Three Centuries of Microbiology

by
Hubert A. Lechevalier
and
Morris Solotorovsky

DOVER PUBLICATIONS, INC.
NEW YORK

Published in Canada by General Publishing Company, Ltd., 30 Lesmill Road, Don Mills, Toronto, Ontario.

Published in the United Kingdom by Constable and Company, Ltd., 10 Orange Street, London WC 2.

This Dover edition, first published in 1974, is an unabridged and corrected republication of the work originally published by the McGraw-Hill Book Company in 1965. Additional bibliographical references have been added at the ends of several chapters.

International Standard Book Number: 0-486-23035-X
Library of Congress Catalog Card Number: 73-91785

Manufactured in the United States of America
Dover Publications, Inc.
180 Varick Street
New York, N.Y. 10014

*"Nothing is more satisfying
for the mind than to be able
to follow a discovery from its very origin
up to its latest developments."*

Louis Pasteur

Preface

Thomas Henry Huxley once wrote that "the thoughts of men are comparable to the leaves, flowers, and fruit upon the innumerable branches of a few great stems, fed by commingled and hidden roots. These stems bear the names of the half-dozen men, endowed with intellects of heroic force and clearness." Microbiology has in recent years given forth so many leaves, flowers, and fruits that the more recent textbooks have, in their attempts to fully cover the expanding disciplines, underemphasized the important stems. The modern student of microbiology is obliged to spend much time in course work mastering the principles and techniques of the active subdivisions of microbiology. Little time is available for leisurely perusal of the older literature associated with the development of microbiology.

We have attempted to reconstruct the growth of microbiology, stressing the main lines of its development. Thus our work has no pretension of giving the reader an encyclopedic coverage. Since we feel that nobody can explain a discovery as well as the person who made it, our text contains numerous, often extensive, quotations from classical papers that outweigh the volume of our own writing. Since many of the papers that we are quoting were not written in English, our work has often been that of translators. In this task, we have taken great liberties that we hope the Masters of the Past will not hold against us, since, in so doing, we are only trying to make their work known and liked by the Masters of Tomorrow.

The reader will find a guide to further reading both in our list of general references and in the few references given at the end of each chapter.

We wish to thank Mrs. M. P. Lechevalier for reviewing our manuscript and Miss E. Gergely and Mr. Francis A. Johns for kindly hunting references for us. The devoted and competent clerical assistance of Mrs. Dorothy Parolise is gratefully acknowledged.

Hubert A. Lechevalier

Morris Solotorovsky

Contents

From Fracastoro to Pasteur

1

Shortly after the discovery of America there appeared in Europe a mysterious disease that was suspected of having been brought back from America by Christopher Columbus's crew. It was universally called the "French disease" (except in France) until Girolamo Fracastoro (1483–1553) published a poem recounting the legend of the shepherd Syphilus, who had been struck with the disease. Fracastoro was a man of many interests. At the University of Padua he was a fellow student of Nicholas Copernicus (1473–1543). He dabbled in astronomy and in all the other sciences of his time, but he is best known to us as the father of the germ theory of disease. In a book published in 1546 he wrote: "Contagion is an infection that passes from one thing to another." There were three types of contagions that he recognized: (1) by contact, (2) by fomites, and (3) from a distance. Concerning fomites, he noted that the "virus" could be preserved in some cases for two to three years. He also warned that contagion at a distance "is not to be imbued with occult properties."

The work of Fracastoro was typical of the state of knowledge at the beginning of the seventeenth century, as science began to emerge as a systematic method of inquiry. The eighteenth century ushered in a new era of scientific development, when men of insight laid the groundwork for modern science. During these two centuries Marcello Malpighi and Antony van Leeuwenhoek made important contributions to the perfection of the microscope. The adoption of the metric system established for most of the world a standard unit of measurement of length and volume, and the invention of the thermometer (and the systems for calibrating it established by Fahrenheit, Réaumur, and Celsius) made possible the precise measurement of temperature. In biology the French naturalist Buffon produced his voluminous encyclopedia, and Lamarck, a botanist turned zoologist, prepared the way for Darwin's theory of evolution. In medicine the English physician Edward Jenner developed the technique of immunization through inoculation; histology became a science, and quinine was discovered for the treatment of malaria. In chemistry Lavoisier refuted the phlogiston theory of combustion, established the chemical composition of water, and was among the first to develop quantitative methods for the study of chemical reactions. At the opening of the nineteenth century, John Dalton published his atomic theory. Coupled with the growth of organic chemistry, these developments set the stage for the work of Pasteur, the first giant of microbiology.

A hundred years after Fracastoro, a German Jesuit, Athanasius Kircher (1602–1680), published a work entitled *Physico-Medical Scrutiny of the Contagious Pestilence, Which Is Called the Plague*. Earlier studies with a microscope had taught him that "every kind of plant is formed of a different and wonderful union of filaments," and when an epidemic of plague struck Rome in 1656, Kircher turned his microscope on the plague victims. His report of seeing "small worms" essentially fortified Fracastoro's ideas on contagion. This work was probably the first time a microscope was used in the study of disease.

The talented Englishman Robert Hooke (1635–1703), another experimenter with the microscope, was one of the last Renaissance men. He made contributions to physics (a law that states that "stress is proportional to strain" still bears his name), architecture, and clockmaking, to name only a few of the fields in which this effervescent, though sickly, man shone. His most important contributions to microbiology were included in his *Micrographia*, published in 1665. It contained an account of what he had seen with the rather primitive microscopes of the time. Most important to us were the cellular structure of cork, the shell of a foraminiferan, and the fruiting bodies of some molds and rusts.

Who invented the instrument that permitted these observations? The answer is not clear. It might have been a Dutch spectacle maker named Janssen, but Marcello Malpighi (1628–1694) of the University of Bologna well deserves the honor of being called the "father of microscopic biology." Malpighi examined the structure of the brain, glands, spleen, liver, and kidneys; studied the anatomy of insects and discovered their tracheas; and followed the development of the chick embryo. He is remembered mainly for the discovery of capillaries in 1661, a discovery that completed Harvey's scheme of the circulation of blood. These observations were made with the crude and rudimentary microscopes of the time; often a simple bead of glass was the complete lens system. Antony van Leeuwenhoek (1632–1723) may have used this system in some of his famous microscopes, but those found after his death were made of a "small double convex glass lens held into a socket between two silver plates riveted together." The lenses were ground from carefully selected glass, and the magnification was perhaps two hundred times at best, but all agreed that the images were very clear.

The life and discoveries of Leeuwenhoek were studied in great detail by Clifford Dobell, who published the now famous *Antony van Leeuwenhoek and His "Little Animals"* in 1932. Leeuwenhoek lived most of his long life in Delft. He was a draper and haberdasher and was obviously on good terms with the local politicians, since he held a sinecure as Chamberlain of the Council Chamber of the Worshipful Sheriffs of Delft. He was also a qualified surveyor and the official wine taster of the town. Among his friends was the physician Reinier de Graaf (1641–1673), the discoverer of the Graafian follicle, who introduced him to Henry Oldenburg, secretary of the Royal Society of London, in 1673. From then on Leeuwenhoek corresponded with the Society, and in more than two hundred letters he described the results of his observations. With his microscopes Leeuwenhoek viewed protozoa and bacteria and used grains of sand as a reference to evaluate the size of some of the microorganisms he saw. He observed protozoa in river water, in rain water that had been collected in a tub, and in infusions that had been standing at room temperature for a few days. He saw what may have been bacteria in the scrapings from teeth. The world had long to wait, however, before his observations were confirmed, for he jealously guarded the secret of the construction of his simple, powerful microscopes and his methods of observation throughout his life. By the time of his death in 1723, only a few men had glimpsed the order of the microscopic world; it remained for their followers to map it.

Eighteenth-century microscopes were primitive instruments, limited in their usefulness by the chromatic and spherical aberration of their lenses. Chromatic aberration derived from the fact that a simple double convex lens does not focus light rays of various wavelengths in the same plane. The rays of short wavelength, such as violet, come to a focus closer to the lens than the rays of longer wavelength, such as red. The resulting blurred, multicolored image enticed many a scientist to believe that the microscope was useful only in creating artifacts.

Chromatic aberration was the cause of a brisk argument between Isaac Newton (1643–1727) and Leonhard Euler (1707–1783). The pessimistic Englishman thought that the aberration was an inherent property of light. The Swiss mathematician reasoned that the aberration would be corrected if lenses were made of two different materials. He thought that glass and water might be suitable, but this idea did not materialize. In 1759, an English artisan, John Dollond, successfully fashioned an achromatic objective for a telescope by combining two lenses made of glass with different indices of refraction. The different refractions of the two glasses canceled the aberration. This practical realization by Dollond, which made possible microscopy as we know it, was based on an earlier observation of an English barrister named Hall, another amateur in optics.

In 1824 a French physicist named Selligue increased the magnification of achromatic objectives by combining several low-power achromatic lenses. However, the manufacture of microscopes remained an art rather than a science until Carl Zeiss of Jena decided to form a partnership with a young physicist called Ernst Abbe (1840–1905). From 1883 to 1895 Abbe solved the basic problems of light microscopy, particularly spherical aberration, which occurs because rays of light passing through the edges of a lens have a focal point different from those passing through the center. Abbe constructed objectives in which a concave lens was added to the basic convex lens system in order to diverge the peripheral rays of light slightly to form an almost flat image. Schott's Optical Glass Works, another firm in Jena, furnished the glasses needed by the Zeiss factory. So the little town of Thuringia, which had witnessed the defeat of the Prussians by Napoleon, became the mecca of microscopy. There, in 1883, were made the first lenses corrected for both chromatic and spherical aberration, and by the turn of the century microbiologists were provided with light microscopes similar to those used today.

The metric system, too, was a product of the intellectual vigor of the eighteenth century. During the French Revolution the new regime became concerned with the scandalous differences in the measuring systems of various parts of France. That, plus a desire to break away from usages of the old regime, led to the formulation of the metric system. The proponents of the reform in weights and measures sought a system based on some natural universal unit that would be acceptable the world over and asked the cooperation of England and the United States. England rejected the offer as impractical, and although Jefferson was enthusiastic about such a change in the United States, he was more interested in creating his own system than in cooperating with French scientists.

The new unit of length, the meter, was defined as one ten-millionth the quadrant of the earth's meridian; that is, one ten-millionth the distance from the pole to the equator. The actual geodesic measurement made was an arc of the meridian between Dunkirk and Barcelona. By 1799 the length of the meter had been measured. One should remember that this operation took place during the revolution, and that at times the poor geometers were harassed by the ignorant and passionate republican population for such trivialities as the white color of their markers. The meter and kilogram standards were made of platinum and deposited in the Archives of the Republic. A law was passed on December 10, 1799, giving a legal value to these standards. Five years previously, Lavoisier had been executed. He had been a proponent of the metric system and had stated: "Never has anything more grand, more simple and more coherent in all its parts, been issued from the hands of men." In fact, despite the care with which the measurements were made, it is doubtful that the meter is exactly one ten-millionth the distance from the pole to the equator, but in practice, the platinum bar deposited in Paris has become the universal reference standard of the metric system.

Another problem was the measurement of temperature. Late in the sixteenth century Galileo (1564–1642) devised a crude alcohol thermometer to measure body temperature and it was perfected by his friend Santorio (1561–1636—also called Sanctorius). (Santorio had a mania for weighing himself before and after all sorts of activities, and, indeed, his observations constitute the first study of human basal metabolism.) The next significant advance came in 1714 when Daniel Fahrenheit (1686–1736), working in Danzig, substituted mercury as the measurable medium. He also favored the world with an odd scale of measurement that bears his name and still persists in English-speaking countries. More logical was the scale proposed by Ferchault de Réaumur (1683–1757) in 1731, which took the temperature of freezing water as $0°$ and the temperature of boiling water as $80°$. (Réaumur, a student of physiology, was by training a physicist and engineer. He studied the nature of the gastric juices of birds, regeneration in crustaceans, and the electric organs of rays. Much of his work also involved the study of insects.) Finally, the centigrade thermometer was proposed by the Swedish astronomer Anders Celsius (1701–1744) in 1742. A vigorous and influential man, Celsius helped the French Academy of Sciences to mount two expeditions to determine whether the earth was flattened at the poles. His uncle, Olaf Celsius, a botanist, published, among other studies, a book on the plants of the Bible. But uncle Olaf is better known for helping and encouraging the young Karl von Linné (1707–1778—generally known by the Latinized form of his name, Carolus Linnaeus) on the road to knowledge.

Linnaeus conceived the idea of using pistils and stamens as the basis of a system of classification. He obtained support for collecting trips that carried him through Lapland, Holland, England, and France, and during his travels he published his *Systema Naturae*. He finally returned home with a solid reputation to become professor of botany at Upsala in 1741. Linnaeus classified plants, animals, and minerals but was mainly a botanist rather than a microscopist. In this he differed from Louis Joblot (1645–1723), was published in 1718 what can be considered the first treatise of protozoology. The first part of Joblot's treatise was devoted to the construction of microscopes, and the second part contained a description of the organisms he observed. Linnaeus did not disregard completely the little beasts of Leeuwenhoek and Joblot; he simply grouped them under the name of "Vermes" in a class which he called "Chaos." In the great organizer's estimation, the world would probably have been better and simpler without them.

One of Linnaeus's best students was the Dane Christian Fabricius (1745–1808), who was responsible for the now classical "Attempt at a Dissertation on the Diseases of Plants," published in 1774. In this work Fabricius noted a close relationship between rusts and smuts and proposed that fungi found in plant lesions were distinct organisms, not just the transformation of diseased tissues of the plant. However, Fabricius's greatest achievements were in the field of entomology. Although he enjoyed a high reputation outside of Denmark, his eventual lot was a poor salary in a professorial chair in Kiel, which was then an outpost of Denmark.

The French naturalist Georges Louis Leclerc, Comte de Buffon (1707–1788), was more fortunate. Reviving the method of Pliny, Buffon produced a voluminous encyclopedia from his office. His work was supported by the court, and he hired many collaborators, one of whom, Étienne de Lacépède (1756–1825), completed the enormous *Histoire naturelle* after Buffon's death. This pompous work comprised fifty-four volumes. Buffon said, "genius is only a great aptitude for patience," and considering that he spent about twelve hours a day writing over forty years, we realize that Buffon must have had strong conviction of his theory. His greatest contribution was to break away from the pious practice of lauding God through the admiration of his work. Buffon wrote a completely laic work ignoring Providence as if "Genesis" had never existed.

This human writing machine was not without interesting ideas. First, Buffon did not believe that there was an absolute boundary between animals and plants. Second, he did not believe in the "preformation" theory that was accepted by numerous naturalists of his time. This theory may date back to Malpighi, who, when he observed fresh chicken eggs, declared that he had "seen" the shape of the chick under his microscope. Following this "basic observation" and that of Leeuwenhoek, who had studied spermatozoa, a lively argument arose to "settle" the question of whether animals were preformed in the eggs or in the spermatozoa. Buffon rejected the theory of preformation on the basis that it presupposes an infinite number of children would be contained in the original parent animal. He considered this an irrational supposition, and instead he proposed a rather vague theory that could be viewed as compatible with present-day knowledge. He believed that nature was filled with living particles that were picked up by the living individual and used, in part, for its own metabolic activities and, in part, for the formation in the genital organs of half its seed. The individual acted as a template: *moule intérieur*. The third important aspect of Buffon's work was his collaboration with Louis Daubenton (1716–1799). This eminent naturalist described the anatomy of individual mammals and compared various structures of one animal to similar structures in others. This may be considered the foundation of comparative anatomy and the antecedent of the concept of evolution. When the Count of Buffon died in 1788, and as candles began to burn continuously for one year around his coffin, the Church's role in science was fixed. Faith and reason were to be divorced, perhaps not officially, but certainly in practice.

It was this same Buffon who first recognized the promise of the young Frenchman Jean Baptiste Lamarck (1744–1829), whose theories of evolution to some extent anticipated those of Darwin. During the Seven Years' War (1756–1763) Lamarck joined the French army at the age of seventeen. During a siege his company lost all its officers and noncommissioned officers, and he, a soldier for only one day, took charge and held out until reenforcement arrived. Oddly enough, this hero was a shy, unassuming man who led a poor, unhappy life. Unhappy in the army, he wrote a book on the flora of France that attracted Buffon's attention. Buffon procured some funds and sent Lamarck to travel throughout Europe with the title of Royal Botanist. During the Revolution there was a lack of zoologists, and at the age of fifty Lamarck found himself the holder of the chair of invertebrate zoology at the Museum of Natural History. Thus the botanist became a zoologist, and during the last thirty-six years of his long life he produced the work for which he is remembered. Though blind during his last years, he was able to continue his work with the help of his daughter Cornélie. When this modest man died he was eulogized by many; one of whom, Georges Cuvier (1769–1832), an eminent zoologist and paleontologist, made it plain that a live Cuvier was more interesting than a dead Lamarck. Cuvier opposed Lamarck's theory of evolution, which was based on the assumption that modifications of an individual caused by adaptation to the environment are passed on to the offspring. Lamarck envisioned the net result as the gradual evolution of one species from another. Cuvier, on the other hand, believed in the stability of species. The perspective of time has shown that, in spite of the opposition of eminent scientists, the mild-mannered Lamarck was a true forerunner of Darwin.

Charles Darwin (1809–1882) was the grandson of Erasmus Darwin, a physician and a poet with a great love for natural history. His father was also a physician, and his mother was the daughter of Josiah Wedgwood, the famous manufacturer of earthenware. Charles was sent to study medicine in the best family tradition. However, anatomy did not please him and he left to study theology at Cambridge. It was then that he started to collect natural history specimens and to study geology. In 1831 the young amateur was offered an unpaid position as a naturalist on the *Beagle,* a ship which was to circle the globe on a cartographic expedition. This trip lasted five years and crystallized his vocation. The eager budding naturalist suffered from incurable sea sickness, but his numerous collections were valuable. In 1839 Darwin married a Wedgwood cousin, and for the rest of his life he was an independently wealthy scholar, living in a Kentish village called Down. His health was miserable, but despite chronic gastric troubles and insomnia he industriously built the theory of evolution. He was a gentle man, and even when his ideas came under the most violent attacks he held himself above personal polemics. He died in 1882 and was buried in Westminster Abbey. Briefly, Darwin postulated that living forms evolved by a process of natural selection that could be called survival of the fittest. Among variant individuals in progeny, those better adapted to functioning and reproduction under the stress of environment tended to prevail. In time, this selection of the more hardy variants would result in the emergence of new varieties and species.

The eighteenth century also witnessed the founding of the science of immunology by the daring English physician Edward Jenner (1749–1823), who devised the technique of vaccination against smallpox. An informative report from Lady Wortley Montagu, wife of the English Ambassador to Turkey, indicates, however, that the inhabitants of Turkey had benefit of a crude form of inoculation at about the same time. In one of her letters, written during her years in Turkey between 1716 and 1718, she reports: "The smallpox, so fatal, and so general amongst us, is here harmless by the invention of ingrafting, which is the term they give it. There is a set of old women who make it their business to perform it every autumn, in the month of September, when the great heat is abated. People send to one another to know if any of their family has a mind to have the smallpox: they make parties for this purpose, and when they are met . . . the old woman comes with a nut-shell full of the matter of the best sort of smallpox and asks what vein you please to have opened. She immediately rips open that you offer to her with a large needle . . . and puts into the vein as much matter as can lie upon the head of her needle, and after that binds up the little wound. . . . The children or young patients play together all the rest of the day, and are in perfect health to the eighth. Then the fever begins to seize them, and they keep their beds two days, very seldom three. They have very rarely above twenty or thirty pustules in their faces, which never mark; and in eight days' time they are as well as before their illness. . . . There is no example of any one that has died in it; and you may believe I am well satisfied of the safety of this experiment, since I intend to try it on my dear little son." Upon her return to England she worked to educate the public in the use of inoculation against smallpox.

Jenner stumbled across the idea in another way. He remembered from his youth the statement of a young girl, who claimed that she could not catch smallpox because she had had cowpox. Acting upon this premise, he began, in 1796, performing numerous human inoculations with cowpox. In 1798 he proposed to present his findings before the Royal Society. The Fellows conservatively advised him that "he ought not to risk his reputation by presenting . . . anything which appeared so much at variance with established knowledge, and withal so incredible." His results were favorable, however, and when he died in 1823, he was a respected and successful man with an international reputation. Immunology had victoriously dominated one of the great scourges of mankind.

The science of histology was founded by the French anatomist Xavier

Bichat (1771–1802) in the eighteenth century; oddly, though understandably, without the use of the microscope. In view of the inadequacies of the microscopes of that time, he claimed that they "gave rise only to fallacies and delusions." On the basis of careful observation and treatment of tissues with physical methods such as heating and drying, and chemical agents such as acid, alkali, and alcohol, he devised a sound classification that depended on a correlation of structure with function. He established twenty-one types that still correspond to modern categories. As examples, "animal musculature" corresponds to our striated muscle; "organic musculature" to our smooth muscle; "tissue of exhalation" to our lung; and "medullary tissue of bone" to our bone marrow. In his concept of life processes, he drew away from the unproductive vitalism of his contemporaries, who claimed that the fluids of the body were the true vital elements. He claimed that the tissue components of the organs were the vital elements and thus the components for basic study. He died in 1802 of an acute infection when he was only thirty-one years old. Bichat left the legacy of a great pupil, René Laënnec (1781–1826), who applied Bichat's approach in the study of tuberculosis. Laënnec recognized the unifying pathohistologic basis of the various types of tuberculosis, then considered to be different diseases.

Laënnec is also known for his invention of the stethoscope. Faced with the embarrassing requirement of examining an obese young woman for cardiac symptoms by direct auscultation, he rolled his notebook into the form of a tube, applied one end to the patient's bounteous chest, and held his ear to the other. His ear received amplified heart sounds.

At about the same time (1820) two pharmacists on the faculty of the University of Paris, Pierre Pelletier (1788–1842) and Joseph Caventou (1795–1877), isolated quinine from the bark of the *Cinchona* tree. They had already isolated strychnine and were to study numerous other alkaloids during their fruitful careers. Knowledge of the beneficial effect of the bark of the *Cinchona* tree in the treatment of malaria had come to Europe from Peru, via Spain and Italy. The inhabitants of Peru had for centuries used the bark in treatment of the dread disease. During the seventeenth century its use was introduced in Europe by the Jesuits and was known as "Jesuits' bark." Linnaeus assigned the generic name *Cinchona* to this tree, after the Countess of Chinchon, wife of the Spanish Viceroy of Peru, who was cured by its bark in 1638. Like the Jesuits, the Count of Chinchon also imported large quantities of the miraculous bark into Europe. It was received with mixed emotions. Its very name, tinted with popery, must have aroused prejudice. An English apothecary, Robert Talbor (1642–1681), who was later knighted, used it in large quantities as a secret remedy to cure fevers; his potion was an infusion of cinchona bark in claret wine. With Pelletier and Caventou, chemistry of natural products replaced such colorful secret concoctions.

Chemistry in the eighteenth century was just emerging from the tradition of alchemy. In 1697 the German scientist Georg Stahl (1660–1734) formulated the "phlogiston theory," which held that combustible substances contained an essence named phlogiston, which left the parent substance during combustion. (The term survives as the root of the adjective "phlogistic," for "inflammatory.") According to this theory, the more combustible a substance, the more phlogiston it contains. The theory was refuted by Antoine Lavoisier (1743–1794), who demonstrated that substances actually *increase* in weight during combustion; in so doing, he introduced quantitative methods in chemistry. He also elucidated the processes of fermentation and respiration by showing the role played by oxygen and carbon dioxide and introducing the use of chemical equations: "Must of grapes = carbonic acid + alcohol." He distinguished between chemical elements and chemical compounds and stated clearly that during a chemical reaction nothing is gained and nothing is lost. He also introduced many words that we still use, such as "hydrogen," "oxygen," and "caloric." The founder of modern chemistry was married to a gifted and devoted lady, who made the illustrations for Lavoisier's publications and helped him with his work. Unfortunately, Lavoisier was a part of the tax-collecting bureau of the *ancien régime* and the Revolutionaries, in a moment of folly, beheaded him in 1794. "It took but a moment," wrote the mathematician Lagrange, "to cut off his head, but another like it cannot be produced in a hundred years."

Atomic theories had been proposed in the seventeenth century by Boyle, Newton, and others, but John Dalton (1766–1844) recast the theory so as to have practical significance in chemistry. He was a color-blind chemist with a keen interest in weather forecasting and in his own affliction, which we still call Daltonism in his honor. He, for the first time, stated clearly that chemical elements are composed of very small indivisible particles of matter called atoms, which maintain their integrity during the course of chemical reactions, and that an atom has a constant mass, different for each element. He proposed his theory in 1808 in "A New System of Chemical Philosophy." At this time Humphry Davy (1778–1829) was discovering new elements with speed and efficiency. Using the Volta battery, Davy electrolyzed a number of substances and in quick succession isolated potassium and sodium in 1807 and calcium, barium, magnesium, and strontium in 1808. He determined the elemental nature of chlorine and invented the miner's safety lamp, with which his name is still associated. Davy first became known in 1795 for his report on nitrous oxide and its potential anaesthetic properties.

Davy's method of isolating the metals of the alkaline-earth group was further perfected by Joseph Gay-Lussac (1778–1850) and Jacques Thenard (1777–1857). Gay-Lussac, a French physicist, formulated the law of combining volumes of gases, with which his name is still associated. At the same time as his countryman, the physicist Jacques Charles (1746–1823), but independently, he discovered that a gas at constant pressure expands, for each increase in one Celsius degree, by 1/273 of its volume at 0°C. In 1804 he made two balloon flights, one to the height of 7,016 meters, to test magnetism and the composition of the atmosphere at high altitudes. Together with Thenard, he isolated boron from boric acid in 1808 and synthesized this acid. Thenard also discovered hydrogen peroxide and observed that the fermentation of gooseberries, cherries, pears, apples, barley, and wheat yielded a deposit which resembled brewers' yeast. We shall return to this observation later.

In 1811 the Italian chemist and physicist Amadeo Avogadro (1776–1856) described the distinction between molecules and atoms. This concept was unacceptable to Dalton, who refused to acknowledge the validity of Gay-Lussac's law and the concept that equal volumes of different gases contain the same number of particles under the same physical conditions. The French physicist André Ampère (1775–1836), in 1814, proposed essentially the same theory as Avogadro's. The investigations of Avogadro, Gay-Lussac, and Ampère were not understood by most chemists of their time. It was not until the first international chemical congress in Karlsruhe in 1860 that a decisive step toward the establishment of a uniform and logical system of chemistry was taken by Stanislao Cannizzaro (1826–1910). At this congress copies of Cannizzaro's "Summary of a Course of Chemical Philosophy Given in the Royal University of Genoa" were distributed. He revived and clearly defined the terms "atom" and "molecule," firmly establishing Avogadro's hypothesis, and he employed hydrogen = one as the universal reference against which the weight of other elements was to be gauged.

Few chemists had a stronger influence on the chemistry of the beginning of the nineteenth century than the Swede Jöns Berzelius (1779–1848). He discovered his share of new elements and developed the modern system of chemical symbols and formulas. On the practical side, he introduced many innovations into the laboratory such as filter paper and rubber tubing. Berzelius was also interested in fermentations and attributed such chemical transformations to the actions of catalysts. He emphasized the importance of catalysis by collecting scattered observations on reactions accelerated by agents which remained unmodified. These reactions included the hydrolysis of starch by acids, first observed by Antoine Parmentier (1737–1813) (the man who introduced the potato to France), and the oxidation of methane on platinum observed by Davy in 1816. The work of Berzelius on catalysts was continued by his student Eilhard Mitscherlich (1794–1863).

Another of Berzelius's students was Friedrich Wöhler (1800–1882), who started his career by publishing analyses of cyanates at the time when one of Gay-Lussac's disciples, Justus von Liebig (1803–1873), was publishing analyses of fulminates. Gay-Lussac, who was then the editor of the *Annales de Chimie*, noted that the analyses of cyanic and fulminic acids were identical. This was the first case of isomerism to be observed, and it was the beginning of a long and fruitful cooperation between Wöhler and Liebig. In the first decade of the nineteenth century, it was thought that organic chemicals differed fundamentally from inorganic substances and could be formed only through the action of a mysterious "vital force." Wöhler did much to destroy this belief when in 1828 he obtained urea from its inorganic isomer, ammonium cyanate. This, although often referred to as a synthesis of urea, is not truly a synthesis. That a vital force was unnecessary for the formation of organic chemicals was gradually accepted, as Hermann Kolbe (1818–1884), in 1844, synthesized acetic acid, and Marcellin Berthelot (1827–1907) synthesized glycerol, organic acids, and hydrocarbons. With his textbook *Organic Chemistry Based on Synthesis* (1860), Berthelot dealt the lethal blow to the theory of vital force. Liebig also investigated a large number of organic chemicals. He was interested in practical chemistry and is credited with the introduction of inorganic fertilizers into farming. Great as these chemists were, they lacked a flair for biological investigations, and, as we will see later, they could not believe that yeasts were living organisms with a metabolism of their own.

The rotation of polarized light was another development important to the growth of organic chemistry. It all started in 1808 one late afternoon in Paris, as Étienne Louis Malus (1775–1812), a former officer of the Napoleonic army, was musing with a piece of Iceland spar in his hand. He was amusing himself by viewing the double image of the setting sun, which had been reflected in the windows of the Luxembourg palace. He was most surprised, as he rotated the crystal, to note that the images of the reflected sun disappeared and, as the rotation was continued, reappeared. Four years later Malus died of tuberculosis, but his discovery was exploited by

Jean Baptiste Biot (1774–1862), who observed in 1815 that certain naturally occurring organic substances rotated the plane of polarized light. Until then optical rotation had been associated only with crystals, their melting or dissolving resulting in a loss of their optical activity. Biot showed that turpentine, camphor, tartaric acid, and solutions of sugar also rotated polarized light. He concluded that this optical activity must be an inherent property of the molecules of these organic compounds and not the result of the alignment of a group of molecules.

Such was the state of science at the close of the eighteenth century, and thus were the tools forged for the work of Louis Pasteur.

References

Dobell, C.: *Antony van Leeuwenhoek and His "Little Animals,"* Dover Publications, Inc., New York, 1960 (republication).

Fabricius, J. C.: *Attempt at a Dissertation on the Diseases of Plants,* Phytopathological Classics no. 1, American Phytopathological Society, 1926.

Major, R. H.: *Classic Descriptions of Disease,* Charles C Thomas, Publisher, Springfield, Ill., 1955.

Rousseau, P.: *Histoire de la science,* Arthème Fayard, Paris, 1945.

Additional Reference

Cartwright, F. F., and M. D. Biddis: *Disease and History,* T. Y. Crowell, New York, 1972.

Pasteur

Molecular Dissymmetry and Fermentations

Louis Pasteur was born in 1822 in Dole, a small town in the Jura. He died in 1895, close to Paris, at Villeneuve-l'Étang, now known as Garches, where the section of the Pasteur Institute dedicated to the preparation of antidiphtheria vaccine was being organized. During the seventy-three years of his life new fields of human endeavor were to be created, mainly because of his work, his vision, and his forceful personality. He started the study of the dynamic chemical reactions that are the bases of life.

Pasteur's father, Jean-Joseph, had been a sergeant in Napoleon's army. After the collapse of the empire, he turned to the modest trade of tanning at Dole. There he married the gentle daughter of a gardener, and embarked on a quiet, simple life based on devotion to family. Louis was the only son among three daughters. This son of simple parents combined genius with homely virtues.

During his early years, Louis showed a great natural talent for art. He made portraits in pencil and pastel and even tried his hand at lithography. Despite his talent, Pasteur decided to study sciences, and the pastels were exchanged for test tubes. During student days, he concentrated single-mindedly on studies and did not participate in any of the gayer and less conventional activities of life in Paris. Louis was a superior student, but in line with the modest position of his family, he had planned to become a secondary school teacher.

The year 1844 was an important one in the life of Pasteur. It was his first year at the *école normale*, and it was the year that the aging Biot, the discoverer of the optical activity of organic substances, presented a note by Mitscherlich at the Academy of Sciences. The essence of this communication was that the paratartrates of sodium and of ammonium and the corresponding tartrates had exactly the same properties except for the fact that the paratartrates were optically inactive while the tartrates had the ability to rotate polarized light. Biot was in perfect agreement with Mitscherlich as to the validity of the reported observation. Pasteur's curiosity was awakened. How could two compounds be identical in every respect and still have a different effect on polarized light? This action must be an expression of their molecular arrangement. If the effect on light is different, the compounds must be different.

Although Pasteur was to solve the riddle of the optical activity of the tartrates and the paratartrates, and although he was aware of the problem as early as 1844, he was not to dedicate his time to its solution until he had finished his doctoral work. His thesis was in two parts: a thesis in chemistry on arsenous oxide, and a thesis in physics on optical rotation of liquids. In 1847 he defended these two theses and received his degree. Pasteur considered these theses schoolboy work. He had in mind greater achievements.

His explanation of the discrepancies observed in the optical activity of the tartrates and the paratartrates (racemates) was impressive in its simplicity. Pasteur said (*Compt. rend. acad. sci.,* **26**: 535–538, 1848):

> By sheer accident, M. Mitscherlich and later M. Biot have been led into error. The paratartrate of sodium and ammonium causes a deviation in the plane of polarized light but, among the crystals of the same sample, there are some which turn the plane of polarization to the left and others which turn it to the right. Consequently, when one has a mixture of an equal amount of both types of crystals, the solution is inactive since the opposed deviations counterbalance each other. The crystallographic difference between these two types of crystals is as follows: They are all hemihedral but some are hemihedral to the right and others are hemihedral to the left. The direction of the deviation of polarized light depends on this dissymmetry. When I want a deviation to the right, I choose the crystals which are hemihedral to the right; when I want a deviation to the left, I choose crystals which are hemihedral to the left. . . . Is it not now quite evident, that the ability of certain molecules to deviate the plane of polarized light, is due to, or at least is linked most intimately to the dissymmetry of these molecules? . . .

Pasteur performed the separation of the racemate into its two optically active constituents in front of Biot, at the Collège de France. The old scientist was much impressed by the young chemist, and for the rest of his life he was to help him in every way he could.

Luck, from the start, was on Pasteur's side, since it was recognized later that the separation achieved by Pasteur was dependent upon a critical temperature. If the racemate had been crystallized from a hot concentrated solution, the crystals would have been all alike, symmetrical and optically inactive. Only when the separation of the crystals occurs at a temperature below 28°C, do the hemihedral crystals occur, forming a mixture of levo- and dextro-tartrates.

In November, 1848, Pasteur left Paris for Dijon, where he had been appointed professor of physics at the *lycée*. Two months later he was transferred to the University of Strasbourg, where he taught chemistry. Within a short time he married Marie Laurent, the daughter of the rector of the University.

Here is how he announced the great news to one of his friends: "I am marrying Miss Marie Laurent. The marriage is settled between my father, her parents, her and me. . . . I think that I shall be very happy. All the qualities that I could desire in a wife, are to be found in her. You will say that I am in love. Yes, but it seems to me that I am not exaggerating anything, and my sister Joséphine is absolutely of my opinion.

"All has been settled eight days ago. Since then, I work but little and I am neglecting completely the laboratory. It is true that we are now on vacation and that I have used well the two months spent in Strasbourg. . . ."

Pasteur was not the man to abandon the laboratory for long. His prediction, however, was accurate. The union was a happy one, as Marie was the proverbial devoted wife. She made few financial demands; she shouldered the major burden of domestic responsibility, and even had time to take dictation of scientific reports in the evenings.

Pasteur continued his study of crystallography up to 1856. He clarified the question of the tartrates, showing that there are four tartrates in all: (1) the dextrorotatory tartrates; (2) the levorotatory tartrates; (3) the racemic form of the tartrates, which is, in essence, an equal mixture of *d*- and *l*-tartrates (paratartrates); (4) the mesotartrates (*i*-tartrates), which have no action on polarized light and which are the optically "internally compensated" forms.

The resolution of a racemate into its optically active components can only rarely be accomplished by crystallization and subsequent mechanical separation of the crystals. Pasteur was to contribute a method of separation which was practical and which has been used extensively. Whereas the salts of *d*- and *l*-tartaric acids with metals or ammonium had identical solubility properties, the salts formed with optically active natural bases such as quinine had very different properties. The differences in the solubility of the *d*- and *l*-salts could then be used to separate the mixture into its components.

Pasteur was fascinated by the concept that dissymmetric forces were the cause of many of the natural phenomena that could be observed in the universe. He tried to induce dissymmetry in crystals by making them grow in magnetic fields. He tried to grow plants under light that had been "reversed" by the use of mirrors, and he studied the growth of plants subjected to artificial magnetic fields. But in these experiments Pasteur found nothing of a fertile nature.

In 1854, Pasteur was appointed dean of the new faculty of sciences of the University of Lille. In his inaugural speech, the young dean stressed the value of theoretical studies: "Without theory, practice is only routine governed by the force of habit. Only theory can breed and develop the spirit of invention." One should not conclude from this statement that Pasteur was afraid to attack practical problems. On the contrary, most of his future career was to be oriented and controlled by the practical problems that he was requested to solve. His eagerness to study such practical problems was an expression of his keen sense of the civic responsibility of the scientist. The fact that he made many fundamental discoveries while solving practical problems is the tangible proof of his towering intellectual stature.

In 1856, an industrialist of Lille, a M. Bigo, requested Pasteur's help in connection with some difficulties that had been encountered in the manufacture of alcohol from beets. By agreeing to help, Pasteur had sealed the fate of his scientific career. His investigations progressively moved from the study of inanimate agents to animate ones. The rest of his life was almost exclusively devoted to the study of microbiology, a science that he essentially created. Before meeting M. Bigo, Pasteur had already observed variations in the end products of a fermentation depending on the nature of the starting materials:

> My investigations [*Compt. rend. acad. sci.*, 41: 296–300, 1855] can be summarized by this rather simple statement: crude amyl alcohol as it is widely available on the market, is mainly formed of a mixture of an amyl alcohol which is active on polarized light, and of an inactive isomeric amyl alcohol. The chemical properties of these two alcohols are identical. Everything that can be done with one can be done with the other with the same ease or with the same difficulty. It would be impossible to differentiate the two alcohols if one's attention has not been sharpened. The active alcohol gives only active derivatives. The inactive one, only inactive derivatives. At first glance all these deriva-

tives are rather similar in odor, solubility, crystalline form, boiling point and specific weight. However differences are there to be found if one uses the utmost exactitude in each and every measurement and determination. The proportion of the two alcohols, the active one and the inactive one, varies greatly depending on the origin of the alcohol. Thus the crude oil from the fermentation of beet juices contains about one-third of the active alcohol and two-thirds of the inactive one, whereas the crude oil obtained from the fermentation of molasses contains an equal amount of the two alcohols. . . .

In his studies on fermentations, Pasteur followed the path opened by Cagniard de Latour (1836), Theodor Schwann (1837), and Friedrich Kützing (1837), all of whom independently came to the conclusion that yeasts were living vegetable matters and were responsible for fermentations. This theory of fermentation was attacked by the great chemists of the time. Liebig, who was very intolerant of any scientific theory that did not meet with his approval, joined forces with Wöhler and published in 1839 an unsigned skit entitled "The Riddle of Vinous Fermentation Solved." The fermentation of sugar was accomplished, according to these great masters, by animalcules that ate sugar and ejected alcohol through their anus and carbon dioxide through their genitals.

When he published his first paper on lactic acid fermentation in 1857, Pasteur was back in Paris as the director of scientific studies of the *école normale supérieure*. The professor of chemistry there at the time was the famous and able Sainte-Claire Deville. The two chemists, both with powerful personalities, did not get along too well at first, and Pasteur, rather than utilize some of the chemistry laboratory space for his work, improvised laboratory facilities in unused nooks and crannies. As the years went by, however, the two men appreciated each other, and Pasteur was deeply moved, in 1881, when he spoke at the funeral of Sainte-Claire Deville. The link between Pasteur the crystallographer and Pasteur the microbiologist is to be found in his studies on the fermentation of tartaric acids (*Compt. rend. acad. sci.*, **46**: 615–618, 1858):

Members of the Academy remember the unusual composition of racemic acid. It is formed of the combination of a molecule of right tartaric acid, which is ordinary tartaric acid, and of a molecule of left tartaric acid, which differs from the right one only by the impossibility of superimposing their shapes, otherwise identical, and by the optical rotations which are identical in absolute value, but one to the right, and the other to the left. . . .

Thus, it was important to determine whether racemic acid would support the same fermentation as dextro-rotary tartaric acid. . . . Ammonium racemate was subjected to fermentation following the method that I have indicated previously for the right tartrate. Fermentation occurred with the same ease, with the same properties and the deposit of the same yeast. But, by following the course of the fermentation with a polarization apparatus, one can see that the process is different. After a few days, the liquid which originally was inactive, becomes levo-rotary and the optical rotation increases until a maximum value is obtained. Fermentation then stops. There are no more traces of the right acid in the medium, which, following evaporation, and mixing with an equal volume of alcohol, yields an abundant crystallization of left ammonium tartrate.

We have here without a doubt an excellent method for the preparation of left tartaric acid. . . .

As for the cause of the difference that I have noticed in the fermentation of the two tartaric acids, it would seem plausible to attribute it to the optical configuration of the compounds which enter into the composition of the yeast. One can understand that if the yeast is itself normally composed of dissymmetric compounds, it will not utilize equally well two nutrients, one of which is dissymmetric in one direction, and the other dissymmetric in the opposite direction. . . .

By 1860, Pasteur's position on the problem of fermentation was well established. This was the year when Pasteur published his impressive "Memoir on Alcoholic Fermentation" [*Ann. chim. et phys.* (3) **58**: 323–426, 1860]. This long paper contains a careful account of the history of the subject and the results of Pasteur's experiments; results that, in great part, he had already published piecemeal. Briefly, Pasteur's position was that alcoholic fermentation is the result of the growth of the yeast, that the yeast needs a source of nitrogen for growth, that alcohol and carbon dioxide are not the only products of the fermentation of sugar, and that a part of the sugar is used to build the carbohydrates and the fats of the yeast. Let us examine this paper to see what was Pasteur's concept of the historical background of this work:

Leeuwenhoek in 1680 studies beer yeast with a microscope and finds it formed of small spherical or ovoid globules. But the chemical nature of this substance is unknown. In a paper on fermentations, which received a prize from the Academy of Florence in 1787, the Italian scientist Fabroni, in the midst of many false concepts, relates and even identifies yeast with gluten. This was progress. It gave an idea of the location of yeast among organic products. . . .

Pasteur, as a footnote, gave this quotation from Fabroni: *"Fermentation is nothing else than the decomposition of one substance by another, such as that of a carbonate by an acid, or of sugar by nitric acid. . . . The sugar-decomposing substance in vinous effervescence is of a vegeto-animal nature. It is located in specialized utricles in grapes and in wheat. In crushing grapes, one mixes this glutenous substance with sugar, as if one were pouring an acid and a carbonate into a vase. As soon as the two substances are in contact, the effervescence or fermentation starts, as in any chemical reaction."*

Pasteur continued:

Fabroni's assertion made the question of the nature of fermentation the order of the day.

It was felt that Lavoisier had solved the problem of fermentation as far as the substrate was concerned, but one had no notion of the nature of the substance responsible for the degradation of sugar. Thus, in the year VIII, one year after the publication in France of Fabroni's work, the Section of Physical and Mathematical Sciences of the Institute offered a prize for the best answer to the following question: "What are the properties which differentiate, among vegetable and animal substances, those which are ferments from those fermented?" Encouraged by this proposition of the Institute, M. Thenard tried to solve the problem, and published in year IX a remarkable memoir in which he deals mainly with the nature of ferment, its origin, and its alteration during fermentation. Here is a summary of his work:

All natural sugary juices which undergo spontaneous fermentation, give a deposit which has the appearance of brewers' yeast and which like the yeast can ferment pure sugared water. This yeast is of an animal nature, that is to say it contains nitrogen and gives much ammonia if distilled.

During fermentation, the yeast progressively loses its nitrogen and it partially disappears as it is transformed into soluble products. . . .

These results, to which I shall soon return, have been the basis of all discussions on the theory of fermentation. . . .

A few years after the publication of M. Thenard's memoir, Gay-Lussac reported a most extraordinary result. While investigating the procedures of M. Appert for the preservation of vegetable and animal substances, he noted that must of grapes which had been preserved unaltered during a whole year would start to ferment a few days after having been transferred to another container. This fact led Gay-Lussac to further experiments and to state that "oxygen is necessary to start fermentation, though not necessary to continue it."

In order to find a new advance worthy of mention in the history of fermentation, after the previously mentioned investigations, and those of M. Colin already referred to,[1] one has to wait until 1835 and 1837 for M. Cagniard de Latour. Following studies with a microscope, unaware that Leeuwenhoek had done so previously in an incomplete way, M. Cagniard de Latour introduced a new concept into the field with which we are dealing.

Before him,[2] yeast had been considered as a product of plants which had the property to precipitate in presence of fermentable sugars. M. Cagniard de Latour recognized that "yeast was a mass of globules capable of reproducing by budding, and not simply an organic or chemical substance as had been suspected."

From his observations, M. Cagniard de Latour concluded "that it is very probably by some effect of their growth that the globules of yeast liberate carbonic acid from a sugared solution and convert it into a spirituous liquid."

This concept immediately found in M. Liebig a powerful opponent.

In Liebig's view, ferment is a very labile substance which decomposes and which excites fermentation as it decomposes. . . . [Pasteur then quotes Liebig]: "Brewers' yeast, and, in general, all animal and vegetable matter in putrefaction, carry over to other bodies the state of decomposition in which they are themselves. The motion which, due to a loss of equilibrium, affects their own components, is communicated also to the elements of substances which are in contact with them."

From the historical studies that M. Chevreul has published recently

[1] Pasteur's previous reference to Colin's work was as follows: "M. Colin is, I think, the first chemist who noticed that yeast washing water was a good source of ferment. In a most interesting paper (1825) . . . he showed that sugared solutions could become alcoholic by addition of flour dough, gluten, beef meat, egg white, cheese, urine, fish glue, blood, etc. But, of the various substances used by M. Colin, it was the waters used to wash brewers' yeast or grapes, or their extract, which were the most satisfactory for fermentation." Pasteur concludes: "All the substances with which M. Colin succeeded to produce alcoholic fermentation of sugar have been in his experiments nothing else but the nitrogenous food for the yeast which had grown spontaneously." Pasteur was, of course, using the word "spontaneously" as we would use it nowadays. His words were to be interpreted differently by the defenders of spontaneous generation.

[2] Pasteur's footnote here draws the attention of the reader to the work of Persoon, who called brewers' yeast *Mycoderma cervisiae* in 1822, and to that of Desmazières, who thought that yeasts were infusoria (1827) because of their motion, which was probably Brownian movement (according to Pasteur).

in the *Journal des savants* (Feb. 1856, p. 99) we learn that Stahl had previously expressed ideas similar to those of M. Liebig on the causes of fermentation.

M. Liebig has expressed his ideas in most of his publications with such persistence and conviction that they have gradually been accepted. . . .

Thus the idea of M. Cagniard de Latour, which at first had been received with a certain amount of attention, was slowly abandoned. Many at least, did not question the idea that brewers' yeast was organized, but it was thought that the yeast partly destroyed itself during fermentation, as had been concluded by M. Thenard, and that, in this respect, it was similar to all other nitrogen-containing substances which could act as ferments. It was to this property that was due its action on sugar. Such was the thinking of M. Liebig.

Berzelius did not share Liebig's views even though he rejected those of Cagniard de Latour and Schwann. To him, fermentation was a contact phenomenon. He did not even believe that there was anything living in yeast. *"It was only a chemical substance which precipitated during the fermentation of beer and which had the usual shape of a non-crystalline precipitate, even inorganic, namely, small spheres which are grouped one after the other like chains of beads."*

In a footnote Pasteur draws attention to variations of the previously described theories on fermentation. For example, Mitscherlich did not hesitate to admit that yeast was truly an organized living being; however, he said: *"The globules of the ferment behave in relation to sugar and water, which contain the elements of carbonic acid and of alcohol, absolutely as the platinum sponge in relation to hydrogen peroxide."* This was also in essence the idea of Berthelot.

The studies on fermentations led Pasteur to the discovery of strict anaerobiosis. He asked Dumas, Balard, and Bernard to come to his laboratory to observe his work. It was convincing and Dumas gladly offered to present the results to the Academy of Sciences to which Pasteur had not as yet been elected. In this paper he said:

We know how varied are the products which are formed during the so-called *lactic* acid fermentation [*Compt. rend. acad. sci.*, **52**: 344–347, 1861]. Lactic acid, a gum, mannitol, butyric acid, alcohol, carbon dioxide and hydrogen appear simultaneously or in succession, in varying and capricious proportions. I have gradually been led to recognize that the plant-ferment which transforms sugar into lactic acid is different from that or those (since there are two) which control the production of the gummy substance, and that these latter ferments, in their turn, do not produce lactic acid. I have also determined that the previously mentioned plant-ferments could not under any circumstances, if pure, produce any butyric acid.

There should thus exist a specific butyric acid ferment. . . . The communication that I have the honor of presenting today to the Academy deals precisely with the origin of butyric acid in the so-called lactic acid fermentation, a phenomenon on which I have long since focussed all my attention.

I shall not enter into the details of this investigation. I shall limit myself by first stating one of the conclusions of my work; that is: *the butyric acid ferment is an infusorian.*

I was so far from suspecting this result that initially, I attempted to exclude these small animals, because I was afraid that they would feed on the unknown plant ferment that I was trying to find in the liquid media that I was using. But eventually I was struck by the constant relationship between the presence of butyric acid and the infusoria, a circumstance that I had attributed at first to the nutritive value of butyric acid for these animalcules.

Since then, numerous tests have convinced me that the transformation of sugar, mannitol and lactic acid into butyric acid is due exclusively to these infusoria, and that they should be considered as being the true butyric acid ferment.

They are small cylindrical rods, rounded at their extremities, normally straight, single or grouped in chains of two, three, four segments, occasionally more. Their average width is 0.002 mm. The length of isolated segments varies from 0.002 mm to 0.015 or 0.02 mm. These infusoria move by gliding. During this motion their bodies remain rigid or undulate slightly. They tumble, rock, or shake briskly. Their undulations become very evident as soon as their length reaches 0.015 mm. Often they are curved at one or both of their extremities. This is rare when they are young.

They reproduce by simple fission, which leads to the segmented chain-like arrangement. The segment which drags others in its wake sometimes vibrates as if to separate itself from the chain.

Although the bodies of these vibrios appear cylindrical, they may be formed of a series of granules or of very short segments. These are without any doubt the first rudiments of these small animals.

One can seed these infusoria as one would seed brewers' yeast. They multiply if the medium is suitable for their nutrition. But remarkably, they will grow in a liquid containing only sugar, ammonia and phosphates, all of which are substances which can be crystallized and are almost all mineral. Their reproduction is clearly correlated with the butyric acid fermentation. The weight of infusoria formed is ponderable, though always small when compared to the total amount of butyric acid produced.

The existence of infusoria having the characteristics of ferments is already a fact which seems to deserve attention, but a characteristic which is even more interesting is that these infusoria-animalcules live and multiply indefinitely in absence of the smallest amount of air or free oxygen.

It would be too tedious to relate how all traces of free oxygen were eliminated from the liquid media. I shall only add that several members of the Academy have witnessed and verified the validity of the experiments.

Not only do these infusoria live without air, but air kills them. Their growth and reproduction is not affected if one passes for any length of time a current of pure carbon dioxide through the medium. If, on the contrary, one substitutes, for one or two hours only, atmospheric air for carbon dioxide, the infusoria die and the butyric acid fermentation stops.

We reach thus this two-fold proposition:

1. *the butyric acid ferment is an infusorian*
2. *this infusorian lives without free oxygen gas.*

This is, I believe, the first known example of animal ferments and also of animals living without free oxygen gas.

The relationship between the mode of life and the properties of these animalcules and the mode of life and the properties of plant ferments, which also live without the recourse of free oxygen gas, is obvious. So are the deductions that can be made in relation to the cause of fermentations. However, I wish to reserve the ideas that these new facts suggest until I have been able to submit them to the light of experimentation.

And Pasteur's work continued along seemingly diverse lines, which had, nevertheless, a certain unity. He studied the basic problem of spontaneous generation, the formation of vinegar (acetic acid fermentation), the diseases of wines, and the diseases of silkworm. To complete the picture of the basic contribution of Pasteur to the theory of fermentation, we will turn to a paper presented in 1872 in which Pasteur describes rather clearly what we now call the Pasteur effect. We will come back later to the problems of spontaneous generation and silkworm diseases. Pasteur states:

For some time I have been led to consider fermentations *stricto sensu,* as chemical phenomena linked to physiological functions of a specific nature [*Compt. rend. acad. sci.,* **75**: 784–790, 1872].

Not only have I shown that ferments are living organisms and not dead albuminoid substances, but I have also demonstrated the fermentation of sugar, lactic acid, tartaric acid, glycerol, and other fermentable substances in exclusively inorganic media. This constitutes incontestable proof that the decomposition of the fermentable substance is correlated with the life of the ferment for which it is an essential nutrient. Under the conditions that I have described, it is impossible that the organic matter of the newly formed ferments contain a single carbon atom which has not been derived from the fermented substance.

The fact that differentiates the chemical phenomena of fermentations from a multitude of others associated with life, is that a weight of fermentable matter has been decomposed which is larger by far than the weight of the acting ferment. I have long suspected that this special characteristic must be linked with the problem of nutrition in absence of free oxygen. Ferments are living beings, but of a special nature, in the sense that they are equipped to perform all functions of life, including reproduction, without atmospheric oxygen. Let us consider, for example, those unique infusoria which conduct butyric acid fermentation, or tartaric acid fermentation, or other putrefactions, and which not only live and multiply in absence of oxygen gas, but even die if this gas is dissolved in their nutritive medium. This is not all. By precise experiments with brewers' yeast, I have demonstrated that although this ferment survived in the presence of some free oxygen, it lost its fermentative abilities in proportion to the concentration of this gas. The weight of yeast which is formed during the decomposition of sugar becomes progressively greater, and tends to approach the weight of the decomposed sugar, as growth proceeds in presence of increasing concentrations of free oxygen.

The ratio of the weight of fermentable matter decomposed, to the weight of ferment formed, will vary with the effect of free oxygen. The maximum, which will also represent the maximum activity of the ferment, will be attained in absence of free oxygen.

Guided by all these facts, I have been led to regard fermentation as a necessary condition if growth is to occur without direct combustion due to free oxygen.

One can foresee, as a result of this theory, that any organized being, any cell, which lives in the absence of the oxygen of the atmosphere or which does not utilize enough free oxygen for all nutritive reactions, must be a ferment for such substances that it uses as a complete or partial source of heat. It seems that these substances must perforce contain oxygen and carbon, since as I pointed out earlier, these are nutrients for the ferment. Indeed all fermentable substances contain these two elements. I am now bringing, to this new theory that I have been timidly proposing since 1861, the support of new facts, which, I hope, will be convincing.

Let us place in a vessel, a sugary liquid capable of supporting the growth of ferments, and arrange this system in such a way that one can seed this liquid with a particular organized being without danger of contamination.

Let us seed a trace of pure *Mycoderma vini* on the surface of the liquid. During the days that follow, the mold will spread over the liquid forming a continuous veil.

It is then easy to observe that the growth of the mycoderma under these conditions absorbs atmospheric oxygen which is replaced by a roughly equivalent volume of carbonic gas, and that, at the same time, no alcohol is formed.

Let us repeat this experiment under exactly the same conditions, but with this difference, that when the growth becomes continuous, we will agitate the flask in order to break up and submerge the veil, since the fatty substances which are associated with the organism prevent it from becoming completely wet. The next day, often after a few hours, when one operates at a temperature of 25 to 30°, one sees small gas bubbles indicating fermentation of the sugary liquid. The fermentation continues slowly during the following days, so that it is easy to observe the presence of a fair amount of alcohol in the liquid. Careful observation of the cells of the mycoderma, under the microscope, shows that they do not reproduce, but become enlarged, and that the internal structure of their *cytoplasm* undergoes profound modification.

If the fermentation stops, it can be started again by breaking up the veil which has re-formed.

The interpretation of these facts does not seem to be open to doubt. In these two comparative experiments, we have cells which acquire or lose the characteristic of ferment as controlled by the experimenter. What is the difference between the growth conditions of *Mycoderma vini* in the two cases? There is only one difference which cannot be refuted. In the first case, growth of the plant occurs at the surface of the liquid in the presence of air, or even better, of oxygen; in the second case, it takes place either in absence of oxygen, or in contact with an extremely small quantity of oxygen, since the oxygen which tends to go into solution is mainly retained by the cells which stay on the surface. Microscopic examination reveals that the submerged cells are viable, less active than surface cells, and are fermentative. . . .

An interesting and remarkable fact is that these same experiments will work with the true molds. *Penicillium glaucum*, for example, which lives in the presence of free oxygen gas, and which utilizes this gas to satisfy all the nutritional activities of its rapid growth, does not produce any alcohol. But, if, while it is alive, it is denied access to this gas, . . . its metabolic activities are accompanied by the formation of quantities of alcohol and of carbonic acid gas bubbles.

Brewers' yeast, this classical example of a ferment, and the other organized ferments that I have discovered, seem to be plants or animalcules which differ from other lower organisms only in their property of living and proliferating normally and for long periods of time without need of contact with air.

The following facts appear to me to be the logical conclusion of these principles.

M. Bérard, in a study which is a model of clear thinking and clear experimental method, showed us that when fruits are placed in air or in oxygen gas, a certain volume of this gas disappears and is replaced by a roughly equal volume of carbonic acid gas. If these fruits, on the contrary, are left in an atmosphere of carbonic acid or another inert gas, carbonic acid gas is still formed in appreciable quantity, as if it were, as M. Bérard remarked, "the result of a kind of fermentation."

Here is, in my mind, the true interpretation of these facts. When a fruit, or in general any organ, is separated from the plant or from the animal from which it was a part, life has not gone from its cells. The ripening of fruits away from the tree which bore them is a tangible proof of this fact. If air is present, . . . nutrition is similar to the nutrition which would take place if the fruit were still on the tree and which is characterized by the fact that the weight of the transformed materials is comparable to that of the materials which serve as food.

Under these conditions, . . . for a volume of carbonic acid gas produced, an about equal volume of oxygen is consumed. This is common respiratory combustion.

If the fruit, on the other hand, is placed in an atmosphere of carbonic acid, life continues by promptly obtaining, from the decomposition of sugar, the necessary heat for its manifestation. The cells are then in the same state as the cells of ferments which live without free oxygen.

In fact, as soon as the fruit is placed in carbonic acid gas, more carbonic acid gas is produced, and also alcohol. Alcohol is formed in small quantities, to be sure, but in large enough quantities that in one of my experiments, twenty-four plums *de Monsieur*, detached from the tree and placed in carbonic acid gas, furnished me, in a few days, with 6.50 gm of absolute alcohol, while they stayed firm and healthy looking. A corresponding amount of sugar had been destroyed. In contrast twenty-four similar plums, left in contact with air, became soft, watery and very sugary.

Grapes, all acidic fruits, melons, etc. behave in a similar manner. In the future I intend to extend my studies to many plants.

I made sure, in these experiments, that neither brewers' yeast, nor any other ferment was present. It is only in very rare cases that cells of yeasts can enter the fruit.

I do not doubt that the phenomena that I have discussed may have practical applications in the art of making wines and liquors. It is conceivable that, by storing grapes under an atmosphere of carbonic acid, it might be possible to obtain products offering new flavors which might be commercially interesting.

Berzelius (*Ann. chim. et phys.*, **61**: 146–151, 1836), with what is now regarded as remarkable foresight, had recognized that there was a certain unifying principle to be found in a number of apparently diverse phenomena. He termed this principle catalysis. He recognized that certain substances are endowed with a special force (catalytic force) that permits them to decompose certain compounds and to carry out rearrangements in the decomposition products of these compounds without being chemically altered in the process. Examples of such catalytic processes would have included the decomposition of starch by wheat amylase, the decomposition of hydrogen peroxide by alkalis, platinum, gold, and other substances, and, of course, alcoholic fermentation. Nevertheless, Berzelius's approach would have led to at least one scientific dead end because a logical but false deduction growing out of his theory was that yeast was a catalyst that was not transformed during fermentation. According to him the yeast globules were to be considered similar in every respect to platinum, rather than organized cells. Pasteur, on the contrary, demonstrated that fermentation was a way of life: life without air.

As is often the case in scientific discussions, there was some truth in every point of view, and, in 1897, the German chemist Eduard Buchner demonstrated alcoholic fermentation without yeast cells (*Ber. deut. chem. Ges.*, **30**, 1: 117–121). In this paper, Buchner, who was to receive a Nobel Prize in 1907, said:

A separation of the fermentative process from the living yeast cells had not been successful up to now. In the following, a procedure is described which solves this problem.

One thousand grams of brewers' yeast, cleaned for the preparation of compressed yeast, but to which no potato starch had yet been added, was carefully mixed and then ground with its own weight of sand and 250 gm of infusorial earth, until the mass became moist and pasty. One hundred gms of water were added to the paste which had been wrapped in filter cloth and subjected gradually to a pressure of 4–500 atmospheres: 350 ml of liquid was thus expressed. . . .

The most interesting property of the expressed juice is that it can cause the fermentation of a carbohydrate. When mixed with an equal volume of concentrated cane sugar solution, within ¼ to 1 hour a regular evolution of carbon dioxide started which continued the entire day. The same reaction occurred with grape sugar, fructose and maltose. In contrast, no fermentation was observed when the expressed juice was mixed with saturated solutions of maltose or mannitol, just as these compounds are not fermented by living cells of brewers' yeast. Mixtures of expressed juice and sugar solution, which had been fermenting for several days, gradually became turbid when standing in the ice box, even though microscopic microorganisms were not detectable. Microscopic examination under a magnification of 700× revealed numerous protein particles that had probably been precipitated by the acids formed during the fermentation. Saturation of the mixture of expressed juice and saccharose solution with chloroform did not hinder the fermentation, but caused the early formation of a small amount of protein precipitate. Similarly, little fermentative ability was lost by filtration of the expressed juice through a sterile Berkefeld-infusorial earth filter, which certainly retained all yeast cells. The fermentation of the mixture of the perfectly clear filtrate with a sterilized cane sugar solution was delayed about a day at the temperature of the ice box. If one suspended, in a 37% cane sugar solution, a parchment paper bag filled with expressed juice, within a few hours the surface of the bag became covered with numerous tiny gas bubbles. Clearly, sugar diffused into the bag with a resulting lively gas evolution. . . . The fermentative power of the expressed juice was gradually lost with time. . . .

Information on the nature of the active substance in expressed juice is limited at present to the results of a few experiments. By heating at 40–50° an evolution of carbon dioxide took place with a gradual precipitation of coagulated protein. After one hour the juice was repeatedly filtered. The clear filtrate was added again to cane sugar and still had a weak fermentative power in one experiment, none in another.

The active substance seemed, therefore, either to lose its action at this strikingly low temperature or to coagulate and precipitate. Further, 20 ml of the expressed juice was introduced into three volumes of absolute alcohol, and the resulting precipitate was filtered off under suction and dried *in vacuo* over sulfuric acid. Two grams of dry substance was obtained which was mostly insoluble in 10 ml of water. The filtrate from this maceration had no fermentative power. . . .

The following conclusions can be drawn concerning the theory of fermentations: (1) Principally, the complicated yeast cell is not required for fermentation.

(2) The causative agent of the fermentative action of the expressed juice is a rather soluble substance and it should be considered a protein. It will be called zymase. . . .

Buchner's discovery eliminated the distinction between organized ferments, such as yeast cells, and unorganized ferments, such as amylase, which had been isolated from malt by Payen and Persoz in 1833. A new concept had been introduced, however, the distinction between extracellular and intracellular enzymes. The word enzyme had been introduced by Kühne in 1878 to refer to extracellular enzymes. The usage of the word naturally spread to all biocatalysts. Of course, the zymase of Buchner was not a single enzyme but rather a mixture of enzymes. It was what we would now call a cell-free enzyme preparation.

Spontaneous Generation

Studies on vinegar, wine, silkworm, and beer

"From ancient times up to the end of the Middle Ages everyone believed in spontaneous generation," wrote Pasteur in his 1861 "Memoir on the Organized Corpuscles Which Exist in Atmosphere." And he continued:

Aristotle said that any dry thing which becomes humid and any humid thing which dries produces animals. Van Helmont described a recipe for the production of mice. Many authors were still describing in the 17th century, the means of producing frogs from the mud of ponds or eels from the water of our rivers. Such errors could not withstand for long the critical spirit which overtook Europe in the 16th and 17th centuries. Redi, famous member of the *del Cimento* Academy, showed that the worms found on putrifying meat were the larva from eggs of flies. His proofs were as simple as they were conclusive since he showed that surrounding the meat with a fine gauze prevented the birth of these larvae. Redi was also the first to recognize that among animals which live in other animals there were males, females and eggs. At a later date, Réaumur noted that it was possible to observe flies depositing their eggs in fruits and that when a worm is seen in an apple, one should know that it hadn't been created by decay, but that on the contrary, it was the worm which had caused the decay of the fruit.

But soon, in the second half of the 17th century and the first half of the 18th, microscopical observations became numerous and the doctrine of spontaneous generation reappeared.

We would interrupt Pasteur's text here to point out that he was not aware of the work of Louis Joblot. In 1718, this microscopist-protozoologist, whom we have previously mentioned, had written: "I boiled some fresh hay in ordinary water for more than one quarter of an hour. Afterwards I put equal quantities of it in two vessels. . . . One of these I closed as well as I could with well wet parchment before it was cooled; the other I left uncovered. In the uncovered vessel I found animals at the end of several days, but I did not find one in the infusion that had been covered." He thought that the air was the carrier of little animals and their eggs. His concept was clear, but his work was unknown since it escaped even Pasteur, to whose text we now return.

Such was the state of the problem when, in London, in 1745, a work was published by Needham, a dextrous investigator and a Catholic priest. . . . The doctrine of spontaneous generation was supported in this work by completely new facts. He was indeed the first one to think of using hermetically closed vessels which had been exposed to the action of heat. . . . It was mainly due to the support that it got from the system of Buffon on generation, that the work of Needham received attention. . . . In the second volume of his *in quarto* edition, which appeared four years after Needham's book, Buffon proposes his system of organic molecules and defends the hypothesis of spontaneous generation. One can presume that Needham's results had a great influence on Buffon's views, since Needham came to Paris and was Buffon's collaborator when the illustrious naturalist was writing the first volumes of his work.

The ideas of Needham and Buffon had their partisans and detractors. They were in disagreement with another famous system, that of Bonnet, on the preexistence of germs. . . . It seems that there were two men of different natures in Buffon. One would admit frankly one day that he was looking for an hypothesis on which to erect a system, and the other, the next day, in a beautiful preface to his own translation of Hales' *Vegetable Staticks* placed the necessity of experimentation in its proper, high place. These two facets of Buffon's genius can be found to different degrees in all scientists of his time. But Needham's conclusions were soon subjected to experimental challenge. There was then in Italy one of the most able physiologists that have adorned science. It was the most ingenious and critical Father Spallanzani. . . .

Spallanzani published in Modena, in 1765, a dissertation in which he refuted the systems of Needham and of Buffon. This book was translated into French, probably at Needham's request, since its 1769 edition is accompanied by notes written by him in which he answers all Spallanzani's objections. Probably struck by the value of Needham's objections, Spallanzani went to work again and soon published the details of his fine work in his *Opuscules of Physics*. . . .

Spallanzani is usually considered as having been the victorious opponent of Needham. If such was the case, should we not be surprised to find out that even today the doctrine of spontaneous generation has so many partisans? . . . An impartial examination of the contradictory observations of Spallanzani and of Needham on the most delicate point of this subject will show us, as a matter of fact, that, contrary to the generally accepted opinion, Needham could not, in all justice, abandon his doctrine simply because of Spallanzani's experiments. . . . Needham's objection can be found in the following remark that he made about Spallanzani's work: "From the way he has treated and tortured his vegetable infusions, it is obvious that he has not only much weakened, and maybe even destroyed, the *vegetative force* of the infused substances, but also that he has completely degraded . . . the small amount of air which was left in his vials. It is not surprising, thus, that his infusions did not show any sign of life.". . . Further advances in science even seemed to legitimate Needham's objection.

Appert applied to home economics the results of the experiments of Spallanzani which had been carried out according to Needham's method. . . . [Gay-Lussac investigated this process and came to the following conclusion (1810):] "One can demonstrate by analysis that the air of the bottles, in which substances (beef, lamb, fish, mushrooms, must of grapes) have been well preserved, does not contain any oxygen, *and the absence of this gas is thus a condition necessary for the preservation of* animal and vegetable substances."

The fears of Needham on the possible alteration of the air in the vases used in Spallanzani's experiments were justified as shown by the lack of oxygen in Appert's preserves.

But an experiment by Dr. Schwann constituted a notable advance in this question. . . . He showed in 1837 that "air which had been heated, then cooled, left unchanged a meat broth which had been boiled." This was a great advance . . . since it conclusively refuted the assertion of Gay-Lussac concerning the role of oxygen in the canning processes of Appert and in alcoholic fermentation. . . . The experiments of Doctor Schwann have been repeated and modified by numerous investigators. MM. Ure and Helmholtz have confirmed his results with similar experiments. M. Schulze instead of heating air before it reached his Appert-type preserves, passed it through chemical reagents: potassium hydroxide and concentrated sulfuric acid. Messrs. Schroeder and von Dusch thought of filtering air through cotton. . . . The first paper of Messrs. Schroeder and von Dusch appeared in 1854 and the second in 1859. These are excellent publications which have also the historical value of revealing the status of the problem, as of 1859. . . .

"It seems from these experiments, said Messrs. Schroeder and von Dusch, that there is some decomposition of organic substances which is spontaneous, which needs only the presence of oxygen gas to get started; for example, the putrefaction of meat without water, the putrefaction of the casein of milk and the transformation of milk sugar into lactic acid. In addition there are other phenomena of putrefaction and fermentation which are associated by error with the preceding, such as the putrefaction of meat broth and alcoholic fermentation which require for their initiation, apart from oxygen, these unknown things which are admixed with atmospheric air and which are destroyed by heat, as shown by the experiments of Schwann, and by filtration of this air through cotton as shown by our own experiments. . . ." Schroeder alone came back to this subject in 1859 when he wondered if the "things" found in air were "organized microscopic germs" or "a still unknown chemical substance."

One can thus state [said Pasteur] that, at that date, those who thought that the problem had been solved did not know its history well.

Thus, in Pasteur's own words, the question of spontaneous generation was not solved in 1858 when Dr. Pouchet, a distinguished naturalist, who was the director of the Museum of Natural History of Rouen, published a paper, "Proto-organisms . . . Born Spontaneously in Artificial Air and Oxygen Gas." On February 28, 1859, Pasteur wrote to Pouchet: ". . . I think, Sir, that you are in the wrong, not because you believe in spontaneous generation, since it is difficult in such matters not to have a preconceived idea, but in stating that spontaneous generation exists. . . . In my estimation the whole question still lacks proofs. . . ."

The French Academy of Sciences was moved by Pouchet's work and proposed a prize for the best work along the following line: *"To try, by well performed experiments, to throw new light on the question of spontaneous generation.*" Biot and Dumas tried to discourage Pasteur from wasting his time on such a controversial subject. However, the origin of microorganisms had to be understood before microbiology, which was emerging from his studies on fermentation, could be established as an experimental science. And so, despite the advice of his friends, Pasteur entered into the controversy. In disproving spontaneous generation, he developed the procedures that are still basic in studies in microbiology.

Pouchet had repeated many of the experiments of Spallanzani, Schulze, and Schwann. His results led him to conclude that spontaneous generation was indeed a fact. However, he believed not that life could be generated from purely mineral constituents, but that in solutions containing organic compounds, previously formed by living organisms, there was a vital force that permitted the formation of living creatures. Pouchet's theory was quite similar to that of Buffon and Needham.

As he had done in the past for other topics that he had studied, Pasteur published a few papers on the subject before publishing a comprehensive memoir. His first publications on spontaneous generation date from 1860, and his previously quoted memoir came out in 1861. It won the prize offered by the Academy (*Prix Alhumbert*–1862).

Pasteur's *mémoire* was a masterpiece of logic. His rigorous experimental approach dealt the death blow to the theory of spontaneous generation. Pasteur demonstrated first that air was filled with microorganisms by filtering air through guncotton. After a large enough sample of air had been filtered through, the guncotton plug was dissolved in a mixture of alcohol and ether. The sediment of this solution was found to contain bodies that could not be distinguished from many of the animalcules previously observed in various natural substrates and infusions. Next Pasteur repeated Schwann's experiment that consisted in sterilizing air by passing it through a red-hot tube. Pasteur then showed that the introduction of the guncotton plug into a heat-sterilized infusion resulted in prompt growth of the animalcules. He also showed that heat-sterilized infusions could be kept sterile in an open flask as long as the open part was tortuous enough to permit the germs in the air to settle on the sides of the tube before reaching the liquid. He performed these experiments with his famous swan-neck flasks. He further showed that the density of microorganisms in air varied from place to place.

Everyone should have been convinced, but as Pasteur had written in 1860: "One cannot expect the doctrine of spontaneous generation to be abandoned as long as one serious argument can be presented in its favor. This is due to its kinship with the unfathomable mystery of the origin of life on the face of the earth. Such a doctrine can be compared with the mythical monster which had many ceaselessly regenerated heads. They must all be destroyed."

Pouchet, for one, was not convinced, and the French Academy appointed a commission in front of which Pasteur and Pouchet would have to carry out experiments on spontaneous generation. Pouchet objected to certain of Pasteur's experiments and withdrew. Pouchet did not want to admit that microorganisms or their seeds were abundantly distributed in atmosphere because: "If panspermism was not just a fiction, atmosphere would be obscured by eggs and seeds, all motion and all respiration would become impossible." He also argued that if panspermism was true, there would not be any microorganisms with very specific habitats. He said that, for example, there is a fungus that grows only on dead spiders, another one that grows only on horse hooves, still another only on the caterpillars of certain moths. "Short of having the imagination of Bonnet," wrote Pouchet in 1868, "is it possible to suppose that nature would uselessly encumber the air of the globe, with the only aim of seeding a few dead spiders or moths? . . ."

The last important heterogenist was Bastian. In 1872 he published a huge book entitled *The Beginnings of Life* in which he stated his belief in spontaneous generation. With unusual determination, he held to his faith in spite of the fact that his arguments were destroyed with monotonous regularity by Pasteur and his collaborators. It was through the objections of Bastian, however, that more was learned about the effect of acidity on the survival of bacteria exposed to heat. The effect of acidity on sterilization was also studied by Roberts, who showed that neutralized hay infusions were more difficult to sterilize than acidic ones (1874). In this same paper Roberts made what might well be the first observation of microbial antagonism:

> The avoidance of air-contamination is important for another reason. The air is admitted, by most observers, to be highly charged with fungoid germs, and the growth of fungi has appeared to me to be antagonistic to that of bacteria, and vice versa. I have repeatedly observed that liquids in which the *Penicillium glaucum* was growing luxuriantly could with difficulty be artificially infected with bacteria; it seemed, in fact, as if this fungus played the part of the plants in an aquarium, and held in check the growth of bacteria, with their attendant putrefactive changes. On the other hand, the *Penicillium glaucum* seldom grows vigorously, if it grows at all, in liquids which are full of bacteria.
>
> It has further seemed to me that there was an antagonism between the growth of certain races of bacteria. On the panspermic theory it

may be assumed that, what takes place when an organic liquid is exposed to the contamination of air or water, is this: A considerable variety of germinal particles are introduced into it, and it depends on a number of conditions (composition of the liquid, its reaction, precedence and abundance of the several germs) which of these shall grow and take a lead, and which shall partially or wholly lie dormant and unproductive. There is probably in such a case a struggle for existence and a survival of the fittest. And it would be hazardous to conclude because a particular organism was not found growing in a fertile infusion, that the germs of that organism were really absent from the contaminating media.

Pasteur was helped in his fight against the theory of spontaneous generation by the studies of the brilliant physicist Tyndall, who showed that the dust and animalcule particles in suspension in air could be seen by passing a ray of light through a dark room. Tyndall built a closed chamber and lined the interior of the chamber with glycerine. After a few days all the particles settled to the bottom, and he was not able to see any particles in suspension when a ray of light was passed through the chamber. Test tubes containing sterile solution were then inserted in the bottom of the chamber. They remained sterile. This experiment was conclusive. Air had been sterilized by sedimentation; no heating had been done, which could have destroyed a "vital force."

Tyndall further showed that various lengths of time were necessary to sterilize various infusions by boiling. Tyndall was led to conclude that bacteria can exist in two forms: a heat-labile form and a heat-resistant form. He found that by heating solutions intermittently it was possible to kill the heat-sensitive forms following their formation from the heat-resistant forms. He thus developed the process of intermittent heating now known as tyndallization. In the same year (1877), Ferdinand Cohn described the spores of *Bacillus subtilis* and demonstrated their heat-resisting properties. The joint effort of all these workers finally killed the chimera of spontaneous generation.

Pasteur's approach to the field of fermentations had many fruitful practical applications. He began to report on his results dealing with the manufacture of vinegar in 1861, and summed up his work on this subject in his comprehensive memoir published in 1864. Pasteur patented his method to prevent minor modifications of his discovery from being exploited by profiteers, and he immediately announced his intention of letting this patent be public domain. His views on the theory of the process of manufacturing vinegar were of course very much opposed to Liebig's. In 1871, Liebig published a long paper attacking Pasteur's ideas. By his own admission he had thought about that paper for ten years before writing it. After such a long period of cogitation Liebig came to the conclusion that vinegar was formed, at least in the best German factories, without the help of Pasteur's *Mycoderma*. This assertion was not made lightly by Liebig, since he had consulted with M. Riemerschmied, the head of one such model factory, and he had been assured that nothing except air and the surfaces of the wood and charcoal could act on alcohol. Besides, M. Liebig had even looked inside one of the production barrels and had not seen any *Mycoderma*. Pasteur answered the same year: "There might be a simple method to convince M. Liebig, although I personally have never tried it; nevertheless it is the property of true theories that they permit logical conclusions to be drawn, the value of which can be stated *a priori*. M. Liebig could ask M. Riemerschmied to take one of his barrels which has long been productive . . . letting it stand with boiling water during half an hour at the most, then, after having removed this water, to start the production process again. According to M. Liebig's theory, the barrel should operate as before, and I state positively that it will not produce vinegar for a long time until new mycodermae will have grown again on the surface of the wood shavings. The boiling water will have killed the old fungus." At this point, the Academy delicately offered to pay for the expenses involved in this experiment. This time, Liebig did not have ten years to think it over, and he never answered.

The success of Pasteur in the field of vinegar manufacture probably led Napoleon III to request his help in the problem of the so-called wine diseases. One of the main problems was to find a method that would permit the shipping of wines over long distances. Pasteur set up a little field laboratory in Arbois, fittingly enough in an unused cafe. He was helped in this by Emile Duclaux, the future director of the future Pasteur Institute. The local artisans produced the primitive instruments used in these studies. In 1866, the first edition of the *Studies on Wine* was published. Pasteur showed that sour wines were caused by contamination of the wine yeast with acid-producing organisms. He also showed that other contaminating organisms were responsible for undesirable secondary fermentations, producing tasteless sparkling wines or oily wines. Pasteur also showed that oxygen played a role in the maturation of wines. Having thus shown that wine was the result of the action of the right yeast and of proper method of maturation, Pasteur pointed out that the application of a modification of Appert's technique, now called pasteurization, was a practical method for killing undesirable ferments.

He added: "One should not confuse the Spanish method of slow heating of wine, which is used only to help oxygen change the color of wine, with my process of heating which aims essentially at destroying parasites."

The famous chemist Dumas, then a senator, was asked to look into the diseases of silkworm, which were causing much concern among the French growers. The interest of Dumas in such problems was natural since he had been born in Alès, in southern France, among the growers of the precious cocoons. He promptly asked for Pasteur's help. France's number one scientific troubleshooter departed for Alès in 1865 and started to study animals, for the first time in his life.

The industry of natural silk production was very important in France. The culture of mulberry bush and the raising of silkworms had started in southern France during the thirteenth century. The industry flourished under the monarchy but declined sharply during the Revolution. The pomp and grandeur of Napoleon's day restored the need for silk. A disease of the silkworm began to undermine French production in 1849, but the cocoon output continued to increase up to the record year of 1853 (26 million kilograms). By 1865 the production had been reduced to 4 million kilograms of cocoons.

As long as sericulture had existed there had been unsuccessful batches of cocoons. In the past it had been possible to attribute such failures to poor handling of the silkworm. However, the new disease could not be prevented or cured by the techniques of sericulture then in use.

As was the case with wines, there was more than one disease of silkworms, and five years passed by before Pasteur published a practical solution suitable for use by the growers. They were five eventful years, which brought both sorrow and joy to Pasteur. He had already lost a daughter in 1859, and two more were to die, leaving only one daughter and a son to grow to adulthood. During that period also, Pasteur had a cerebral hemorrhage, which left him partially paralyzed. As a further trial, the book containing the detail of his research came out in 1870 as the war with Germany started. But these years also brought their share of joy as Pasteur and his collaborators, Gernez and Maillot, elaborated a satisfactory method for curbing the destruction of the valuable insect.

The *Studies on Silkworm Diseases* (2 vols., Gauthier-Villars, Paris, 1870) contained the following preface:

"I should begin by apologizing for having ever started this study. I was so badly prepared for the investigations which are the subject of this book, that in 1865, when the Minister of Agriculture asked me to study the diseases which were decimating the silkworm, I had never even seen this valuable insect. I hesitated to accept this delicate mission. Not only did I lack the hope of carrying it to a logical conclusion, but I also regretted having to abandon, for an obviously long period of time, investigations which were dear to me. . . . This was when the results of my investigations on organized animal and plant ferments were opening to me a wide field. As applications of these studies, I had recognized the true cause of vinegar formation and discovered that wine diseases were caused by microscopic fungi. My experiments had thrown a new light on the question of so-called spontaneous generation. If I can be permitted this antithesis, the role of infinitely small beings appeared to me infinitely large, either as cause of various diseases, especially contagious diseases, or as contributors to the decomposition and to the return to the atmosphere of everything which has lived."

However, the studies on silkworm diseases were to fit harmoniously into the large mosaic of knowledge that Pasteur was elaborating. It had been observed previously that silkworms (*Bombyx mori*) that suffered from the so-called pebrine disease carried microscopic bodies, which were called corpuscles. It is known now that pebrine is caused by an intracellular protozoon (*Nosema bombycis*), which can infect the insect either through infected eggs or by ingestion with food.

Pasteur recognized the infectious nature of the disease and noted that adults without corpuscles never had corpuscular eggs. However, corpuscular moths could produce eggs free of corpuscles. From this basic observation, Pasteur devised a method of curbing the disease and for rearing healthy worms. The moths were mated, and no attention was paid to the males after copulation. Each individual female was enclosed with the eggs that she produced. If the female was found by microscopic examination to contain corpuscles, the eggs were discarded. This simple method, which required only the use of a microscope, permitted the silk growers to obtain robust broods, which were kept healthy by the prevention of infection.

Pasteur also recognized that flacherie was a different disease from pebrine. He thought that flacherie was caused by a vibrio (*Bacillus bombycis*). The modern view (E. A. Steinhaus, *Insect Microbiology*, Comstock Publishing Associates, Ithaca, N.Y., 1947) is that flacherie is caused by a virus which predisposes the silkworm toward infection by the *Bacillus*.

A third disease of the silkworm, muscardine, had been studied previously by Bassi, who had shown, in 1835, that it was caused by a mold, called *Botrytis bassiana* during Pasteur's days, and now placed in the genus *Beauveria*. Bassi is considered by many as the founder of the parasitic concept of infection.

The defeat of France had a profound influence on Pasteur. He, who in the past had been an admirer of German scientific institutions, became a strong germanophobe. He had forewarned that France was misdirecting its energies and not giving adequate support to the development of science and technology. In 1868 he had written a critical article, "Le budget de la science," that succeeded in stimulating the interest of the Emperor. Before Napoleon could act, however, the war broke out and France experienced a crushing defeat. Commenting after the war on reasons for the defeat, Pasteur wrote: "Whilst Germany was multiplying her universities, establishing between them the most salutary emulation, bestowing honors and consideration on the masters and doctors, creating vast laboratories amply supplied with the most perfect instruments, France, enervated by revolutions, ever vainly seeking for the best form of government, was giving but careless attention to her establishments for higher education. . . ." This germanophobia was to give impetus to Pasteur's study of the manufacture of beers. He wanted French beers to emulate the preferred German brews. Whether he succeeded or failed in this patriotic endeavor is purely a matter of one's personal taste, but in his studies, Pasteur extended to brewing principles which he had already applied to vinegar making, oenology, and to sericulture. He demonstrated the role of microorganisms in the manufacture and the spoilage of beers. Once the causative agents were known, it was only a question of studying their properties in order to devise methods of brewing and methods of storing beers that would be rational and successful.

The studies on beer were conducted first in the laboratories of the faithful Duclaux, then professor of chemistry at the University of Clermont-Ferrand, continued in Paris, and concluded in industrial breweries. Altogether, Pasteur received three patents on his brewing process. He summed up his results in a book published in 1876 (*Studies on Beer. Its Diseases, the Causes of the Diseases and Methods of Preservation, Together with a New Theory of Fermentation, Gauthier-Villars, Paris*). With the studies on beer a chapter in the scientific work of Pasteur came to a close. Pasteur had concluded his work in the field of industrial microbiology; he had introduced the laboratory into the factory and the microscope to the farm. From then on, he was to apply his talent to the study of contagious diseases of vertebrates.

Studies on Diseases of Vertebrates

If we think of medicine as it prevailed at the beginning of the nineteenth century, at a time when some of the former surgeons of the Napoleonic armies were remembering with nostalgia the magnificent thighs of guardsmen that they used to amputate, and if we compare it with medicine at the beginning of the twentieth century, the medicine that was practiced on men that held the trenches of Verdun, we note two major differences: the introduction of anesthesia and an understanding of the nature of infectious diseases.

We have already mentioned the pioneering observation by Davy (1798) of the analgesic effect of nitrous oxide. His disciple, Faraday, showed in 1818 that ether had "effects very similar to those occasioned by nitrous oxide." A few experiments with animals followed, but no interest was shown by the medical profession, and, in 1839, a leading French surgeon summed up the general feeling of the profession by stating that "to escape pain in surgical operations is a chimera. We cannot expect such an eventuality in our time." Nevertheless, the solution was around the corner, and in the studies that interest us, we have to cross the ocean to America, for the first time, to find an important discovery. Ether had been used previously, in America on a small scale to relieve pain during the extraction of teeth, but the first use in surgery was performed in 1842 by Dr. C. W. Long of Jefferson, Georgia. When Dr. Long had been a student, he had attended gay little parties called "ether frolics." He had noted after such orgies that he had bruises on his body, even though he could not remember having felt any pain at the time the injury occurred. This gave him the idea of using ether during minor operations. He published his results in 1849. In the meantime, there was in Boston a dentist by the name of Morton who used ether successfully in his work and who also experimented with it on puppies. He wrote to the professor of surgery at Harvard and offered to administer ether to a patient before an operation. A demonstration, which was most successful, was carried out in 1846 at the Massachusetts General Hospital. The surgeon summed it up by stating: "Gentlemen, this is no humbug." The use of this method rapidly became widespread; the very same year it was used in Paris and London. Morton later became involved in a series of lawsuits and died, bankrupt, in 1868. The only thing wanting in the new method was a name. It was suggested by Holmes in 1846.

Oliver Wendell Holmes was a law student when he wrote "Old Ironsides," a poem that saved the famous ship *Constitution* from being scrapped. He then turned his attention to medicine and studied at Harvard and in France. Back home, he taught at Harvard and eventually became dean of the Medical School. He warned the students: "Do not dabble in the muddy sewer of politics, nor linger in the enchanted streams of literature, nor dig in the far off fields of alien sciences. The great practitioners are generally those who concentrate all their powers on their business." Considering Holmes's literary activities, it seems that he had disqualified himself from ever becoming a great practitioner. His paper "The Contagiousness of Puerperal Fever" was published three years before he christened anesthesia. In it, he stated: "The disease known as puerperal fever is so far contagious as to be frequently carried from patient to patient by physicians and nurses. . . . The woman about to become a mother, or with her new-born infant upon her bosom, should be the object of trembling care and sympathy whether she bears her tender burden or stretches her aching limbs." What practical methods did Holmes suggest to curb the spread of puerperal fever? (1) A physician involved in midwifery should not take any active part in post-mortem examinations of cases of puerperal fever. (2) If a physician is present at such autopsies, "he should use thorough ablution, change every article of dress, and allow twenty-four hours or more to elapse before attending to any case of midwifery." (3) If a physician should note, within a short period of time, two cases of the fever in his practice, he should relinquish obstetrical practice for at least a month. This last conclusion must have caused more than one of his colleagues to feel a pinch in the pocketbook! He concluded his paper with a statement that having puerperal fever cases in one's practice was not "a misfortune but a crime." Holmes thus suggested thorough ablutions but did not urge the use of any special disinfectant. At the time of the publication of that paper (1843), was there any knowledge of disinfectants? Pasteur was only twenty-one years old then, but mankind had unknowingly seen microorganisms in action ever since the first man existed. The decomposition of organic matter in the primeval forests, the foul odors coming from the sewage and dumps of the cities, the stale odor of stagnant water, and the putrid smells of dead men and animals had been associated with the revolting odors that sick men could also produce. It is therefore not surprising that even in the nineteenth century, foul odors and disease were associated. The first disinfectants were thus effective deodorants.

A French druggist, Labarraque, sponsored the use of "chlorides of sodium and calcium oxides" (1825) as a means of preventing putrefaction of corpses in the Paris morgue. So striking was the deodorizing effect of Labarraque's method of wrapping corpses in a sheet soaked with sodium hypochlorite, that it received the full endorsement of the Paris prefect of police.

Thus at least one disinfectant was available; what was needed to bring about its effective application was an understanding of the concept of sepsis. Such an application had already been advocated by Bassi, in order to control the parasitic fungus that was responsible for the muscardine disease of silkworm. He had described a method of preventing the disease by segregation and had found that certain chemicals, such as calcium chloride and potassium nitrate, could be used to disinfect contaminated places. Indeed, in 1844, he had postulated that not only muscardine but other diseases, including syphilis and plague, were caused by living parasites. Bassi was aware that "perhaps some of his readers would respond with a smile to his doctrine of living contagions." For the first time, the contagion theory of Fracastoro and Kircher was receiving solid scientific support. Botanists and agronomists were interested by Bassi's work, and Pasteur paid due credit to him in his *Studies on Silkworm Diseases*. The medical profession was not overly impressed even though Bassi's work was mentioned by Henle in a paper that was published in 1840. Henle was a professor of anatomy, specializing mainly in pathology. He advocated strong support of the germ theory of diseases, and at a later date he was one of the teachers of Koch in Göttingen.

The routine use of an antiseptic, to control a human septic disease, was introduced by Semmelweis. Like Holmes, he first studied law before indulging in medicine, and also like him, he studied puerperal fever. In 1846 he became assistant at the obstetric clinic of the Allgemeines Krankenhaus of Vienna. There were two maternity wards in this hospital. One was used to train medical students; the other was used only to train nurses and midwives. Semmelweis noted that the death rate was higher in the ward where medical students were admitted, and that the death rate went down during the summer, when the medical students were absent. He also noted that the medical students were bringing with them the odor of the autopsy room. His fertile mind made the link between the putrefaction of the autopsy room and the disease. By forcing the students to wash their hands and arms in a solution containing chlorine before being admitted into the ward, he was able to reduce the death rate caused by puerperal infections close to the level of the nurses' and midwives' ward.

The head of the clinic was not impressed. He thought that the improvement was due to a change in the *genius epidemicus*. Some of Semmelweis's colleagues were more responsive, and one of them published Semmelweis's results in 1847, giving him full credit. A great controversy followed this publication, and Semmelweis left Vienna for his native Hungary. In Budapest he published the full account of his work on puerperal fever. Disturbed by a lack of recognition, he died, insane, in 1865.

Two years later, Joseph Lister, a surgeon in Glasgow, wrote the following lines in *Lancet* ("On a New Method of Treating Compound Fractures . . . ," *Lancet,* 1867, pp. 364–373 and 418–420):

The frequency of disastrous consequences in compound fracture, contrasted with the complete immunity from danger to life or limb in simple fracture, is one of the most striking as well as melancholy facts in surgical practice.

If we inquire how it is that an external wound communicating with the seat of fracture leads to such grave results, we cannot but conclude that it is by inducing, through access of the atmosphere, decomposition of the blood which is effused in greater or less amount around the fragments and among the interstices of the tissues, and, losing by putrefaction its natural bland character, and assuming the properties of an acrid irritant, occasions both local and general disturbance. . . .

Turning now to the question how the atmosphere produces decomposition of organic substances we find that a flood of light has been thrown upon this most important subject by the philosophic researches of M. Pasteur, who has demonstrated by thoroughly convincing evidence that it is not to its oxygen or to any of its gaseous constituents that the air owes this property but to minute particles suspended in it, which are the germs of various low forms of life long since revealed by the microscope and regarded as merely accidental concomitants of putrescence, but now shown by Pasteur to be its essential cause resolving the complex organic compounds into substances of simpler chemical constitution, just as the yeast-plant converts sugar into alcohol and carbonic acid.

A beautiful illustration of this doctrine seems to me to be presented in surgery by pneumothorax with emphysema, resulting from puncture of the lung by a fractured rib. Here, though atmospheric air is perpetually introduced into the pleura in great abundance, no inflammatory disturbance supervenes; whereas an external wound penetrating the chest if it remains open infallibly causes dangerous suppurative pleurisy . . . in case of puncture of the lung without external wound the atmospheric gases are filtered of the causes of decomposition before they enter the pleura, by passing through the bronchial tubes, which by their small size, their tortuous course, their mucous secretion, and ciliated epithelial lining seem to be specially designed to arrest all solid particles in the air inhaled. . . .

Applying these principles to the treatment of compound fracture . . . it appears that all that is requisite is to dress the wound with some material capable of killing these septic germs, provided that any substance can be found reliable for this purpose, yet not too potent as a caustic.

In the course of the year 1864 I was much struck with an account of the remarkable effects produced by carbolic acid upon the sewage of the town of Carlisle, the admixture of a very small proportion not only preventing all odor from the lands irrigated with the refuse material, but, as it was stated, destroying the entozoa which usually infest cattle fed upon such pastures. . . .

My first attempt [to use carbolic acid] was made in the Glasgow Royal Infirmary in March, 1865, in a case of compound fracture of the leg. It proved unsuccessful, in consequence, as I now believe, of improper management; but subsequent trials have more than realized my most sanguine anticipations.

Carbolic acid proved in various ways well adapted to the purpose. It exercises a local sedative influence upon the sensory nerves; and hence is not only painless in its immediate action on a raw surface, but speedily renders a wound previously painful entirely free from uneasiness. When employed in compound fracture its caustic properties are mitigated so as to be unobjectionable by admixture with the blood, with which it forms a tenacious mass that hardens into a dense crust which long retains its antiseptic advantages, as will appear from the following cases, which I will relate in the order of their occurrence. . . .

The method consisted of wrapping the wound with various fabrics impregnated with carbolic acid diluted either in water or in oil. Oiled papers and tin sheeting were used to keep evaporation down and to prevent wetting of the patient's bed. Splints were used, of course, to permit proper knitting of the patient's bones.

Joseph Lister was born in London, son of a Quaker family. He could not go to Oxford or Cambridge since he was not a member of the Church of England. Instead he attended the nonsectarian University College. He started his scientific career with studies of muscles of the eyes and skin. His first post was in Edinburgh, where his chief was James Syme, a pioneer in modern clinical education. Lister spent much time with the Symes and married Syme's daughter in spite of the fact that she was not a Quaker. As the years passed Lister published papers on the coagulation of blood, inflammation, and gangrene. In 1860 he was appointed at the University of Glasgow and was elected to the Royal Society. A local professor of chemistry introduced Lister to the work of Pasteur and that, together with the circumstances described in his above quoted paper, led him to the use of antiseptics in surgery. Senior surgeons opposed the new method, at least in Great Britain, but it was highly successful on the Continent. Lister continued to work. He introduced the use of ligatures soaked in disinfectants and started to perform operations under a spray of phenol. After seventeen years he finally admitted that the spray was not necessary. It is certain that his work had a tremendous influence on surgery, but he probably was not the first one to use phenol as an antiseptic in conjunction with surgery. Enrico Bottini, an Italian surgeon, started using phenol in surgery two years before Lister. Lister is also credited with having obtained the first pure culture of a microorganism, but this priority may be questioned. In 1877, Lister was appointed professor in London. He faced opposition from his colleagues, and students did not dare to study under him until after they had graduated. But even conservative Englishmen had to recognize the value of his work, and from 1883 to his death in 1912, he received numerous honors.

As Pasteur's "philosophical studies" were transforming surgery from a hazardous affair to a safe, predictable procedure, the great man himself was starting to apply his knowledge of microorganisms to diseases of vertebrates. In these studies, which were to occupy him up to his death, he had several collaborators. Among his early coworkers were Roux, Chamberland, and Joubert, whose names we find associated with his in some of the papers that we will now quote.

Pasteur in collaboration with Joubert and Chamberland wrote in 1878 (*Compt. rend. acad. sci.*, **86**: 1037–1043):

> In order to prove experimentally that a microscopic organism is truly an agent of disease and contagion, I do not see any other method, in the present state of science, than to submit the microbe (New and fortunate expression proposed by M. Sédillot *) to the method of successive cultures outside the living organism. . . .
>
> This is precisely the technique to which M. Joubert and I submitted the anthrax bacterium. We noted, after having grown the bacterium through a large series of such cultures, each fresh culture being inoculated with a droplet from the previous culture, that the last culture of the series was able to multiply and act in the body of animals in such a way that the animals developed anthrax with all the symptoms typical of this affection.
>
> Such is the proof, which we consider flawless, that *anthrax is caused by this bacterium.*
>
> As far as the sepsis vibrio is concerned, the results of our investigations have not been so convincing. As a result, we have intensified our

* *Compt. rend. acad. sci.*, **86**: 634–640, 1878.

investigations on this subject. We have tried to grow the vibrio from an animal which died of septicemia. All our experiments failed in spite of the variety of media used: urine, beer yeast aqueous extract, meat broth, etc.

Our liquids did not stay sterile, but we usually obtained a microscopic organism which had no relation to the sepsis vibrio. . . .

The idea occurred to us that the sepsis vibrio might well be an exclusively anaerobic organism, and that the sterility of our seeded liquids might have been due to the fact that the vibrio had been killed by the atmospheric oxygen which was dissolved in these liquids. Members of the Academy may recall that in the past I have made similar observations with the vibrio responsible for butyric fermentation which not only lives without air, but that air kills.

We then had to try to grow the sepsis vibrio *in vacuo*, or in the presence of inert gases such as carbon dioxide gas. The results fulfilled our expectation: the sepsis vibrio grew as easily in a high vacuum as in the presence of pure carbon dioxide.

It was logical to think that by putting a liquid containing many sepsis vibrios in contact with pure air, one should be able to kill the vibrios and suppress all virulence. This is exactly what happens. . . .

But if oxygen destroys the vibrios, how can septicemia exist since the air of the atmosphere is present everywhere? How do these facts fit into the germ theory? How can the exposure of blood to the dust of air make it septic?

Everything is occult, obscure, and open to discussion when one does not know the cause of phenomena. Everything is clear when it is known. What we have just said is true only for a septic liquid loaded with adult vibrios which are reproducing by binary fission. Things are different when the vibrios have transformed themselves into their germs, that is to say into these shiny corpuscles, described and drawn for the first time in my "Studies on the disease of silkworms." I was at that time referring to vibrios isolated from worms which had died of a disease known as the *flacherie*. Only the adult vibrios disappear, burn up, and lose their virulence in contact with air. The germ-corpuscles are preserved under these conditions; they remain always ready for new cultures and new inoculations.

We can see, at this point, that Pasteur considered it was he, and not Cohn or Koch, who had discovered the endospore of bacteria. Pasteur had an excellent illustration of spore-containing bacteria in his booklet on the diseases of silkworm. It is not quite sure, however, that he attributed the longevity of the bacterium that he studied to the spore. In other words, he observed the spores and he observed also that the bacterium of the flacherie was not easy to kill, but he did not clearly indicate that the spore was the organ responsible for the toughness of the bacterium.

Let us return to Pasteur's text:

> All this does not yet solve the problem at hand. How can there exist septic germs on the surface of objects which are floating in air and in water? Where can these corpuscles be generated? The fact is that nothing is easier than to produce these germs, in spite of the contact of air with septic liquids.
>
> If abdominal serosity, filled with dividing sepsis vibrios, is exposed to air in a layer of at least 1 cm, within a few hours one witnesses a strange phenomenon. On the surface, the presence of oxygen can be observed by a change in the color of the liquid. There, the vibrio dies and disappears. On the contrary, in the deeper layers of the liquid, the vibrios, which are protected from oxygen by their brothers which are dying above them, continue to multiply by binary fission. Slowly they transform themselves into germ-corpuscles by resorbing the rest of their filiform bodies. One no longer sees the moving threads of various lengths, some of which extend beyond the field of the microscope; one sees only a dust of shiny dots, isolated or surrounded by an amorphous matrix which is scarcely visible. Thus, germs are formed which are not susceptible to the destructive action of oxygen, and they constitute the septic dust. We can now understand what seemed previously obscure. We can understand how putrescible liquids are seeded by dust particles which are present in the atmosphere. We can understand why putrid diseases can exist as a permanent phenomenon on the surface of the earth.

One of Pasteur's main concerns in 1878 was still the elucidation of the germ theory of disease. He obviously felt that he had to have an answer to, and ever foresee, the objections of the enemies of panspermism. His studies were becoming more and more physiologically oriented, and, as we can see in the same paper, Pasteur was already examining the causes of immunity:

> Having noted that the anthrax bacterium does not grow to any extent at temperatures of 43–44°, we thought that such might possibly be the explanation of the well known but mysterious fact that certain animals are not susceptible to anthrax. We have observed that hens cannot be infected with anthrax. Could it be that the 42° body temperature of these birds, added to physiological resistance, would prevent the development of the anthrax bacterium in the body of these animals? If

this conjecture is well founded, we should be able to give anthrax to hens easily by lowering the temperature of their body. This experiment was successful. If one inoculates a hen with the anthrax bacterium and submerges its legs in water at 25°, enough to reduce the temperature of the whole body to 37–38°, the temperature of normally susceptible animals, the hen dies in 24 to 30 hours of invasion by the anthrax bacterium. Converse experiments have already given us favorable results. By raising the temperature of animals which are normally susceptible to anthrax we have been able to protect them against this horrible disease which is today without remedy.

And, in the same paper, Pasteur, who was then aware of the work of Lister, dropped this practical remark:

If I had the honor of being a surgeon, since I am convinced of the dangerous conditions which can be caused by the germs of microbes which are to be found everywhere, especially in hospitals, not only would I use only instruments in a perfect state of cleanliness, but also after having cleaned my hands with the greatest of care, I would flame them rapidly, a practice not much more dangerous than a smoker passing a hot charcoal from one hand to the other, and I would only use bandages, cloths and sponges which would have been exposed to air heated at 130 to 150°. I would use only water which would have been heated at 110 to 120°.

In 1880, Pasteur, in collaboration with Chamberland and Roux, discussed the etiology of anthrax (*Compt. rend. acad. sci.*, **91**: 86–94, 1880):

One of the most deadly diseases of livestock is anthrax. All parts of France are affected to different degrees, and the losses are well in the millions. Especially affected is the district of Eure-et-Loir. Of the numerous flocks of sheep which are raised there, the chances are that not a single one is spared by the disease each year. Every farmer considers himself lucky when the death rate does not exceed 2 to 3 per cent of the total number of his animals. All countries are familiar with this disease which is at times so devastating in Russia that it has been named the *"Siberian plague."*

Where does this disease come from? How is it propagated? Is it not possible that the exact knowledge of its etiology could lead to prophylactic measures which would be easy to apply and could rapidly squelch the dreaded disease?

For a long time it was thought that anthrax was produced spontaneously under various conditions. A multitude of predisposing factors have been considered, such as the nature of the soil, the water, the food and the method of rearing the livestock. But the investigations of Davaine and of Delafond in France, of Pollender and of Brauell in Germany, have focussed attention on the presence of a microscopic parasite in the blood of animals which have died from this infection. . . .

This opinion was further strengthened when, in 1876, Dr. Koch of Breslau demonstrated that the bacterium could be transformed from a vibrio or bacillus-form into germ-corpuscles or spores.

Two years ago, I submitted to the Minister of Agriculture and to other officials a project for investigation of the etiology of anthrax which was well received. . . .

The experiments were started in early August 1878. Initially, certain groups of the sheep were fed with alfalfa which had been sprinkled with artificial cultures of the anthrax bacterium.

In spite of the large numbers of spores of bacteria ingested by all the sheep of a given group, many of them escape death, often after having been obviously sick. A smaller number die with all the symptoms of *spontaneous* anthrax after an incubation period of eight or ten days. In its last phase the disease is fulminating. In the past, this aspect of the disease had led to the belief that its incubation period was short.

One can increase mortality by mixing, with the contaminated food, prickly objects such as the end of thistle leaves and barley spikes cut in small fragments about one centimeter long.

It was important to establish by autopsy whether the animals which died under these conditions would show lesions similar to those observed in animals dying spontaneously in stables or in outdoor grazing areas. The lesions were identical in both cases and by their nature they lead to the conclusion that the site of invasion is in the mouth or in the back of the throat. . . .

The question of the origin of the germs of the bacterium still had to be elucidated. If one rejects the idea of the spontaneous generation of the parasite, it is natural to focus one's attention first on the animals which are buried under ground.

To demonstrate the presence of corpuscles having a diameter of one to two thousandths of a millimeter in a sample of soil collected in the fields of Beauce and to show that they can cause anthrax is an arduous problem. It would be comparatively easy if the corpuscles of the anthrax parasite were alone in the soil, but soil contains an infinite multitude of microscopic germs of various species. These organisms hinder each other's growth both in living animals and *in vitro* cultures. . . .

We added blood from animals which died of anthrax to soil which was sprinkled with an aqueous yeast extract or with urine. This was done at summer temperatures and at the temperature that fermenting cadavers impart to the surrounding soil. In less than twenty-four hours multiplication of the bacteria occurred and germ-corpuscles were formed.

These germ-corpuscles remain dormant, but they can germinate and transmit anthrax not only for months, but for years.

These are only laboratory experiments. We still had to find out what happens in the field. In August, 1878, in a garden of the farm of M. Maunoury, we buried, after autopsy, a sheep from his herd which had died of anthrax. Ten and fourteen months later we collected soil from the grave and were easily able to find the anthrax bacillus. When inoculated into guinea pigs this organism caused fatal anthrax. Even more important was the fact that this same search was successful in soil samples collected from the surface soil. . . .

I would not be surprised if at the present moment the Academy doubted the veracity of these facts. The soil is a powerful filter. How could it let germs of microscopic organisms rise to the surface? . . .

Earthworms are the messengers of the germs and it is they, which from the depth of the burial, bring back to the surface of the soil the terrifying parasite.

The germs of anthrax in addition to a multitude of other germs can be found in the cylinders of finely particulate soil deposited by the worms on the surface of the soil. . . .

And now, is not the prophylaxis of anthrax disease obvious? One should not bury the animals in fields which will be used to grow crops or in fields which will be used for grazing. Whenever possible one should choose for burial, poor soils which are not suitable to the life of earthworms.

In conclusion, I daresay that if the farmers wish it, the disease of anthrax will soon be only a memory. . . .

Pasteur was thinking of the control of the disease by preventing infection. This was still the idea expressed in our next selection, which discusses *fowl cholera*. This paper (Pasteur, *Compt. rend. acad. sci.*, **90**: 239–248, 1880) also indicated that new methods of control were in the making.

At times chicken yards are infested with a devastating disease which is commonly called *fowl cholera*.

The affected animal is weak, walks with difficulty and its wings drag flaccidly. The body feathers stand away from the body so that the animal is shaped like a ball. It wants to sleep. If one forces it to open its eyelids it seems to emerge from a deep slumber into which it soon sinks again. Most often the animal dies, without stirring, in silent agony. At most the wings are moved for a few seconds. Internal damage is extensive. The disease is caused by a microscopic organism,[1] which was first implicated by M. Moritz, from high Alsace, and better described in 1878 by Perroncito, from Torino. It was finally isolated in 1879 by M. Toussaint, at the veterinary school of Toulouse, who cultivated the small organism in neutral urine and showed that it was the virulent agent of blood. . . .

An excellent medium is a broth of chicken muscle neutralized by potassium hydroxide and sterilized by heating at 110–115°. The multiplication of the microscopic organism in this culture medium is fantastically rapid. . . .

A few drops of a culture of our microbe, deposited on bread or on meat fed to the chicken, are enough to cause the disease by intestinal invasion. The small microscopic organisms multiply abundantly in the intestine so that the excrements of diseased animals are deadly. One can thus understand the manner of natural dissemination. Evidently the excrement of sick animals is important in contagion. The spread of the disease could be arrested by isolating the affected animals for a few days. The chicken quarters could be washed with water, especially water acidified with a little sulfuric acid, which easily destroys the microbe and prevents its growth at a concentration of less than a gram per liter. . . .

Certain modifications in the method of culture of the infectious microbe can decrease its virulence. However, I request the Academy to permit me the liberty for the moment of not disclosing the methods which enable me to produce the attenuation that I just mentioned.

The same year Pasteur continued:

The time has come for me to explain the statement that I made previously, that there are various stages of virulence in fowl cholera [*Compt. rend. acad. sci.*, **91**: 673–680, 1880].

This, at first, sounds strange when one realizes that the virus causing this disease is a microscopic organism which can be handled, as a pure culture, like beer yeast or the mycoderma of vinegar. However, it is logical to suspect that variable virulence is probably a common property of the causative agents of various virulent diseases. . . . For example, do we not witness extremely serious epidemics of smallpox as

[1] Now called *Pasteurella multocida*.

well as very mild ones? Do we not see the great epidemics disappear slowly only to reappear later and vanish anew? . . .

Let us begin with the cholera virus in a state of high virulence. This can be done by collecting the virus from a chicken which just died of the chronic form of the disease rather than of the acute form.

Let us make successive pure cultures of this virus, in chicken muscle broth, by taking each inoculum for a new culture from the old one, and let us test the virulence of these various cultures. There is no detectable change in virulence. . . .

However, in my previous statement, I did not mention the length of time elapsing between two successive transfers of the culture. If this interval is one day or one week, the virulence does not change. If the interval is two weeks, it is the same. . . . If we wait three, four, five, or eight months or more before we study the virulence of the cultures, then the whole picture has changed. Differences in virulence are now important.

When there is such an interval between the successive transfers, virulence, which was initially such that ten chickens out of ten died, decreases steadily. Only nine, eight, seven, six, five, four, three, two, then only one chicken in ten dies. Sometimes mortality may be nil and all the inoculated chickens may be sick, but they all recover. In other terms, by a simple change in the mode of cultivation of the parasite, by a mere increase in the length of time elapsing between the successive transfers of the virus, we can obtain a true vaccinal virus, which does not kill, but gives a benign form of the disease and protects from the fatal disease.

One should not assume that attenuation proceeds with mathematical regularity [stressed Pasteur cautiously].

Pasteur, Chamberland, and Roux the next year wrote (*Compt. rend. acad. sci.*, **92**: 429–435, 1881):

We have turned our efforts to the possible application of the action of the oxygen of air to the attenuation of viruses.

The anthrax virus being one of the best known, first attracted our attention. . . .

In neutral chicken broth the bacterium does not grow at 45°. Its growth is excellent at 42° and 43°, but spores are not formed. As a consequence one can maintain spore free mycelial cultures of the bacterium in contact with air at 42° to 43°. One can then note the following facts which are most remarkable. After about a month, the culture is dead. One or two days before the death of the culture, abundant growth can still be obtained by subculture. As far as virulence is concerned, it is lost after eight days at 42–43°; as judged by lack of virulence for guinea pigs, rabbits and sheep. . . . How is virulence lost during these eight days at 43°? Let us recall that the microbe of fowl cholera also dies when it is maintained in contact with air. In that case, however, the length of time required for death is much longer and in the interval the microbe slowly becomes attenuated. Would we not be justified in thinking that the same transformations are undergone by the anthrax microbe? Experimentation confirms this hypothesis. Before the extinction of virulence, the anthrax microbe passes through various degrees of attenuation and, as was also observed for the microbe of fowl cholera, each one of these states of attenuated virulence can be reproduced by cultivation. Each one of our attenuated anthrax microbes constitutes a suitable vaccine for protection against a less attenuated strain. In this gamut of anthrax microbes, are there viruses capable of causing non-fatal anthrax in sheep, cows and horses but which would make them refractory to the fatal disease? We have been successful with sheep. As soon as the grazing season starts in the Beauce we will test our vaccine on a large scale. . . .

So, the large-scale experiments in the Beauce were performed, but, as Pasteur explained, under the auspices of an independent impartial organization.

Last April, the Society of Agriculture of Melun, through its president, the Baron of la Rochette, proposed a decisive experiment to evaluate the results that I had just announced to the Academy [*Compt. rend. acad. sci.*, **92**: 1378–1383, 1881]. I accepted promptly and the following plan of action was agreed to on April 28:

1. Sixty sheep will be put at the disposal of M. Pasteur by the Society of Agriculture in Melun.

2. Ten of these sheep will not be treated in any way.

3. Twenty-five of these sheep will receive two inoculations of vaccine twelve or fifteen days apart, with two anthrax viruses attenuated to different degrees.

4. Twelve or fifteen days after the last inoculation of vaccine, these twenty-five sheep and the twenty-five remaining sheep, will be inoculated with a highly virulent strain of anthrax microbe.

The twenty-five unvaccinated sheep will all die, the twenty-five vaccinated animals will live. The ten untreated sheep will be used for comparative purposes in order to show that vaccination does not prevent sheep from returning to normal.

5. After the inoculation of the very virulent virus, the fifty sheep will remain in the same stable. The two groups of animals will be distinguished by a hole which will be punched in the ear of the vaccinated sheep.

6. All the sheep which die of anthrax will be buried side by side in an enclosed field.

7. During the month of May, 1882, twenty-five fresh sheep will be sent to graze in this field. This will prove that contamination can occur by means of anthrax germs brought to the surface of the soil by earthworms.

8. Twenty-five fresh sheep will be sent to graze in a field, a few meters away, in which no animal dead of anthrax has ever been buried. None of them will die of anthrax.

The experiments were started on May 5, in Pouilly-le-Fort, close to Melun on a farm belonging to M. Rossignol. . . .

It was decided to replace two sheep by two goats and to add to the group ten bovidae: eight cows, a steer and a bull.

On the fifth of May, 1881, twenty-four sheep, a goat and six cows received five drops of attenuated anthrax virus, using a syringe of Pravaz. On May 17, the process was repeated using a more virulent virus.

On May 31, the very virulent strain was used to inoculate the thirty-one animals previously vaccinated and also the twenty-four sheep, a goat and four cows which had received no previous treatment.

The very virulent virus had been grown from germ corpuscles of a strain of the anthrax parasite which had been kept in my laboratory since March 21, 1877.

On June 2, visitors came to witness the results. They were amazed. The twenty-four sheep, the goat and the six cows which had received the attenuated virus were all obviously in good health. On the contrary, the twenty-one sheep and the goat which had not been vaccinated were already dead of anthrax. Two other non-vaccinated sheep died under the very eyes of the spectators. The last one passed away within the day.

The unvaccinated cows were not dead. We have already proved that cows are less susceptible to die of anthrax than sheep. However, all had copious oedema, at the point of inoculation. . . .

Pasteur was a great fighter. Whenever the truth, as he saw it, was attacked, he would counterattack, until the last opponent was either convinced of the value of his arguments or unable to strike back.

As an example of his polemical style, here is what he wrote to another great bacteriologist, Robert Koch.

To Monsieur Koch
Privy councillor in Berlin

Paris, this December 25, 1882

Sir,

In 1881, you hastily and carelessly attacked my work in the first volume of the reports of the *Imperial Public Health Office of Germany*. In Geneva, on September 5, 1882, I refuted your erroneous statements. It is unfortunate that you shied away from a public discussion at that time. Even though you chose not to debate with me face to face in front of competent judges, I accept the challenge.

You said, at the Geneva Congress, that I have contributed nothing new to Science. Really, Sir! A general method of attenuation of viruses by a simple exposure to the action of oxygen, the knowledge of new microbes, the investigation of the factors permitting their attenuation, all this to you has no novelty! Of course, it is true that in the German publication that I just referred to, you have implied that the attenuation of viruses was a fable, the effect of some contamination of my cultures or of the presence of a foreign germ on the needle used in vaccination.

Even though I am accustomed to contradictions of all sorts, I must admit that I was bewildered when I read in your article that:

In the study of a disease, I do not search for the causative microbes, that I do not care where they are located, and that I do not try to demonstrate their parasitic properties in every single case.

It is indeed necessary to have these lines before my eyes to accept the fact that they have been written. . . .

You go back to the development of anthrax in chickens by the mere cooling of the subjects. . . . In 1881, in the reports of the *Imperial Public Health Office of Germany* you doubted the veracity of this experiment. Today you are more reserved and you admit its veracity. I am grateful to you for this change. However, you do not accept my interpretation of the results. The manner of fixing the wings of the hens on small boards does not meet with your approval. . . . According to you, 33 per cent of the hens which are inoculated with the anthrax microbe under normal conditions, develop the disease.

German chickens may very well be more cooperative than French chickens. Personally I have never been able to give anthrax to hens which had not been cooled, be they fixed on boards or not.

Your first publication dealt with *Anthrax* or *Milzbrand* and appeared, as you recall in 1876. Here is how I referred to it, on April 1877, in front of the Academy of Sciences:

"In a remarkable Memoir, doctor Koch observed that the small filiform bodies discovered by M. Davaine can transform into shiny corpuscles. . . ."

As you can see, sir, I was one of the first to recognize the value of your work. . . .

Will you say that you did not know of my memoir on the diseases of silkworms? . . . You have persisted in ignoring my work simply because you did not want to admit that your study on the anthrax bacillus should be considered, in spite of its intrinsic value, as a new application of principles which I had already established.

In summary, it is not you, sir, who has found the mode of generation of bacilli and vibrios by spores; it is not you who has noted their unusual mode of formation; it is not you who has recognized that they survive as a dry dust and that their viability persists over long periods. The precision with which I described and illustrated the formation of these cysts, germ-corpuscles, or spores is such that you could have copied the plate which is present on page 228 of my Memoir of 1869–1870 and introduced it into your paper of 1876. It would have illustrated what you were saying on *Bacillus anthracis*. . . .

You may recall that this letter was dated 1882. Pasteur's greatest success was still to come. In October, 1885, Pasteur addressed the Academy of Sciences to announce that a method had been tried, apparently with success, to prevent the development of rabies in persons who had been bitten by rabid animals. Pasteur said (*Compt. rend. acad. sci.*, **101**: 765–773, 1885):

This method essentially rests on the following facts:

The inoculation into a rabbit, by trepanation, under the dura mater, of medulla from a dog suffering from street rabies, always produces rabies after a mean incubation time of about fifteen days.

If the virus is passed from this first rabbit to another, and from the second to a third, and so on, using the same method of inoculation, then with each successive passage the incubation time of rabies, in the rabbits, decreases.

After twenty to twenty-five passages from rabbit to rabbit, the incubation period is eight days. After another twenty to twenty-five passages the incubation period is seven days. The incubation period remains seven days up to the ninetieth transfer, the last one made.

This type of experiment, which was started in November 1882, has continued for three years without any interruption in the chain. . . . Rabies virus of perfect purity and homogeneity can be maintained easily. This is the *practical* basis of the method.

The rabid medulla of these rabbits is always uniformly virulent throughout its whole length.

When sections of these medullae a few centimeters long, are excised aseptically and suspended in dry air, the virulence decreases slowly and eventually the medullae sections become avirulent. The duration of this extinction of virulence varies somewhat with the thickness of the pieces of medulla, but mainly with the temperature. The lower the temperature, the longer the virulence remains. This is the *scientific* basis of the method.

With such medulla preparations, a dog can be rendered refractory to rabies, in a relatively short period of time.

In a series of jars, one suspends daily a piece of fresh medulla from a rabbit which has died of rabies seven days after it had been infected with the virus. The jar atmosphere is kept dry with potassium hydroxide. The dog is daily inoculated subcutaneously with a syringe-full of medulla suspended in broth. One starts with a medulla which has been drying for a relatively long time, and continues with fresher pieces of medulla, each one having been dried two days less than the previous one. The desiccation period of the medulla, for safety at the first injection, has to be determined by pilot experiments.

When the last injection has been made with medulla which has been drying for only two days, the dog has become refractory to rabies. The virus can be inoculated with safety under the skin or even in the brain of such a dog.

Using this method, I had succeeded to make fifty dogs of all sorts refractory to rabies without a single failure, when on Monday, July 6th, three persons who had come from Alsace, presented themselves at my laboratory:

Théodore Vone, grocer at Meissengott, bitten on the arm, on July 4th, by his own dog which had become rabid;

Joseph Meister, nine years old, bitten also on the 4th of July at 8 o'clock in the morning by the same dog. The child had been overpowered by the dog and had numerous bites on the hand, the legs, the thighs. A few were deep and he could walk only with difficulty. The most prominent bites had been cauterized on the fourth of July at 8 o'clock at night by Dr. Weber of Villé;

The third person, who had not been bitten, was the mother of the young Joseph Meister.

The autopsy of the dog, which had been killed by his master, revealed that his stomach contained hay, straw and wood fragments. There was no doubt that the dog was rabid. Joseph Meister had been picked up from under the dog covered with saliva and blood.

M. Vone had many contusions on the arms, but he assured me that his shirt had not been penetrated by the teeth of the dog. Since he had nothing to fear, I told him to go back to Alsace, the same day, which he did. But I kept with me the little Meister and his mother. . . .

The opinion of Dr. Vulpian and Dr. Grancher was that, considering the intensity and the number of the bites there was little doubt that Joseph Meister was to become rabid.

Since the death of the child was almost certain, I decided in spite of my deep concern, to try on Joseph Meister the method which had served me so well with dogs. . . .

The 6th of July at 8 o'clock at night, sixty hours after the bites of the 4th of July, in the presence of Drs. Vulpian and Grancher, the little Meister received the first injection. . . .

I decided to give a total of 13 inoculations in 10 days. Fewer inoculations would have been sufficient, but one will understand that I was extremely cautious in this first case.

The virulence of the various pieces of medulla used, was carefully determined by inoculation into the brain of fresh rabbits.

This method showed that the medullae used in the first five days were not virulent, whereas those used in the last five days of the treatment were increasingly virulent. . . .

During the last days, I had inoculated Joseph Meister with the most virulent virus. . . .

Joseph Meister escaped not only the rabies that he might have received from his bites, but also the rabies which I inoculated into him. . . .

Today, three months and three weeks after the accident, Joseph Meister's health does not leave anything to be desired. . . .

Last Tuesday, October 20th, with the kind cooperation of Messrs. Vulpian and Grancher, I started the treatment of a young man of fifteen years of age, who had received unusually severe bites, six full days before the beginning of the treatment. . . .

The Academy will probably not hear without emotion about the courage and the resourcefulness of this child that I have just started to treat. He is a shepherd of Villers-Farlay (Jura) by the name of Jean-Baptiste Jupille. He saw a large dog of a suspicious appearance attacking a group of six young children.

Armed with only a whip and his wooden shoes, the child Pasteur mentions had fought with the dog and killed him, but he was badly bitten.

So it was that Pasteur and his collaborators—Chamberland, Roux, and Thuillier—had found a way of preventing rabies. It was a magnificent culmination to a career during which enough great discoveries were made to have filled the life of several masters of the science.

Money came from all ends of the world, and with it a building was erected to house the laboratories of Pasteur and his coworkers. It was dedicated in 1888 and christened l'Institut Pasteur. The master by then was a tired man, and the future of the Institute was placed in the hands of his collaborators. The first aim of the new Institute was to furnish the proper facilities for the manufacture of vaccines. As it still stands now, the Pasteur Institute of Paris and its branches throughout the world are institutions devoted partly to research and partly to the manufacture of various products, including laboratory media.

Everybody was not happy however, and "anti-Pasteurian" lectures were given by what Lister called "a few ignoramuses." An International Society of Antivaccinators was founded. In the opinions of some people, everything that Pasteur had done was wrong. We simply can quote this humoristic passage from Paul Boullier ("La Vérité sur M. Pasteur," Paris, 1887): "Do you know what is the new treatment of M. Pasteur?"

"It consists in making from sixteen to *forty inoculations* depending on the gravity of the bite."

"Is that called *inoculate?*" "I call that *tattoo.*"

Few shared that point of view, and France heaped on Pasteur's shoulders all the honors that could be given a scientist.

In 1892, an international celebration was held in Paris in honor of his seventieth birthday. As Pasteur watched the delegations of school children who had come to honor him, he probably was thinking of his daughters who died so young. Thanks to his work and ideas, many parents have since been spared the helpless suffering that he had to endure.

References

Bassi, A.: *Del mal del segno* (English translation), Phytopathological Classics, no. 10, American Phytopathological Society, 1958.

Bottini, E.: Dell'acido fenico nella chirurgia pratica e nella tassidermica, *Ann. univ. di med.*, **198**: 585–636, 1866.

Dubos, R. J.: *Louis Pasteur*, Little, Brown and Company, Boston, 1950.

Duclaux, E.: *Pasteur, The History of a Mind,* English translation by E. F. Smith and F. Hedges, W. B. Saunders Company, Philadelphia and London, 1920.

Eve, A. S., and C. H. Creasey: *Life and Work of John Tyndall,* Macmillan & Co., Ltd., London, 1945.

Nicolle, J.: *Louis Pasteur*, La Colombe, Paris, 1953.

Pasteur, L.: *Oeuvres, réunies par Pasteur Vallery-Radot* (7 vols.), Masson et Cie, Paris, 1922–1939.

Roberts, W.: "Studies on Biogenesis," *Phil. Trans. Roy. Soc. London,* **164:** 466, 1874.

Vallery-Radot, R.: *The Life of Pasteur*, English translation by Mrs. R. L. Devonshire, Dover Publications, Inc., New York, 1960 (republication).

Additional References

Gortvay, Gy., and I. Zoltán: *Semmelweis, His Life and Work*, Akadémiai Kiadó, Budapest, 1968.

Théodoridès, J.: *Un grand Médecin et Biologiste, Casimir-Joseph Davaine (1812–1882)*, Pergamon Press, Ltd., London, 1968.

Vandervliet, G.: *Microbiology and the Spontaneous Generation Debate during the 1870's*, Coronado Press, Lawrence, Kansas, 1971.

Koch

3 Pasteur had now built the foundations on which microbiology could be erected as a creative experimental science. In Germany, microbiology was maintained as an adjunct of botany before the advent of Pasteur. Ferdinand Cohn was the foremost German microbiologist when Pasteur appeared on the scene, and Cohn's institute was one of the first laboratories to accept and extend Pasteur's advances. Cohn's first publication on microbiology, "The Life Cycles of Microscopic Algae and Fungi," appeared in 1853. Reminiscing in 1872, he said: "The study of the smallest of all organisms, . . . was then overwhelmingly directed toward morphology and life cycles . . . all newer investigations must be based on Pasteur's work." Appropriately, Cohn introduced Koch to the scientific world.

Robert Koch was born in 1843 in the town of Clausthal, in the Harz mountain area of Germany. His parents were diligent and well disposed toward education, though limited in their financial resources. His father was a mining inspector for the government and eventually rose to a high position in the service. The slow steps up the ladder were accompanied by a continuous increase in young Kochs, up to the number of thirteen. Robert was the third-born. As a child he was intense and energetic with a wide variety of interests. He was active in nature study and collected plants, beetles, caterpillars, butterflies, and minerals. He was also caretaker of the family's chickens, cow, pig, and horse, augmented by rabbits and guinea pigs of his own. There still, apparently, was time for the piano, zither, chess, and the newly discovered art of photography. With all, he found time for emotional involvement, and at age fifteen he was deeply in love with his cousin, Agathe Goedicke. Aunt Marie, Agathe's mother, tried firmly but kindly to dampen the seriousness of Robert's ardour, but Agathe finished the job by discontinuing correspondence. This was probably from loss of interest and the considerable effort demanded to reply to his long emotional letters that were much in advance of the average early adolescent. Robert always remained fond of Agathe. Years later, as Mrs. Kolle, she wrote Robert to request help for her son, Wilhelm, who was interested in a career in bacteriology. Koch gave Wilhelm Kolle continuing guidance and friendship; in time, Kolle succeeded Paul Ehrlich to the important directorship of the Serum Institute at Frankfurt. One of the last letters Koch ever wrote was to Agathe, thanking her for flowers sent him on what was to be his last birthday.

His record, at gymnasium, was between satisfactory and very good. Although the Kochs were not in easy circumstances, it was decided to provide Robert with a university education in medicine. After full family conference, Göttingen was selected. The Georgia-Augusta University in Göttingen was then known as the "big school in the small town." In the year 1862 Göttingen was a rural community. In the morning, cows, sheep, and pigs were herded through the streets, past the windows of the University, to pasture, and evenings the same procession moved back to town. The villagers maintained reverence and respect for the University, the "aura academica." The students were not tempted by the distractions of city life. The medical faculty included Wöhler, in chemistry, who was the first to synthesize an organic compound found in the body; Krause, in pathology, who discovered the specialized sensory nerve endings; Hasse, in medicine, who was among the earliest to help establish it as a scientific discipline; Meissner, in physiology, who was a gifted experimentalist; and Henle, in anatomy, who was the subject of earlier comments. Each was a master in his field, and all were inspiring teachers.

The students were encouraged to try their hands at research, and Koch was among the industrious participants. Koch said later that "at the university I had not been stimulated directly toward my eventual field of research, simply because bacteriology was not taught at that time. However, I owe a debt of gratitude to my former teachers, Henle, Hasse and especially Meissner, who awakened my interest in research." Koch investigated problems under Krause and Meissner and won the university prize of 30 ducats for a study called "Ganglion Cells in the Nerves of the Uterus." He used the money for a trip to a scientific meeting of the Society of German Naturalists and Physicians. His early devotion to research had its humorous touches. The problem with Meissner involved the formation of succinic acid in the human body. Koch was the experimental animal. In line of scientific duty he consumed such items as 20 to 30 grams of potassium malate in a single dose, a pound of asparagus at single sittings, and a half-pound of butter per day for five days. The last feat he declared "was not very feasible because of digestive disturbances."

At about the time that Koch graduated from Göttingen, Metchnikoff arrived to work under Henle and recalled: "When, in 1866, I worked under Henle, in Göttingen, at a time when there were serious investigations on the microscopic agents of infectious disease he remained indifferent and . . . at no time did the question of contagious diseases come up in his laboratory." Yet, in his "Pathological Investigations," published in 1840, which he admitted were more theory than facts, Henle had presented a clear exposition of the foundations of the germ theory of disease. The style was prophetic of Koch's postulates. Henle believed that contagion was due to living organized matter that could be transmitted through the air or by contact. The agent entered the body and multiplied during a latent period. To establish the etiology of a specific disease (1) the agent must be found regularly in the host during the disease, (2) the agent must be isolated, (3) the isolated agent must be shown to be capable of producing the disease.

Koch left the University of Göttingen, in 1866, with his fine new *eximia cum laude* diploma. Twelve eventful, unsettled, and strenuous years passed before the publication of his work on anthrax, in 1878, a period of *Sturm und Drang*.

He turned first toward Berlin, which was then a mecca of medicine, to improve his clinical background. But he found the facilities "organized on so large a scale that the individual, with his special needs, is lost. For example, of the 4,000 cases in the Charité Hospital, the professor selects one or two for presentation before a class of 200. The students sit removed from the patients." A hope to study under Virchow did not materialize. He hesitated in choosing a career as a rural practitioner, a military surgeon, or a ship's doctor. Through the good offices of an uncle he obtained a temporary assistantship at the General Hospital of Hamburg. While home, in Clausthal, and waiting for the disposition of his application to Hamburg, he became engaged to Emmy Fraatz, the daughter of a local official. The courtship was not overly emotional.

The appointment at Hamburg was uneventful at first; but before he left, the city was struck by one of its periodic cholera epidemics, and Koch had an opportunity to whet his appetite for research. He applied himself energetically to the problem. A relative coming to visit him at the hospital remarked: "I recall with horror seeing him occupied with intestinal specimens while his breakfast 'mush' was sitting on a nearby table." Records of his microscopic studies at that time reveal a careful description of "S shaped and gently bent or wavy whip-like structures"; it was his first viewing of the organism that he identified later as the cholera vibrio.

The position in Hamburg carried too small a salary; so, he left and, after some difficulty, was appointed house physician in an institution for idiot children, at Langenhagen. The situation appeared to be satisfactory. In addition to his official duties, he was permitted to practice privately. The practice soon prospered, and in July, 1867, he married Emmy, who had been waiting patiently. The marriage festivities were not protracted; on the day of the wedding, bride and groom had to rush back to Langenhagen so that Koch could minister to a critically ill patient. He soon attained to an expanding practice requiring a horse and carriage, but the deep interest in research continued despite the busy life. Emmy remarked how, on his calls, he would stop to collect samples of brackish water, for microscopic examination.

The apparently good turn in fortune was short-lived. In 1868, his job was abolished by an administrative change, and the Kochs were again on the move. They experienced a number of brief unsuccessful starts, moving continually eastward into the hinterland. Daughter Gertrud, his only child, was born in 1868 in Niemegk, Potsdam. By 1869, they were settled in Rakwitz in the province of Posen. It was a strange place for middle-class Germans. Koch's mother, writing to another son, Adolph, regarding Robert's new situation, stated: "In Rakwitz much Polish is spoken, there are many Jews, one is addressed as gracious sir and there is bowing and hand-kissing." Practice again was exhausting and arduous but interesting and remunerative. In addition to practice, and the never-abandoned microscopy, he raised a variety of animals and plants.

The outbreak of the Franco-Prussian War, in 1870, gave Koch an opportunity to indulge his suppressed love for adventure and travel. Despite legal exemption from service, on the basis of both nearsightedness and Hannoverian citizenship, he enlisted in the medical corps and saw service in fixed field hospitals. He cared competently for the wounded and the more numerous typhoid and dysentery cases. The shells occasionally came close, but the hospitals, to which he was attached, were never caught in the line of fire. He offered the usual complaints regarding monotony of diet and requested from his parents such items as "woolen underwear, not too itchy, and, if possible, a cheap, durable raincoat." Toward the end of 1870, Rakwitz began to feel the absence of his service, and the mayor officially requested his release from military service. Fearing the loss of his practice, he accepted release and returned. The urge for change and improvement was again upon him, and he now sought appointment as a district health officer. Favorable recommendations and very good performance in the examinations gained him certification, despite his youth. The district councillor soon requested his appointment for Wollstein, a town near Rakwitz.

The new position was not a bed of roses, but it was a distinct improvement over the previous situations. The post carried with it an adequate house, supervision of a small hospital, and the possibility of private practice. However, the salary was inadequate for reasonable subsistence. He again settled down to the combination of official duties and exhausting rural practice. Night calls were the rule because the poor peasants could not spare the daylight hours so necessary for economic survival. Koch soon improvised a rudimentary laboratory. In a small room adjoining his consultation office he installed an incubator, a sink, a small darkroom, and a work bench, which was soon crowded with glassware. As we learn from his wife, the overflow soon found its way into the consultation room. In addition, a household chest of drawers was frequently filled with potato slices for cultivation of bacteria and fungi as recommended by Cohn. A tile stove was used as an animal autopsy table. The garden was also called into service and a net for trapping birds helped to furnish experimental animals. Larger animals, guinea pigs, rabbits, and even apes were kept in the district hospital. All expenses were borne by Koch himself from his very modest resources.

Koch's interests in science were broad and included such activities as archaeology and pigeon breeding. Understandably, in his approach to infectious diseases, he did not differentiate sharply between those of man and of animals. He began his studies on anthrax in 1873. Anthrax was one of the major subjects of study and dispute in this early phase of the germ theory of disease. Fortuitously the disease was rampant in the Wollstein area and was in the scope of Koch's administrative concern.

Briefly recapitulating the status of the germ theory of disease at that time, we recall that Pasteur had recognized the possible application of the principle of specific fermentation by specific microorganisms to the etiology of infectious diseases. Lacking the necessary medical background, he was slow to enter into the latter phase of the problem. Lister, then professor of surgery at Glasgow, appreciating the analogy between fermentation and putrefaction in wounds and recognizing the role of bacteria entering wounds from air, developed the antiseptic treatment of wounds with carbolic acid dressings. The great success of the Listerian procedure led to its wide adoption and fortified the case for the germ theory of disease. Investigators began the search for pathogens. Klebs, for example, had identified a variety of bacteria from diseased tissue under the microscope, to which were assigned such names as *Micrococcus septicus* and *Microsporon diphtheriae*. But because science lacked a method of pure culture, the germ theory of disease had yet to be proved. The theory was given additional support, in 1873, by young Otto Obermeier's observation of the presence of threadlike structures in the blood of relapsing fever patients during attacks of fever and their absence during periods of defervescence. Weigert confirmed these observations and demonstrated them to Cohn. Weigert also suggested that these organisms could be classified under the generic name of *Spirochaete* and that the species designation be *obermeieri*, in honor of Obermeier, who had just fallen victim to cholera, while investigating intestinal contents from cholera cases. On the other hand, many of the leaders of German medicine, including Virchow, were strongly opposed to the germ theory of disease, and a running debate continued from one scientific meeting to another. At the Congress of German Surgeons in 1875, Klebs presented an impassioned case for the germ theory. But the famous surgeon Thiersch presented a sound case for those opposed to the theory: "It is still undecided whether putrefaction follows from bacteria or bacteria simply appear where there is putrefaction." Yet recalling the advances in surgical practice achieved with Lister's antisepsis, Thiersch also said: "My heart draws me toward the bacterial theory, but my mind says, wait."

By spring of 1876, Koch felt that the major problems concerned with the etiology of anthrax had been solved, although a marginal note written in his notebook in April listed no less than thirty-two items for further study. Impatient with the continuing appearance of erroneous observations in the literature on anthrax, he decided to publish his findings. Concerned over the possibility of hidden errors, and with doubts regarding his self-designated "primitive technics," he turned to Ferdinand Cohn for criticism. With a unique combination of humility and confidence he wrote:

> Esteemed Professor:
>   Stimulated by your bacterial investigations which were published in *Contributions to the Biology of Plants*, and having ample access to the necessary source material, I have been investigating the etiology of anthrax for considerable time. After many failures, I have finally succeeded in completely elucidating the developmental cycle of *Bacillus anthracis*. I believe I have amply confirmed my results. Nevertheless, esteemed Professor, I would be most grateful if you, as the leading authority on bacteria, would give me your criticism of my work before

I submit it for publication. Since my demonstration material cannot be preserved, I seek your kind permission to demonstrate the critical experiments for you over a period of several days at the Institute of Plant Physiology. Should you decide to grant this favor, please suggest the most convenient time for my trip to Breslau.

Years later, Cohn described his reaction to Koch's letter: "At that time I had already been active in bacterial research for a period of years, and therefore, I had been receiving such communications from dilettantes, regarding alleged discoveries, in this yet undeveloped field. I anticipated little of value from this request from a completely unknown physician from a rural Polish town. Naturally, however, I wrote him that I would be pleased to see his material."

Koch lost little time in making the trip to Breslau, and he arrived heavily laden with equipment, animals, and a microscope. On the first day, Cohn had invited one observer, Traube, one of the early investigators of enzymes. Both were amazed by the simple perfection of the demonstration. Of his reaction, Cohn wrote, "Within the very first hour I recognized that he was the unsurpassed master of scientific research." For the second day of the demonstration, Cohn asked the staff of the Pathology Institute to attend. Weigert was scheduled for autopsy duty, but Professor Cohnheim, director of the Pathological Institute, came to the demonstration. Cohnheim was no less impressed than Cohn. When Cohnheim returned to the Pathology Institute he said to Weigert: "Drop everything and go see Koch; this man has made a great discovery, all the more amazing is that he has developed simple, precise and definitive methods entirely on his own. There is nothing to add. I consider this the greatest discovery in the field of bacteriology and believe that Koch will continue to surprise us and shame us with further discoveries." By the third day the audience included Cohnheim, Weigert, Lichtheim of the Internal Medicine faculty, Auerbach, the histologist, and Traube, in addition to Cohn.

Koch returned home in an elated state. He resumed his experimentation and rapidly prepared his report for publication in Cohn's journal *Contributions to the Biology of Plants*. In three weeks the manuscript of "The Etiology of Anthrax Based on the Developmental Cycle of *Bacillus anthracis*" was submitted. The paper appeared in *Beiträge zur Biologie der Pflanzen* (2: 277–310, 1877). In it, Koch wrote:

Davaine has demonstrated decisively that anthrax could be transmitted with fresh or dried blood from infected animals only when the blood contained characteristic rods, and that these rods were bacteria. He attributed the transmission of anthrax in man and animals to the dissemination of these bacilli, which remain viable in the dried state for long periods of time. However, a variety of objections have been raised to Davaine's conclusions. Some investigators have produced anthrax by injecting blood that contained bacilli but have failed to demonstrate bacteria in the animals so infected, and conversely others believed that they transmitted the disease with bacteria-free blood. Still others have noted that the transmission of anthrax depended not only on a contagious agent, which may be present in the upper layers of soil but that soil conditions also had an influence.

Access to cases of anthrax in animals offered me the opportunity to clarify these questions. I rapidly reached the conclusion that Davaine's principles were only partly correct. The bacilli in anthracic blood are not resistant, as indicated by Davaine's experiments. As I shall demonstrate later, blood containing bacilli maintains its infectivity in dry state only for several weeks and in wet state only for several days. How then does such a susceptible organism survive a winter season and how does it continue to be a source of contagion for years? We can explain this if the organism can change to a form that is resistant to alternate drying and wetting. The present investigations have revealed a spore stage for the anthrax bacillus, similar to the spores demonstrated by Cohn for other bacteria.

According to Cohn's classification of schizomycetes, the anthrax organism belongs to the genus *Bacillus*, and is here designated *Bacillus anthracis*.

1. In the blood and in tissue fluids of living animals, these bacilli multiply by fission. This process has not been observed directly, but can be established from animal passage experiments. Mice were both convenient and available. At first, I infected these animals on the ear or in the middle of the tail, but I found these methods uncertain because animals could remove the inoculum by licking or rubbing. Subsequently the base of the tail or the back was used. The mouse was grasped by the tail, an incision was made in the skin and a small drop of inoculum was inserted into the wound. The results were regularly positive when fresh anthracic material was used. This procedure can therefore be used as a test for viability. This passage technique also assured a continuous source of fresh anthracic material. Even after 20 passages, the spleen was always enlarged and showed numerous, non-motile-non-spore bearing rods of uniform size, which were also found in the blood, but in smaller numbers. Multiplication by fission after elongation was indicated by the observation of forms approximately twice the size of

ordinary bacilli, with indentations in the middle. Also shorter forms, contiguous as pairs, could be seen. The results of this long series of mouse passages indicated little chance of change in morphology or alteration of generation.

The distribution of bacilli in the body of injected animals is not always the same. In guinea pigs, large numbers of bacilli are found in the blood, often equaling or exceeding the number of red blood cells. The blood of the rabbit contains much fewer bacilli, frequently less than one per microscopic field. In mice, there are so few bacilli in blood that one frequently may fail to observe them. On the other hand, in the rabbit there are many bacilli in the lymph glands and spleen, and in mice, astounding numbers in the spleen. Marrow in infected mice contains only occasional bacilli.

2. In the blood of dead animals or in other suitable nutrient liquids, the bacilli grown, within a specified temperature range and with access to air, form unusually long, unbranched, *Leptothrix*-like threads, and many spores. For example, a very fresh drop of beef serum or beef aqueous humor is placed on a glass slide and a fragment of fresh infected spleen is centered in the drop which is covered with a coverslip. The slide is placed in a moist chamber and transferred to an incubator. The humidity of the moist chamber must be so regulated that moisture does not penetrate under the coverslip, and the serum at the border of the coverslip does not evaporate. In the first instance the bacilli move from under the coverslip and in the second instance, the drying of the serum deprives the bacilli of air, and thus hinders growth.

These preparations are incubated for 15 to 20 hours at 35 to 37°C. In a typical trial, many unaltered bacilli are found in the center of the preparation; going toward the edge, one sees bacilli which are 3 to 8 times longer than the original ones with occasional angles and bendings. As the edge is approached, the threads grow longer, attaining 100-fold the length of the original bacilli. The threads lose their uniform transparency, become finely granulated, and light refracting bodies appear at uniform intervals. In the threads nearest the edge, where the exchange of gases is most favorable, and development most advanced, completely formed ovoid refractile spores are present at frequent intervals, suggesting chains of pearls. Many threads are in the process of disintegration and their original structure is suggested only by a mucous substance which holds the spores together. Some spores are free while others occur in clusters. In a successful preparation, there are therefore all stages from the short bacillary rods to the long spore-bearing threads and free spores. Since the non-motile bacilli were originally only in the center of the preparation, it was possible that the threads could have developed from air-borne contaminants penetrating at the edge. The preparations were not protected and indeed colonies of micrococci and other bacilli were frequently seen near the threads. It was important not to consider contaminants as transition forms of the organism selected for study. Since the nutrients, temperature and air-requirements of *Bacillus anthracis* were known, I sought to establish cultures on the microscope stage to facilitate direct observation. After many unsuccessful experiments, I found the following method satisfactory. A petroleum lamp was used as the heater for the microscope stage. The microscope was placed on a stand in order to set the heating cylinder of the lamp under the microscope stage. A small flame was adequate to maintain the stage at the desired temperature. The moist chamber was replaced by a hollow ground slide. Although there was only a small amount of air in the slide preparation experience showed that it was adequate for the test cultures. As a temperature indicator, a drop of tallow was placed on the inner surface of the coverslip and the coverslip was sealed with a ring of oil. The tallow would melt if the temperature of the stage rose to 45°C. A stage temperature of 40°C. was satisfactory for my experiments.

A thin drop of fresh beef serum, or better, beef aqueous humor, was placed on a coverslip and seeded at the periphery of the drop with infected spleen. The coverslip was placed on a glass slide and the preparation was sealed with oil. Initial evaporation from the drop was so slight that only occasional bacilli at the outermost edge dried up. The preparation was placed on the heated stage and after stabilization, the location of bacilli was noted and the microscope tube was then raised to minimize cooling of the preparations. Observations were made at 10 to 20 minute intervals.

In the first two hours the bacilli become thicker. Growth then begins and after three to four hours the bacilli elongate ten to twenty-fold. They then bend and form a network. After several additional hours the individual threads stretch across several microscope fields and appear like a mass of wavy parallel glass threads. Soon the threads become irregular, form knotted masses and it is impossible to follow individual threads. When one observes the free end of a thread for a long time—fifteen to twenty minutes—one can directly observe growth as revealed by elongation. After ten to twenty hours the contents of the more vigorously growing threads become finely granulated. The granules enlarge and become light-refracting spores. The threads then disintegrate, the spores are freed and sink to the lower layers of the drop where they collect in masses. If the development of the bacilli fails or if they grow only sparsely and do not attain spore formation, experimental error should be suspected. For example, at one point I was unable to obtain adequate spore formation although the bacilli elongated in a satisfactory manner. After attempting to implicate the heating apparatus, the nutrient medium, etc., it finally occurred to me that the oil used for sealing had a rancid odor. When I substituted fresh oil for rancid oil, good spore formation was again obtained. Thus it seems that small amounts of vapor of volatile fatty acids could inhibit spore formation.

3. The spores of *Bacillus anthracis* develop into the original bacillary form, as seen in blood, under specific conditions of temperature, nutrient medium and access to air. That these refractile bodies are spores, and not disintegration products of the bacillus, could be assumed with certainty from the analogous developmental stages of other organisms such as fungi and algae. Animal passage experiments, to be described later, showed that spores produced anthrax and confirmed this hypothesis. To obtain complete information on the life cycle of *Bacillus anthracis*, it appeared pertinent to induce spore germination under controlled conditions that could be followed with microscopic examination.

All efforts to induce germination of spores in distilled water and well water, at room temperature and at 35°C., failed. Germination in blood serum and aqueous humor, in sealed slide preparation, was irregular. I finally devised a procedure which solved the problem. Pure cultures of *Bacillus anthracis* were held until spores formed from the threads. Drops of such cultures, containing masses of spores, were placed on coverslips and dried rapidly. The rapid drying favored cohesion of the spores. The preparations were kept dried for several hours to several days. They were then inverted on a drop of aqueous humor placed on a conventional glass slide. The preparations, which were not sealed with oil, were placed in a moist chamber at 35°C. Germination of spores was evident after 3 to 4 hours and the process started at the periphery of the preparation. The peripheral areas were occupied by bacillary threads and few spores while dense concentrations of ungerminated spores were still present at the center.

The sequence of germination is as follows:

The spore, which is initially oval and surrounded by a narrow ring of transparent material, elongates and appears distended. During this elongation, the spore loses refractility, disintegrates and disappears. When the preparation contained few spores, the conversion to bacilli was followed by a reconversion into spores. These observations give the complete developmental cycle of *Bacillus anthracis*. . . .

The demonstration of conversion of bacilli to spores and spores to bacilli, under experimental conditions, still does not elucidate the relationship of these developmental stages to the naturally occurring disease. Anthrax can be disseminated when present in dry or wet carrier substrates. The organism remains viable in dry substrates but there is disagreement on the length of viability under these conditions. Pieces of spleen and lymph gland the size of pea, millet seeds or larger, and blood, obtained from infected mice, rabbits, and guinea pigs were spread and dried on coverslips. The preparations were kept in a shaded, ventilated location.

Samples of dried tissues removed after one day, and then every two to three days, were suspended in aqueous humor, injected into mice and cultured in moist chambers. The samples which were spread very thinly, lost infectivity and viability after 12 to 30 hours. After suspension in aqueous humor the bacilli appeared the same as in the fresh state but they disintegrated rapidly. Thicker layers remained infective and viable for 2 to 3 weeks. The thickest layers maintained infectivity and viability for 4 to 5 weeks. Contrary to Davaine's reports, freshly dried material did not remain infective for longer periods of time, despite a variety of modifications in experimental procedure. Only those dried tissue samples which yielded spore bearing threads in culture, were capable of producing anthrax. This establishes that the direct transmission of anthrax is dependent upon the presence of viable bacilli.

In the next phase of the study, I investigated the temperature limits for spore formation. Paraffin sealed slide cultures were prepared and maintained at various temperatures. Since this experiment was being performed during the winter it was easy to maintain preparations at 5°C. Temperatures above 40° were obtained with the heated microscope stage. The bacilli grew most rapidly at 35°C. and elegant spores were seen after 20 hours at this temperature. At 30° the spores appeared after approximately 30 hours. At still lower temperature, the formation of the spores was correspondingly more languid: at 18 to 20°, approximately 2½ to 3 days were required, under 18°C. sporulation was rare, and spores were not formed below 12°. Above 40° the development of bacilli was retarded and appeared to stop at 45°.

We will now observe the behavior of bacilli, in various liquids, under conditions simulating natural environment. Freshly removed spleen, from infected mice, was triturated with fresh beef blood, or preferably aqueous humor, in proportions which simulated the consistency of serous or mucous liquids from cadavers of animals killed by anthrax. The blended liquids were placed in well-stoppered glass containers and placed in the incubator. The liquids putrefied and the bacilli were destroyed after 24 hours. Anthrax, therefore, cannot be transmitted under such conditions. As we will see, the destruction of the bacilli is due to the absence of oxygen and not to the action of the putrefactive gases. A slide preparation of anthracic blood is prepared without an air bubble and sealed air-tight with oil. The slide is then placed on a heated microscope stage. Initially the sample of blood shows the two spectroscopic bands characteristic for oxyhemoglobin and in 3 hours the bacilli increase 4 to 5 times in length. As the available oxygen is consumed, the hemoglobin is reduced as indicated by spectral examination, and growth of the bacilli ceases. Putrefactive bacteria are not seen and there is no odor of putrefaction. The sequence of events is entirely different if small amounts of oxygen are available. These changes can be easily observed when approximately 10 to 20 grams of the infectious suspension are placed in a watch glass partially covered with a glass plate and kept at room temperature. A penetrating putrefactive odor develops after 24 hours and micrococci and other bacteria appear in large numbers. Nevertheless, *Bacillus anthracis* multiplies, the bacilli of that species elongate after 24 hours and many spores are apparent at 48 hours. After sporulation, the threads disintegrate and the spores sink to the bottom. The other flora grows abundantly for several days, disappears gradually, and the putrefactive odor abates. A slimy sediment forms, the supernatant becomes free of organisms, and the preparation attains a glue-like or cheesy odor. Moderate dilution with distilled water or well water does not deter spore formation, but high dilution with distilled water leads to death of the bacilli after 30 hours. Specific concentrations of salts and protein are therefore required for spore formation. These findings indicate that, in practice, conditions for spore formation are often as favorable as in the above described experiments.

The lack of agreement among previous investigators regarding the duration of viability of dried anthrax blood could be explained. If fresh infected spore-free blood was dried rapidly, no spores would have been present and infectivity would have been maintained for 5 weeks. If the blood was dried slowly, at room or summer temperatures, spores would have developed and viability would have been prolonged beyond 5 weeks. I had a small collection of anthrax tissue specimens which were dried in open bottles under various conditions and at various times. These tissues were cultured in aqueous humor and tested for infectivity in mice. Small pieces of tissue, which dried rapidly, did not contain spores. These yielded no viable bacteria and were non-infectious. On the other hand, large pieces of sheep spleen which were dried slowly, contained spores, gave viable cultures and were infective. All spore bearing substances produced anthrax and in nutrient solutions developed into spore bearing threads of *Bacillus anthracis*. The viability time of dried spores has not been determined accurately. It is likely that it will be a span of years. I have found a sample of sheep's blood, at present still viable and infectious after 4 years.

The possibility remained that the disease, produced with putrefied preparations containing anthrax bacilli or spores, was due to putrefactive bacteria. However, preparations of normal blood and aqueous humor allowed to putrefy, and injected into mice was fatal only for 2 of 12 animals. The mice that died showed enlarged spleens, but their spleens and blood were free of bacteria. Spores from hay infusion, cultured by Professor F. Cohn, did not produce any disease in mice. Similarly, putrefied aqueous humor, containing a spore bearing bacillus resembling *B. anthracis*, did not produce anthrax in mice. Thus, only one type of bacillus can produce the specific disease.

Mice and rabbits fed for several days with spleens from rabbits and sheep that had died of anthrax remained free of disease.

Mice were injected with anthrax, they were sacrificed, by chloroform anesthesia, at 2, 4, 6, 8, 10, 12, 14 and 16 hours and their blood and spleen were tested for infectivity. Bacilli were first demonstrable 14 hours after injection. By 16 hours, the spleens were enlarged, and after 17 hours, the spleens contained dense masses of bacilli. Thus the bacilli reach the blood stream slowly, but having reached a preferred site they multiply very rapidly. In addition to mice, I have also tested rabbits, guinea pigs, dogs, chickens, sparrows and frogs. The latter 4 species were resistant.

The experiments with the frog merit further discussion. Fragments of injected mouse spleen were inserted under the skins of the test frogs and the frogs were sacrificed 48 hours later. The blood was free of bacteria. The fragments of mouse spleen were loosely bound by ahhesions to the surrounding frog tissues. The color of the spleen had changed from dark-brownish-red to a lighter greyish-red. Microscopically, there were large numbers of morphologically unaltered bacilli in the center of the spleen fragments. At the periphery, there were thickened and elongated bacilli but predominantly the bacilli were coiled, and, at some points, enclosed by thin-walled capsules. The coiling of the bacilli was explained as follows: as a reaction to the spleen fragments, there was deposition of a dense liquid within which were pale, turgid cells, the size of erythrocytes, containing prominent nucleus and nucleolus and cytoplasmic granules. In most of these cells there were one or more short straight bacilli, but in some, the bacilli were bent or coiled. They were forced into these shapes by the enclosing cells. The cytoplasmic granules, of these cells, first enlarged and then disappeared followed by disintegration of the nucleus. The cells exploded, freeing the coiled bacteria and leaving empty cell membranes. This appeared to be the first clear demonstration of ingestion of schizophytes by ameboid white blood cells. It had been believed previously that these cells ingested micrococci, but it had been difficult to differentiate between the micrococci and the cytoplasmic granules.

Let us recapitulate the facts that have been established and relate them to the etiology of anthrax, without forgetting that there are still gaps to be filled before the etiology can be fully proven. At the outset we must remember that the experiments were performed with small rodents. Even though it is unlikely that ruminants would react very differently from rodents to the anthrax bacillus, one can note that the experimental disease in the small animals is fatal in 24 to 30 hours but that the natural disease in large ruminants is not fatal before several days. . . . The results of feeding experiments in rodents may not be applicable to ruminants because of the great differences in the digestive processes of the two types of mammals. Inhalation experiments are still lacking. The nature of spore formation in cadavers at various soil depths, temperatures, and in various types of soil, has not been studied even though such data would have practical value. Factors concerned with hindering or preventing penetration of the bacilli and spores into the blood and lymph vessels should also be investigated. Despite these unsolved questions we have now described the developmental cycle and determined the etiology of the disease. Anthrax material, regardless of the length of time it has been aging, can produce and disseminate anthrax only if it contains viable bacilli or spores of *Bacillus anthracis*.

The transmission of the disease by moist bacilli in fresh blood occurs only infrequently, and is limited to people employed at slaughtering, disjointing and skinning animals that may be infected with anthrax. The infected blood or tissue juices gain access through wounds. The disease is transmitted more frequently by dried bacilli which may remain viable up to 5 weeks. The dried bacilli adhering to wool and similar materials may be carried to wounds by insects. Water is an unlikely source of infection because the infectious material loses viability rapidly when diluted. The most usual means of dissemination is through spores. The bacilli in dried state can remain viable only for short periods of time and they cannot survive in wet state or in rain or frost. The spores, on the other hand, are incredibly resistant, surviving year-long dryness, moisture, or alternate wetting and drying. If sporulation has occurred in a particular locale, anthrax will not be eliminated from that area for a long time. A single cadaver, disposed of in an improper manner, can furnish innumerable spores. Even after long sojourn, the spores attached to dust formed from soil, hair, horns, etc., can infect animals through skin lesions. It is also likely that they can reach the blood and lymph vessels from the air passages or digestive tract.

Having determined the manner and conditions for dissemination of anthrax, it should be possible to hinder or prevent the development of *Bacillus anthracis* and reduce or eliminate the disease.

The existing regulations require that the cadavers be disinfected, and buried in moderately deep graves and that the affected area be barred to grazing. The incidence of the disease is highest among sheep and in their case improper disposal of cadavers is frequent. Burying the cadavers in shallow wet graves promotes rather than hinders the development and dissemination of spores. The most certain method to combat anthrax would be to destroy all materials containing *Bacillus anthracis*. However, the large number of cadavers renders chemical, steam or incineration treatment impractical. If we could eliminate or reduce spore formation to a minimum, the disease would be reduced and disappear. We have seen that the bacilli require air, moisture, and a temperature above 15°C. to form spores; if one of these requirements were not satisfied, spore formation would be prevented. The rapid drying or incineration of large cadavers would be difficult operations. On the other hand, the cadavers could be cooled below 15°C. with little trouble and expense, even in summer, and access to air could also be prevented without difficulty. In Germany the average soil temperature at a depth of 8 to 10 meters is below 15°C. Thus, burying cadavers in trenches of this depth would lead to destruction of the bacilli and render the cadavers harmless. The diggings should be made at adequate distances from farm buildings and be covered appropriately. The deep burial would also discourage the operation of thieves who at night reopen burial trenches to recover the cadavers for use as meat. Spore development would also be hindered where the soil is dry and the water table is low. Buhl has reported that anthrax was eliminated from a horse stud farm at Neuhof by draining so as to lower the water table. Appropriate experimental stations should be established to investigate control procedures.

Another prophylactic measure should be considered. Cases of anthrax among sheep herds occur throughout the year, claiming small numbers of animals at intervals of several days or weeks. Cases among beef cattle are fewer than among sheep and are more widely spaced, with remission periods of several months to half a year. Few cases occur among the horses. It appears therefore that the sheep is the usual reservoir for *Bacillus anthracis* and that transmission to other species occurs only under special conditions. Thus Leonhardt has observed that anthrax was eliminated among steers, when they were separated from sheep. It also follows that preventive measures should be focused to eliminate the disease among sheep.

We have now, for the first time, established the etiology of anthrax and determined the soil conditions which facilitate its transmission. Typhoid fever and cholera are two diseases that are similar to anthrax in mode of dissemination. . . . The infectious agents of these diseases may be spherical bacteria or similar schizophytes. . . . However, should organisms be found in typhoid fever and cholera, we would still be hampered by the fact that these diseases do not occur in animals.

Despite these obstacles we should not be deterred from proceeding as far as available methods can carry us. One should first investigate the problems with attainable solutions. With the knowledge thus gained, we can proceed to the next attainable objectives. Diseases such as diphtheria, which can be transmitted to animals, appear immediately amenable to successful investigation. With a knowledge of comparative etiology of infectious diseases we can learn to hold at bay the epidemic diseases of man.

Koch's position as a major force in the newly emerging science of bacteriology was well established with the publication of his studies on anthrax. Between 1876 and 1880, although still limited to his personal laboratory facilities and hampered by the need to support his family with private practice in addition to his job, he continued to establish the basic principles of medical bacteriology. His successive investigations rose steadily, reaching a climax in the "Etiology of Tuberculosis." In 1877 he published a paper on the methodology of the art, including the application of photography, in which he noted that a proper technique of fixation and staining would permit investigators to exchange preparations: "how many incomplete and false observations might have remained unpublished instead of swelling the bacterial literature into a turbid stream, if investigators had checked their preparations with each other!"

He thus introduced the rapid air drying of thin films. To stain the films, he used the aniline dyes that had been introduced by the work started by Paul Ehrlich, in 1875. In addition, he developed chemical fixation techniques and mounting methods. These were sufficiently sensitive to demonstrate delicate bacterial flagella. Photomicrography, a subject long close to his heart, was developed to an exact procedure that could be described for use by others. Here, as for other advances in science during that time, German industry gave its full and expert cooperation.

Koch's rapid sequence of advances was shown by a monograph entitled

Investigations into the Etiology of Traumatic Infective Diseases, which was published in 1878 and soon translated into English (1880). In this monograph we read:

> The frequent demonstration of microorganisms in traumatic infectious diseases renders their parasitic nature probable. However, the proof will be complete only when we succeed in demonstrating the presence of a given type of parasitic microorganism in all cases of a given disease and when we can further demonstrate the presence of such organisms in such numbers and distribution that all the symptoms of the disease may be explained.

Koch then noted that the best available method for demonstrating bacteria in tissues, at that time, was that of von Recklinghausen wherein tissues were treated with acid and alkali to which bacteria were more resistant than other cells. After discussing the inadequacies of the method, he continued:

> An attempt has been made to obtain better results by the use of staining fluids. Hematoxylin has been most recommended, but it does not stain rod-shaped bacteria, and only lightly, the spherical forms. Staining with aniline dyes has given better results than those with hematoxylin. Weigert, in 1878, was the first to use aniline staining for demonstrating bacteria in tissues. The specimens are hardened in alcohol, stained in aqueous methyl-violet, treated with acetic acid, dehydrated with alcohol, cleared with oil of cloves and mounted in Canada balsam. Other aniline dyes such as fuchsin and aniline-brown may be used in place of methyl-violet. In such preparations, only nuclei of cells and bacteria are stained. Individual bacteria are more easily recognized than after hematoxylin treatment. Isolated large bacilli are seen readily but the method yields uncertain results with smaller bacteria and is quite useless with the smallest bacterial forms. . . .
>
> In attempting to photograph bacteria embedded in Canada balsam . . . I tried without success numerous lenses and condensers until I encountered the Abbe illumination apparatus made by Carl Zeiss. The apparatus consists of a combination of lenses with a focal point which coincides with the object. The light is conveyed to the lenses by a movable mirror. The beam of light can be altered by diaphragms introduced between mirror and lenses. Lateral displacement of the diaphragms provides oblique light and, by obstructing the center of the aperture, the middle portion of the beam can be eliminated. In thin tissue sections with a wide aperture diaphragm and light reflected from

white clouds, it becomes easy to distinguish bacteria among the stained components.

Having discussed questions of methodology, Koch then described animal experiments:

Mice are especially adapted for experiments on infective diseases. Accordingly, putrefying blood, putrefying meat infusions, etc. were injected subcutaneously. The results varied with the nature and quantity of the putrid fluid. Blood and meat infusion putrefied for a long time are less injurious than those putrefied for a few days. Five drops of the latter, injected subcutaneously, kill the animal in 4 to 8 hours. In such a case the injected area contains a mixture of bacteria. There is no inflammation or changes in the internal organs, and bacteria cannot be found in the organs or the blood. Thus the animal has not died from an infective disease, but simply from the effects of a chemical poison.

When less putrefied fluid is injected (1 or 2 drops), only a third of the animals become sick with characteristic symptoms. There is first an increased secretion from the conjunctiva, and the lids become glued with a whitish mucus. Lassitude ensues with loss of appetite, respiration slows and the animals die without convulsive episodes in a sitting position, rather than on their back with extremities stiffened and extended, as in anthrax. If a small quantity of the edema fluid or blood from such a mouse is injected into another one, the same symptoms develop and the second mouse dies in about 50 hours. Seventeen such passages have been performed in series. Material taken from any organ can be used for successful inoculation whereas with anthrax the inoculum from the spleen was the most reliable material for passage.

The high virulence of the blood of septicemic mice leads us to suspect that bacteria must be present in the blood in great numbers. However, in my early examinations, I failed to observe bacteria in septicemic blood and it was not until I used the Abbe condenser that I succeeded in demonstrating their presence with complete certainty, in spite of their minute size. Blood was examined by drying on a coverslip and staining with methyl violet. Large numbers of small bacilli were seen singly or in small groups among red blood corpuscles. The bacilli were 0.8 to 1 micron long and approximately 0.1 to 0.2 micron thick. The bacilli were often in pairs, attached longitudinally. When septicemic blood is placed on a concave slide and incubated the bacilli grow in dense masses.

Their relation to the white blood cells is peculiar. They penetrate into these and multiply. Hardly one is without bacilli. Many contain only isolated bacilli, others have thick masses of bacilli with a recognizable nucleus, in others the nucleus can no longer be distinguished, and finally some cells may appear as a cluster of bacilli about to fragment.

Rabbits and field-mice could not be infected with the blood or organs of septicemic mice. . . .

Koch continued with a description of his experiments on gangrene in mice, spreading abscesses in rabbits, pyemia in rabbits, septicemia in rabbits, and erysipelas in rabbits. He finally concluded:

The conditions, which must be satisfied before the parasitic nature of experimental traumatic infective diseases is established, were completely fulfilled in the first five diseases and partially fulfilled in the sixth. Infection was produced with such small amounts of fluid, that the effects could not be attributed to chemical poisons. Bacteria were always present in the materials used for inoculation, and a different organism could be demonstrated for each disease. At the same time, the number of bacteria in the bodies of animals which had died of the experimental traumatic infectious disease were adequate to account for the symptoms and for death. Furthermore, the bacteria recovered from the dead animals were the same as those originally injected. The experimental traumatic infectious diseases resembled human traumatic infectious diseases with regard to origin from putrefying substances, and character at post mortem examination. We can anticipate that human traumatic infectious diseases will be proved to be parasitic when investigated by the improved methods used in this study.

The morphological differences among the bacteria studied are as great as can be expected among particles that border on the invisible. I refer to size, shape and conditions of growth. Examination must include groups of bacteria as well as individual bacteria. Attention must also be paid to physiological effects such as character of spread in tissues, types of cells attacked and toxic effects. We must regard the different forms of pathogenic bacteria as distinct and constant species. This view is contrary to Nägeli's who asserts there is no necessity for separating bacteria into species. Great stress should be laid on pure culture. If, in serial cultivation, the same form of bacterium is obtained, this organism must be accepted as a species. Some bacteria, such as *B. anthracis*, which is easily recognized, can be maintained successfully in pure culture. With very small bacteria, the occurrence of contamination in culture cannot always be determined because distinct morphological characteristics may be lacking.

My studies demonstrate that pure cultures can be maintained in the animal body. For example, a great variety of bacteria are present in putrefying blood, but only two, a small bacillus causing septicemia and a coccus in chains causing gangrene could survive and grow in the living laboratory mouse. Further, these two could be separated by inoculation into field mice where the bacilli disappeared while the cocci multiplied. In my experiments the form and size of bacteria passed through the animal body have remained constant.

Despite all this fine work, Koch still did not have suitable laboratory facilities. Koch's friends, especially Cohn and Cohnheim, tried diligently to obtain for him a suitable position that would permit him to devote all his time to research. Efforts toward an appointment on the faculty of Breslau were unsuccessful, but in 1880 Koch was appointed a visiting member of the Imperial Health Office in Berlin. The same year he was made a full resident or ordinarius, finally attaining a position fitting to his stature and a laboratory eventually adequate for his scope of research. In the preliminary interview with Dr. Struck, Director of the Office, Koch did not request a specific salary, but asked for and received one that would be adequate to maintain his family in a manner fitting to the position.

In July, 1880, when Koch joined the Imperial Health Office, the chemical and hygienic sections had acquired nearly all the laboratory space, and Koch's bacteriology division was assigned a small single-windowed room. He was soon joined by Löffler and Gaffky, and the department moved to a three-windowed room. With the addition of Hueppe, Proskauer, and Von Knorre a third expansion was required. Research advanced energetically from the very start. Within the first year of activity, the institute had so much to report that a new journal, the famous *Mitteilungen aus dem kaiserlichen Gesundheitsamt,* was founded to meet this need. In 1881 the journal appeared with Koch's report "On the Study of Pathogenic Microorganisms" as the first article. The report outlined advances in procedures for the study of infectious disease along the lines now known as Koch's postulates. The most important topic in the report was the preparation of pure cultures. Koch observed that this phase of bacteriology was most urgently in need of improvement. Pure cultures could be isolated and maintained by using slices of cooked potato, but many bacteria did not grow on them. He investigated substances that could be added to liquids to render them transparent and solid. He recommended the addition of 2.5 to 5 per cent gelatin to the commonly used infusions, such as meat or hay infusions. It was not the complete answer, because gelatin liquefied at 37° and some bacteria could hydrolize it enzymatically at room temperature. The definitive solution came by way of a household kitchen. Frau Hesse, the New Jersey-born wife of one of the investigators at the Institute, suggested the use of agar-agar that her mother had been using to make jellies. Her mother had received the formula from a Dutch friend who had lived in Java.

Koch's demonstration of his methods during the International Medical Congress in London in 1881 was the high point of the congress. The demonstrations were held in Lister's laboratories at King's College. On this occasion Pasteur remarked to Koch: *"C'est un grand progrès, Monsieur."* Pasteur's remark was not the beginning of a beautiful friendship; contact between the two great men remained infrequent, and on some occasions they disparaged each other's work.

After returning from the international congress, Koch began his studies on tuberculosis. The first guinea pigs were inoculated on August 18, 1881, with tissue from an ape that had died from the naturally acquired disease. The work was conducted in strictest secrecy; apparently, not even Ferdinand Cohn, Koch's scientific father confessor and a visiting member of the Institute, was aware of these investigations. The work was conducted at a feverish pace, and progress was so satisfactory that Koch decided to present a preliminary report in May of the next year before the Berlin Physiological Society. The meeting, directed by Emil Du Bois-Reymond, was held in a small reading room, 30 by 45 feet in size. A total of seventy-two chairs were available, and table space for 200 preparations was also required. One of the audience remarked that the elaborate demonstration table "appeared like a cold buffet." The room was filled to the last available seat. Virchow was conspicuously absent. He had recently taken a strong stand against the "juvenile work" of the "youngsters" of the Imperial Health Office. Because of Virchow's dominating influence in the Berlin Medical Society, Koch did not attempt to offer his report to one of their meetings. Koch's words at first came slowly and haltingly, but he gradually recovered confidence as he proceeded step by step with his masterly solution of the riddle of the etiology of tuberculosis. The audience was left spellbound, and for a time after he had ended the presentation not a word was uttered. The audience recovered, and after enthusiastic applause, many rushed up to congratulate him. Paul Ehrlich stated that "all who were present were deeply moved and that evening has remained my greatest experience in science."

That was in 1882. Two years later there appeared in the *Germany*,

Reichsgesundheitsamt Mitteilungen (**2**: 1–88, 1884) a long memoir, "The Etiology of Tuberculosis," which reads essentially as follows:

The successful demonstration of bacteria as the etiologic agents of many infectious diseases has led to the search for a microorganism which might cause tuberculosis. Initially one must determine if foreign structures can be demonstrated in the diseased area. If so, they must be investigated for evidence of organization and viability independently of the host. Further, their distribution in the body, their appearance at various stages of the disease, and other circumstances which might reveal a direct connection between these structures and the illness should be studied.

To obtain a complete proof of a causal relationship, rather than mere coexistence of a disease and a parasite, a complete sequence of proofs is necessary. This can only be accomplished by removing the parasites from the host, freeing them of all tissue elements to which a disease inducing effect could be ascribed, and by introducing these isolated parasites into a healthy animal, with the resulting reproduction of the disease with all its characteristic features. An example will clarify this type of approach. When one examines the blood of an animal that has died of anthrax one consistently observes countless colorless, non-motile, rod-like structures. These were often considered inanimate crystalline structures. But, after they were observed to grow and form spores from which rods developed again, their affinity to lower plants left no doubt. When minute amounts of blood containing such rods were injected into normal animals, these consistently died of anthrax and their blood in turn contained rods. This demonstration did not prove that the injection of the rods transmitted the disease because all other elements of the blood were also injected. To prove that the bacilli, rather than other components of blood, produce anthrax, the bacilli must be isolated from the blood and injected alone. This isolation can be achieved by serial cultivation. A small amount of blood containing bacilli is transferred to a solid culture medium, for example, nutrient gelatin or cooked potato. The bacilli multiply profusely while the other blood components remain static. After two or three days the bacilli have formed a dense mass of spore bearing threads and a minute amount of whitish growth can be transferred again to nutrient gelatin or cooked potato. The serial transfers can be continued for 3 or for as many as 50 passages and in this manner the other blood components can be eliminated with certainty. Such pure bacilli produce fatal anthrax soon after injection into a healthy animal, and the course of the disease is the same as if produced with fresh anthrax blood or as in naturally occurring anthrax. These facts prove that anthrax bacilli are the unique cause of the disease.

This procedure, successfully used in demonstrating the parasitic nature of anthrax, has formed the basis for my investigations of the etiology of tuberculosis.

Pathogenic organisms of the size of anthrax bacilli, appearing in the blood in large numbers, or structures like the spirochete of recurrent fever that strikes the eye with its characteristic motility, are not difficult to demonstrate with ordinary optical aids. When the organism is small and infrequently present in tissues, and the tissue has become necrotic, examination becomes difficult. It is therefore necessary to use more precise microscopy involving fixation and differential staining, and to examine the preparations with oil immersion systems and Abbe condensers. An especially difficult problem was anticipated with tuberculosis in view of the consistent failure encountered by other workers in attempting to demonstrate this elusive organism.

The investigations were initiated with material in which the infectious agent could be anticipated with certainty. For example, newly formed tubercles from the lungs of animals sacrificed three or four weeks after inoculation. Sections from such lungs, fixed in alcohol, were examined by the best available methods. Tubercles were also crushed and streaked on glass slides, dried and then examined under the microscope. All attempts to demonstrate bacteria in these specimens were unsuccessful. Alkaline stains were next investigated because alkali has been reported to intensify staining reactions. One cc. of concentrated alcoholic methylene blue solution was added to two-hundred cc. of 0.01% KOH solution. When streaked preparations were treated for 24 hours with this staining solution, fine rod-like forms could for the first time be recognized in the tuberculous masses. Further investigation proved that these rods were able to multiply and form spores, and therefore belong to the same group of organisms as the anthrax bacillus. In sections, it was unusually difficult to demonstrate these bacilli among densely packed nuclei and debris. It was therefore decided to follow Weigert's technique of differential staining that has been used successfully in anthrax. The preparations were stained first with alkaline methylene blue and then with concentrated aqueous Bismarck brown until the preparations assumed a brown color. Under microscopic examination, tubercle bacilli remained blue and contrasted with the tissue elements that stained brown. The methylene blue did not, however, give a very intense color, and practice was required for consistent recognition of the bacilli.

Ehrlich has developed a method, modified by Weigert, which stains the tubercle bacilli more intensely and I now use this method exclusively. A saturated aqueous solution of aniline is prepared by vigorously shaking five cc. of pure aniline with one-hundred cc. of water for one half hour. The mixture is then filtered until clear. A saturated solution of methyl violet is prepared by mixing 20 gm. of dried methyl violet with 100 to 150 cc. of absolute alcohol with vigorous shaking and the residue permitted to settle for 24 hours. One-hundred cc. of aniline water are mixed with 11 cc. of methyl violet solution. I add 10 cc. of absolute alcohol to this mixture to improve stability. Glass slides which are to be used for stained preparations are first cleaned by washing in nitric acid followed by alcohol. The specimen is then smeared on the glass slide with a scalpel or needle, carefully crushing the caseous masses and spreading them thinly. In the case of sputum, care should be taken to select the yellowish masses of pulmonary origin that are distributed in the frothy, slimy liquid. The yellowish masses are crushed and spread thinly, the slides are dried thoroughly in the air and then fixed with heat, either by being placed in a drying oven at 110°C. for 20 minutes, or by slow passage through a flame three times with smear surface held upwards. With careful heat fixation the morphology of the bacterial cell is not altered. The slide is laid face downward in a dish of staining solution which is then heated until it just begins to bubble. The flame is removed and the stain allowed to act for about 10 minutes. Better results are obtained by staining in the absence of heat for two to 12 hours. To examine tissue sections for tubercle bacilli, pieces of the organ are fixed in absolute alcohol. Thin sections are not necessary because the counter-staining permits recognition of the baccilli even in thick sections. The preparation of large sections is desirable because the distribution of bacilli is often very irregular. The sections are placed in the stain for at least 12 hours but can be left safely in the staining solution for several days. To render the preparation suitable for microscopic study, excess stain must be removed.

For decolorization, the slides are transferred from the staining solution to dilute nitric acid, prepared by diluting the concentrated acid with 3 to 4 parts of water. The slides are then agitated in dilute acid until a greenish-blue color appears, followed by treatment with 60% alcohol for 10 to 15 minutes and finally counterstained.

In order to obtain good contrast between the bacilli and the cellular elements a yellow or light-brown is chosen for the counterstain when the bacilli are stained blue, and a blue counterstain when the bacilli are stained red. In the first case, Bismarck brown is recommended; and in the second, methylene blue. Both counterstains must be used in dilute solution, and not for a long time. The counterstaining should render the nuclei evident but not be so intense as to mask single bacilli. I use freshly filtered Bismarck brown, which is barely transparent in a depth of 2 cm. The slide preparations are laid downward in the Bismarck brown solutions. Tissue sections are kept in the Bismarck brown for several minutes, then transferred to 60% alcohol, followed by absolute alcohol. Cedar oil is recommended for clearing because it does not remove the aniline. For mounting, Canada balsam diluted in turpentine is used. The Abbe illuminating apparatus is the best device for microscopic examination of tissue sections under oil immersion. Smear preparations should be so thin in density that the elements are distributed in a single layer. Under such conditions the preparations can be mounted in water for microscopic examination, and a condenser will provide adequate light. When methyl violet is used as a primary stain and Bismarck brown as counterstain, the tissue elements are completely differentiated from the tubercle bacilli.

With this staining procedure the tubercle bacilli retain the dark blue stain, while the nuclei of tissue cells, necrotic cellular components and granules of plasma cells are stained brown. The demonstration of tubercle bacilli is further facilitated by the fact that all bacteria now known, except the leprosy baccilus, take the counterstain when stained by Ehrlich's method. Tuberculous sputum is regularly contaminated with organisms from the buccal cavity, but I have never seen any of these stain like the tubercle bacillus. The same observation is valid for smears of intestinal organisms. Spores of one type of large bacilli which can be found there stain blue while the remainder of the cell will be stained brown, but this will be true only in the early developmental stage of such spores. The other bacterial spores in the intestinal contents, as observed by Gaffky, remain unstained. On the contrary, Gaffky has found that fungal spores are stained blue.

Recently I have examined many types of bacteria-containing substances with Ehrlich staining technique, these included putrefying meat infusion, decomposing urine, blood, milk, decaying infusions of vegetables, mud from swamps, gutter slime, etc., but I have never found bacteria which stained the same way as the tubercle bacillus. I must, therefore consider as erroneous, claims for bacteria found in sputum, putrefying liquid, intestinal contents of healthy man, and in mud, which would stain like the tubercle bacillus.

The leprabacillus is the only other one which stains similarly by Ehrlich technique. This relationship is notable since there are also anatomical similarities between the two diseases. To be sure, the staining reaction of the two types of bacilli are not identical. For example the tubercle bacillus will take Weigert's nuclear stain but the lepra-bacillus will not. However, the case of the leprabacillus indicates that in time other types of bacteria might be found which possess the same staining recation as the tubercle bacillus. Such discoveries need not discredit the demonstration of the tubercle bacillus as the agent of tuberculosis since the tubercle bacillus possesses other biological characteristics which reliably differentiate it from other bacteria.

It appears likely that other methods will be found for staining the tubercle bacilli. Ehrlich's method has already been modified. Among others, the modification developed by Ziehl in which phenol or resorcin can be used in place of aniline appears worthy of mention. One of the difficulties in the staining of the tubercle bacillus is the instability of the reaction, in presence of Canada balsam. Preparations stained with methyl violet and gentian violet are decolorized in as little as two days and the bacilli eventually become invisible. Preparations stained with fuchsin or alkaline methylene blue are much more durable. However, decolorized preparations can be restained: the Canada balsam is lique-fied by heat and the section transferred to turpentine, after 24 hours it is transferred to absolute alcohol and finally restained. The tubercle bacillus will again take the blue color intensely.

An acceptable explanation for the differences in staining character-istics between the tubercle bacillus and other bacteria cannot be offered because of inadequate knowledge of the finer structure and chemical composition of bacteria. There are many reasons for believing that the tubercle bacillus, like other bacteria, is covered with an outer layer, or capsule, that functionally is different from the inner substance. The bacilli stained with methylene blue appear thinner than those stained with methyl violet or fuchsin. When methyl violet-stained preparations are decolorized, an outer layer decolorizes first, so that a thin stained thread remains, which is about as thick as the bacillus stained with methylene blue. Furthermore the tendency of the bacilli in culture to clump is compatible with the presence of a capsule.

To determine whether the bacilli are constantly associated with tuberculosis, I have examined material from two hospital sources. The cases will not be presented individually but will be grouped according to an anatomical classification. When a tuberculous nodule is examined, in unstained sections, and without the diffuse light of the Abbe con-denser, it appears to consist of densely packed cellular elements, and is therefore opaque. As soon as the center of a tubercle becomes caseous, the cells are changed into a finely granular opaque mass in which fine details cannot be distinguished. A different picture is ob-tained with stained sections, placed in a strongly refractive medium, and examined under diffuse illumination. The earliest tuberculous nodules then appear to consist of masses of stained nuclei. Nevertheless, a section of ordinary thickness appears transparent enough to allow the reliable recognition of the most delicate organized elements among the nuclei. The caseous centers of the tuberculous nodules, however, appear very different because the cells have died, and they contain the remains of disintegrated nuclei in the form of collections of stained granules.

In stained preparations, the tubercle bacilli appear as rods one-fourth to a half the diameter of a red blood corpuscle in length. As mentioned previously, their thickness is dependent on the staining technique. The bacilli are usually bent and sometimes approximate a spiral form. These departures from linearity help distinguish them from other bacteria of their size. The distribution of the bacilli in tuberculous tissue is highly variable. Occasionally, they may be collected in dense masses, but most often they are present only in small numbers. The bacilli are most constantly present in early or rapidly progressive lesions. At first they are found in moderate numbers in the cytoplasm of epithelioid cells, near the nuclei. As the bacilli multiply in the cells, they become arranged in dense groups of parallel rods. At this stage, the cells are in the process of destruction. The nuclei begin to disintegrate forming irregular granules of varying size. The granules, in turn, disintegrate gradually, leaving a homogeneous material which will not accept nuclear stain. This material becomes the caseous center of the tubercle. In general the caseous material contains few tubercle bacilli. Large numbers of tubercle bacilli are seen only when necrosis is occurring rapidly. The bacilli may die but retain their staining affinities longer than the cells, or they may form spores, thereby losing their ability to take stain. In such cases, only the spores remain and since spores cannot as yet be stained, their presence is revealed only by the infectivity of the caseous material in which they are present.

The giant cells are frequent components of tuberculous tissue but are not specific for the disease. Since the giant cells are seen often in the centers of lesions, it has been suggested that the tuberculosis germ should be present in these cells. This view has been substantiated. The giant cells are more frequent in slowly developing types of tuberculosis, such as scrofula, or tuberculosis of the joints. In these affections, few tubercle bacilli are evident, and when present, they are found in giant cells. Generally, only a small number of bacilli will be present in these cells. A single bacillus within a giant cell may be easily missed because the rod may not be placed in the plane of the prepared section. The tubercle bacilli may appear black against the brown background of the giant cell when stained with methylene blue and Bismarck brown. The nuclei of the giant cell are often pushed toward one side and the bacilli are usually first found in the area away from the nucleus. As the numbers of bacilli increase, they approach the nuclear area and eventually cause destruction of the giant cell.

The first stage in development of the tubercle is the appearance of one or more bacilli in the interior of cells which become epithelioid in character. It is difficult to explain how the bacilli, that are non-motile, gain entrance into the cell. However, they may be taken up from other tuberculous areas by wandering cells in blood, lymph, or in tissue itself. The bacilli produce harmful effects in the wandering cells and stop their movement. Further studies will determine whether the wandering cells disintegrate and the bacilli are taken up by other cells which are converted to epithelioid form, or, as appears to me more probable, whether the wandering cells transporting the bacilli change into epithelioid cells and thence into giant cells.

Direct examination also supports the view that the tubercle bacilli are taken up and transported by wandering cells. In rabbits, sacrificed soon after intravenous injection of large numbers of tubercle bacilli, many circulating leucocytes take up one or several of the bacilli. In addition true round cells with single or divided nuclei, much like white blood cells, and containing tubercle bacilli are seen in the lungs, liver, and spleen. These cells must be considered to have wandered into the tissues from the blood. The same sequence was observed following the intraperitoneal injection of tubercle bacilli into guinea pigs.

After the wandering cells, which have carried the bacilli, have been altered to sessile epithelioid cells, the pathogenic effects of the bacilli appear to spread to the neighboring cells. All cells situated within a definite range of an infected cell change into epithelioid cells. The cells containing bacilli undergo further changes, they grow larger, their nuclei increase, and finally become giant cells. The subsequent fate of the mature giant cells varies with the rapidity of the disease. When the progress of the disease is slow, there are few bacilli within a giant cell, generally one or two. The bacilli in giant cells tend to die as indicated by decreased intensity in staining. In the same giant cell, there may be seen an intensely stained bacillus together with a pale one. The giant cell appears to be an enduring structure. The tubercle bacilli may form spores within the giant cells and leave behind them the germs for future propagation, but frequently the bacilli die off and the empty cell then remains as a residuum of the presence of the tubercle bacillus. One could compare this with a volcanic region in which, in addition to individual active volcanoes, there are a great number that are slumbering or extinct, but the extinct ones show marks of prior eruptive activity. The giant cells in which bacilli have multiplied rapidly may be disrupted by the bacilli and disintegrate into small granules.

The ability to form spores is an important property of the tubercle bacillus. . . . The spores are oval and there will be 2 to 6 in a single bacillus.

Following these observations on the general properties of tubercle bacilli, I shall proceed with a description of their behavior in various tuberculous processes.

First Koch discusses miliary tuberculosis in man.

A total of 19 cases were examined in which the lesions were miliary and sub-miliary gray tubercles with whitish or light-yellowish centers, disseminated through lungs, brain, liver, spleen and kidney. Bacilli were present in all cases but the smaller and newer tubercles contained the greater numbers of bacilli, whereas, as soon as the nuclei of the central cells disintegrated, caseation was accompanied with a decrease in the number of bacilli. In the larger nodules, where caseation was far advanced, few bacilli were seen and then only near nuclei of epithelioid cells at the periphery of the nodules. Occasional bacilli were also seen in giant cells at the border of the caseous areas. . . . In many of the older nodules the bacilli appeared to have vanished completely. However, the examination of a few sections of a nodule is not an unequivocal proof of the absence of tubercle bacilli. When an adequately large number of sections are examined, areas containing many bacilli are found. In the livers and spleens of miliary cases, bacilli were seen only in giant cells. The tubercles of meninges contained large numbers of bacilli and they were found bordering small arteries. In many areas there were young tubercles consisting of round cells with dense intracellular masses of bacilli. Occasionally there were accumulations of bacilli within blood vessels. . . .

After presenting the detailed observations of the nineteen individual cases, Koch continued with human lung tuberculosis:

Twenty-nine cases were examined and tubercle bacilli were present in tissues from all of them. The bacilli were most abundant in fresh caseous infiltrations and cavities where necrosis was proceeding rapidly. They were less abundant in cavities containing dense fibrous walls and most scarce in scarred, contracted, strongly pigmented lung tissue. As the number of bacilli decrease, they tend to be confined to giant cells. The distribution of bacilli is irregular, one tuberculous lung may show a large number of bacilli and another only occasional bacilli. In the same lung, some areas may be free of bacilli, and other areas will contain masses of them. . . . Thus, a variety of locations must be examined to observe bacilli in the walls of cavities.

The sequence of invasion appears as follows: a small number of bacilli reach the lung and are soon localized by infiltrating cells. They are not destroyed, on the contrary, they produce necrosis and caseous degeneration in the center of the cell mass. The nodule gradually increases in size and eventually penetrates into a bronchus forming a cavity. Progression of the cavity is irregular, dependent upon the rate of growth of the tubercle bacilli in the various parts of the cavity. In general, the cavity retains the characteristics of the caseous focus. At the inner border, there are necrotic masses followed outwards by epithelioid cells interspersed with giant cells, frequently containing tubercle bacilli. They are more frequent in the caseous areas of cavities than in closed caseous nodules. It would appear that closed caseous nodules are no longer adequate to support the growth of the tubercle bacilli, but that, in cavities, the disintegrating wall tissue presents fresh nutrients for the bacilli. The distribution of bacilli in cavities is irregular, due to differences in the suitability of local areas for growth. Conditions for suitability of growth in a single area may vary with time. If spores are present nearby, an unsuitable area may again become the site of an active lesion. When bacilli or spores are not invading from a neighboring area, contraction, scarring and healing may occur.

The tubercle bacilli may spread from the original focus of infection in several ways. They can reach the larger blood vessels of the lungs and be disseminated over the entire body, causing miliary tuberculosis, apparently spread through the lymph to the bronchial glands. Dissemination through the air passages appears to be frequent with preferential seeding of the larynx. The sputum may also be swallowed producing foci in the intestinal wall.

The bronchial spread of large masses of tubercle bacilli contained in cavity exudate may produce a rapid necrosis in large areas of the lung. When the structure of the alveoli is still evident, though filled with caseous material, the process is described as caseous pneumonia. The various combinations of the two described processes namely lesions arising from a single focus and spreading slowly, and caseous infiltration resulting from a massive contact with infectious material, produce the varied picture included under the general category of pulmonary tuberculosis. Caseous pneumonia may be produced in other ways. Lesions in the larynx, throat and mouth, as well as caseous bronchial glands may empty their contents into the bronchi, producing massive caseous infiltrations in the lung.

The demonstration of tubercle bacilli in sputum is of great diagnostic significance since only tubercle bacilli can cause tuberculosis. Using Erlich's staining method, tubercle bacilli were demonstrated in a considerable series of patients. Large numbers of bacilli were seen in sputa of patients with a rapid dissolution of the caseous lung tissue. The caseous fragments found in tuberculous sputum consisted almost entirely of tubercle bacilli. Since sputum was mixed with saliva, there was also other types of bacteria which tended to overgrow the tubercle bacilli. With Ehrlich's stain, the tubercle bacilli retained the methylene blue, whereas other bacteria took the counterstain. Bacteria other than tubercle bacilli may penetrate and multiply in tuberculous lesions. These may generally be considered non-pathogenic, but some, such as a special type of micrococcus, may participate in the destructive process. This micrococcus had certain similarities to sarcinae, but differences have been described by Gaffky.

Among the 29 cases of pulmonary tuberculosis, the specimens of small intestine from 8 were ulcerated with tuberculous lesions and caseous mesenteric lymph nodes. Occasionally the ulcers were surrounded by fresh tuberculous foci, grouped along the lymph channels. In the material here available, the tubercle bacilli appeared to grow better in the intestines than in the lung. It should therefore not be surprising that the feces of patients suffering from lung tuberculosis would contain tubercle bacilli. Among the innumerable rod-shaped bacteria seen in the intestinal contents, the tubercle bacilli may be seen only by virtue of their special staining properties.

After a description of tuberculosis of various organs, scrofulous glands, joints, bones, and skin, Koch turns his attention to tuberculosis in animals:

Tuberculosis varies greatly in its manifestations among different animal species, but the tubercle remains the basic pathological unit. The differences are attributable to secondary manifestations, among which are coagulation necrosis without caseous degeneration, as in the liver and spleen of guinea pigs; caseous softening of pus-like exudation, as in tuberculosis of monkeys; slow deterioration with the caseous material giving a dense lesion, as in tuberculosis of man; caseous degeneration and calcification, as in tuberculosis of cattle; and formation of tumors with deposition of lime, as in tuberculosis of birds.

After a description of tuberculosis in the cow, horse, pig, goat, sheep, fowl, and monkey, Koch describes spontaneous tuberculosis of guinea pigs and rabbits:

Among the many hundreds of rabbits and guinea pigs that were purchased for experimental use, and that were eventually submitted to autopsy, none were initially tuberculous. But after normal animals had been kept for 3 to 4 months in the same rooms with experimentally infected animals, cases of spontaneous tuberculosis developed among them even though they were in separate cages. The incidence of the spontaneous disease varied directly with the number of experimentally infected animals in the animal quarters. The spontaneous disease differed from the experimental one. Animals dying from spontaneous tuberculosis showed one or few large areas of caseous necrosis in the lungs and caseous bronchial lymph nodes. Occasionally, the lesions in the bronchial glands were much larger than the lesions in the lungs. The lesions in other organs were less advanced than those in the lungs. The type of disease in experimentally infected animals varied with the route of inoculation. Most often, animals were injected on one side of the abdomen, whereupon the site of inoculation and the local lymph nodes became caseous, while the bronchial glands remained unaffected. In these animals the livers and spleens showed severe tuberculous involvement while the tubercles in the lung were still small. In animals infected by inhalation, there were large numbers of small tubercles in the lungs. Comparisons suggest that the disease in spontaneously infected animals resulted from inhalation of one or few organisms. In the larger lung lesions the central destruction tended to advance to cavity formation; this picture was suggestive of human phthisis. The infection did not remain localized long enough to form true cavities, but spread to other organs and so cause death.

The dissemination to other organs differed in the guinea pig and the rabbit. Initially, the tubercles in the liver and spleen of both species were like those in the lungs: miliary, greyish-translucent, with yellowish centers and of dense consistency. In the guinea pig the spleen enlarged significantly and became dark red in color. Soon, tubercles fused to form whitish-gray islets. With continuing confluence of lesions, the spleen became mottled with light greyish areas and dark red areas. Finally the lighter areas predominated. Ruptures of small vessels in the pulp led to punctate hemorrhages. The spleen showed coagulation necrosis which contrasted with caseation necrosis in the lymph nodes. The sequence of changes in the liver was similar to that in the spleen. Under microscopic examination, the lightly colored areas revealed coagulation necrosis with few scattered bacilli. Only in occasional cases was characteristic multiplication and grouping of bacilli seen. At the borders of the necrotic areas, the bacilli were more frequent and were often within giant cells. Grossly visible tubercles were never seen in the kidneys of guinea pigs. In rabbits, the spleen and liver were enlarged to a lesser degree than in the guinea pig, and the tubercles in these organs always remained small. The kidneys almost always showed small whitish nodules attaining pea size and containing numerous tubercle bacilli. The urethra was sometimes involved.

Experimentally produced tuberculosis behaves in general like the spontaneous disease. The disease shows species specific characteristics. Obviously, the disease is also modified by the route and the size of infecting doses. In the present study, 273 guinea pigs, 105 rabbits, 3 dogs, 13 cats, 2 hamsters, 10 chickens, 12 pigeons, 28 white mice, 44 field mice, and 19 rats were infected. Without exception, tubercle bacilli were found in all tubercles that were examined. The large number of animals used rendered it impossible to examine all organs that showed tubercles. In most cases, I limited myself to the examination of several tubercles in the lung or spleen. The tubercle bacilli are evident with the first cellular changes in the tissues. The bacilli induce the accumulation of epithelioid cells, giant cells, and later the especially characteristic caseation. The progress of the disease is correlated with the presence and the number of tubercle bacilli. Thus, in chronic tuberculosis only few and scattered bacilli are found, but many densely packed bacilli are seen in rapidly progressive disease. When the tuberculous process is arrested, the bacilli disappear. Three points, namely that the tubercle bacilli occur regularly in tuberculosis, that they precede the pathological changes in time and place, and that their appearance and disappearance occur in direct relation to the course of the disease, permit us to conclude that they are not an accidental accompaniment of tuberculosis, but are in direct causal relation to the disease. For incontrovertible proof, the tubercle bacillus must be isolated from the disease process, cultivated outside of the body in pure culture, and finally the disease must be produced experimentally by the bacilli rendered free of material from the disease processes.

It was anticipated that pure cultures of tubercle bacilli would be difficult to obtain. Cultivation on solidified medium was the method selected at the outset because it is more reliable and convenient than all other methods. Nutrient peptone gelatin cultures seeded with crushed lung tubercles were unsuccessful. These trials were performed at room temperature because gelatin liquefies at higher temperatures. Solidified blood serum was next tried because it would appear to possess the necessary nutrients, it would not liquefy with prolonged incubation at body temperature, and it is transparent. The blood of slaughtered animals was collected in vessels washed in bichloride of mercury and rinsed in alcohol. The first blood to emerge from the wound was discarded. After the blood was collected, the vessel was closed with a glass stopper and refrigerated for 24 to 30 hours. The serum was removed with pipettes and transferred to culture tubes plugged with cotton. Pipettes and tubes were sterilized by heating for at least 1 hour at 150 to 160°C. in a hot air oven. The culture tubes were filled to one-third capacity with serum and the stopper replaced immediately. Despite all this care, bacteria from the air and animal hair would have soon caused putrefaction of the serum if they had not been destroyed. This was done by Tyndall's method of repeated heating. Experimentation revealed that heating for one hour at 58°C. for five successive days nearly always rendered blood serum sterile. The heating could be performed on an open water bath, but a tin vessel with water jacketed walls gave more uniform heat. The blood serum sterilized in this manner was then solidified by heating at 65°C. for 30 to 60 minutes. In order to obtain the maximum possible surface for growth, the culture tubes were held in a slanted position. Covered tin boxes with a slanted jacketed bottom were practical for this operation. The water of condensation which collected in the tubes during sterilization was of value for cultivation. When the bacteria to be grown were spread on the solidified slant, up to the edge of the water of condensation, they developed both on the solidified surface and in the water of condensation so that the mode of growth in liquid and on the solidified medium could be observed simultaneously. Culture tubes were incubated for several days to eliminate the possibility of contamination.

For inoculum, soft tissue as free as possible of putrefying bacteria is desirable. When the putrefying bacteria are confined to the surface of the organ, it is still possible to obtain pure cultures of tubercle bacilli, but as soon as foreign bacteria have penetrated into the deeper layers of the organs, all attempts to cultivate the tubercle bacillus in the absence of these organisms will be unsuccessful, because the putrefying bacteria overgrow the tubercle bacillus. Also, cultures are difficult to initiate when the tissue samples are dense and contain few bacilli. In the case of the guinea pig, pure cultures are best obtained from lung tubercles or caseous lymph glands. Surgical tools are sterilized by direct heating in a flame. The skin of the animal is thoroughly moistened with 0.1% bichloride solution. The skin is removed with hot scissors and forceps, exposing, but not touching, the lymph glands of the axillary and inguinal areas. The lungs are exposed with another pair of hot scissors.

Tubercles are then excised with still another set of instruments which have been heated and cooled. The tubercles are crushed between two sterilized scalpels and spread on the surface of the culture medium with a platinum wire which has been heated and cooled. The culture tube is held almost horizontally between thumb and forefinger and the cotton stopper held with the other fingers of the hand. The inoculation is performed rapidly to avoid contamination from the air. It is also advisable that the room be free of dust and air currents. Despite all precautions, the entry of occasional foreign bacteria carried on dust cannot always be avoided. It is therefore necessary to inoculate a series of 5 to 10 culture tubes to assure successful pure cultures. The cultures are grown at 37°C in incubators designed to give even distribution of heat. And, since growth is slow, evaporation should be minimized. The d'Arsonval incubator is recommended. Growth appears after 10–15 days as pale whitish mat, dots and flakes; they resemble dry scales which are lightly attached to the surface of the serum which is never liquefied. With a small inoculum the flakes remain separated. With a heavy inoculum, the flakes coalesce, forming a thin greyish-white membrane. The growth spreads horizontally. This is most striking in liquid medium, where a surface pellicle is formed which moves up the wall of the tube for a distance of several millimeters. When a liquid culture is disturbed the membrane ruptures into clumps which sink slowly to the bottom, leaving the medium clear. This observation confirms the lack of motility suspected from direct microscopic observation. On soft serum, the growth is not as uniform as on hard serum, and accumulates in compact masses which adhere tenaciously to the substrate.

After 5 or 6 days, examination at 80 × magnification reveals the growth of fine, wavy, undulating threads. The growth slowly assumes the form of flakes with wavy markings indicating the original membranes covering the culture surface. To study the arrangement of bacilli in the colony, a cover slip is applied to the surface of the culture removing a portion of the colony, the preparation is then dried, fixed with heat and stained by Ehrlich's method. The bacilli are arranged end to end parallel to the long axis of the colony, and appear to be separated by a very slight space. This characteristic suggests that the organisms are surrounded by sheaths that render the colony cohesive. Maximum growth is usually attained in four weeks. In mature colonies, almost all bacilli contain spores.

Most cultures were discarded after study of pathogenicity and growth characteristics. Only a few cultures were saved. Cultures from various sources maintained their pathogenicity and growth characteristics for as long as 22 months, through as many as 34 transfers.

Other types of media were next investigated. Cultures on liquid blood serum formed thin, whitish, brittle, friable membranes. When the cultures were disturbed, the membranes broke up and sank to the bottom, but the serum remained clear. Growth could not be initiated from colony fragments which had dropped below the surface of the medium. Sheep, beef, and calf sera supported growth best, but the sera of horse and pig also gave good growth. Growth was also satisfactory on dog serum, although this species is apparently resistant to tuberculosis. Solidified egg white did not support growth. Liquids other than blood serum at first appeared unsatisfactory. In a neutral meat infusion broth slight growth was obtained in 4 to 5 weeks, but the growth was better when the inoculum was triturated and the culture frequently shaken. Shallow layers of broth up to one cm. thick in wide bottom containers, such as Erlenmeyer flasks, supported optimal growth. The broth remained clear, but in 4 to 5 weeks the organisms grew as a finely granular layer on the bottom of the flask. When one compares this slow growth with the results of earlier culture studies by Klebs, Schüller, or Toussaint, where turbidity was observed after one to three days, it appears that these investigators had contaminated cultures. Infusions from the meat of various species of animals were equally satisfactory. Meat infusion agar did not support growth with the characteristic membrane, a compact mass was formed. Cooked potato was an unsatisfactory medium. The limits of temperature for growth were 30 to 42°C.; growth was optimal at 37 to 38°C.

The transmission of the disease to experimental animals was the most important phase of the investigation. The experimentation was extensive in scope, but the presentation will be restricted to a limited series of experiments where precautions in technic eliminated reasonable basis for criticism. Three possible sources of error must be considered in animal studies. First, mistaking spontaneous tuberculosis for experimentally produced tuberculosis. Second, mistaking the effects of true tuberculosis for other pathological changes which are similar in appearance. Third, producing accidental tuberculous infection with contaminated instruments.

How may one avoid these mistakes? With regard to spontaneous tuberculosis, absolute elimination of error is not possible. The investigator should be alert for the symptoms discussed earlier. Tuberculous animals should be kept in individual cages. Their care should include frequent ventilation, cleaning, and disinfection of cages. Normal rabbits and guinea pigs should not be kept for long periods of time in the same rooms with tuberculous animals. Communal residence beyond eight to ten months leads to frequent cross infection. Regarding the second source of error, true tubercles, are infectious and contain tubercle bacilli as can be demonstrated by animal inoculation and microscopic examination. For example, if inhalation of some inoculum by a dog has led to grey nodules in the lungs, the nodules must be shown to be infectious. The spread to other organs would be an indication, but transmission to other animals should be attempted. If inoculation of another animal with material from a nodule does not produce progressive infection, it would appear unlikely that the inoculum was tuberculous. The third source of error, infection from contaminated instruments, etc., was frequent in previous investigations on tuberculosis. Instruments should be carefully disinfected and inoculating loops should be heated red-hot. Conventional syringes cannot be disinfected with certainty, because they are damaged by high heat, and cannot be sterilized properly by liquid disinfectants. Syringes fabricated from glass and metal are necessary. The needle end of the syringe is inserted into a piece of cork and the remaining portion is wrapped in cotton wool. The syringe so wrapped, can be sterilized by heating at 150–160°C. The hands of the experimenter and the operating area can be disinfected with 0.1% bichloride of mercury solution. All these precautions were taken by us and in addition, freshly purchased animals were used in each experiment. The animals were kept in separate cages and used quickly to reduce the possibility of complications from spontaneous infection.

There were two types of experiments, those performed with tuberculous tissue and those performed with pure cultures.

Infected tissues were used both for direct animal inoculation studies and as sources of pure cultures. These included organs from human miliary tuberculosis, tuberculous lungs, various localized tuberculous lesions, fungoid joints, scrofulous glands, skin from lupus, and tuberculous organs from various animals. Inocula were always examined for the presence of tubercle bacilli. A half centimeter-deep incision was made into the abdomen of guinea pigs and a fragment of tissue the size of a millet to mustard seed was inserted into the pocket. The wound was closed after one day and showed no reaction. Generally, after two weeks a swelling appeared at the site of injection and the nearby lymph glands, usually the axillary glands on the injected side. At the same time, a hard nodule developed at the site of the healed wound. The lymph gland enlarged rapidly to hazel nut size and the nodule opened and became covered with a dried crust under which was formed a flat, caseous ulcer, producing little discharge. The animal began to lose weight, the fur became rough, respiration labored, and the animal died or was sacrificed usually between the fourth and eighth weeks. Inoculation of rabbits through a skin pocket led to a slower and more irregular disease than was observed in guinea pigs. Subsequently, the anterior eye chamber was the only site of inoculation used for rabbits. Inoculation into the eye led to a characteristic tuberculosis of the iris, which has been described frequently in the past. . . .

Tuberculosis was produced in all of the animals used for these passage experiments; there were 179 guinea pigs, 35 rabbits and 4 cats. A progressive infection was produced, characterized by swelling of lymph glands, caseous ulceration at the site of injection, emaciation, respiratory difficulty. At autopsy the disease was observed to have spread from the site of injection to the neighboring lymph glands, lungs, spleen and liver. Microscopic examination revealed characteristic tubercle formation and the presence of tubercle bacilli.

Even though there have been unsuccessful attempts, by other investigators, to perform animal passages, my results will not appear unexpected if one recalls that I only injected materials in which tubercle bacilli were demonstrable, and that I used highly susceptible animals. Controls injected with non-tuberculous tissues did not appear necessary since in our institute, hundreds of guinea pigs and rabbits were being injected with a large variety of non-tuberculous tissues. In no instance was tuberculosis produced with non-tuberculous tissue.

My experiments lead to the conclusion that true tuberculosis can be produced only by injection of material containing tubercle bacilli.

In the group of experiments now presented, pure cultures of tubercle bacilli were used.

If the bacilli freed from tissue produced tuberculosis, they would clearly be its etiological agent. . . . The cultures were first checked for purity and removed from the solidified blood serum with heated platinum needles, avoiding admixture with serum. In several experiments, control animals were injected with sterilized blood serum; none of these animals developed tuberculosis. One may therefore conclude with certainty that if true tuberculosis is produced with pure cultures of tubercle bacilli that have been transferred a number of times, it is a result of the activity of this organism.

A pure culture from miliary tubercles of human lung, subcultured 5 times, totaling 54 days of cultivation, was used. This culture was injected subcutaneously into 4 guinea pigs. Two additional uninfected animals were kept in the same cage. After 14 days, the inguinal glands of the injected animals were swollen, ulcers developed at the sites of inoculation and the animals began to lose weight. One of the inoculated animals died after 32 days and the others were sacrificed on the 35th day. The 4 inoculated guinea pigs showed advanced tuberculosis of the spleen, liver and lungs. The inguinal glands were severely swollen and caseous, more so on the side of injection than on the other. The bronchial glands were slightly enlarged. The uninjected animals did not show any traces of tuberculosis.

Koch continued for a total of eleven experiments with pure cultures isolated from a variety of infected tissues. He summarized:

In this series of 11 experiments, the animals were injected subcutaneously. In general, the effects were the same as with subcutaneous injection of tuberculous tissue. The small skin wound closed and healed within a few days. Followed by swelling of the lymph glands, loss of weight and death. Autopsy revealed a severe tuberculous process involving lung, spleen and liver with subsequent characteristic changes in these organs. The only difference was that the disease followed a more rapid course after injection with pure cultures than with tuberculous tissues. Among guinea pigs the difference in time was approximately two weeks. This result is probably due to the fact that, in tissues, the organisms are embedded and cannot become infective until the tissue is resorbed, while with pure cultures the organisms immediately enter into action. The same sequence is even clearer following injection into the anterior eye chamber of the rabbit, because the development of the tubercle can be followed visibly.

Microscopically, tubercles produced by the inoculation of pure cultures were similar to tubercles produced by injection of tuberculous tissue and to those occurring spontaneously. The virulence of the organisms was apparent from dissemination to all susceptible organs. In addition, passage to other animals uniformly produced tuberculosis.

The injection of pure cultures was without effect in resistant animal species while they produced tuberculosis without exception in other animals. All control animals remained well. Thus, it seemed that tubercle bacilli should be considered the only cause of tuberculosis.

Nevertheless, it appeared important not to stop at this point, but to continue to produce tuberculosis in experimental animals with pure cultures administered by other routes. These were inoculation into the anterior eye chamber of the rabbit, injection into the peritoneal cavity, injection into large veins and inhalation of pure cultures.

For intraocular injection, an incision was made into the upper edge of the cornea and a piece of culture was inserted with a blunt forceps into the anterior eye chamber through the incision. Subsequently, suspension of culture was injected using a syringe with a fine needle, thus permitting control of size of inoculum. . . .

Where small inocula were used, the course of the disease was the same as observed following implantation of tuberculous tissue, but more rapid. Isolated tubercles developed in the iris, led to caseation of the bulbus, and finally to generalized tuberculosis by spread from iris to lymph glands and thence through the blood stream. With heavier inocula of pure culture, there was a rapid dissemination through blood, bypassing the local lymph glands, leading to generalized tuberculosis, without the prior involvement of the iris and bulbus.

Pure cultures suspended in serum or distilled water were injected intraperitoneally into guinea pigs, rats, mice, cats and other animals. This procedure could be accomplished without perforation of the gut or producing traumatic peritonitis. Rabbits were not suitable in view of their heavily filled caeca. The reaction to intraperitoneal injection was similar to that following injection into the eye chamber. After injection of few bacilli a disseminated infection with tubercles in the omentum and spleen was produced. When, however, large numbers of tubercle bacilli were injected intraperitoneally in guinea pigs, the organisms mainly were taken up in the large omentum, which retracted to form a horizontal sausage-like swelling. In cut section this swelling appeared like a freshly caseated lymph node. Masses of tubercle bacilli, mostly in spore stage, were present in dense nodules in the omentum. In addition, the spleen, the liver, and the peritoneum were heavily invaded by tubercle bacilli, but death occurred so rapidly that there was insufficient time for the development of macroscopically visible tubercles. An exudation of fluid was not seen in guinea pigs, but was present in dogs and cats. On the other hand, in guinea pigs, the pleurae frequently were so filled with a faintly yellowish fluid that the resulting compression of the lungs led to death of the animals. Generally, guinea pigs died 10–20 days after injection. When a smaller amount of culture was injected, survival was longer and countless tubercles developed primarily in the peritoneum, omentum, spleen and liver. Species less susceptible to tuberculosis, dog, rat, white mouse, succumbed only after injection of large numbers of organisms and only after several months. They then showed an unusually heavy tuberculous involvement of the abdominal organs, with a smaller number of tubercles in the lungs. . . .

Infection was most rapidly achieved by the intravenous route. Through the blood channels, the host is at once overwhelmed by the infecting agent. The organisms do not have to overcome barriers, such as lymph glands, which are encountered with other routes of inoculation, and they are disseminated immediately to all organs, causing an extensive and uniformly distributed formation of tubercles. This is most comparable with miliary tuberculosis of man, where likewise the tuberculous virus is channeled into the blood stream and transported to all parts of the body. Finely divided suspensions of pure culture were filtered through gauze, to eliminate coarse particles, and then injected into the jugular vein or, in rabbits, into the ear vein. . . .

The tuberculous nodules produced by injection into the blood stream, like those in other infections produced with pure cultures, were similar to nodules observed in spontaneous tuberculosis. The lesions contained varying numbers of tubercle bacilli. These were virulent and produced tuberculosis that was indistinguishable from the disease produced with injection of tuberculous tissue and from the spontaneous disease.

To deliver tuberculous material to the lungs of experimental animals, material was injected into the bronchi by tracheotomy or animals were exposed to a spray of a liquid suspension of infectious material. The second method was favored, because it was more closely comparable to the natural mode of transmission, but the atomization procedure was not without danger to the experimentor.

A large chamber, with an opening for an atomizing apparatus, was placed in the garden far from occupied rooms. The orifice of the atomizing apparatus was set within the chamber. A tube was passed into the laboratory room through a tight wooden window shield and attached to a rubber balloon. The suspension could then be atomized from a remote location.

The experiment involved eight rabbits, ten guinea pigs, four rats, and four mice, and was terminated twenty-eight days after exposure. The results of this experiment were as follows:

All rabbits and guinea pigs showed many tubercles in the lungs, and their size increased with the length of survival. In the animals that died late or that were eventually sacrificed there were also tubercles in the liver and spleen. The tubercles observed following inhalation of pure cultures were identical with those observed after inhalation of sputum suspensions. The tubercles were not sharply round and delimited, but encompassed roughly the center of a lobulus. The individual alveoli were filled with caseous material thus producing finely granular white areas against a dark greyish red field. The largest tubercles filled entire lobuli and sometimes coalesced with adjoining lesions giving the picture of caseous pneumonia. Such alveolar dissemination is also seen in the primary nodules of spontaneously occurring tuberculosis. Thus spontaneous tuberculosis in these animals is almost always acquired by inhalation.

The lungs of sacrificed rats and mice revealed numerous grey nodules up to the size of millet seed. Many of the nodules contained white yellowish centers, but the caseation was by no means as advanced as in the lungs of guinea pigs and rabbits. There were occasional nodules in the spleens of rats and mice. These animals, as already noted, were much less susceptible than other species to tuberculosis. The individual tubercles developed slowly and the spread to other organs was not rapid.

Microscopically, the tubercles produced by inhalation of pure cultures were identical to those in natural tuberculosis with regard to the arrangement of epithelioid cells, giant cells and content of tubercle bacilli. Tubercles from the various organs of several guinea pigs and rabbits and from the lungs of one rat and one mouse were injected intraperitoneally into a total of 22 guinea pigs. These guinea pigs showed early swelling of the axillary glands on the injected side, lost weight and died of tuberculosis in 5 to 8 weeks.

Summarizing the results of pure culture studies: Guinea pigs, rabbits, field mice and cats were highly susceptible; all of 94 guinea pigs, 70 rabbits, 9 cats and 44 field mice succumbed to infection. None of the control animals injected with suspending medium alone acquired the disease. Among the less susceptible animals, only half of the chickens given a single subcutaneous injection, became tuberculous. Dogs, rats, and white mice were even less susceptible but succumbed to infection with large inocula. The disease produced with pure cultures was the same as that produced by inoculation with infected tissue but was more rapid. Infection with pure culture completes the proof that the tubercle bacillus is the only possible etiological agent of tuberculosis.

We have proven that tubercle bacilli are the cause of tuberculosis and we are now faced with the problem of the origin of the bacilli. Can they be found outside the animal body as are the anthrax bacillus and the erysipelas micrococci? This is important not only for etiologic studies but even more so for prophylaxis. For, if tubercle bacilli occur, grow and form spores in putrefying animal and plant material, it would be impossible to avoid contact with them. Fortunately, this is not the case. The tubercle bacillus grows much more slowly than other organisms, grows only in serum and meat broth, and most important, requires an incubation temperature above 30°C. Even when all these conditions are satisfied, if the tubercle bacilli are not protected against the overgrowth of other rapidly growing bacteria, they are overcome and killed. They are therefore true parasites, dependent upon the animal body for survival. They are not facultative parasites, like the anthrax bacillus, which does not sporulate in the animal body and passes through its life cycle in free nature.

Another question, is whether or not usual ubiquitous bacteria can, under favorable conditions, be converted into tubercle bacilli. Conversely, can tubercle bacilli, while in the body or after leaving the body, revert to non-pathogenic forms. If so, invasion by the specific bacterium would not be required to produce tuberculosis. There are no proven examples of conversion of non-pathogenic bacteria into pathogenic ones and there is no reason for believing that tubercle bacilli may develop from indifferent bacteria. Such a change has not been observed in the guinea pig or rabbit that are unusually good media for tubercle bacilli. The possibility of a decline in virulence of the tubercle bacillus is a different problem. The loss of virulence by the anthrax bacillus is a possible precedent. However, the decline in virulence of the anthrax bacillus can be produced only artificially; it does not occur in the animal body or by simple sojourn outside the body. Tubercle bacilli have been cultivated outside the animal body for as long as two years without the slightest change in various characteristics, including virulence.

Fischer and Shill, as cited previously, have shown that the tubercle bacillus does not lose virulence when kept in decaying tissue for six weeks. All these considerations do not support the possibility of readily induced alterations in the virulence of the tubercle bacillus. It is most likely that the tubercle bacillus has, at some time, arisen from another organism. Having once become a true parasite, it appears to have the property, in common with other parasites, of tenaciously preserving its parasitic characteristics. The only reservoirs are the animal and human body. In these hosts, the organism has ample opportunities to multiply profusely, develop resistant forms, be disseminated and attack other victims.

The pulmonary form, which is the commonest form of the disease, is the most efficient source for dissemination. The other forms play almost no role in transmission, since the organisms are present in small numbers and remain so contained that they rarely have the opportunity for dissemination. A seventh of all humans die of pulmonary tuberculosis and most cases of pulmonary tuberculosis discharge large amounts of sputum containing great numbers of spore-bearing tubercle bacilli, for weeks to months. The infectious germs are thus disseminated on floors, lint, etc. A high proportion of these organisms die without reaching new hosts, but a considerable number remain viable and virulent for many days, or even months.

There is no uncertainty regarding the medium and manner in which phthisis patients can transmit the disease. By coughing, sputum particles are divided, discharged into the air and atomized. Many investigations have demonstrated that the inhalation of atomized tuberculous sputum is infectious, not only for highly susceptible species, but for relatively resistant ones as well. If a healthy person inhales particles of freshly discharged tuberculous sputum, he may become infected. However, infection, in this manner, does not occur often, because sputum particles generally are not small enough to remain suspended in the air. A more likely medium is dried particulate sputum, which can remain suspended in the air. Sputum that has been expectorated and that has dropped to the floor, dries, and by trampling is resuspended as dust. Sputum is also often discharged on bed linen, clothing, and handkerchiefs. Even tidy patients, by wiping their mouths with handkerchiefs after expectorating, contaminate handkerchiefs which may then become sources of infected dust.

Recovery of viable bacteria from the air has revealed that the bacilli are not suspended as isolated form, but that they are covered with the medium in which they have grown. These droplets dry on surfaces where they have been deposited. The organisms are resuspended in the air when the dried masses are divided into fine particles which can be set into motion by weak air currents. Dust from fragments of plant fibres, animal hair, epidermal scales and similar materials can function as carrier. But the contaminated dust from bed linens, bed-covers, clothes and handkerchiefs is most infectious. Dried expectorated sputum on floors is broken only into coarse particles and these are not easily raised into air. On the other hand, rapidly drying discharges on fabric produce fibrils which can remain suspended in air for long periods of time, and, when they finally reach the ground, they can be resuspended by weak air currents.

When tubercle bacilli are inhaled in dust form, they can lodge in the upper air passages or penetrate into the alveoli. The depth to which the particles penetrate depends upon the type of breathing of the host. With deep mouth breathing, the particles penetrate deeply. Nasal breathing offers a certain amount of protection against penetration of the infectious material, since a large amount of dust is retained by the mucous membrane of the nose. Whether or not the bacilli that have been carried to the bronchi and alveoli will gain a foothold, depends upon a variety of circumstances. The slow growth of the tubercle bacillus will be an important factor. Other factors may be damaged respiratory epithelial tissue, or deposition of exudate as a result of infectious diseases such as measles. Structural defects, such as adhesions and defective chest bone cage formation, hindering lung motion, may promote accumulation of bronchial secretion that might serve as loci for tubercle bacilli. Since tuberculous persons discharge large numbers of organisms, there is opportunity for contact of infectious material with parts of the body other than the respiratory tract. Tubercle bacilli may thus invade superficial lymph nodes through scratch wounds, skin irritations, etc. Less frequently feces of patients may be a source of infectious material.

Tuberculous domestic animals, are only of minor etiologic importance. Animals do not discharge sputum and tubercle bacilli are found only rarely in their feces. However, the milk of tuberculous animals can be a source of infection. Except for that, animals can be a source of infection only after death, through ingestion of infected meat. In that case, the infectious agent will invade through the digestive organs and the first evidence of disease will appear there. Primary tuberculosis of the intestines is rare as compared with primary lung tuberculosis. One may therefore conclude that infection from eating tuberculous meat is infrequent. Cooking is a factor in this regard and in addition tuberculosis of cattle tends to be localized in the lungs and lymph glands. Secondary intestinal tuberculosis is fairly frequent in phthisis patients as a result of swallowing sputum. However, the intestine is a poorer locale than the lungs for growth of the tubercle bacillus and spore forms are more likely to colonize the intestine than vegetative forms. Regarding milk as a source of infection, organisms are infrequently found in the milk of tuberculous cows. Tubercle bacilli occur in milk only when the mammary glands are infected, but tubercles do not develop often in the udders.

In all forms of tuberculosis, foci of epithelioid cells containing giant cells, appear at the site of infection and coagulation necrosis is first seen in the center and then moving outwards. Subsequently foci appear in adjoining areas of the primary lesion. The organism is not motile and is carried by moving components in tissue. Since the tubercle is avascular, wandering cells must be the vehicle of transport. The wandering cell moves until motility is lost as a result of activity of the bacilli and a new tubercle develops where it comes to rest. When the lymph stream is reached the wandering cell will move to the next lymph gland where it will then produce tubercles. In this manner the thoracic duct may be reached by parasitized cells which will then enter the blood stream. The blood stream may also be reached through rupture of tubercles in the walls of veins or by lesion proliferation into arteries.

A variety of factors influencing the type and rapidity of development of tuberculosis has been discussed. Among these were the size of infecting dose, the route of inoculation, and predisposing factors such as prior damage to respiratory epithelium, structural and concomitant functional defects. Intra-uterine transmission to the foetus is rarely observed and does not occur experimentally in guinea pigs. Thus, if a hereditary factor is to be accepted, it would be due to inheritance of characteristics, which would be a predisposition that would become evident upon later contact with the germ.

The investigations that have been conducted until now, have not led to improved therapy. At this time emphasis must be placed on prophylaxis. Such measures must be directed, in part toward suitable direct disinfection and, in part toward prevention of contact of healthy persons with tubercle bacilli. Such measures will have to include a consideration of sociological factors. The question of prophylaxis will be discussed on another occasion.

Koch's pupils rapidly discovered the etiologic agents of other diseases. In 1884, Gaffky published his studies on the etiology of typhoid fever; and Löffler, on the etiology of diphtheria. By 1892, members of the Koch school had isolated also the agents of erysipelas, tetanus, glanders, acute lobar pneumonia, and epidemic cerebrospinal meningitis.

Koch's basic interest remained tuberculosis, and he started next on studies directed toward specific therapy. But there were other demands for the new discipline of bacteriology. In 1883, cholera appeared in Egypt, and the Egyptian government called on France and Germany for assistance. Pasteur sent Nocard, Roux, Thuillier, and Straus; he himself, being partially paralyzed, did not participate. Koch led the German group, comprising Gaffky, Fisher, and Treskow. The epidemic had begun to abate by the time the foreign commissions had arrived. The death of Thuillier from cholera demoralized the French group, which returned home without conclusive accomplishments. Koch's group accumulated data implicating a comma-shaped organism and then went on to continue these studies in India, where cholera was endemic. A vibrio, apparently identical with one seen in Egypt, was isolated consistently from Indian patients in Calcutta. The organism was cultivated on gelatin dishes, but it did not produce infections in laboratory animals. Although the fit to Koch's postulates was not complete, the evidence was highly convincing. Koch returned home triumphantly and was decorated by the Prussian crown.

In 1885, Koch accepted a call to the newly established chair of hygiene at the University of Berlin. Gaffky succeeded him at the Public Health Laboratories. Munich had established such a chair in 1878 with Pettenkofer as the first director, and was followed by Göttingen in 1881 with Flügge as director. Such chairs were subsequently added in the next few years to other

Prussian universities, and all but one were directed by Koch's pupils. Gaffky went to Giessen, Löffler to Greifswald, and Fränkel to Königsberg. The chair at Halle went to Renk, who was not a disciple of Koch.

The staff allotted to Koch at the University Institute consisted of two assistants, a curator, and three military surgeons. The teaching laboratory accommodated twenty-two students. The flow of able bacteriologists continued from the Institute of Hygiene. The first pupil was Shibasaburo Kitasato, who joined Koch in 1885; others important in the advancement of bacteriology were Richard Pfeiffer in 1887 and Emil Behring in 1889. Ehrlich and Uhlenhuth were others from this group. Koch concentrated on his major problem of tuberculosis, seeking a therapeutic agent. By 1890, he was prepared to make his first public statement regarding such an agent. The occasion was the Tenth International Medical Congress in Berlin. Germany played host in the grand manner, and for the joint sessions the largest auditorium, the Busch Circus, seating 8,000 was used. The auditorium was decorated to simulate the Temple of Zeus, substituting a statue of Asclepius in place of Zeus. The rumors were that Koch was to announce a world-shaking discovery. Actually, according to one of Koch's confidants, Koch was hesitant about communicating his findings at this time, but was acceding to the urgings of Von Gossler, the Minister of Culture. The title of Koch's address was "Bacteriological Research." Stating first that he would not be presenting anything new to those active in bacteriology, he discussed the manner in which investigations for therapeutic agents should be conducted, going from studies involving pure cultures to those involving animals and finally man. Proceeding from generalizations to tuberculosis, he stated that after many failures, he had found a substance that could inhibit growth of tubercle bacilli in cultures and in animals. This substance, injected into normal guinea pigs, rendered them resistant to tuberculosis, and injected into guinea pigs with advanced tuberculosis, it arrested the disease. He expressed the hope that it might be useful for therapy in man and justified communication at this preliminary stage with his desire to encourage others, the world over, to participate in the problem. Three months later, toward the end of 1890, he published the following paper ("Progress Report on a Therapeutic Agent for Tuberculosis," *Deut. med. Wochschr.,* **16:** 1029, 1890):

> In a paper that I presented before the International Medical Congress several months ago, I described an agent that renders the experimental animal resistant to infection with tubercle bacilli and arrests the disease in infected animals. This agent is now being tested in man.

I originally intended to conduct comprehensive studies on its preparation and use in medical practice before publication. Despite all precautions, so much inaccurate information regarding this agent has been disseminated that it appeared obligatory to avoid further misinformation by summarizing the immediate situation. . . .

A description of the origin and preparation of the agent must be delayed until the pertinent experimental studies are completed. The agent is a stable, clear, brown liquid that must be diluted prior to use. To avoid bacterial contamination, the preparations, diluted in distilled water, must be sterilized by heat or the addition of 0.5% phenol and kept in cotton stoppered vessels. The diluted agent may in time lose potency and should therefore be diluted as shortly as possible prior to use. The agent is effective by subcutaneous route and ineffective orally. We use a syringe devised for bacteriological procedures, it is fitted with a rubber bulb rather than with a rubber gasket and it can be kept aseptic by rinsing with absolute alcohol. More than a thousand injections have been administered with this type of syringe without producing a single abscess. The area between the shoulder blades and the lumbar region are the preferred sites of injection because pain and local reaction are minimal in these areas.

Man was found to be more sensitive than the guinea pig to this material. This is an additional demonstration of the finding that the results of animal experiments cannot immediately be applied to man. A healthy guinea pig can tolerate more than 2 cc. of the undiluted agent without untoward effects but 0.25 cc. is sufficient to cause an intense reaction in a healthy man. On the basis of body weight, the guinea pig can tolerate 1,500-fold the reactive dose for man.

A 0.25 cc. dose in man produced the following symptoms: 3 to 4 hours after the injection there developed lassitude in the limbs, a tendency towards coughing and difficulty in breathing. The symptoms increased in intensity and at the 5th hour chills and trembling set in for about 1 hour, accompanied by malaise, vomiting and temperature rise to 39.6°C. After 12 hours the symptoms began to recede, the temperature dropped and was normal the next day. Lassitude in the limbs and pain at the injection site persisted for several days. On the basis of repeated trials, the lowest effective dose for a healthy man is approximately 0.01 cc. Most people reacted to this dose only with lassitude but some experienced a slight rise in temperature.

Despite the difference in the minimal effective dose for guinea pig and man, the effects on the tuberculous processes in both species are similar. I shall therefore confine myself to a description of the effects in man.

A healthy person hardly reacts to 0.01 cc. but this dose, in the tuberculous patient produces a generalized reaction evidenced by a rise in temperature that may reach 41° and a local reaction. The generalized reaction is accompanied by pain in the limbs, coughing, lassitude and vomiting. Occasionally there is a slight icterus and a measles-like exanthem on the chest and neck. The reaction begins 4 to 5 hours after the injection and lasts for 12 to 15 hours. There are few residual effects and generally the patient will feel better than he did prior to injection. The local reaction can be observed best in forms such as lupus. Within a few hours after injection, the areas of lupus become swollen and erythematous. During the period of fever the swelling and redness increase and at some points the lupous areas become brownish red and necrotic. Sharply circumscribed lesions may be surrounded by a blanched ring 1 cm. wide, in turn surrounded by a second area of erythema. After the fever abates, the swelling of the lupous areas recedes. The lupous areas become covered with serous exudate which encrusts and forms scabs. The scabs drop off in 2 to 3 weeks leaving a smooth red scar. This reaction occurs in all the areas of active lupus, regardless of size, while the scar tissue, within the lupous areas, remains unaffected.

The reaction in lupous patients demonstrates convincingly the specific nature of the agent. Investigators desiring to study this agent should begin with lupus. Less striking but still visible and palpable are the local reactions in tuberculosis of the lymph glands, bones and joints where swelling, pain and erythema also occurs. The reaction in the internal organs, mainly in the lungs, escape observation. However, the increased coughing and discharge of sputum by pulmonary patients after the first injection of the agent suggest a local internal reaction. It must be assumed that changes similar to those seen in lupus occur in lesions in the internal organs.

The generalized reaction to injection of 0.01 cc. is uniformly elicited if any tuberculous process is present. Thus, it is not too presumptuous to assume that in the future this agent will become an indispensable diagnostic aid. Questionable early cases will be detectable where sputum examination and physical examination may still be negative. Glandular effects, inapparent bone tuberculosis, questionable lupus and similar conditions will be recognized as tuberculosis. It will be possible to determine whether apparently regressed tuberculosis has healed completely or lesions remain like smouldering ashes to light up anew.

Applications to therapy are more important than to diagnosis. In the description of the reaction of the lupous areas it was seen that lesions regressed as a result of sloughing of diseased tissue. In some areas, a single treatment was adequate to produce healing, but, in others, repeated treatments was required. The inflammatory changes occur only in tuberculous tissues and produce destruction and sloughing of the diseased areas but tubercle bacilli are not killed. The action is limited specifically to living tuberculous tissue, caseous and necrotic tissue are not affected. Tubercle bacilli in the dead tissue remain viable and if such tissue is not eliminated successfully, the tubercle bacilli spread to neighboring areas. The possibility of spread resulting from the rapid release of tubercle bacilli during the destruction of living infected tissue must be assessed in the process of treatment.

The next step would involve removal of the dead tissue, as, for example, by surgery. Where procedures such as surgery are not possible, and the body itself must effect a slow elimination of the dead tissue, continuing treatment with the agent is indicated to protect threatened living tissue. The restriction of activity of the agent to living tissue explains why rapidly increasing doses can be given. As living infected tissue is killed off, less living infected tissue remains to produce the reaction and more of the agent must be used to attain the same effect. The tolerance rises too rapidly to be considered as a process of adaptation, a 500-fold rise in dosage can be attained in three weeks. Even when the patient has progressed to the point when he is reacting weakly to high doses, the treatment should be continued to prevent the establishment of new areas of infection.

We may now direct our attention to specific recommendations for treatment. Lupus is again selected as the simplest type with which to start. Here, treatment can be instituted with the full dose of 0.01 cc. Two cases of facial lupus regressed to formation of smooth scars after administration of 3 to 4 doses. The other patients in the group improved after a longer series of treatments administered at 1 to 2 week intervals. Glandular, bone and joint tuberculosis required a similar regimen, namely large doses with long rest periods. As with lupus, the earlier and less severe cases responded more rapidly than the older and more severe cases.

Phthisis was more sensitive to the action of the agent than lupus, glandular and bone tuberculosis. The initial dose of 0.01 cc. was here found too high, and had to be reduced to 0.002–0.001 cc. We then initiated treatment of phthisic cases with 0.001 cc. and continued daily with this dose until no elevation of temperature was produced. The dose was then raised in increments of 0.001 cc. to 0.01 cc. or more. This mild treatment was especially necessary for patients in poor condition. Proceeding in this manner, one could attain high doses with little febrile reaction. Several phthisic patients, in good general condition, tolerated

treatment initiated with higher doses, and in such cases favorable results were attained more rapidly. In phthisis the treatment first resulted
in increased coughing and expectoration. Coughing and expectoration
then decreased progressively and were entirely eliminated in favorable
cases. The character of the sputum changed from purulent to mucous,
with a decrease in the number of tubercle bacilli. Night sweats ceased
and the patients gained weight. Early cases were freed of symptoms in
4 to 6 weeks and could be considered cured. Patients with moderate
or small cavities also showed symptomatic improvement and could be
considered as approaching cure. Only patients with large and multiple
cavities failed to show objective improvement, although subjective improvement was noted.

The experience, up to the present, indicates that in phthisis the
treatment cures early cases with certainty, and that moderately advanced
cases may also be cured. Advanced cases with large cavities, complicated by secondary infection, may be temporarily improved but durable
improvement is rare. In such advanced cases, the tuberculous tissue is
killed, but the killed tissue cannot be eliminated in the presence of
secondary pus producing invaders. If treatment is improved to a point
where the purulent component might be eliminated, as by surgery, advanced cases might also be controlled by such a combined treatment.

I must emphasize that the agent should not be used indiscriminately.
It can be given most easily in early phthisis and when surgery can be
used such as in cases of lupus, glandular, bone and joint tuberculosis,
but careful judgement is required for application to other forms of the
disease. Supporting therapy is also important in the treatment.

A physician who does not use all available diagnostic aids, including
sputum examination, is guilty of neglect. In doubtful cases, injection
of the agent should be used to determine with certainty the presence
or absence of tuberculous infection. The new agent will become a
benefaction to ailing mankind when tuberculosis is diagnosed and
treated early. The number of neglected advanced cases, which form
the inexhaustible reservoir for new infections, will then decline.

Despite Koch's cautious statements on his therapeutic agent, tuberculous
patients from all countries flocked to Berlin for treatment, and physicians
injected the agent indiscriminately, sometimes causing fatal reactions. Injections were given even in hotel rooms. It was not long before public opinion
turned unreasonably against Koch. Contrary to what he thought, the diagnostic value of the agent was more important than its therapeutic effect. The
following year an accumulation of data on approximately two thousand

cases failed to indicate efficacy, and Koch's agent, designated tuberculin in 1891, was gradually abandoned as a therapeutic agent.

In the midst of these difficulties, there were changes in Koch's personal life. For years Koch and his wife had been estranged and lived apart. The separation was eventually completed by divorce. There was always a strong romantic and emotional component in Koch's complex character. While sitting for his official portrait in an artist's studio, he noticed a portrait of an attractive young girl. He fell in love immediately with the subject, who turned out to be Hedwig Freiberg, a student of the artist, and a minor actress at the Lessing Theatre. Koch began an ardent courtship that led to marriage in 1893.

In the spring of 1891 he took a vacation alone in Egypt, to recover from the strife surrounding the work on tuberculin. And he wrote letters to Hedwig that were no less ardent than those he used to send to Agathe when he was an adolescent. He ended a letter from Cairo: "As long as you love me, I cannot be beaten down by the vicissitudes of fate. Do not abandon me now for your love is my comfort and the star to which upward I gaze." The marriage to Hedwig proved happy and durable. His wife accompanied him even on the demanding research expeditions to the tropical areas of Africa. The desire for adventurous travel remained with him, and such trips became more frequent as the years passed.

The demands of research began to conflict with academic activities. The solution was to organize an institute that would be devoted solely to research; this was the Institute for Infectious Diseases, established in 1891. Max Rubner replaced Koch at the University. The staff roster at the new institute included Behring, Ehrlich, Pfeiffer, Kitasato, and Wassermann. One year after organization of the institute, an epidemic of cholera broke out in Hamburg, and Koch organized the campaign for control under the official responsibility of Gaffky, who was assigned as Federal Commissioner to the distress area. The role of carriers in the spread of the disease and Pfeiffer's specific bacteriolytic test were discovered during this operation. Koch's coworkers made other important discoveries. Ehrlich, by 1890, established the scientific bases for immunization procedures, and Behring and Kitasato, late in 1890, developed antitoxin for treating diphtheria and tetanus. The diphtheria and tetanus serum therapy was submitted to clinical trial in 1891, and its effectiveness was immediately apparent. This rapid application of a basic discovery gave Behring a degree of fame matching and possibly exceeding Koch's momentarily tarnished greatness. Within the institute, a circle of coworkers gathered under Behring. This soon reached the proportion of an institute within an institute. Behring left in 1894 to become professor of hygiene at the University of Halle. Koch continued persistently at attempting to improve tuberculin, as we can note from a report he published in 1897: "On New Tuberculin Preparations." Koch also tried a variety of emulsions containing bacillary material rather than soluble culture products alone, as in the original tuberculin.

Behring also tried his hand at tuberculous immunity but with an aggressiveness that led to a complete rupture with Koch, following a patent dispute. Since therapeutic and prophylactic agents were patentable discoveries, Behring patented diphtheria antitoxin and Koch patented tuberculin, but Behring also submitted a patent application for an immunizing agent against tuberculosis. His application was disallowed in favor of Koch's. In the race for honors, however, Behring moved faster than Koch. The first Nobel Prize for medicine was awarded to him in 1901, whereas Koch did not receive this award until 1905.

Other important problems in addition to tuberculosis continued to interest Koch. We recall the expedition to Egypt and India in 1883. In 1896, Koch was asked by the English government to help overcome rinderpest, which was ravaging the cattle in the Cape Colony, in South Africa. Koch lost little time in facing this new challenge and was at Cape Town in less than two months. An etiological agent was not to be seen microscopically or cultured, but Koch determined that the infectious agent was present in the blood and bile. He was able to devise procedures for active immunization. A combination of infectious blood and serum from convalescent animals was effective; also the subcutaneous injection of infected bile produced immunity. Koch got Kolle to continue the study, and he rushed to Asia, where bubonic plague was then spreading. Four years later, rinderpest was to break out in Egypt, and the team of Koch and Kolle was to be called to arrest the epidemic.

Coming back to 1896, we note that a German commission led by Gaffky and staffed with Koch's pupils, Pfeiffer, Stacker, and Dieudonné, had been called to control the Indian epidemic. Koch visited the German commission in India and then proceeded to German East Africa, where the plague had spread. After investigating some animal diseases, he started studies on malaria. In this domain he found tough competition in Ross, who had just described the role of the *Anopheles* mosquito in the life history of the parasite. Koch related the quartan, tertian, and tropical forms of the clinical disease to different varieties of parasites. He also emphasized the importance of latent human carriers in the spread of the disease and suggested active measures for control, such as microscopic examination of blood smears, prophylactic ingestion of quinine, and use of netting to avoid mosquito bites. The first monument dedicated to Koch was erected on the Italian island of Brioni for his help in eradicating malaria there in 1900.

Continuing study of tubercle bacilli led to the recognition of various types. The separation of bovine from human strains was made by Theobald Smith. An avian type was also characterized. These results were accepted and extended by Koch, who abandoned his view of a single variety, as expressed in the 1884 paper, and they were important in the formulation of public health control measures.

Epidemics of typhoid fever were still being encountered in Germany, and Koch participated in the planning of control measures. A commission under Frosch started a control campaign based on early diagnosis, isolation of cases, and the control of water. The commission had eleven laboratory stations for this study. Support was available for such public health campaigns since the minister for culture, Kirchner, in charge of health activities, was a former pupil of Koch.

The older generation of antibacteriologists was beginning to disappear. Symbolically, Koch replaced Virchow as foreign member of the French Academy of Sciences, after his death in 1902.

The same year, there was again a call from the Rhodesian government to investigate a new cattle disease. Koch thought that it was caused by *Trypanosoma theileri* but was unable to devise an immunization procedure. However, dipping in an arsenic salt solution, as advocated by Theiler, killed the transmitting tick and was effective. Koch named the disease African Coast fever. On his sixtieth birthday, which should have been a much celebrated occasion, Koch was still in Rhodesia, mixing amateurish archaeological trips with microbiology. A work day began with a hasty breakfast at 7 A.M. and continued until sundown. His wife accompanied him as usual. Upon his return he announced his decision to retire from the directorship. ". . . But this step does not mean that I shall abandon our science. No! on the contrary I shall serve science so long as my strength permits. . . ."

In 1904 Koch and his wife presented themselves at the Pasteur Institute, and Metchnikoff attempted to serve as his guide and host. He approved of Koch's interest and taste in paintings during a visit to the Louvre, but was less charmed by the visitors' indulgence in gourmet cooking and the Montmartre cabarets. Koch mocked Metchnikoff's strict adherence to frugal hygienic living. The conversation even reached the level of philosophical discussions. Metchnikoff was well impressed by Koch's acquaintanceship with many fields of knowledge, and they parted good friends. In earlier years Koch had treated Metchnikoff unkindly, but the years had chastened Koch, and Metchnikoff was too saintly a man to hold a grudge.

Retirement did not mean rest, and the next year, 1905, Koch was again in Africa working on three diseases—African Coast fever, relapsing fever, and trypanosomiasis—rather than on one. The transmitting agent for relapsing fever was presumed to be the bedbug, but Koch believed that the tick was the intermediate host. He was able to demonstrate that apes could be infected through the bite of naturally and artificially infected ticks and that ticks could transmit spirochetes to offspring. But the chain of transmission had just been demonstrated by Dutton and Todd, and Koch gracefully acknowledged the priority.

In 1905 he returned in time to receive the Nobel Prize, and he addressed the Stockholm audience on the control of tuberculosis, still allowing a place in therapy for tuberculin treatment. Other honors followed the Nobel Prize; the Order of Merit for Art and Science and the Order of Wilhelm. In 1906 and 1907 he was back in Africa, heading a German sleeping sickness expedition in East Africa. The expedition headquarters rapidly acquired a patient colony of 1,000 persons. Three stages of the disease could be differentiated from observation of *Trypanosoma,* first in blood, then in blood and lymph vessels, and then in the brain and cerebrospinal fluid. Koch's group also showed that the trypanosomes underwent developmental stages in the *Glossina.* Control was attempted through thinning trees, cutting back brush, and eradicating crocodiles, the source of blood for *Glossina palpalis.* Some patients were even treated with atoxyl.

He returned to Berlin in November, 1907, to accept further honors. These included a rise to the rank of "Exzellenz" and the acknowledgment of the Robert Koch Grant Fund for control of tuberculosis. The fund attained resources in excess of 1 million marks, a large sum for 1908. This fund was wiped out in the German inflation following the First World War. In 1908, he left with his wife for a trip that he hoped would circle the globe. The trip included a visit to the United States, where he was to see his brother and other relatives. In the United States he was honored in the grand manner that included a banquet at the Waldorf-Astoria Hotel. Andrew Carnegie was among the active participants in this celebration and had been one of the major donors to the Koch Grant Fund. Closer to his heart was the next major stop, Japan, where he visited his old pupil Kitasato. The Kochs were enchanted with Japan and spent forty days there. Writing from Kyoto to his daughter, he said: "I have been travelling around this wonderland for almost

2 months and every day brings new and interesting experiences. I have been received with great regard and everything is being done to make my trip pleasant. Professor Kitasato and two assistants are travelling with me and are tireless in their attendance. . . . I would have desired to visit China after Japan and return to Berlin in the spring of next year. But circumstances have altered these plans. I have received a telegram informing me that I have been delegated to the International Congress in Washington and unfortunately, I will have to return eastward. . . ." At the Washington meeting, as at previous meetings during the past few years, he gave a presentation on the relationship between human and bovine tuberculosis. Back at Berlin, Koch returned to the study of tuberculosis with Bernhard Möllers as his assistant. He investigated tuberculin prepared on protein-free substrates. This material was designated TAF. He was also studying the significance of complement-fixing antibodies in tuberculous immunity.

It was now February, 1910, and Koch had the happy occasion to honor the sixtieth birthday of his first staff member and pupil, Georg Gaffky. Two months later, in April, Koch experienced a severe heart attack. He had had mild heart attacks previously. In March of 1910 the attacks became more severe, but he chose to work until complete collapse occurred. His physicians diagnosed accurately and prescribed as best could be done, but the prognosis was poor. By May Koch managed to travel to the Frey-Dengler sanatorium in Baden-Baden, where he passed away peacefully on the twenty-seventh of May. Koch wished to be cremated. His ashes were deposited behind a simple white marble slab showing a relief of his profile, name, and dates of birth and death. The edge of the slab carries a short chronology of his scientific triumphs. The memorial sits in a mausoleum within the Institute for Infectious Diseases.

References

Bochalli, R.: *Robert Koch*, Wissenschaftliche Verlagsgesellschaft mbH, Stuttgart, 1954.

Hitchens, A. P., and M. C. Leikind: "The Introduction of Agar-Agar into Bacteriology," *J. Bacteriol.*, **37**: 485–493, 1939.

Koch, R.: *Investigations into the Etiology of Traumatic Infective Diseases*, Translation by W. Watson Cheyne, The New Sydenham Society, London, 1880.

Additional Reference

Barlow, C. and P.: *Robert Koch*, Heron Books, Geneva, 1971.

Bacteria as Agents of Disease

4 Koch had demonstrated that anthrax and tuberculosis were caused by specific bacteria. He had isolated the causative agents in pure cultures and had rigorously demonstrated their etiologic role by applying the principles that we now call "Koch's Postulates." Before we follow the road traced by Koch further, we must remember that he himself had followed the path of some enlightened precursors. In this respect the work of Davaine on anthrax was outstanding, and in the field of tuberculosis, there had also been a great forerunner, a French army surgeon named Jean Antoine Villemin. Villemin had studied medicine in Strasbourg and had later become a professor at the Army Medical School of Val-de-Grâce in Paris. Being in the army, he was familiar with horses, and he had noticed a similarity between a disease of horses, called *morve* in French, and human tuberculosis. In a book entitled *Studies on Tuberculosis* that he published in 1868, he wrote:

> There is no other affection which has been the object of as many studies as tuberculosis. . . . Constant and numerous efforts having been unable to throw light on this dangerous affection, one has tried to analyze it, to divide it into distinct species, in the hope that this might permit the capturing of truth by fragments. But these attempts, far from simplifying the question, have only complicated it even more and have thrown the whole history of tuberculosis into hopeless confusion.

This conviction that there was a unity in tuberculosis was based on a detailed study that Villemin summarized in the book that we are quoting. He had carefully investigated the anatomical and pathological aspects of the various forms of tuberculosis, and he had critically questioned the role of heredity and environment. He had reviewed the distribution of the disease among various animal species. All these studies were already an imposing contribution to the knowledge of the disease, but, in addition, Villemin furnished the conclusive proof that tuberculosis was a contagious disease that could be transmitted from man to rabbit, from cow to rabbit, from rabbit to rabbit, from man to guinea pig, from man to dog, etc. Many passage experiments were performed with various animal species. Some, such as sheep, seem to be refractory to tuberculosis. Various tuberculous materials were used for animal inoculation, varying from human sputa to caseous materials. The description of these inoculation experiments read much like those of Koch with the major difference that the etiological agent was only suspected instead of seen. Villemin concluded:

> The inoculation of tuberculous material does not act by virtue of the visible and tangible matter, but because it contains a more subtile principle which escapes our senses. . . . Tuberculosis is inoculable, this is an uncontestable fact. From now on this affection must be classified as a virulent disease, such as *morve-farcin* [glanders], which is its closest relative, nosologically speaking. . . . We would be wrong to think that the affected organism has made the virus, since if we transfer from one organism to another a drop of vaccinal serosity, a drop of variolar or syphilitic pus, a fragment of tuberculous matter, etc., one reproduces in the inoculated subject a multitude of lesions which are similar to those found in the subject from which the inoculated material had been taken. . . . But the organism plays only the role of a medium in which the virus multiplies as a parasite. . . . We must establish a fundamental distinction between the virus and the substance that contains it. The latter is made by the organism under the prodding of the virus. The variolar virus is contained in the pus of the pustule but the pus is not the virus.

These were very profound concepts! Their author was born in the Vosges Mountains in 1827. He died in 1892. His work, the best of pre-Kochian pathology, became the beacon that Koch followed to greatness. And there were those who, in their turn, stimulated by Koch, advanced in this direction even further. We will now turn to some accomplishments of those he had inspired.

When Koch established the bacteriological division of the Imperial Health Office, he was assigned two professional assistants from the Army Medical

Corps; these were Georg Gaffky and Friedrich Löffler. The latter was assigned to study glanders, and in the same year that Koch described the etiologic agent of tuberculosis, Löffler described and isolated the *Actinobacillus* of glanders. This problem finished, he turned to the more difficult and more important problem of diphtheria. These studies matched in comprehensiveness that of Koch's work on tuberculosis. Accordingly, Löffler's report on diphtheria required eighty pages of the Reports of the Imperial Health Office (*Germany, Reichsgesundheitsamt Mitteilungen,* **2**: 421–499, 1884).

This long report can be summarized as follows:

Sixty years have elapsed since Bretonneau presented his classic description of diphtheria. Despite many studies, no universally acceptable explanation of the etiology of this disease has been reached. The divergence of opinions arises from the characteristics of the disease. The appearance of individual cases varies with the age of the patient, the severity and the stage of the disease. Especially significant is the frequent difficulty in deciding whether the observed lesions are due to the frequent complications or to the disease proper. Furthermore, many inflammatory diseases other than those caused by the virus of diphtheria produce pharyngeal lesions which cannot be differentiated from diphtheria. The only significant pathognomonic differential factor is the etiological agent.

This disease is localized on a mucous membrane which is exposed to the extensive bacterial flora of the exterior world. The numerous organisms from food and drink reaching the rugose mucous membrane of the upper respiratory tract find there a favorable medium for growth. These favorable conditions are further improved by the protein-rich inflammatory exudate induced by diphtheria. It will therefore be difficult to differentiate the primary etiological agent from the proliferating multitude of saprophytes present in the normal mouth and pharynx. The failure to define the etiological agent of diphtheria has been due to the difficulty in recognizing the essential organism in the presence of many others and to the inadequacy of methods previously available for differentiation and isolation. The inadequacies of the experimental methods have now been overcome by Koch.

Earlier investigators sought the specific agent in the mucous exudate of the pharynx rather than the internal organs. Experience from accidental infections justified this choice in source material, since physicians and nurses aspirating exudate by mouth suction through tracheal cannulae would acquire the disease if they came in contact with it, but the disease was not acquired by contact with material from other organs and body fluids. . . .

If, then, diphtheria is a disease caused by a microorganism, three postulates must be fulfilled:

1. The organism typical in form and arrangement must be consistently demonstrated in the diseased area.

2. The organism which by its behavior appears responsible for the pathological process, must be isolated and grown in pure culture.

3. A specific experimental disease must be produced with the pure culture. . . .

An evaluation of the results of the past investigations reveals that these postulates have not been fulfilled. Previous investigators saw bacteria in pseudomembranes, but few gave detailed data regarding the types of bacteria present. The earliest investigators stated merely that "molds" were present. The status of bacteriology at the time did not permit more precise differentiation, technique being limited to smear preparations. Subsequently, sections were prepared which demonstrated the relationship of "molds" to tissue. "Molds" were more frequent in the upper layers of pseudomembranes. Micrococci predominated and were located in the lymph vessels. The demonstration of bacteria in internal organs was uncertain and depended upon resistance of tissue structures to the types of reagents used, acetic acid, alkali, ether, etc. Reliable investigations, utilizing stains, Abbe condensers and oil immersion, revealed the absence of bacteria in internal organs and their presence in pseudomembranes. Attempts to culture the micrococci resulted in mixed growth, since the source material was tonsillar overlay in which other bacteria were always present. Cultures of tissue from internal organs were negative.

Attempts to transmit diphtheria to animals, using a variety of infecting procedures did not lead to a truly typical diphtheria. Intramuscular injection of diphtheritic material produced a hemorrhagic inflammation. Corneal injection led to a keratitis and intratracheal injection led frequently to a pseudomembranous tracheitis. Most investigators could not produce these reactions with disrupted organic materials that were used as controls, but some investigators could produce identical changes with non-diphtheritic material. . . .

It seemed indicated, using Koch's recently developed methods for isolation and cultivation of bacteria on solidified culture media, to reinvestigate types of bacteria associated with diphtherial tissue, to determine which might be of etiological significance, to grow these in pure culture, and finally to perform inoculation experiments with pure cultures in the largest possible number of animal species. . . .

In most instances the source specimens at my disposal were not accompanied by comprehensive clinical data; acquisition of such data would have required full attendance in the hospital. However, I attempted to determine the clinical course, and especially the duration of illness, of source cases. Even though scarlatinal diphtheria was differentiated from diphtheria, the frequency of the combined disease led me to include such cases in the scope of this study.

For examining microscopic sections it was necessary to use a staining method which would be applicable to a wide variety of bacteria. None of the available methods, namely the Weigert nuclear stain, the picro-carmine gentian-violet differential stain and the Koch-Ehrlich differential stain for tubercle bacilli were satisfactory. While studying sections of syphilitic sclerae, I found that a more intense and rapid staining was achieved with a mixture of 1 cc. of concentrated alcoholic methylene blue in 200 cc. of water when 0.2 cc. of 10% potassium hydroxide was added than when such a mixture lacked potassium hydroxide. Since this weak methylene blue solution did not stain some types of bacteria adequately, I raised its concentration, and the most rapid and intense staining was achieved with 30 cc. of concentrated alcoholic methylene blue in 100 cc. of a 1–10,000 dilution of potassium hydroxide in water. The sections were immersed in the stain for several minutes and then rinsed in 0.5% acetic acid to remove excess stain and achieve differentiation of nuclei. They were then dehydrated in alcohol, treated with cedar oil and finally mounted in Canada balsam. Anthrax, rabbit and mouse septicemia bacilli, typhoid and glanders bacilli, erysipelas micrococci, *Micrococcus tetragenus*, spirochetes of relapsing fever, fungal mycelia, etc., were all equally well stained. This technique thus approached a universal staining method. . . .

Löffler examined clinical materials from twenty-seven patients. Anatomical diagnoses were followed by a description of sections from various tissues and organs, stained by the alkaline methylene blue method. He concluded:

A consideration of the case material reveals that 3 types of cases may be recognized. In one type, chain-forming micrococci appear to play a major role. The mucosal surfaces are denuded of epithelium, are greyish yellow in color and there has been a loss of tissue as a result of necrosis. Pseudomembranes are absent. The micrococci are present on the surface and penetrate into the tissues forming wedge- or tongue-like areas of necrosis margined by a narrow unstained border followed by a layer of proliferating cells. The micrococci may thus penetrate into the lymph vessels and then spread to other organs on the body where they form plugs in the capillaries. A similar picture is seen accompanying other diseases such as smallpox, typhus fever, puerperal fever, etc., where chain-forming micrococci are known to produce complications accompanying the primary disease. Similarly these organisms are most likely a complication of primary diphtheria, thus the frequency of the occurrence of micrococci on and in mucous membranes in diphtheria should not lead to confusion regarding the true primary agent. . . .

Another type of case offers more direct evidence of the secondary invasive role of the micrococci. True pseudomembranes are present and in addition to micrococci, a second type of organism is seen. Too few micrococci are present to account for the observed damage, but the other organism may readily account for the pseudomembrane formation. The second type of organism is Klebs' bacillus. Unlike the micrococcus, Klebs' bacillus alone may be present in typical cases that are characterized by the presence of a thick pseudomembrane in the pharynx, larynx and trachea. A variety of bacteria are irregularly distributed on the surface of the pseudomembrane. Below the surface are small masses of bacilli which stain intensely with methylene blue. . . .

The third type of case strongly supports the hypothesis that the bacilli occurring just below the surface are the etiological agents. It can be illustrated by the case of a child who died on the third day of the disease. The pseudomembrane in the trachea, which undoubtedly was the last such element to be formed, was a thick membrane containing masses of the characteristic bacillus alone. Another child who also died on the third day of the disease with dyspnea and symptoms of lung inflammation had not only a pseudomembrane in the trachea but also masses of rods in the alveoli as well. . . .

From the pathologico-anatomical studies it could not be determined whether the micrococcus or the bacillus was the etiological agent. The only remaining course was to obtain pure cultures of each of the 2 types of organisms and, by animal inoculation, to determine which could produce a disease analogous to human diphtheria.

The cases chosen for cultural study of the coccus were those showing chain-forming micrococci either solely or in overwhelming proportion in microscopic smear preparations of tonsils and internal organs. Meat broth-peptone-gelatin was used as culture medium. Fragments from infected tonsils were incorporated into the liquefied medium on slides, and the mixtures were covered with glass cover slips. The gelatin was then allowed to solidify, permitting the development of isolated colonies. The preparation of pure cultures from internal organs was simpler. To eliminate surface contaminants, the organs were washed in 5% phenol for 10 minutes to kill vegetative forms of bacteria and in 1% bichloride of mercury for 5 minutes, to kill spores. The organs were dried on blotting paper. After the surfaces were dried, the organs were cut with a hot glowing knife and hot glowing forceps were used to tease portions of tissue for culture study. With these methods pure cultures of chain-forming micrococci were obtained from 5 cases: 2 of scarlatinal diphtheria and 3 of typical diphtheria.

The chain forming micrococci grew slowly in gelatin. After 3 days of growth, examination with direct light revealed small, round, greyish translucent colonies. . . . In reflected light the colonies were white and granular, indistinguishable from erysipelas cultures. The micrococci grew luxuriantly in meat infusion medium (meat infused with 1% peptone, 0.5% NaCl and 1% dextrose). After 24 hours of incubation at 37°C. finely threaded floccules appeared. These floccules were formed by long interweaving chains comprising up to 100 cocci. The organism grew on meat infusion broth solidified with 1% agar, but growth was more luxuriant when serum was used as gel. Pure coagulated serum was inferior to a mixture of 1 part meat infusion medium and 4 parts serum. . . . Growth on cooked potato was slow. In the various culture media, bacterial cell division was observed in both the horizontal and longitudinal plane.

Animal passage studies were performed with 4 pure cultures of chain-forming cocci. . . .

After extensive animal experimentations with mice, guinea pigs, rabbits, birds, dogs, and apes, Löffler was finally convinced that the micrococcus could not produce a diptherial disease in animals. He concluded:

Since, (1) the chain-forming cocci did not produce a diphtheria-like disease in any experimental animal species, (2) the cocci were seen only in a limited proportion of diphtheric cases in man, (3) they occurred in diphtherial lesions only in association with a type of bacillus to be described more fully below, and finally (4) they were also found in internal organs in other diseases, we are justified in concluding that the chain-forming cocci are only accidental secondary invaders in diphtheria. The chain-forming micrococci may, however, produce a diphtheria-like disease when the pharynx is invaded and the organisms spread through the lymphatics to the trachea and lungs. Such was the case of an infant who had a nasal discharge and a slight pseudomembrane in the throat. After several days the infant became hoarse and a dyspnea of increasing severity set in followed by death. At autopsy the pharynx was clear, but there was a thin grey pseudomembrane on the epiglottis, trachea, and bronchi, and areas of hemorrhagic bronchopneumonia in the lungs. In sections the epithelium of the epiglottis was still intact, but the lymph vessels in the mucous membrane below were filled with micrococci. The trachea was denuded of epithelium and covered with an exudate of inflammatory cells and micrococci. Micrococci were also seen in the mucous membrane and the lungs. All other organs were normal. The disease could be interpreted as a diphtheria-like mucous membrane erysipelas.

The characteristic rods, which [had been observed previously by Klebs] were slightly bent and enlarged at the poles, and were frequently seen in smears and sections, could not be cultured on meat infusion-peptone-gelatin. . . . It was therefore decided to culture the material on coagulated blood serum with incubation at body temperature. Despite the possibility of rapid overgrowth with putrefying contaminants, the very first trial using this technique met with success:

Segments of organs from a dead patient were sent to me in a glass jar. The organs lay for several hours in blood that had seeped from the cut surfaces of the organs. Liver, heart and kidney which in smears yielded only micrococci, were washed in 5% phenol followed by 1% $HgCl_2$. Fragments of these organs were cultured on meat infusion-peptone-gelatin incubated at room temperature, coagulated beef serum and coagulated sheep serum incubated at 37°C. In addition, scrapings from areas of pseudomembrane of the pharynx, showing both micrococci and Klebs' bacilli in smear preparations, were cultured on coagulated sheep serum incubated at 37°C. On the following day discrete translucent colonies were seen in all serum cultures. Most of these colonies consisted of micrococci, but occasional ones were of rods identical with the ones seen in pseudomembranes. After three days of incubation the colonies of micrococci were small, yellowish and translucent, whereas the colonies of rods were large, whitish and opaque. No Klebs' bacilli were recovered from gelatin cultures. Occasional colonies of rods were now seen in cultures of liver and kidney, which until then had revealed only micrococci. However, the presence of the rods on the peripheral areas of the internal organs is, in my estimation, a post-mortem change caused by the organs and pharynx being held in the same fluid.

For animal passage experiments loopfuls of growth from colonies of the characteristic rods were suspended in 10 cc. amounts of sterile water and loopfuls of those suspensions were cultured on fresh serum slants. The organisms grew as isolated colonies. Third passage cultures prepared from isolated colonies, were used for animal inoculation. Such pure cultures were certainly free of original tissue. The medium consisted of 3 parts calf or sheep serum and 1 part veal bouillon containing 1% peptone, 1% dextrose and 0.5% NaCl. Growth on this medium was luxuriant; in 2 days a 1 mm. thickness of culture was evident and isolated colonies attained a diameter of 0.5 cm. This medium was used for subsequent cultures. Another pure culture of rods was obtained from the tracheal pseudomembrane from another case, which in smear preparations revealed many rods.

Following the isolation of rods from autopsy material, recovery of pure cultures of rods from living patients presented no difficulties. Pure cultures were obtained from 4 successive living cases. Portions of membrane were removed for microscopic section and for pure culture isolation as described above. In sections, micrococci predominated on the surface, but rods were abundant in the cellular layer immediately below the surface. The fibrinous layer was free of bacteria.

In culture, the rods appeared first followed by the micrococci. Pure cultures of rods were obtained as before. The rods were non-motile and stained intensely with methylene blue. They were straight, or slightly bent, about as long as and twice as thick as tubercle bacilli. The larger ones were segmented with thickening at their points of junction. The ends were sometimes enlarged. Barred staining and intensified polar staining were frequent.

Treatment of methylene blue smears with dilute iodine intensified the stain in the polar granules and decolorized the remaining areas of the cells. The polar granules did not appear to be spores. They did not glisten in reflected light, they stained readily, and rods with polar granules were destroyed by heating for ½ hour at 60°C. Cultures were viable for 3 months. The rods required temperatures above 20°C. for growth and cooked potato medium did not support their growth.

Of major interest was the effect of pure cultures of the rods in various animal species. Six pure cultures were tested in mice, rats, guinea pigs, rabbits, apes, small birds, pigeons, and chickens. The animals were inoculated by subcutaneous route, application to traumatized and intact mucous membranes, and by inhalation.

After having shown that mice and rats were refractory to any pernicious effect of the rods, Löffler found that the bacillus was highly pathogenic for guinea pigs. Löffler performed pathogenicity trials, both with tube culture inocula and tissue from infected animals. Twenty-eight guinea pigs were used and all were injected by subcutaneous route. He presented the observations and results in considerable detail, and he summed up as follows:

Pairs of guinea pigs were inoculated subcutaneously with each of a series of cultures. All guinea pigs were sick on the day following injection; they were listless, coats were ruffled, and there was swelling at the site of injection. The animals died within 2 to 7 days. There were greyish white membranes at the sites of injection, edema in the surrounding subcutaneous tissue, bloody serous exudate in the pleural cavities, and brownish red dense and atelectatic areas in the lungs. The remaining internal organs were normal. Bacilli were recoverable only at the site of injection.

I have presented the results with this series of guinea pigs in detail, first because they demonstrate that bacilli isolated from the various human cases produced the same pathology, second because they demonstrate conclusively that death resulted not from a generalized dissemination of bacilli, but from an effect induced by the bacilli at the site of injection. The hemorrhagic edema, the pleural exudate, the brownish red areas of consolidation in the lungs, where bacilli could not be demonstrated, were conclusive indications that a toxin generated at the site of injection and transmitted through the blood stream, induced severe damage to vessel walls. The toxin generated by the bacilli in the guinea pig is undoubtedly similar to the toxin of human diphtheria. The toxin in man, like that in the guinea pig, appears to act primarily on blood vessels.

Other noteworthy observations in the guinea pig were the formation of greyish white pseudomembranes at the sites of inoculation, and the disappearance of bacilli in lesions after several days, despite the lethal outcome in the animals.

The canary, finch, siskin, etc., were more susceptible than the guinea pig, when inoculated intramuscularly. . . . All birds died by the third day following infection with the symptoms described above. Bacilli were recovered only from the site of inoculation and all cultures from internal organs were sterile. When birds were inoculated with nutrient broth-peptone-gelatin cultures, survival could be prolonged up to 5 days, but the pathological changes were the same as those obtained after injection with cultures on coagulated serum and all cultures from internal organs were sterile. As in guinea pigs, the animals died from the effects of the localized infection.

The responses of rabbits were more variable than those of guinea pigs and small birds. Cultures of the bacillus were applied on the scarified cornea or conjunctiva, injected intramuscularly in the thigh or applied on the trachea by means of a tracheotomy. Where tracheotomies were performed, both the muscle and skin wounds were closed with sutures. A small number of intravenous inoculations were also performed.

Löffler performed an extensive series of rabbit inoculations. Twenty-five rabbits were injected with three different strains.

The various isolates of the bacillus induced the formation of dense fibrinous membrane on the conjunctival and tracheal mucous membranes of a high proportion of the test rabbits. The pseudomembrane formation must be attributed to the bacilli because the operative procedure alone and other bacteria isolated from diphtheritic material injected in the same manner did not produce these effects.

The bacilli induced the formation of pseudomembranes in experimental animals, but the bacilli in the experimental pseudomembranes were not as numerous as in human cases and were distributed differently. The membranes in animals were essentially limited to the site of inoculation and contained few bacilli. Despite the limited site of membrane formation, as in guinea pigs, the infections could lead to death. Where death occurred following intratracheal inoculation, the mechanical obstruction caused by the intratracheal pseudomembrane possibly could be implicated as the primary cause of death but deaths following conjunctival or intramuscular inoculation must be attributed to the formation of toxin at the site of inoculation and its subsequent dissemination through the blood stream. The toxin is not formed in the blood stream; intravenous inoculation of large numbers of bacilli in rabbits did not lead to death. The organisms seen in rabbit tissues stained poorly and showed aberrant giant forms. These forms and staining reactions were encountered in the unfavorable nutrient gelatin cultures and imply that the rabbit body is not a favorable environment for the development of the bacilli.

Since there are reports of many cases of transmission of diphtheria from man to larger birds and *vice versa,* it appeared especially interesting to study the effect of the bacilli on pigeons and chickens.

The third serial transfer of bacillar strain No. 1 was injected under the tongue and into the gums of 3 pigeons; one of the three pigeons was also injected in the breast muscle. A yellowish exudate developed at the sites of injection and spread over the beak cavity. The infiltration in one animal interfered with food intake and led to death on the 11th day. In both other animals, the exudates were resorbed and the animals recovered.

Pigeons are useful for such studies because the laryngeal and tracheal mucosae are easily injected without traumatization. These areas are exposed when the beak is opened. It was thus possible to see if introduction of the rods into the untraumatized respiratory tract could induce pseudomembrane formation.

The trachea of a pigeon was streaked with the 13th serial transfer of strain No. 1 and remained well. The fifteenth passage of the same bacillus was then applied to the trachea through tracheotomy. The soft tissues of the throat became edematous on the following day and a pseudomembrane was formed on the larynx. The animal died of respiratory failure on the 7th day. At autopsy there was a yellowish fibrinous pseudomembrane at the tracheotomal wound, larynx and buccal cavity, containing a variety of bacteria including the characteristic rod.

He then repeated the experiment with some variations, using various strains that had gone through five to twenty-three serial transfers in vitro. The variations in the route of infection involved intratracheal application through the beak cavity and through tracheotomy, injection into multiple sites in the beak cavity, and intramuscular injection. Application of cultures through tracheotomy led to death. Injection into beak cavity or muscle was not fatal; a local lesion developed but healed in approximately two weeks. Löffler then demonstrated a similar range of pathogenic reactions in chickens with six cultures injected in a total of eleven animals. Summarizing:

> In general, these experiments demonstrated that pigeons and chicks are not nearly so susceptible as small birds, namely finches, sparrows and canaries.
>
> It is important at this point to discuss observations on two pigeons and one rooster. In one of the pigeons injected with strain No. 1, a weakness in the limbs and inability to fly was evident after 4 weeks. This progressed to severe paralysis and then to eventual recovery. This animal, which had been injected in the beak cavity with a pure culture of rods isolated from a fatal human case, had developed a pseudomembrane that regressed. The paralytic episode in the bird was reminiscent of the transitory paralysis that occurs sometimes in human diphtheria, and might be attributed to the specific organism.
>
> This conclusion was strengthened by a similar observation on a rooster that had been infected by tracheal application of the 25th serial transfer of the same strain. After transitory pseudomembrane formation, the animal developed paralysis of the legs and wings beginning 4 weeks after injection. The paralysis did not regress. Since such a symptomatology for chickens is not described in the literature, I believed that the paralysis was due to diphtherial infection. Paralysis was also encountered in a rabbit that had received the very same inoculum.
>
> The pigeon that had been paralyzed and had recovered was reinjected in the beak cavity and breast with the 30th transfer of strain No. 1. A transitory pseudomembrane formation was again followed by

paralysis. The animal died 23 days after the second injection. Autopsy revealed pneumonia to be the cause of death and there were masses of uric acid crystals around the joints and connecting tendons. This paralytic episode therefore was not of diphtherial origin but due to a uratic arthritis.

The rooster was then sacrificed. The entire rump musculature was atrophic, the sternum and ribs were distorted, the junctions of the ribs with the sternum were swollen, and the vertebrae softened. Thus the animal was suffering from a rachitis of the rump, bones and muscles, that accounted for the paralysis. Again diphtheria did not seem to be directly involved.

Similarly, for the second pigeon in this group of three large birds that developed paralysis, the symptoms were nondiphtheritic in origin. At autopsy a myxoma was found in the lower spinal cord.

I considered it necessary to present the information in these three birds in order to spare subsequent investigators the possibility of being misled by such observations.

There remains to be described two experiments in an ape. A long-tailed Java ape was infected by streaking and traumatizing the pharynx with a stiff brush infected with the 9th serial transfer of strain No. 3. A small transitory ulcer resulted. The animal was then injected in the conjunctiva and subcutaneously in the axilla. A transitory swelling occurred in the conjunctiva and an edematous infiltration followed by ulceration and healing occurred in the axilla. These experiments indicated that apes were not susceptible to diphtheria.

In the experiments so far discussed natural modes of infection were not used. A more natural method of infection was then tried. Three chickens, 3 pigeons, 3 rabbits and 3 guinea pigs were placed in a chamber 1 meter × 0.5 m. × 0.5 m. Three hundred cc. of a densely turbid suspension of the 6th serial transfer of bacillar strain No. 3 were then atomized into the chamber. An ape was placed in a cylindrical chamber 0.75 meters high and 0.5 meters in diameter. The cylinder was equipped with the necessary food, water and bedding. One hundred and fifty cc. of the same bacterial suspension were sprayed into the cylinder. Even though the animals were exposed to infection both by inhalation and ingestion of contaminated food, none became sick. Similarly repeated attempts to infect rabbits and guinea pigs by application of cultures to undamaged mucous membranes were unsuccessful.

Löffler decided to try one more route of infection, the vaginal inoculation of guinea pigs. He noted that: "*Recovery occurred in most animals despite severe initial symptoms of intoxication."* This recovery was due, he thought, to the fact that the animals could remove the inoculum and the membranes that might form by licking. He continued:

Such rapid removal and subsequent recovery could not be expected to occur following subcutaneous injection of organisms.

Before further discussing the evidence regarding the implication of the bacillus as the etiological agent of diphtheria we should determine whether or not the bacillus can be found in the oral and pharyngeal secretions of healthy individuals. Children were used for this study because they are the most susceptible age group. Cultures on broth peptone sugar serum medium were taken from the oral mucus of 20 children, ages 1 to 8, and cultures from 10 adults served as controls. The cultures were examined after 3 days of incubation and methylene blue smears were prepared from all greyish white or white colonies. These colonies were found to consist either of micrococci or of short ovoid bacilli that were not even remotely similar to our specific rod. Pure cultures isolated from three colonies of the short bacilli were each injected into two guinea pigs and were found to be avirulent.

In one culture from a child, bacilli morphologically similar to the diphtheric rods were observed. They were slender, the size of tubercle bacilli, showed intense polar staining and polar clubbing. After four in vitro transfers, the culture of this organism was injected subcutaneously into two guinea pigs. Both animals were sick on the day following injection and died on the third day. In one there was a greyish pseudo-membrane at the site of inoculation and edema in the surrounding area. The axillary glands were swollen and hemorrhagic, the kidneys and adrenals were engorged with blood. The specific bacilli were observed only at the site of inoculation. The findings in the other guinea pig were complicated by tuberculosis. Fifth passage culture of the same organism was then injected subcutaneously into 3 guinea pigs. All three animals died after 2 days with the typical diphtheric syndrome. There could be no doubt that this organism isolated from a healthy child was the same as the bacillus isolated from cases of diphtheria.

The following facts favor designating the bacillus as the etiological agent of diphtheria: The rods were found in thirteen of twenty-seven typical cases of diphtheria with fibrinous pharyngeal exudate. The rods were present in the oldest areas of the pseudomembranes and were deeper than other organisms. Cultures of the rods were lethal when inoculated subcutaneously into guinea pigs and small birds. Whitish and hemorrhagic exudates developed at the sites of injection with diffuse edema in the surrounding tissues. As in humans, the internal organs were free of lesions. The bacilli produced pseudomembranes on the exposed tracheae of rabbits, chickens and pigeons, on the scarified conjunctivae of rabbits and on the vaginal vulvae of young guinea pigs. Another characteristic effect was severe lesions of the blood vessels as evidenced by the bloody edema, hemorrhagic lymph glands and pleural exudate. As in humans, younger animals were more susceptible than older ones.

The following points may be made against the conclusion that the bacillus is the etiological agent of diphtheria:

1. The bacilli were absent in a number of typical cases of diphtheria.

2. The bacilli in the pseudomembranes of rabbits and chickens were not arranged as in the pseudomembranes of man.

3. The bacilli did not produce disease on the untraumatized pharyngeal mucous membranes of animal species which were susceptible when the mucous membranes were traumatized.

4. Animals surviving experimental infection did not become paralyzed.

5. Finally, a typical virulent bacillus was found in the throat of a healthy child.

The proof that the bacillus is the etiological agent of diphtheria is thus not complete. However in typical human cases where the bacilli were not found, they might have been recently eliminated, as was indicated in the experiments on vulval infection in guinea pigs. The third objection may not be important since it has been frequently observed in man that diphtheria is preceded by catarrh of the throat and air passages. The effect of sharp north or northeast winds on the incidence of such catarrhs is well known. In addition one should note that the experimental animals used in these studies did not possess an organ comparable to the human tonsil. The tonsil with its many crypts and folds offers a favorable site for growth of the organism. Paralysis was not observed in the susceptible animal species because few survived infection and, in addition, its frequency even in human diphtheria is relatively low. A maximum incidence of 11% has been reported. As far as the last objection is concerned, it is conceivable that agents which are rarely infectious could occasionally be found in healthy subjects.

In my estimation, the noted objections are not major ones. The future lines of study seem clear. Special emphasis should be put on the characterization of the toxin. It should be possible, in view of our knowledge of the nutritional requirements of the organism, to produce large amounts of the chemical substances that it forms. If the same specific compound occurs in the culture medium, the infected guinea pig and the diphtheria patient, it would be an important argument in support of the bacillary etiology of the disease.

Löffler thus showed that the principles of etiological proof that Koch had taken six laborious years to establish through his studies on anthrax, traumatic infestious diseases, and tuberculosis were not always applicable. The organism seen by Klebs and cultured by him did not, because of its limited distribution in the affected organisms, seem to account for the observed symptoms. Fortunately for Löffler it produced a disease in animals that was reasonably similar to that in human beings. Despite his great importance in bacteriology, Friedrich Löffler does not seem to have attracted the attention of any devoted biographer. However, we know that he came from a family with a history of distinguished service in the Prussian Army Medical and Sanitary Corps.

In 1852 at the time of Friedrich's birth in Frankfurt, his father was a regimental surgeon, who was eventually elevated to the rank of general and who became the associate director of the Friedrich-Wilhelm Institute of Military Medicine. Friedrich attended the French *Lycée* in Berlin, where, besides receiving a fine classical education common to good gymnasia of the time, he learned to speak French with great fluency. He began his medical studies in Würzburg, but, in good family tradition, he soon transferred to the Army Medical School. Before he could finish his studies, the Franco-Prussian War broke out, and he was sent into service in the field. He returned to Berlin after the war and finally passed the State Medical Examinations in 1874. After having filled two vacancies in Hanover and Potsdam, he was assigned to the Imperial Health Office in Berlin. This, at first, did not seem to be such an exciting appointment as Löffler was asked to devote his energy to the study of the flash points of petroleums. The big opportunity came when, in 1880, he and another military physician, Georg Gaffky, were transferred to the newly formed laboratory of bacteriology, which was, as we recall, under the supervision of Robert Koch.

Koch opened a new world to his two military assistants. Löffler recalled: "The days when we worked side by side with Koch, almost daily learning new wonders of bacteriology, have remained unforgettable. We worked from early morning to late evening, hardly finding the time for the needs of physical existence. This was when we learned the meaning of precise work and how to pursue an objective."

After studying disinfection with Koch, Löffler, together with Schütz, discovered the etiologic agent of glanders in 1882 and the etiologic agent of hog cholera soon after. He then turned his attention to diphtheria, which was a new kind of disease, a disease the symptoms of which were caused in part by a toxin, as we have seen from his paper of 1884. Soon after this monumental publication, Löffler was appointed to the laboratory of the First Garrison Hospital in Berlin. While at this post, he qualified as a *Privatdocent*, in 1886. Two years later, he was offered the professorship of hygiene both at Giessen and Greifswald. He chose the latter post, and he remained there for twenty-five productive years. He had always been a skilled technician, and in 1890 he developed a staining method for bacterial flagella. He was soon faced with another great challenge when he was teamed with Frosch and asked to solve the etiology of foot-and-mouth disease. On this occasion, one more weakness of the principles of Koch was exposed, when these two investigators demonstrated that the causative agent was invisible and filtrable. It was the first rigorous demonstration of the role of filtrable viruses in animal diseases. With their brilliant work, these two German investigators showed that Pasteur's work on rabies might not have been as idiotic as the dogmatic Koch might have been inclined to think. The study of foot-and-mouth disease was continued with the added cooperation of Uhlenhuth. These three investigators were to devote years of study to the development of prophylactic and therapeutic methods for control of this malady.

In 1913, Löffler left Greifswald to head the Koch Institute in Berlin. Soon after, the war broke out, and he found himself back in the Army Medical Corps and its oldest general. His age did not prevent him from participating in field service. His two sons were also army physicians. In January, 1915, he had to return home because of poor health. Surgery was of no avail, and he died in the month of April, 1915.

Throughout his fruitful life, Löffler had been a highly critical investigator who would not publish a paper unless he was convinced that a significant advance had been made. Considering the greatness of his contributions, one is surprised that he did not receive a large number of honors.

Löffler was a remarkable technician who had introduced new methods of staining. The most important contribution in this field was, however, the work of the Danish physician Christian Gram. The method of Gram is a simple method for separating bacteria into two basic groups. Using the Gram stain one can, for example, differentiate *Klebsiella pneumoniae* from pneumococci. However, at the time Gram published his most important paper, the main interest of bacteriologists was simply to see the difficult-to-detect bacteria and to differentiate them from mammalian nuclei. Gram was working in Friedländer's laboratory at the morgue of the City Hospital of Berlin. In 1884, Gram published a paper in which he described his staining

method, which had been referred to in passing by Friedländer the year before. It was entitled "The Differential Staining of Schizomycetes in Sections and in Smear Preparations" (*Fortschr. Med.*, **2**: 185–189, 1884). Gram described his method as follows: "*After having been dehydrated in alcohol, the preparations are immersed in the aniline–gentian violet solution of Ehrlich for 1 to 3 minutes (tubercle bacilli are immersed for 12–24 hours). The preparations are then placed in an aqueous solution of iodine–potassium iodide (iodine-1 gm., potassium iodide-2 gm., water 300 gm.) directly or after a rapid rinsing in alcohol. They are allowed to remain there for 1 to 3 minutes, during which time the color of the preparations changes from dark blue-violet to deep purple-red. The preparations are then completely decolorized with absolute alcohol. Further clearing is achieved with clove oil. . . . Bacteria are stained intense blue while the background tissues are light yellow. . . .*"
After indicating that counter staining was possible, Gram reported the most crucial part of his observations:

> **I.** The following forms of schizomycetes retain the aniline–gentian violet after treatment with iodine followed by alcohol: (*a*) cocci of croupous pneumonia (19 cases) . . . (*b*) cocci of pyemia (9 cases) . . . (*k*) tubercle bacilli (5 cases) . . . (*l*) anthrax bacilli (from 3 mice) . . . (*m*) various putrefactive bacilli and cocci. . . .
> **II.** The following schizomycetes are decolorized by alcohol subsequent to treatment with iodine: (*a*) encapsulated cocci from croupous pneumonia (1 case) . . . (*b*) non-capsulated cocci from croupous pneumonia (1 case) . . . (*c*) typhoid bacilli (5 cases). The bacilli from typhoid fever are readily decolorized by alcohol either with or without prior treatment with iodine. Decolorization occurs even after the preparations have been immersed in stain for 24 hours.

Gram must have been much chagrined to see that certain bacteria refused to keep his stain. He wanted to stain all bacteria and not the surrounding mammalian tissues. However, what he thought was a defect of his method soon became a great contribution to bacteriology. It is not definitely known who thought of utilizing the method of Gram the way we do today, but in a textbook published by Flügge in 1886, two years after the paper of Gram, one can read: "The method of Gram is mainly useful for the differential staining of bacteria in tissues and for the diagnostic differentiation of species." Gram, who had concluded his paper with the modest hope that "the method would be useful to other workers," died in 1935 without having studied the method further. Not only had it been of great practical value as a diagnostic tool, but it had also raised profound questions about the chemical composition of bacteria and about the mode of action of the staining-decolorizing process.

From Löffler's paper on diphtheria, it was obvious that he strongly suspected that a toxin was formed by Klebs's bacillus. This point was clarified by two coworkers of Pasteur, Roux and Yersin. They published, in 1888, a paper which was simply entitled "Contribution to the Study of Diphtheria" (*Ann. inst. Pasteur*, 2: 629–661), in which, after having confirmed in general the observations of Löffler, they wrote:

> Löffler reported that he had not seen diphtheritic paralysis among animals which had survived inoculation with Klebs' bacillus. . . . This constant absence of such a characteristic phenomenon as paralysis in the experimental infection, could have led to doubts that the bacillus of Klebs and Löffler was the cause of diphtheria. Some paralyses are nevertheless observed in animals inoculated with diphtheria in the trachea or under the skin. It is even very frequent when the experimental animals are not subjected to too rapid an intoxication.
>
> The first example of experimental diphtheritic paralysis we observed was in a pigeon inoculated in the pharynx with a pure culture of Klebs' bacillus. After the formation and regression of a characteristic diphtheritic membrane, this pigeon appeared to have recovered, but after three weeks it became so weak it experienced difficulty in standing and fell forward when prodded to walk. . . . The muscular weakness continued for a week and was followed by partial recovery. Nevertheless, the animal died five weeks after inoculation. At autopsy the animal was emaciated but there were no lesions of the joints or nervous system which could explain the motor difficulty. Is this not an example of characteristic diphtheritic paralysis following an experimentally produced angina?
>
> When rabbits survive the immediate effects which are provoked by diphtheria in the trachea, they often present symptoms of paralysis. As an example: a rabbit inoculated intratracheally through a tracheotomy developed dyspnea and rasping respiration as early as the second day. The symptoms abated, but on the sixth day, paralysis developed in the hind legs. . . . The paralysis progressed rapidly and the animal died. The cervical lymph nodes were enlarged and edematous, the trachea was congested, but no longer showed a diphtheritic membrane. The lungs were edematous.
>
> We have already stated that intravenous injection of one cc. of diphtheria culture was frequently fatal in less than 4 days. Most often

the disease terminates in a total paralysis which precedes death only by several hours. When death does not occur in so short a period of time, paralysis is easier to observe. It usually begins at the posterior part of the body, and sometimes it is so rapidly progressive that within one or two days it involves the entire body and the animal dies with respiratory and cardiac failure. At other times, the paralysis is limited temporarily to the hind limbs. It begins with a muscular weakness which gives the gait a peculiar appearance. . . . The disease is almost always progressive and the neck and front limbs become involved. It is not rare to see death without convulsions suddenly overcoming the animal in a position in which one had seen it shortly before. . . . At autopsy of such paralytic rabbits, when the disease has not been of too long duration, the lymph nodes and various organs are congested and the livers are fatty. Sometimes the spinal marrow is softened. Pigeons recover from paralysis more frequently than rabbits. . . . The paralysis following inoculation with the Klebs-Löffler organism completes the resemblance of the experimental disease to the natural disease and establishes firmly the specific etiologic role of this bacillus.

The authors then described their main discovery, the proof of the production of a diphtheria toxin:

We have already seen that the diphtheric bacillus is not disseminated through the organs of humans or animals suffering from diphtheria, but is localized in the diphtheritic membranes or at the site of inoculation. How can a culture restricted to one location in the body give rise to a generalized infection and vascular lesions in all organs? It has been postulated that a very active poison was elaborated at the site of localization of the bacilli, and from that point spread throughout the body. Löffler and Oertel, among others, believe in the existence of this toxin. Baumgarten, on the other hand, believes that it is unnecessary to invoke the action of a hypothetical chemical product and he claims, erroneously, that the microbes of diphtheria can invade the organs.

The diphtheritic poison exists, and can be demonstrated in cultures of the Klebs bacillus. Let us filter a 7-day veal bouillon culture through porcelain. The liquid filtrate is completely clear, slightly acidic, and free of living organisms as determined by culture. Injected subcutaneously into animals in doses of 2 to 4 cc. the liquid is innocuous. It is not innocuous in larger doses. A dose of 35 cc. injected intraperitoneally in a guinea pig, or intravenously in a rabbit is fatal. After 2 or 3 days the hair of the guinea pig becomes ruffled, the appetite declines and the urine is sometimes bloody. The respiration then becomes irregular, and the animal dies on the fifth or sixth day after injection. At autopsy the axillary and inguinal lymph nodes are congested. All the blood vessels, especially those of the kidneys and adrenals, are dilated. There are ecchymoses along the blood vessels, and the pleural cavities contain a serous exudate.

The severity of the symptoms vary with the dose of the poison. In a guinea pig we have seen dyspnea develop on the fifth day after injection and persist for an entire week. Respiration was only diaphragmatic and irregular and the animal became very weak. The symptoms abated slowly and the guinea pig recovered. Later, it succumbed to a subcutaneous inoculation of the bacillus of diphtheria. Are not these symptoms similar to those seen in man, where paralysis of certain respiratory muscles occurs after an attack of diphtheria?

Paralysis appears rapidly in the rabbit, following intravenous injection of 35 cc. of culture filtrate. On the fourth or fifth day, sometimes later, muscular weakness begins in the posterior limbs, soon extends to the entire body, and the animal succumbs rapidly. When the intoxication is less acute, the paralysis can be limited for a time to one group of muscles. In one case, the extensor muscles of the hind limbs lost function and the voice became hoarse. Improvement began after two days but the animal later relapsed with complete paralysis, and death followed respiratory failure.

The diphtheritic poison is more abundant in older cultures. A rabbit that is injected intravenously with 35 cc. of filtrate from a 42-day culture dies within 6 hours after injection. . . . The muscles of the abdomen become flaccid and the rabbit shows muscular weakness. It passes liquid feces, the respiration becomes labored and irregular, hiccupping develops, and the animal dies without convulsions. At autopsy, the blood is poorly clotted and the intestines are distended by diarrheal liquid. A guinea pig receiving 35 cc. of the same liquid intraperitoneally dies of respiratory failure within 10 hours or so. Autopsy shows the characteristic congestion of the viscera, especially the kidneys and adrenals, and often a pleural exudate.

The presence of the diarrhea in rabbits which have received large doses of diphtheritic poison led us to investigate the occurrence of diarrhea during the toxemia of human diphtheria. Although this fact has hardly been noted, diarrhea is very frequent in fulminating diphtheria, according to Mlle. Daussoir, a head nurse in a diphtheria ward. Her position makes her especially competent in this domain. . . .

With such potent cultures, such large doses are not necessary. . . . Let us inject a series of guinea pigs subcutaneously with quantities of filtrate, varying from ⅕ of a cc. to 2 cc. and let us compare the effects with those following injection of fresh culture of Klebs' bacilli. With both the filtrate and living culture, there is edema at the site of injection, ruffled coat and irregular respiration. Both groups of animals die with similar symptoms. The lesions in both are the same except for the presence of diphtheritic membrane in the group inoculated with living culture. The guinea pigs receiving the largest dose of toxic liquid die in less than 24 hours, the others in 48 or 72 hours according to the dose received. In all, there is edema and induration of tissue at the site of injection, hemorrhagic congestion of internal organs especially kidneys and adrenals, and pleural exudate. Guinea pigs injected with small doses of toxin survive. Thus, subcutaneous injection of ⅟₁₅ of a cc. of filtered liquid led only to edema and necrosis.

Rabbits, pigeons and small birds die like guinea pigs following injection of soluble diphtheritic products under the skin. . . . Animals which are resistant to the subcutaneous injection of Klebs' bacilli, are likewise remarkably resistant to the injection of diphtheritic poison. . . .

The injection of animals with varying doses of soluble poison of diphtheria has shown us the various forms of diphtheritic intoxications varying from those which lead to death in several hours to those in which there is recovery after an extended period of time. The delayed and not the acute manifestations of bacterial diseases are the most important. . . . The future will show us without doubt that a number of organic diseases of unknown etiology such as some nephritides and nervous diseases are perhaps the result of a microbial infection which has escaped detection.

What is the nature of the diphtheritic poison? Is it an alkaloid or a diastase? Since the answer is as yet unknown, we shall content ourselves with reporting a few facts which tend to clarify the problem. The toxin is heat-labile. A filtrate which is fatal for a rabbit at a dose of 2 cc. administered subcutaneously, is no longer fatal at a dose of 35 cc. administered intravenously if the filtrate has been heated at 100°C. for 10 minutes. . . . Diphtheria toxin loses activity more rapidly when exposed to the atmosphere than when kept in a sealed tube protected from air and light. Filtration through porcelain is the least deleterious procedure for separating the toxic liquid from the bacillus. Although we have not isolated the active substance of diphtheria cultures, lability to heat and air indicates that it is a diastase-like substance.

Is it possible to acclimate the animal to diphtheritic toxin and to produce an immunity against diphtheria by this method? This question will be the subject of a later report. . . .

Pierre Paul Emile Roux, the senior author of this report, had been nurtured in an academic environment from the time of his birth in 1853. His father was the principal of the secondary school of Confolens (France). At the time of his birth, there were five living brothers and sisters, two having died while still very young. His parents were worried by his frail appearance. Nevertheless, Emile, though sickly, managed to eke out a long and active life.

The Confolens secondary school dominated the river Vienne and was built on the remains of an old abbey. Emile's father was the monarch of this little domain. He was dreaded by the children because his temper was short, a characteristic that he passed along to his son. This disagreeable disposition did not improve with age, since, soon after the birth of Emile, he began to suffer from gout. His affliction, however, did not prevent him from fathering two more children by his hard-working wife, who was responsible for all nonacademic aspects of life at the school. She supervised the maintenance staff, nursed the sick, and had also to comfort her irascible husband. As efficient as she was, there remained little time for Emile. He was adopted, so to speak, by a maid, who raised him with truly maternal love but frightened him with tales of the world of spirits.

Emile was nine years old when his father died. The old man had left behind him a family that was described as "blessed because all the boys were intelligent and all the girls were pretty." The eldest daughter married the professor who replaced Roux *père*. When her husband moved to the directorship of a bigger school, in Aurillac, the birthplace of Duclaux, Emile followed in order to continue his studies under the supervision of his brother-in-law. The child was highly intelligent and performed outstandingly. He was an ardent reader, and soon a skeptic who began to doubt the dogmas of the Catholic Church and the integrity of its priests. Later in life, Dr. Roux fondly recalled that one of his teachers had told him: "Roux, Emile, you will come to no good."

At the beginning of the war of 1870 Emile was in the Puy, where his brother-in-law had just moved. Two of his older brothers fulfilled their obligation to their country and never returned. Like Monsieur Pasteur, Dr. Roux would have no love for Germany. He was, however, always very correct with German scientists; and with Behring he conducted a friendly correspondence. On a visit to Germany, he stayed with Behring to act as godfather to one of his sons.

After a brief sojourn in the army, Emile obtained his baccalaureate degree, and the next year he entered the School of Medicine of Clermont-

Ferrand. He supported himself by assisting in the teaching of physics at the local *lycée*. His teaching was competent, and the next year he also was asked to help the new professor of chemistry at the *lycée*. This new teacher was Emile Duclaux, a graduate of the famous *école normale supérieure* and a disciple of Pasteur. Duclaux, thirty-two years old, was to have more influence on the future of his eager assistant than the professors at the school of medicine. At that time, Duclaux, who was to help establish biochemistry as a separate discipline, was already probing the chemistry of life's processes. The appeal of the unknown in science must have been irresistible to Roux, who had all the characteristics of a great mystic, save faith. Although he continued his medical studies, he became less a healer and comforter, and more a scientist trying to understand the mechanism of disease.

Roux had found his vocation, but having no money, he had first to insure himself of some kind of income in order to complete his studies. The School of Medicine of Clermont-Ferrand provided instruction only for the first three years of studies. He had to go elsewhere. Surely, he would have gladly forgotten about medicine to study chemistry, which was now his mistress, but how? Finding no answer, he at least managed to continue his medical studies free of charge by attending the Val-de-Grâce Army Medical School. There life was not inspiring for the individualistic Roux, but the warm friendships that he formed with other cadet students, such as Nocard, kept up his spirits. As the end of their studies approached, a thesis had to be presented. The general in command of the school was not in search of the brilliant and novel. He simply wanted a routine performance that would pass the minimum requirements. He was not interested in any "academic nonsense." Roux and Nocard failed to see the light and insisted on pursuing their theses seriously. This led to difficulties. After an exchange of harsh words between the troublemakers and the general, Roux and Nocard were first ordered into confinement and then expelled for breach of military discipline.

Roux was soon in Paris, without degree and penniless. He registered at the school of medicine in Paris and obtained a minor job in a secondary school. In a fit of his infamous temper he almost strangled one of the juvenile miscreants under his jurisdiction. He left and found an equally humble, but somewhat more medically oriented, job at the Hôtel-Dieu Hospital. During this time, Duclaux had returned to Paris, as professor at the Institute of Agronomy, and was only too glad to rescue his former assistant. Roux found the time to attend many of Pasteur's lectures. The great master, who must have had a nose for talent, took notice of the young man and asked Duclaux whether he could have Roux as an assistant. Roux felt that it would be ungrateful to leave his benefactor, but Duclaux encouraged him to follow his destiny. Thus, in November, 1878, at the age of twenty-five, Roux started his life work, which was to be intimately associated with that of Pasteur.

He joined Chamberland and Thuillier in the laboratories on Ulm Street. A little nook over the laboratories was put at his disposal, and the frugal Roux started a life of full and complete integration with the laboratory. As monastic as his inclinations may have been, Roux did not neglect to take his vacation the next summer. When he returned in September, 1879, and resumed work with cultures that had been dormant since July, he noticed a loss of virulence. It was this observation that was to lead Pasteur to the development of his method of vaccination using bacterial strains of reduced virulence.

During part of the years 1880 to 1882, Roux and Chamberland were installed in the town of Chartres, the center of the rich farming region of the Beauce, in order to carry out field studies on anthrax. Pasteur came once a week to review the experiments and give fresh instructions after having made his own observations. Roux recalled how Pasteur would linger in the fields: "one had to remind him of the time, to show him that the spires of the cathedral were fading in the night, before he could be induced to leave the experimental fields. . . . He liked to question farmers and farmhands and he used to attach much importance to the opinions of shepherds, who, because of their solitary life, were often keen observers. . . ." The climax of this work in Beauce was the famous public demonstration of the vaccination of sheep against anthrax. "That day," said Roux, "among all the people who were flocking to Pouilly-le-Fort, there were no more doubters, only admirers."

Back in Paris, Roux was delegated to study rabies. This work became the subject of his doctor's thesis, which he defended in 1881. The work on prevention of rabies occupied most of the attention of Roux, Chamberland, and Thuillier. Behind all their experiments lurked the fear that one of them would be accidentally inoculated with the dread disease. Fortunately, this never happened, although Thuillier was spared only to die of another disease, a victim of his vocation.

In 1883, a cholera epidemic erupted in Egypt, and Pasteur decided that a French expedition should be sent to study the etiology and methods of control of the disease. His proposal was approved by the government, and on August 15, Roux, Straus, Nocard, and Thuillier arrived in Egypt. Koch was already there, heading a German mission. The French mission, which had, in Nocard, a distinguished scholar in veterinary medicine, did not limit its observations to human cholera but also studied bovine plague. On September 21, Roux wrote Pasteur a letter in which he explained what had happened to Thuillier:

"Thuillier and Nocard went to witness the autopsy of an animal which had died of bovine plague on Wednesday the 14th. . . . Thuillier was in fine spirits all day, he went swimming in the sea and in the evening he went for a drive in a carriage. He ate his dinner with a good appetite and retired at 10:30. He fell asleep easily, but at 3 A.M. he was awakened by abdominal pain and, entering my room to tell me that he felt sick, he collapsed on the floor. Straus and I carried him to his bed . . . we first thought it was indigestion . . . after taking an opiate medication, he went back to sleep. I remained in his room. At 5 A.M. he had an attack of severe diarrhea and vomiting. He went back to sleep after taking additional medication. At 7 A.M. he became worse. Straus and I sustained him with the most potent medication. He showed all the symptoms of the dreaded cholera. From then on, all the French and Italian physicians in town attended him. Everything possible was done by all of us with the energy and the faith of those who have decided to repulse death. . . . With all our strength and efforts we managed to extend his agony up to Wednesday morning, the 19th, seven o'clock. . . . He was buried Wednesday afternoon at four o'clock in an imposing and moving ceremony. . . . M. Koch and his collaborators came as soon as they received the news. They found beautiful words to honor the memory of our dear friend. . . . They placed wreaths on the coffin . . . and M. Koch was one of the pall bearers. . . ."

Louis Thuillier was twenty-seven years old. It was not his first trip outside of France, having previously worked in Hungary and in Germany. Before departing for Egypt, he had gone to visit his family in Amiens without telling his parents of his plans. From Marseille he had written them to let them know that he was with the French mission. How he contaminated himself, nobody knows. Roux considered him to be by far the most careful worker on the French team. Pasteur described him as follows:

"Thuillier entered my laboratory after having placed first in the examination in physical sciences at the Ecole Normale. He was, by nature, deeply meditative and quiet. He was extremely energetic, hard-working and ready for any sacrifice."

Straus summed up the results of the mission as follows: "During the two months of the study, it has not been possible to ascertain the etiology of cholera. It is hoped that the data collected will help the future solution of this problem." Straus's hope was not in vain. Koch had had indications that a certain comma-shaped bacterium was involved. He later proved that it was indeed the pathogen that had killed Thuillier. But for Roux, it had been his first trip outside of France, his first mission, and it had ended in disaster.

The work on rabies continued, and the day came in 1885 when the first patients were treated. Pasteur, with his customary faith in himself and with his usual penchant for prolific writing, was preparing a communication and wished to include Roux's name. The latter considered the timing of publication premature. Pasteur would not wait and was thus the only author. During the first year that the rabies treatment was made available to the public, of 2,682 persons subjected to the Pasteurian inoculations, only thirty-one died. Some of the failures were due to excessive delay before starting treatment. The results appear to have justified Pasteur's enthusiasm. Interestingly, this was the only serious dispute between the two men that has filtered down to us.

With the press coverage that the rabies treatment received, and the subsequent flood of financial support for the work, came the foundation of the Pasteur Institute. From its inception, Roux exerted a profound influence on the role the Institute was to play in the development of microbiology as a science. He and Yersin made important contributions to the planning of the laboratories located on the block of Dutot Street, now known as the *rue du Docteur Roux*. After the opening of the Institute in 1888, Roux, again with Yersin, started to teach a course in microbiology, which was to be called "Roux's course" for the next twenty-five years. Shortly thereafter, they announced their discovery of the bacterial toxin that we have previously described. Thus, no sooner had the Institute begun to operate than the value of the work it could produce was demonstrated by a basic discovery. This finding was completed by Behring and Kitasato, who soon discovered that the injection of the modified toxin into animals induced them to form a neutralizing substance or antitoxin that rendered them and others that were injected with it immune to the action of the toxin. The problem of producing these antitoxins on a large enough scale for use in human treatment was solved by Roux. Here, the help of Nocard, by then professor at the Veterinary School of Maisons-Alfort on the outskirts of Paris, was most vital. In 1892, inoculation of horses with increasing doses of diphtheria toxin was begun. With the help of two other physicians, Louis Martin and A. Chaillou, the antidiphtheric sera were first tried in two children's hospitals. The results were highly encouraging, and Roux presented the preliminary findings at the

International Congress of Hygiene, in Budapest, in 1894. Interest in and support for this new method of therapy were enthusiastic. Once again, money poured in and the former imperial estate of Villeneuve-L'étang in Garches was remodeled to accommodate the many horses required to produce the therapeutic sera.

Roux's health now began to deteriorate; a quiescent tuberculosis was reactivated. From time to time he required nursing care, and this was a duty for the Pasteurians themselves. Among these were the Metchnikoffs. However, Roux was not the only sick one; there was Pasteur. During the first days of 1895, Pasteur was taken to one of the laboratories so that he could see the bacillus of plague that had been isolated by Yersin. Pasteur enjoyed discussing the achievements of all his disciples—Calmette, Nicolle—all those who were spreading his teachings in faraway places. At the beginning of the summer he was taken to Garches, where Roux was a most frequent visitor, trying to comfort both the master and his wife. Pasteur died in September.

Duclaux became the director of the Pasteur Institute and his former assistant, Roux, the associate director. Life continued much as before. Roux still spent much time in Garches, often with Nocard. They were working on the practical aspects of the production of sera. Soon, Roux had to take long periods of rest. His tuberculosis was progressing, and his health declined to a low point during the years 1897 to 1899. However, Roux was determined to resist the microbe that was afflicting him. With sheer stubborn determination, he survived. He planned the construction of the Pasteur Hospital with the help of Louis Martin, who was to be its first director, just as a few years previously he had designed the Pasteur Institute with the help of Yersin. The hospital was dedicated in 1900. Conceived as a center for treatment of contagious diseases, it was, for that time, a model of perfection.

When Duclaux died in 1904, Roux replaced him as director of the Pasteur Institute, and Chamberland became the associate director. From then on Roux was only concerned with the direction of the Institute, and he abandoned personal research.

With the war of 1914, long-range research at the Pasteur Institute stopped, and the production of vaccines and sera was increased. Roux took an active part in the inspection of sanitary installations for the army. His numerous field trips were debilitating, and his health declined further. During the war Roux left an apartment that he had occupied with one of his sisters and went back to live permanently at the Pasteur Institute. His living quarters were located at the hospital, permitting the good sisters to keep a sharp eye on him. He had two rooms, both small, that had been intended as quarters for an intern, one for his office, the other for his bedroom. He referred to this setup as his "pigeon hole."

Emile Roux was ascetic in his lack of interest in money, in his devotion to science, and in his monastic simplicity. His scientific vocation overshadowed his other sentiments, although he was receptive to feminine love. Handsome and romantic, as well as famous, he formed the inevitable attachments. Besides a mistress of many years, there were others with whom he had affairs, including a blond English girl scientist who died young of the consumption that Roux was to fight most of his life. Mme. Metchnikoff was also a favored friend. Other conquests included the rich ladies who were charmed by his presence and who allowed sentiment to flow over in the form of financial support to the Institute. He once said: "Women are like drugs, when they no longer act, one must change." His wit was apt. It ranged from an explanation given to a child as to why water made noises when it boiled—"It is the crying of scalded microbes"—to the answer he gave a lady who was being treated for rabies and who wanted to pay him: "I can assure you, dear Madam, that all my services are free, including autopsy."

In 1919, the exhausted Roux was relieved of some of his work load by the appointment of Calmette as associate director. In his monastic cell at the Institute he would do much reading and much listening. All the members of the Institute confided their problems to him, and he became marvelously well informed on all aspects of the Institute's life. As the years passed, he was witness to scientific advances made in the laboratories of the Pasteurians, including the discovery of anatoxins by Ramon, the development of the BCG vaccine, and its setback by the Lübeck disaster. The impact of all these events with their joys and sorrows was felt in the little intern room where Roux spent much of his time when he was not in his modest directorial office. He died on the third of November, 1933, a few days after the death of Calmette.

Nowadays, when the noontime weather is pleasant, the *laborantines* of the Pasteur Institute sit and chat under big trees that cast their soft shade, not only on the green benches on which they cackle, but also on an austere slab of polished stone under which rests the dedicated, monastic, yet sensual Roux. A few yards away, in sculptured stone, the fight of Jupille against the rabid dog is forever frozen.

The man who had inspired and guided Roux along the austere paths of science was Emile Duclaux, born in 1840 in Aurillac, in the French province of Auvergne. Duclaux's father was a rather grave man who was usually lost in a dreamy world of his own. His mother had a happy and affectionate disposition. There were other children in the family, but it was on Emile's frail little shoulders that his father elected to heap all his dreams of intellectual grandeur. While the other children were permitted to play the games suitable to their age, Emile had to sit for long hours listening to his father dispense knowledge that would have been far better understood by an older child. Emile had a sweet disposition and endured these intellectual orgies patiently and with profit. When he attended the local secondary school, his devoted father would rise every morning at 5 A.M. in order to keep up with his son's lessons. Under such guidance Emile became a true scholar, not only in the humanities, but also in the sciences. Together they undertook numerous outings during which Emile learned a great deal of natural history. This fatherly attention was mainly explained by the desire of M. Duclaux Sr. to see his son enter the Paris Polytechnical School, for it is said that the secret desire of all French parents is to see their sons enter the *École polytechnique*. In 1859, Duclaux was admitted to both the Polytechnical School and the *école normale supérieure*. But by then his devoted father was dead, and the growing fame of Pasteur influenced Duclaux to enter *normale* rather than *polytechnique*. After graduating, Duclaux became Pasteur's assistant, a post that he retained for three years. In 1865, he left Paris to teach in various secondary schools, first in Tours, then in Clermont-Ferrand, where he met the young Emile Roux, then in Lyon. He was finally appointed professor of meteorology, of all things, at the Institute of Agronomy in Paris. No matter what his official function was, Duclaux's main interest was in biochemistry. In 1882, he published a book on *Ferments and Diseases,* which he dedicated to his first wife, "the innocent victim of the infinitely small," who had died of puerperal fever. The book was aimed primarily at enlightening the physicians of the time in the hope of reducing the number of senseless deaths caused by sepsis. From 1877 to 1896, the personal research of Duclaux dealt with the application of microbiology to the dairy and cheese industry. In 1898, the first tome of his *Treatise of Microbiology* was published. The second volume, which appeared one year later, was devoted, in great part, to enzymes.

As the Pasteur Institute was to open, Duclaux was made its associate director. He began by founding the *Annales*, the official publication at the Institute. Following Pasteur's death, he became the director of the Institute. He wrote a biography of Pasteur that was published one year later, in 1896. He himself died in 1904. His biography was written by his second wife, Mary, nee Robinson.

Duclaux, in spite of his careful preparation for intellectual life, is not the famed discoverer of any specific important phenomenon. He was, first and foremost, a great teacher, and second, he wrote important books and contributed enormously to the dissemination of microbiological knowledge with his *Annales*. Nevertheless, his work on enzymes, his chemical prying into the living world, and his teaching made him one of the founders of biochemistry.

The life of Alexander Yersin, the codiscoverer of the diphtheric toxin, started in Switzerland, far from Indochina, where he spent most of his long life. He was born in Aubonne, in 1863, a few days after the death of his father, who was a professor of natural history. He showed an aptitude for sciences that finally found expression as an interest in medicine. The young man began his medical studies in Lausanne, continued them in Marburg, and finished them in Paris; thus culling his formal training from three countries. Inordinately shy, he chose pathology as a field for specialization because contact with people would be minimal. He became an assistant to Cornil, the pathologist at the Hôtel-Dieu Hospital in Paris. While in this capacity he cut himself during the autopsy of a former patient who had died of rabies. Fortunately, this was in 1886, and the "tatoo" method for treatment of those threatened with this dread disease was available in Pasteur's little laboratory on Ulm Street. The young physician was so impressed by the fact that he survived both his wound and the treatment, and by the sympathetic attitude of the Pasteurians, that he decided to join the group.

Before doing so, the restless young pathologist went, in 1887, to the laboratory of Pasteur's great competitor, Robert Koch. Armed with his up-to-date knowledge of the new pathogenic bacteriology, as practiced by Koch, he returned to Paris to help Roux, for whom he had much affection, give a course in bacteriology. We can thus see that the influence of Koch's methods must have made its way into Pasteurian teaching through Yersin. Yersin's choice of subject for his initial investigations (the effect of antiseptics and heat on tubercle bacilli) also reflects the profound impression that Koch must have made on the young man. Following this, he did other work on experimental tuberculosis. Good as these studies may have been, greater achievements were in store for Roux's young collaborator. Roux had decided to solve the perplexing problems posed by the recently discovered diphtheria bacillus.

The paper previously quoted on this subject was the first of three papers on diphtheria that were published one year apart by the Roux-Yersin team. This was a period of hard work and long hours in the laboratory during which Yersin discovered the great adventure that is scientific research. However, Yersin was a man of paradox, having not only a strong vocation for science but also an irresistible desire for physical adventure. He was a small man with great physical endurance. Shy and retiring, he was nevertheless strong willed and restless. One day, when Roux was absent, Yersin left the Institute to take up a new vocation. The young doctor, who was to be known to all during the years to come as Monsieur Yersin, had decided to explore the tropical jungles. He left a long letter for Roux, gave instructions to Haffkine regarding experiments in progress, collected a month's wages, and accepted a position as physician on one of the ships of the Messageries Maritimes that traveled between Saigon and Manila. He was fascinated by Indochina, and it was there he was to spend most of his adventurous life. After a period at sea he decided to explore the unknown interior of this country with the aim of aiding its development. To this end, he made three trips between 1892 and 1894, each one successively more ambitious and of greater value.

He was first entrusted with a small expedition to explore the mountainous region bordered by north Cochin China, south Annam, and Laos. Two Annamese boys and a collection of trinkets gave the expedition the proper colonial flavor. Everything went as well as could be expected until Yersin was shot in the leg at the entrance of a village. He did not look very bellicose however, and he was nursed by the very people who had shot him. As soon as he could walk, he continued on his journey. Upon his return to Saigon he was severely debilitated for months.

At the time Yersin was beginning his new life in Indochina, Calmette had been sent to establish a Pasteur Institute in Saigon, in 1891. He prevailed upon Yersin to accept an appointment with the Colonial Sanitary Corps. Yersin promptly used his position to arrange for two more exploratory expeditions of greater scope and importance than the first ones.

He collected data on topography, flora, fauna, climate, possible sources of ore, sites for settlements, and routes for roads. He also collected anthropological and sociological data on the native tribes. In Yersin's words, "Here is a vast country, placed under our protectorate, which is nevertheless almost totally unknown to us." The trips were replete with hazards; Yersin acquired malaria on one trip and severe dysentery on another, and at one point his camp was attacked by a tiger that severely injured two members of the party. But the greatest danger was from hostile tribes. Yersin wrote, "One quickly forgets the fatigue, misery and privations experienced during an exploration as soon as one returns to civilization. Before long one misses the adventurous life where the unknown is so inviting and it is a rare explorer who is content with a single trip." Within his lifetime, Yersin saw these areas settled, economically productive, and traversed with railroads, telephone lines, and automobile highways.

Plague, that great scourge of the Middle Ages, brought Yersin back to his bacteriological studies. As he explained, in the paper that follows, he was sent to Hong Kong to study plague and the methods to curb it. Not mentioned in this paper was the rather small size of the staff of his expedition. Besides himself, it was composed of two untrained persons, one of whom promptly absconded with Yersin's money. In contrast there was a well-staffed Japanese mission in Hong Kong headed by Kitasato.

Upon his arrival in the plague-ridden city, Yersin paid a visit to Kitasato and his staff, but, even by speaking German, they could not communicate. Philosophically Yersin noted: "I shall avoid their society; if one of us goes off on a false track, the other will not imitate or follow him." At first Yersin was unable to obtain cadavers for autopsy. He soon found "that a few piastres judiciously distributed, and the promise of a good tip for each body supplied, had an immediate effect." In an incredibly short period of time, Yersin was able to publish a paper, in which he described his work in Hong Kong (*Ann. inst. Pasteur*, **8**: 662–667, 1894) as follows:

At the beginning of last May there erupted in Hong Kong an epidemic of bubonic plague which was taking a heavy toll of the population. The disease had been endemic for a very long time on the high plateaus of Yünnan, and had appeared from time to time very close to the border of our Indochinese territories, in Mong-Tse, Lang-Tcheou and Pakhoï. In March, this year, plague appeared in Canton and in a few weeks caused the death of 60,000 persons in the city alone. The great commercial traffic which exists between Canton and Hong Kong and also between Hong Kong and Tonkin has worried the French authorities that Indochina might be struck by the epidemic. This fear is well founded since it is difficult to establish an efficacious quarantine in these regions.

I received from the Ministry of Colonies the order to go to Hong Kong to study the nature of this affliction, the conditions under which it spreads, and the methods to curb it, in order to prevent its reaching our territories.

When I arrived in that city, the 15th of June, more than 300 Chinese had already died. As rapidly as possible, temporary structures were erected, since the hospitals of the colony were overflowing.

I settled with my laboratory equipment in a straw hut that I had built with the permission of the English government, on the grounds of the main hospital.

The disease which was almost exclusively in the Chinese boroughs of the city, had all the symptoms and all the characteristics of the ancient bubonic plague which, in the past centuries, had decimated many times the populations of occidental Europe as well as those of the East. The famous epidemic of Marseille, in 1720, was the last one that affected France. Since that time the plague has been limited to a few foci in Persia, Arabia and in the Chinese province of Yünnan.

Here are the symptoms of the disease:

The onset is rapid, with an incubation period of 4½ to 6 days. The patient is prostrated.

Abruptly a high fever sets in, often accompanied by delirium. On the very first day a discrete bubo usually appears. In 75% of the cases it is located in the inguinal region, in 10% of the cases in the axillary region, and occasionally at the back of the neck or in other regions.

The nodule rapidly reaches the size of an egg. Death occurs after 48 hours and often sooner. If the patient manages to survive 5 to 6 days, the prognosis is better, the bubo softens and one can operate to drain the pus.

In a few cases, the bubo does not have time to form, and one will note in such cases hemorrhages in the mucous membranes or petechial spots on the skin.

Mortality is high; about 95% in the hospitals.

In the infected boroughs, many dead rats are found on the ground. It is interesting to note that in the part of the town where the epidemic started, a new sewage duct had just been installed. The pipes are much too narrow and are difficult to clean. They constitute a permanent focus of infection. One does not understand why there are two different sewage networks in Hong Kong. One is narrow and chronically clogged and is used for the disposal of water from the houses. The other is large and well planned, and is used for the disposal of rain water.

The toilets are mobile and are emptied every day, the contents are used, after a certain amount of treatment, as fertilizer for the innumerable gardens which border the Canton River opposite the island of Hong Kong.

The lodging quarters of the poor Chinese are often so revoltingly dirty that one scarcely has the courage to enter them. In addition, the number of occupants is unbelievable. Many of these slums do not have windows and are below the level of the ground. One can imagine the havoc that can be caused by an epidemic on such a terrain and the difficulty involved in its control. The only remedy was to burn down the Chinese town. This was proposed but was unfeasible for budgetary reasons.

Few Europeans have been struck by the disease, thanks to the better conditions under which they live. European houses are not fully safe, however, since even there one can encounter dead rats.

The physicians of the Chinese Customs, which had the occasion to observe the epidemics of Pakhoi and Lien-Chu, in the province of Canton, and Mr. Rocher, French Consul in Mong-Tse, had remarked that the scourge starts by severely affecting mice, rats, buffaloes and pigs before striking men. . . .

The fact that certain animals were naturally susceptible to the disease, gave me the proper material for its experimental study.

It seemed logical to start first by looking for a microbe in the blood of patients and in the pulp of the buboes. The pulp of the buboes always contains masses of short, stubby, bacilli which are rather easy to stain with analine dyes and are not stained by the method of Gram. The ends of the bacilli are colored more strongly than the center. Sometimes the bacilli seem to be surrounded by a capsule. One can find them in large numbers in the buboes and the lymph nodes of the diseased persons. They are seen in the blood from time to time, but less abundantly than in the buboes and lymph nodes, and only in very serious and rapidly fatal cases.

The pulp of buboes, seeded on agar, gives rise to transparent, white colonies, with margins that are iridescent when examined with reflected light.

Growth is even better if glycerol is incorporated into the agar. The bacillus also grows on coagulated serum.

In broth, the bacillus has a very characteristic appearance resembling that of the erysipelas culture: clear liquid with lumps depositing on the walls and bottom of the tube.

The most favorable medium is an alkaline solution of 2% peptone to which 1 to 2% gelatine has been added.

Microscopical examination of the cultures reveals true chains of short bacilli interspersed with larger spherical bodies. . . . As the cultures get older, swollen and abnormal forms become more common, and these do not stain easily.

If one inoculates mice, rats or guinea pigs with the pulp from buboes, they die, and at autopsy one can note the characteristic lesions as well as numerous bacilli in the lymph nodes, spleen and blood. Guinea pigs die in 2 to 5 days, mice in 1 to 3 days. In the first passages, the microbes are mainly in the mononuclear leucocytes.

In the guinea pig after a few hours one can already detect edema at the point of inoculation, and the surrounding lymph nodes become palpable. After 24 hours, the hair of the guinea pig is ruffled, it will no longer eat, and it is suddenly seized by convulsive crises which become more and more frequent until death.

At autopsy there are hemorrhages in the abdominal cavity. The site of inoculation is edematous and the edema extends to the regional lymph node, which is enlarged and full of bacilli. The intestine is often hyperemic, the adrenal capsules are congested, the kidneys are violaceous and the liver is enlarged and red. The spleen is also enlarged and often covered with an eruption of small miliary tubercles. When the disease has lasted somewhat longer than usual, there are sometimes abscesses in the abdominal wall.

In the pleura and the peritoneum, there is a small amount of serous liquid that contains the bacillus. The organism can also be found in the blood, where it is more elongated than in the lymph nodes. The liver and the spleen are also very rich in microbes.

Passages in guinea pigs are easily achieved with macerated spleen or blood. Death is more rapid after a few passages.

Pigeons do not die after having been inoculated with a moderate dose of either the pulp of the bubo or a culture of the plague bacillus.

The first culture from a bubo is difficult to establish on peptone agar. It grows, however, and kills as fast as the pulp of the bubo.

One notes, after a few days, that in these cultures, certain colonies grow faster than others. Examined with the microscope, they all contain the pure bacillus. . . . If one inoculates the faster growing strains into animals one notes that their virulence is strongly diminished. They kill the guinea pigs only after a rather long time or they do not kill them at all, but they are still fatal for white mice.

I have noted that on agar, the less virulent colonies grow faster than the others and tend to predominate. The net result is that upon successive transfers the cultures rapidly lose virulence.

One can often kill mice, and almost always rats, by feeding them fragments of spleen or liver from animals which have died of plague. At autopsy, one recovers the bacillus from the blood, the liver, the spleen and the lymph nodes.

The mice that have survived many contaminated meals die when they are injected subcutaneously.

Dead rats that one finds in houses and in the streets almost always harbor large quantities of the microbe in their organs. Many have real buboes.

I have placed healthy mice and inoculated mice in the same cage. The inoculated ones died first, but within a few days, all the others die of invasion with the plague bacillus.

Plague is thus a contagious and transmissible disease. It is probable that rats are the major vector of its propagation, but I have also noted that flies can acquire the disease and die. Thus flies may also be an agent in its transmission.

I had noted that there were many dead flies in the laboratory where I was performing my autopsies. The body of a fly, was freed of wings, and the legs and head ground in broth and inoculated into a guinea pig. The inoculum contained a large number of bacilli which were similar to the plague organism and the guinea pig died in 48 hours with the specific lesions of the disease.

I have been able to isolate the plague bacillus from soil which had been collected 4 to 5 centimeters under the surface in a house where disinfection efforts had been made. It was in all respects similar to the one isolated from buboes, but it was not virulent.

I have mentioned previously that in cultures from blood or from a bubo, one could isolate several varieties of the bacillus which differed in their virulence to animals, and that some colonies had even lost all virulence for the guinea pig. From a bubo of a patient who had recovered and had been convalescing for three weeks, I was able to isolate a few colonies which completely lacked virulence to mice. . . .

These very telling facts permit me to suppose that the inoculation of certain strains or varieties with little or no virulence, would be able to give animals immunity against plague. I have started some experiments in this direction and I shall publish their results in the future.

The paper that we have just read was not the initial announcement of the discovery of the plague bacillus. This announcement had been made, as is the custom in France, in a note that had been presented at the Academy of Sciences. During this time Kitasato also thought that he had discovered the organism responsible for plague; however in his first publication that appeared in *Lancet* he admitted that he could not determine whether the organism that he was working with was Gram-positive or negative. His description was not as good as Yersin's, and the purity of his cultures was in question. The purity and specificity of Yersin's cultures were not subject to doubt.

We may recall that Calmette had founded the Pasteur Institute of Saigon in 1891. In 1895 Yersin was asked to create a second Indochinese Pasteur Institute for the manufacture of sera and vaccines. This he did at Nhatrang, a beautifully located fishing village. Lyautey, the future organizer of Morocco, wrote in 1896: "Nhatrang is a paradise of tropical vegetation and temperate climate which is located at the end of a bay protected by cliffs which seem to have been put there very specifically to break the wind of the sea and the heat of the land. . . ."

"There Doctor Yersin, a student of Pasteur and of Doctor Roux, a young Navy physician (he is not yet 30 years of age) has devoted his life to science and to the search for vaccines against the diseases of the Far East. In this he has the faith, the stubbornness and the passion of the great. . . ."

Soon after, Yersin became the head of the two Indochinese Pasteur Institutes. His main interest, however, changed with the years. He became extremely conscious of the importance of developing the agriculture of the Far East. With his usual drive, he became an agronomist, dedicated to all the aspects of the agricultural expansion of Indochina.

Nhatrang remained his home during his long productive life. His house was perched on the edge of a cliff that overlooked the river and the sea. It was full of fine instruments because at heart Yersin was a physicist. He was interested in astronomy, meteorology, radio, and photography, and he provided himself with proper instrumentation in all these fields.

Even though Yersin had led an active and fruitful life, which was highlighted by the development of many practical applications in agronomy, exploration, and bacteriology, he remained shy and introverted. Solitude was his most precious possession. Elected a member of many societies, he would only attend the occasional meetings at which he thought he might be useful. After the death of Roux, in 1933, he was named a member of the Scientific Council of the Pasteur Institute in Paris. Every year, he went by plane to fulfill his administrative obligation. His last trip was made in 1940. It was Monsieur Yersin's farewell to Paris; he was then seventy-six years of age. He took the last plane that left France for Indochina at the beginning of the Second World War. There he died, three years later, far from his native land and from France, the country he had adopted as his own and had served so devotedly.

A discussion of the work of Yersin naturally brings in the name of the pioneer Japanese bacteriologist, Shibasaburo Kitasato. He was born in 1852, in Ogunigo, a mountain village of southern Japan. He studied medicine in Kumamoto and at the Imperial University in Tokyo. After graduation in 1883, he joined the Sanitary Bureau of the Department of the Interior. Two years later he was sent by the government to study under Robert Koch. After having been involved in a number of minor problems dealing with typhoid fever, cholera, and glanders, he had acquired a facility in bacteriology that he put to good use when he was assigned the problem of elucidating the etiology of tetanus. This was a project on which Gaffky and Löffler had spent some unfruitful time. Kitasato reported his results in 1889 (Z. *Hyg. Infektionskrankh.*, **7**: 225–233) in a paper entitled "On the Tetanus Bacillus." It read as follows:

A soldier in the Berlin garrison hospital died of tetanus. Microscopic investigation of the pus from the wound revealed Nicolaier's club-shaped spore-bearing bacillus in addition to other organisms. Mice injected subcutaneously with this material died in 2–3 days of typical tetanus. At the site of inoculation, but not elsewhere, were club-shaped spore-bearing bacilli in addition to other organisms. Cultivation was attempted. As found by Nicolaier, these organisms could grow on coagulated serum, but always with other organisms. Similar results were obtained with agar and gelatine media.

On the other hand, the following procedure gave positive results. When I streaked tetanus pus on coagulated serum or agar and incubated the cultures at 36 to 38°C., all organisms grew within 24 hours. After 48 hours there was an increase in the number of club-shaped bacilli. The cultures were then placed in a waterbath at 80°C. for ¾ to 1 hour. Only spores remained viable in such cultures. All mice injected with these cultures died of tetanus. After I had thus confirmed that the cultures contained spores, I mixed a loopful of such culture fluid with nutrient gelatine and poured part of the mixture into dishes and part of the mixture into glass vessels. . . . Hydrogen was passed into the vessels and the openings were sealed. The culture vessels and culture dishes were then incubated at 18–20°C. After one week, colonies were seen in the vessels supplied with hydrogen, while there was no growth on the conventional dishes. After ten days the sealed hydrogenated culture vessels were opened and the growth was examined in smear preparations. There were rods, smaller than the bacillus of malignant edema, arranged singly and in threads. Since these bacilli unquestionably were anaerobes, they were cultured in broth in the hydrogenated vessels and in deep agar tubes. All cultures grew in 30 to 48 hours. The organisms were bacilli with refractile terminal spores. Mice injected with growth from the agar and broth cultures, died of typical tetanus in 2 to 3 days.

This procedure for isolation of pure cultures of tetanic bacilli was repeated successfully a number of times. I should like to note that fortunately the other types of spore forming bacilli in tetanic pus did not withstand the heat treatment of 80°C. for 30 minutes.

Properties of the tetanic bacilli are as follows: They are obligate anaerobes. They grow well under hydrogen, but not under carbon dioxide. The bacilli grow in weakly alkaline peptone agar and gelatine.

Gelatine is liquefied gradually with slight gas formation. The addition of dextrose to 2% concentration, accelerates growth . . `.` and so does 5% litmus. Cultivation under hydrogen imparts a burnt odor to the growth. The bacilli could be subcultured repeatedly without loss of virulence.

The colonies in gelatine, under hydrogen, had a dense center with radiating thread-like growth. When the gelatine was liquefied, the growth became entirely thread-like.

The tetanus bacilli grew best at temperatures of 36 to 38°C. At 20 to 25°, gelatine cultures began to grow after 3 to 4 days. In sealed gelatine cultures under hydrogen, at 18 to 20°C., growth was evident after one week. There was no growth under 14°C. Spores were formed after 30 hours at 37°C., and after one week at 20–25°C. The tetanic bacilli could be stained with analine dyes. They retained the Gram stain. The Ziehl double stain method could be used to define the spores.

Spores applied to silk threads, then dried over sulphuric acid and then kept in air, were virulent after several months. Spores mixed with sterile soil also retained viability for months. The spores were killed after steaming at 100°C. for 5 minutes. The spores were resistant to chemicals. They were viable after 10 hours in 5% carbolic acid, but were killed after 15 hours. They were killed after two hours in 5% carbolic acid containing 0.5% hydrochloric acid. They also died after 3 hours in 1% bichloride of mercury or after 30 minutes in 1% bichloride of mercury containing 0.5% hydrochloric acid. The spores were not killed by chloroform.

Mice inoculated by stabbing with a platinum needle dipped in pure culture, developed tetanus after 24 hours and died in 2 to 3 days. Rats, guinea pigs and rabbits could be infected with larger amounts of culture, such as 0.3 to 0.5 cc of broth culture. Rats and guinea pigs became sick after 24 to 30 hours, while the incubation time in rabbits was 2 to 3 days. Pigeons were more resistant. With pure cultures, infection was achieved without the aid of foreign bodies such as cotton, wool and wood splinters that were necessary for infection with mixed cultures.

The symptoms of tetanus begin in the area of inoculation. If the animal was injected in the hind portion of the body, the first spasms were in the hind limbs. At autopsy there was hyperemia without pus at the site of inoculation. There were no changes in the internal organs. Neither bacilli nor spores could be seen in smears from the site of injection or the various tissues and organs throughout the body. Tissues from the various organs did not induce tetanus in other animals and did not yield the bacilli in culture. Two rabbits were injected in the dura mater and two were injected intravenously with broth culture. They developed tetanus after 30 hours and died in 48 to 60 hours. Tetanic bacilli could not be seen in smear and sections from brain or spiral cord and cultures from these organs were negative.

To see whether the tetanic bacilli produced a specific toxin in the animal body, I injected culture into the base of the tail, removed the tissue at the site of injection after one-half, one, two, three, etc. hours, and cauterized the dissected sites. The mice submitted to excision after one hour developed fatal tetanus, but those submitted to excision before one hour, remained alive. Tetanic bacilli were seen in smears from the site of injection until 8 to 10 hours after injection; they were not seen in the internal organs at any time.

Since the tetanic bacilli disappeared rapidly after injection but still produced typical tetanus, it seemed that the bacilli might produce an active toxin. This possibility is now being investigated by Weyl and myself. However, the possibility still exists that the organism may be demonstrated by improved methods of detection.

The following conclusions can be drawn: (1) Tetanus is an infectious disease caused by a specific bacillus. (2) The causative agent of human tetanus and experimental tetanus is the bacillus described by Nicolaier. (3) This bacillus is present in the pus of man and experimental animals suffering from tetanus. The bacilli form spores in the pus. (4) The bacilli can be cultured from cases of tetanus and such pure cultures can produce tetanus in experimental animals.

Kitasato's success in solving this rather difficult problem was a pleasant surprise for Koch. It was not the first time that an anaerobic bacterium had been reported to be involved in an animal disease, however. We will recall in this regard the paper by Pasteur, which was quoted previously, in which, in 1878, he had reported, in collaboration with Joubert and Chamberland, a strictly anaerobic "sepsis vibrio." This early work of the French workers did not, nevertheless, have the polish of Kitasato's demonstration.

In the study of the suspected tetanic toxin, Kitasato joined forces with Behring. Together, as will be seen in Chapter 6, they worked not only on the tetanic toxins but also on the diphtheritic toxin. This led to the study of anti-toxins. In the domain of diphtheria and toxins a healthy competition existed between the German and French workers, which culminated with the discovery of the anatoxins by Ramon. Kitasato, who was involved in some of the work that demonstrated a therapeutic effect of tuberculin in tuberculosis, remained with Koch until 1892. The time had come to go back home to the land of the chrysanthemums. Even though his future there was uncertain and despite offers of professorial positions in England and the United States, Kitasato chose to go back to his country and to help his people profit from the vast experience that he had accumulated in Europe. The Japanese government was not yet ready to spend money on the lowly bacteria, and Kitasato had to start the development of bacteriology in Japan with the help of a wealthy benefactor, Yukichi Fukusawa. The value of a bacteriological laboratory became obvious when, in 1894, plague was almost at the door of Japan. Kitasato went to meet the enemy with the limited success that was previously described. However, his time had not been wasted, and when a second epidemic of bubonic plague broke out in Manchuria in 1911, his previous experience was invaluable in helping break the scourge.

In 1899, the government finally assumed the support of Kitasato's laboratories. These were soon expanded to become a sort of Pasteur Institute, which was charged with the manufacture of vaccines and antisera. The staff of this institute, which grew in number, kept in close touch with the developments in Europe not only by reading the scientific literature but also by sending young investigators for periods of study in the best laboratories. Thus one of Kitasato's assistants, Hata, went to Paul Ehrlich's laboratory in Frankfurt, and history repeated itself. Hata, too, made good in Germany, but he did it by playing a key role in the discovery of Salvarsan. Kitasato kept in contact with Europe by taking a trip abroad himself, and in 1908 he acted as Koch's zealous guide when the great German bacteriologist visited Japan with his wife. Kitasato and one of his assistants devoted an entire month to showing the land of the rising sun to the Kochs. The climax of this visit was the erection of a Shinto shrine for Koch. Knowing that Koch trimmed his own beard, Kitasato secretly arranged with the maid serving the Kochs to obtain some of Koch's hair clippings for preservation in the ceremonial vessels in the shrine.

Following his studies on bubonic plague, Kitasato relinquished laboratory work in favor of administrative responsibility. In 1917 he was nominated to the House of Peers and was ennobled in 1924. In the political sphere, his interest remained medical, and he helped establish many of the laws regulating medical and public health practices. He died in 1931, and following his wishes, a shrine in his honor was erected not far from the one honoring Koch on the grounds of the Kitasato Institute.

One of the assistants of Kitasato, Kiyoshi Shiga, also had a brilliant career. He went to Germany and worked with Paul Ehrlich from 1900 to 1903, but before his stay abroad, he had already published a most significant series of papers entitled "The Agent of Dysentery in Japan" (*Zentr.*

Bakteriol., Parasitenk. Abt. I, **23**: 599–600; **24**: 817–828, 870–874, 913–918, 1898), which he referred to as "The Dysentery Bacillus or *Bacillus dysenteriae*." Shiga started by pointing out:

> Dysentery is widely prevalent in almost all provinces of Japan and causes many thousand fatalities. Last year, from June to December, there were 89,400 cases with a 24% fatality rate . . . I have conducted a thorough etiological study on a series of 36 cases. Previous investigators have been unsuccessful in determining the etiology of the disease because culture methods were inadequate and experimental animals do not give pathological reactions similar to human dysentery. The typhoid bacillus and cholera vibrio, likewise do not produce disease in animals similar to that seen in man. There is no doubt that the agent of dysentery, like the cholera and typhoid bacillus, must be present in the intestines; this is indicated by clinical symptoms and anatomical changes.

Shiga showed that he had learned the principles of Koch well, and like so many investigators of the time, he carefully repeated:

> To determine an etiological agent the following conditions should be satisfied: (1) the organism must be present in all cases; (2) it must not occur in other diseases or in normal people; (3) it must produce a specific disease in animals, possibly similar to the disease in man. . . .
>
> Not infrequently there are diseases where the three conditions are fulfilled yet the evidence remains unconvincing. Dysentery is the best example of such a situation. Widal's recent discovery of the agglutination reaction in typhoid fever is an important advance toward the solution of this problem. With this principle we can extend the above conditions to include, (4) if the serum from an actively sick or convalescent patient afflicted with the disease in question agglutinates a specific organism, the organism has an intimate association with the disease.
>
> I have found a specific organism in dysentery that, in the Japanese variety at least, satisfies all 4 conditions. I have named this organism *Bacillus dysenteriae*.

Shiga gave the following description of *Bacillus dysenteriae*, which is now appropriately called *Shigella dysenteriae*:

> The dysentery bacillus is a short rod with rounded ends much like the typhoid bacillus and most types of coli. The organisms most frequently occur singly and occasionally in pairs. With methylene blue, the rods are more intensively stained at the ends. They are decolorized by the

method of Gram. They are moderately motile but flagellae have not been demonstrated with certainty. No spores are formed. Gelatine is not liquefied. Growth is optimal at body temperature, but occurs at room temperature. Alkaline media are recommended for cultivation. . . .

Shiga noted that the feces of five of his thirty-four living patients contained amebae, and he considered it possible that amebae might have been present in more patients. However, his attention was directed to bacteria, and he described as follows his attempts at culturing bacteria from the dysenteric stools:

Alkaline media were most useful. The bacillus was difficult to isolate early in the disease, but nearly pure cultures were readily obtained during the height of the diarrhea. As the stools became normal, the number of other bacteria increased and the number of dysenteric bacilli decreased. I could cultivate 10 types of bacteria from dysenteric stools, but the specific bacillus predominated. The dysenteric bacillus was cultivated from all cases in the series and was the only one that was agglutinated by the serum of dysenteric patients. The dysenteric bacillus was not found in the stools of healthy persons or in patients suffering from typhoid fever, acute diarrhea, chronic diarrhea, tuberculous diarrhea, and beri-beri.

Few or no dysenteric bacilli could be recovered after clinical symptoms had abated, when the stools were back to normal or yellow and soft. The bacilli that were recovered at the height of the disease were highly virulent and agglutinable but both characteristics declined in intensity with the abatement of symptoms. I hope to perform more precise experiments on this point when immune serum of sufficiently high potency will be available. In two dead patients the organism was recovered in almost pure culture from the deeper layers of the intestinal wall, while in the superficial layers, coli bacilli and other forms predominated.

The detection of the dysenteric bacillus, described by Shiga in the last two paragraphs quoted, was done with methods that were already classic. As he stated, he also used the newly discovered method of agglutination:

On August 8, 1897 the serum of one patient was found to agglutinate one type of organism from agar cultures. It was then that I named this organism *Bacillus dysenteriae*. The agglutination tests were performed as follows:

1. *Hanging drop.* Serum was diluted ten-fold with broth. A drop of

the diluted serum was mixed with bacilli from an agar culture on a cover slip which was inverted over the depression in a hollow ground slide, sealed and incubated for 24 hours. Higher dilutions were also prepared.

2. *Macroscopic observation.* Three loopfuls of growth from agar culture were suspended in 2.9 cc of broth and 0.1 cc of serum was then added. Higher dilutions were also prepared.

The sera from 25 patients were tested. All sera were positive, but to varying degrees. The sera of 11 normal persons did not agglutinate the dysenteric bacillus, neither did the sera from 3 cases of typhoid fever, 3 cases of intestinal catarrh, or 3 cases of beri-beri. The reactions with sera of various animal species were negative. Various immune sera, for typhoid fever, cholera, tetanus, diphtheria and tuberculosis, did not agglutinate the dysenteric bacilli. Typhoid and coli bacilli were not agglutinated by dysenteric serum.

On rare occasions, another bacillus was agglutinated by dysenteric serum. This organism was thinner than the dysenteric bacillus, gram negative, it did not liquefy gelatine, growth on agar was almost transparent and bluish white in color. It formed gas when stabbed into glucose agar. Milk was coagulated. It produced indol in peptone water. The organism was more virulent than the dysenteiy bacillus for the guinea pig.

Rarely, in dysenteric stools, coli types were found that were agglutinated by dysenteric serum and by normal serum. Consequently they cannot be considered significant in the etiology of dysentery.

Shiga performed a number of animal experiments that showed that pure cultures of his dysenteric bacterium could kill mice, guinea pigs, rabbits, and cats when injected subcutaneously. It was also fatal to mice and guinea pigs when injected intraperitoneally. Orally, it could induce diarrhea and kill guinea pigs, cats, and dogs. Rectal application was not a satisfactory route for establishing a fatal disease. Birds seem to be unaffected by subcutaneous or oral administration of the pathogen. Shiga concluded:

These animal experiments showed that with pure cultures, I was unable to produce the intestinal changes seen in man. Hyperemia and hemorrhagic areas were the maximal amount of pathological alteration seen in experimental animals.

He then proceeded with a colorful, if not impressively thorough study of vaccination and immunization:

Using the method introduced by Kolle for cholera and typhoid fever, I tested a vaccine in animals and in myself. A 24 hour agar streak culture was suspended in 10 ml. of broth and heated for 20 minutes at 60°C. After sterility of the suspension was verified, I injected 0.1 ml. intraperitoneally into a guinea pig. On the next day there was a slight rise in temperature and slight infiltration at the site of inoculation. Ten days later the blood serum produced weak agglutination. I then tested the vaccine on myself. My serum taken prior to vaccination did not produce agglutination. I administered ½ of an agar culture subcutaneously in the rump. Several hours later I developed a headache, and tenderness at the site of injection. At 8 hours there were chills and fever followed by lassitude and pain in the knee joint and calves. The local tenderness increased and on the next day, the swelling reached the size of a dish, the local axillary glands were swollen and sensitive to pressure and there was slight rise in temperature. On the second day my temperature returned to normal and local symptoms began to abate. I bathed. On the third, 4th and 5th day I still had a headache. On the fifth day, the local area again became painful, and on the eighth day, the local swelling increased. The lesion was incised. The more superficial layers were hard and thickened. The deeper layers were pussy and infiltrated. The pus was sterile. The local changes were therefore due to toxemia. The symptoms abated on the ninth day. Blood serum taken on the 10th day agglutinated dysentery bacilli at a 1 to 10 dilution. This vaccination test shows that the toxin of the dysentery bacillus is more virulent than that of the typhoid or cholera bacillus in man. I was previously injected with cholera vaccine produced by the same method and experienced only a transitory rise in temperature. Appropriate immunization studies are in progress in order to develop a safe vaccine. . . .

The role of bacteria in the etiology of some diseases of man and animals had been suspected from the work of Pasteur, Villemin, and Davaine, to name only a few, and it was established with certainty by Koch and those who followed his methods. Shiga had introduced the last refinement into the postulates of Koch by using a serological method as one of the criteria to prove the etiological role of the causative agent of a bacterial dysentery. The total number of bacterial species that cause human and animal diseases is not very high, a little more than a hundred perhaps, but still these few organisms play a very important role in our health and that of animals. Bacteria can also attack plants. As a matter of fact, even though their role in phytopathology is less important than that of fungi, there may be more than

200 bacterial species that are able to cause plant diseases. Whereas the majority of bacterial troublemakers for animals are Gram-positive bacteria, the reverse is true in the case of bacterial plant pathogens, and the Gram-negative rods lead the way.

Bacteria had been seen associated with plant abnormalities ever since 1858, when Lachmann observed bacteria in root nodules of leguminous plants. The same observation was made in 1866 by Woronin, who considered the nodules of lupine a manifestation of a disease since they were infested with bacteria. As we recall, Hellriegel and Wilfarth, in Germany, demonstrated the bacterial nature of the nodules of leguminous plants and showed that without the bacterial nodules the plants were unable to fix nitrogen. Beijerinck, in 1888, isolated a bacterium responsible for the formation of the nodules in pure culture and named it *Bacillus radicicola*. The role of bacteria in the formation of root nodules is not to be confused with their role as etiological agents of diseases *stricto sensu*. The role of bacteria as disease producers in plants was first indicated clearly by Thomas J. Burrill in Illinois. During the years 1878 to 1884, Burrill correlated the transmission of specific plant diseases with specific bacteria. He did not completely fulfill the requirements of the postulates of Koch since he did not cultivate the responsible agent in pure culture far from the tissues of the host. Thus his role in the elucidation of the etiology of bacterial plant diseases was closer to that of Davaine in the study of animal diseases than to that of Koch. As we will see further on, Erwin F. Smith was the Koch of phytopathology.

For the first time in the history of microbiology, American investigators were playing a pioneering role. This may be explained by the importance of large-scale agriculture in the development of the economy of the United States and by the articulate role of the farmer in American life.

Thomas J. Burrill was born in Pittsfield, Massachusetts, in 1839. The unproductivity of the New England soil led the family to migrate to Illinois when Thomas was nine years old. The demands of pioneer farming permitted only sporadic schooling. After attending high school at an older age than usual, Burrill entered the State Normal University and finished when twenty-six years old. He had developed an interest in botany, but economic circumstances led him to accept a job as superintendent of schools at Urbana, Illinois. While there, he continued to study botany, and, in 1867, he joined an exploratory expedition to the Colorado Rocky Mountains as botanist. Burrill became assistant professor in natural history at the Illinois Industrial University in 1868, and by 1870 he was promoted to professor of botany and horticulture. Burrill recognized the importance of the study of the diseases of plants. By 1871 he had incorporated such subject matter in a new course that he designated "cryptogamic botany." Soon he regarded, in contrast to many European botanists, fungi in plants as "the cause, not simply the result, of disease." For microscopy, he purchased oculars and objectives and had the shops of the University machine the stands and bodies. With these instruments, Burrill began his studies on bacteria as agents of disease in plants. These studies were pursued diligently through 1887 and reported first in the *Transactions of the Illinois State Horticultural Society,* beginning in 1877.

Simultaneously, Burrill was devoting considerable effort toward developing small Illinois Industrial University into an effective university. He relinquished research to devote his time toward development of what is now the University of Illinois, and he was appointed its vice-president in 1879. He was one of those who catalyzed a strong interest in plant pathology in American universities and at the U.S. Department of Agriculture.

Burrill remained with the University of Illinois for his entire career. He died in 1916, well-honored for his role in the development of the science of plant pathology and his service to higher education.

The following are excerpts from a paper entitled "Anthrax of Fruit Trees; or the So-called Fire Blight of Pear and Twig Blight of Apple Trees," which Burrill presented and which was reported in *Proceedings of the American Association for the Advancement of Science* (**29**: 583–597, 1880):

> The widespread and disastrous disease of the pear tree, commonly called Fire Blight and that no less prevalent, at least in the western states, usually known as Twig Blight of the apple tree, are due to the same immediate agency. They are identical in origin and as similar in their pathological characteristics as are the trees themselves. . . .
>
> The immediate and exciting cause of the disease is a living organism, belonging to a group of minute beings called bacteria which produce butyric fermentation of the stored carbonaceous compounds in the cells of the affected plants, and especially in those of the bark outside of the liber or fibrous inner layer. This organism, if indeed specifically different, is closely allied to the *"vibrion butyrique"* of Pasteur and the *Bacillus amylobacter* of Van Tieghem, now known to be identical with the preceding. The proof of these assertions is the burden of this paper. . . .
>
> In 1877, the writer presented to the Illinois State Horticultural Society the results of microscopical observations, in the account of which occurs the first published notice of the minute moving bodies always present in the portion of trees suffering from the malady of which we

treat. I quote: "The cambium [a term here too loosely used] of the blighted branch, when trouble first shows itself, and for some days thereafter, is filled with very minute moving particles. . . . Not unfrequently a thickish, brownish, sticky matter exudes from affected limbs, sometimes so abundant as to run down the surface or drop from the tree. This proves to be identical with that notice in the cambium and unquestionably has the same origin. The sticky, half-fluid substance thus exuding is entirely made up of these minute oscillating particles." In a subsequent discussion in the same society I am credited with the following: "If we remove the bark of a newly affected limb and place a little of the mucilaginous fluid from the browned tissues under our microscope, the field is seen to be alive with moving atoms, known in a general way as bacteria. . . . A particle of this viscous fluid introduced upon the point of a knife into the bark of a healthy tree is in many cases followed by blight of the part, but with me not in every instance. . . . If we look once more to the affected branch, we find the disease spreads more or less rapidly from the point of origin, and upon examination the moving microscopic things are discovered in advance of the discolored portions of the tissues, but not very far ahead—an inch perhaps. [They are now known to advance several feet in certain cases, ahead of the black colored parts.] Does it not seem plausible that they cause the subsequently apparent change? It does to me, but this is the extent of my own faith; we should not say the conclusion is reached and the cause of the difficulty definitely ascertained. So far as I know, the idea is an entirely new one—that bacteria cause disease in plants—though abundantly proved in the case of animals. . . ."

Then came the description of more recent experiments:

I began, July 1, 1880, a series of experiments with a view to determine whether these organisms were really active agents in the observed changes, or simply accompanied other causes of destruction. . . .

The term inoculation is used to denote the introduction of the virus by a wound in the bark. External applications are not included under this designation.

The inoculations of July 1st and 10th were made by cutting pieces of diseased bark freshly taken from the tree, and inserting them after the manner of budding, as practised by nurserymen, without, however, taking the precaution of tying. These pieces of bark were about three-sixteenths of an inch by two-thirds of an inch. Those made after the date mentioned were performed with a sharp-pointed knife, or needle dipped in the exuding virus of diseased trees. This was usually collected in the morning and placed in a vial with a little distilled water. Requisite care was always taken to cleanse the instrument thoroughly or to choose a new one when changing the infecting material. . . .

Of the pear trees inoculated with virus from diseased pear, fifty-four per cent received the disease, while of those inoculated from blighting apple, seventy-two per cent became as thoroughly infected, and the parts as speedily died. This greater per cent of infection from the apple virus does not necessarily prove that it is more destructive, for another set of experiments may reverse the percentages; but I think the result may be taken as showing the identity of the disease in the two trees. All other experiments agree in this respect. . . .

The organism to which we attribute the death of our fruit trees is so minute that a magnifying power of two hundred diameters is necessary to make out its outline at all, and one thousand diameters are required for careful study. During the course of its development, it assumes various shapes, the different forms usually appearing together in the field of the microscope. What seems to be the most characteristic form consists of two oblong joints with rounded ends, which have a transverse diameter of about $\frac{1}{1000}$ of a millimetre. The two articles usually seen together have a length of about $\frac{3}{1000}$ of a millimetre. They are comparatively thicker than *Bacterium termo* and their motions are less rapid. They slide forward with a slightly undulating motion, they turn over and on end, but never glide across the field. It is impossible specifically to identify these creatures by form alone. So far as we now know, this may be the same as a common omnivorous little agent which converts sugar, amylaceous matter, lactic, tartaric, citric, malic and mucic acids and albuminous substances into carbon dioxide, butyric acid and hydrogen, whenever and wherever the conditions permit. Should this prove to be the case, it would not necessarily, if presumably, invalidate its agency in producing this disease of the pear; but it might render less hopeful the discovery of remedial treatment. . . .

There is absolutely no trace of other fungous growth in the tissues examined by me until after death has taken place; neither have any been found for several days after this in sections of twigs or diseased parts of trees left in the meantime out of doors. . . .

Burrill concluded: *"The evidence now given of disease in plants produced by bacteria contributes something to the germ theory of disease in animals and may lead to very important scientific and practical results."*

In Europe also there was an interest in bacteria as the cause of plant diseases. Edouard Prillieux, in France, observed, in 1879, the presence of micrococci in wheat kernels affected with rose-red disease. He did not attempt to cultivate the pathogen, nor did he try to prove pathogenicity with inoculation experiments. Prillieux also studied other bacterial plant diseases, but his major interest was in mycology. A professor at the Institute of Agronomy in Paris, he died in 1915 at the ripe old age of eighty-seven.

In Italy, Orazio Comes, at about the same time, also started to associate bacteria with plant diseases. He described a certain *Bacterium gummis*, which he thought was an abundantly distributed plant pathogen. Like Prillieux, he was a mycologist who worked mainly on diseases produced by fungi. He did not obtain pure cultures of his *Bacterium*, and he did not carry out any conclusive inoculation experiments. Despite this lack of initiative when it came to bacteria, Comes was one of Italy's greatest phytopathologists.

More important in this field was the work of the Dutchman J. H. Wakker, who was working in the laboratory of Hugo de Vries in Amsterdam, with the aid of a grant received from the hyacinth growers of Haarlem. He demonstrated that the disease known as the "yellow disease" of hyacinths was a bacterial infection caused by a yellow organism that could be inoculated via the leaves of the plant. The first publications of Wakker on this subject were published in 1883. His papers were mostly in Dutch and did not attract much attention even though the work was excellent. He left Holland for Java, where he headed a sugar experiment station. There he worked mainly on diseases caused by fungi. No matter how good his work was, proper promotions did not come. Eventually, Wakker settled down to teaching mathematics in a secondary school.

With Wakker being ignored in Europe, the leadership in the study of bacterial plant diseases remained in America. In this field Erwin Frink Smith was the most outstanding scholar. He was born in 1854 in the small village of Gilbert's Mills, located not far from Syracuse, in the state of New York. There Erwin spent the first fifteen years of his life in serene contact with nature for which he had a strong love. At the village school his intellectual potential was suspected, and the teachers helped by providing him with books and periodicals that ranged in scope from Hawthorne's stories to Gray's botany and from the *Atlantic Monthly* to the *Journal of Applied Chemistry*. He was talented in all sorts of ways, and as he matured he started to muse on the profession at which he should aim. For a while he would think of literature as a potential outlet for his intellectual fervor. Then he would consider science. This was a difficult choice indeed, since science had so many ramifications. However it seems certain that natural history was always one of his great favorites.

In 1870, the Smiths, as the Burrills had done, went west. The productivity of the soil of Michigan had attracted Erwin's father, who bought there an 80-acre farm in Clinton County. Another attraction was the good quality of the educational institutions. The state of Michigan was rightfully proud of its fine university in Ann Arbor and a good college of agriculture in Lansing. However, formal education, even though respected by the Smiths, was not actively pursued, in part because of the necessity of farming the land, and in part because of the individualistic attitude of Erwin. He preferred to indulge in botanical studies with the cultured local druggist, Charles F. Wheeler, than to attend school with a too monotonous regularity. He graduated from high school at the mature age of twenty-six. At about the same time, in 1881, he published a book in collaboration with Wheeler on the flora of Michigan. The principal of the high school remembered him as follows: "Wearing a full beard . . . he entered school the second week of the fall term. . . . His unusual intelligence, courteous bearing and evident acquirements were such that I fell in love with him. . . . Before I knew him he had acquired a fine knowledge of the French language. . . ." Wheeler had been his tutor in the language of Pasteur, and it is not surprising that later, as we may recall, he translated Duclaux's book on the life of Pasteur into English.

After graduation from high school Smith still lacked the money to attend college. During the summer of 1880 he attended the Michigan Agricultural College and then obtained a job as prison keeper at the Ionia Reformatory. A Unitarian by confession, and believing in public service, he used some of his spare time to prepare a study on prison reform. The major extracurricular activity was writing articles on hygiene and public health for the *Michigan School Moderator,* a journal for high school teachers. The diligent flow of good articles led the editor to employ Smith to take charge of a "scientific and sanitary department" for the journal. In addition to this editorial position, the Michigan State Board of Health appointed him as a correspondence clerk. The income from the two positions allowed him to move to Lansing in 1882.

Erwin Smith's articles in the *School Moderator* gave an accurate and exciting picture of the new frontiers of science. The May 18, 1882, issue carried the following account on Pasteur:

". . . M. Pasteur has opened a great field for study. . . . What if we shall be able to discover and stamp out the germs of yellow, typhoid, and miasmatic fevers, by and by, and to reduce the dreadful mortality of diphtheria and scarlatina. It now seems possible! . . ."

The May 31, 1883, issue carried an article on Koch's discovery of the tubercle bacillus:

"The researches of the German savant, Dr. Koch, concerning the nature and cause of consumption, have attracted world-wide attention. As a result of experiments upon lower animals, and of exhaustive microscopic study, the Doctor concludes: 1. That consumption is contagious, and 2. That it is caused by a *bacterium*. . . ."

Smith was an academic anomaly. He was a very competent botanist, and his knowledge of the developments in the field of hygiene was second to none. By 1885, a number of the professors at the University of Michigan had become acquainted with Erwin Smith and were prepared to grant him a degree after one year of resident study if he could pass the appropriate examinations. He resigned from his position with the Board of Health in order to enter the University.

According to plan, he emerged with a bachelor's degree after one year of residence. He was then past thirty. Soon an offer for a position came from Dr. F. Lamson Scribner, the first head of the Mycological Section of the U.S. Department of Agriculture. Dr. Scribner's financial situation was not too bright, and he could only muster 500 dollars for a six-month period during which Smith could investigate some of the fruit diseases that Scribner did not have time to study himself. Smith accepted in September, 1886, and left for the capital.

In Washington, Erwin Smith first roomed in the same house as Theobald Smith. This was a lot of different Smiths for the same house. They were to become, in their respective fields, the most brilliant bacteriologists of their time in the U.S.

The facilities at the disposal of Scribner and his associates were less than modest: "We had no laboratory facilities. . . . We had very few books and nothing in the way of apparatus beyond the simplest sorts of microscopes. . . ." Erwin Smith recalled at a later date. The total budget for the year 1887 was 5,000 dollars. Smith's arrival in Washington was contemporary with Millardet's description of the Bordeaux mixture. The great things that were happening in phytopathology contrasted with the inadequacy of the government's setup. It was in a way a stroke of bad luck for Smith to start his work in the Department of Agriculture with a study of peach yellows. Scribner's new division needed something spectacular to get off the ground; the glitter of some practical results was needed to untie the strings of the frugal Congress. It turned out that peach yellows was a "beast" of a disease. The careful Smith was able to confirm that it was a contagious disease, but he was unable to find the etiological agent. He, of course, thought of bacteria, which had become rather fashionable in America since the work of Burrill, but he could implicate neither bacteria nor fungi. As we will see in Chapter 8, Smith's work was good enough to lead Beijerinck to the tentative conclusion that peach yellows was a disease of the same type as tobacco mosaic. This might eventually have been a great consolation to Smith, but, in the meantime, Scribner was running out of money and had to dispense with Smith's services. Smith was not alone in the world, since everybody who had had contact with him was ready to sponsor him, but unfortunately not much of interest turned up. At this point, Scribner left for the University of Tennessee, and his successor, B. T. Galloway, found 1,800 dollars a year to reappoint Smith and get him back on the track of the etiological agent of the peach yellows. By 1889, he had done a considerable amount of work on this disease, and he presented a thesis on this subject at the University of Michigan. The complexity of the first problem that Erwin Smith tackled will be illustrated by the fact that the viral nature of the disease and the exact mode of transmission of the virus by a leafhopper were only elucidated during the 1930s by L. O. Kunkel, who worked first at the Boyce Thompson Institute, and later at The Rockefeller Institute in Princeton. [1]

Smith found that plant diseases caused by bacteria were a more rapidly rewarding field of activity than the yellows. So rewarding was it that his work was soon noticed on the European continent. Professor A. Fischer of Leipzig, who did not believe in bacteria as the cause of plant diseases, did not hesitate to suggest that the lack of proper workmanship might be the cause of the "odd" results obtained in the new lands across the ocean. Smith did not appreciate the comments of the German "Master," and he felt obliged to reply. Smith recalled in 1922 that his reply "silenced all the critics, and won over the doubting European public. Fischer never forgave me, but I could not do otherwise . . . it cleared the air and advanced the science. . . ."

Between 1905 and 1914 Smith published a three volume book entitled *Bacteria in Relation to Plant Diseases.* In this he wrote:

> The writer's studies of the bacteria themselves and of the diseases which they cause . . . began in 1893. At that time there was very little reliable information on the subject. . . .
>
> It was with the hope of making useful discoveries and clearing up contradiction and uncertainty that the writer began his study of this class of diseases. His first effort in the way of preparation was to supplement his botanical training with a knowledge of bacteriological methods. . . . His second effort was to gather together and properly digest

[1] Peach yellows is actually caused by a mycoplasma-like organism.

all literature relating to the subject. . . . His third endeavor was to carefully work over, in the laboratory, field and greenhouse, as opportunity offered, all of the so-called bacterial diseases of plants, submitting each supposed parasite to all the tests of modern pathology. . . .

Wilt of cucurbits was the first disease he studied exhaustively, and the specific agent, *Bacillus tracheiphilus* (now in the genus *Erwinia*), was so named because it localized in the vessels of the affected plant. He isolated it in 1893. After defining the disease, listing host plants, and describing geographical distribution and signs of the disease, Smith proceeded to prove the etiological role of the bacterium. Approximately forty experiments were presented in detail. An abridgment of one of the critical trials follows:

Inoculations of July 16, 1896

The plants were in a hothouse and the bacteria used were from . . . an 8 day old culture. . . . Well-developed, young, healthy, and rapidly growing cucumber plants (*Cucumis sativus* variety White Wonder), were inoculated. . . . Many delicate pricks (40 to 70) were made in the apical part of one leaf-blade of each plant, covering an area of not more than one sq. cm. The pricks themselves did the plant no injury. The platinum loop and the steel needle used in the operation were flamed and cooled each time before using. A big loop of the fluid, containing many thousand bacteria (some of which were motile, as determined by examination under the microscope) was put on the clean surface of the leaf, spread a little, and then rapidly pricked in, taking special care to make the needle holes as small as possible. . . . The plants were examined every day for the first 8 days and frequently after that. Twenty-four plants were inoculated.

(Plant 355) This plant was 18 inches and very thrifty. The inoculation was made on the sixth leaf 9 inches away from the stem. The pricked leaf-blade was 5 inches broad. Up to the morning of July 21, there was no trace of disease but at 3 PM of the same day about 0.5 sq. cm. on one side of the pricks was wilted. . . . By noon of the seventh day, the wilt covered about 10 sq. cm., and reached half-way down the blade. The leaf was now cut off close to the stem with a hot knife. Four days later the vine was normal, apparently except for a droop of the first two blades below and a fainter one of the first two above the node which had borne the pricked leaf. I filled the pot several times with water, but an hour later the absorption of the water had not relieved the droop of the foliage. The next day in the afternoon, the first two leaves below were cut away. They had not recovered their turgor. Three days later . . . the blades of the next four up showed the wilt. The eighteenth day the blades of the second and fourth leaves up were shriveled but the petioles were turgid. The fourth leaf was on the same side as the second. The blade of the third leaf which was on the opposite side was flabby but had not yet shriveled. The blades of the fifth, sixth, seventh and eighth leaves were drooping. The others were turgid. The twenty-third day after inoculation all the leaves were shriveled. . . .

Similar results were obtained with the twenty-three other plants inoculated, and Smith continued:

Every one of the twenty-four plants contracted the disease, and in each case it first appeared in the pricked area. Nineteen of the plants subsequently developed constitutional signs and died of the disease. No general signs appeared in the other five plants, i.e., the disease was stopped by the removal of the affected leaf. . . . In eighteen cases the amputation of the affected leaf did not check the spread of the disease. . . . The bacteria . . . pass down through the vessels of the leaf at the rate of about 0.75 inch to 2 inches (2 to 5 cm.) a day. . . .

Between August 5 and 8 numerous freshly wilting leaves were cut from these plants and fixed in strong alcohol to determine whether the bacteria are actually in the vessels of the leaf at the time the secondary wilt appears or whether this wilt is due simply to the plugging of the vessels of the stem. . . . Thin microtome sections were made from the basal part of the petiole of 66 leaves . . . staining in carbol fuchsin. Bacteria can not be demonstrated in every one, but they occur in 61 of them; no fungi are present, neither are there any insect-injuries. In most cases the bacteria are confined strictly to the spiral vessels of these petioles, and they do not occur in all of these, nor in all of the bundles. They are not present in the phloem, the cortical parenchyma or the tissues between the bundles. Summarized, the amount of bacterial infection in the basal part of these petioles is as follows: (1) In a few petioles nearly every bundle is occupied and bacteria occur in many vessels . . . (2) in 5 no bacteria detected; (3) in by far the greater number the bacteria are confined to a few vessels of a few bundles. . . .

After a description of the "Morbid Anatomy," Smith proceeded to "The Parasite," *Bacillus tracheiphilus.* His summary reveals the thorough nature of his investigation:

Résumé of Salient Characters

Positive

A bacillus in the vascular bundles of cucurbits causing a wilt-disease; short rods (single, paired, in fours end to end, or in small clumps); motile, peritrichiate; capsules; pseudozoogloeae; involution-forms; stains readily; smooth; white; viscid; glistening; slow grower on media; surface colonies small, round discrete; no growth at 37°C. or at 6°C. (16 days); aerobic; facultative anaerobic (with grape-sugar, cane-sugar or fruit-sugar); from these sugars a non-volatile acid, soluble in ether; grows only in open end of F-tube with dextrine or glycerine, acid from glycerine; slime on steamed potato is same color as the ungrayed substratum; usually it grays potato after a time; clouds peptone-bouillon and Dunham's solution thinly; growth retarded in acid juice of cucumber-fruits; also retarded or inhibited by juice of many other vegetables, e.g., table beet, sugar-beet, turnip, etc.; grows on many media at 25°C., carrot, coconut, Fermi, Uschinsky, etc.; asparagin as carbon food (?); thermal death-point 43°C.; optimum for growth 25° to 30°C.; maximum, 34° to 35°C. (?); minimum (?) 8° or below; easily killed by dry air, sunlight, or freezing (50 per cent or more); ammonia-production (moderate); feeble production of hydrogen sulphide; in litmus-milk persistent growth without reduction or distinct change in color of litmus; short-lived on many media; killed readily by acids, but lives long in cane-sugar-bouillon with carbonate of lime; grows on some media in hydrogen and carbon dioxide; dissolves middle lamella (cucumber-parenchyma); distributed by insects, especially by *Diabrotica vittata*. Group No. 222.2322023.[1]

Negative

Mealy or dendritic surface growths; Gram's strain; endospores; chains; filaments; growth not yellowish, piled up or wrinkled; pellicle on bouillon; liquefaction (gelatin, blood-serum, egg-albumen, etc.); lactose and pure maltose in closed end of fermentation-tube; lab ferment; acid (in milk); gas (all media); pigment (except gray stain on potato); indol (?); nitrite from nitrates; starch-splitting; cellulose-dissolving

[1] In 1895 Wyatt Johnston had suggested that all important characteristics of a species could be expressed by numbers arranged in a definite order. This system was variously modified. For example, 200 would mean "not sporulating."

(except possibly in host); asparagin as nitrogen food; ammonium salts as nitrogen food; steamed turnip, and cauliflower; Cohn's solution; acid bouillon (+33); acid gelatin; nearly odorless; not a soft rot; not infectious to tomato, potato, etc. On steamed potato liable to be confounded with a non-infectious coccus (follower) which reddens litmus milk.

Any organism which reddens or blues litmus-milk decidedly, reduces the litmus, throws down the casein, or clears litmus-free milk without precipitation may be set down at once as something else.

It was during his study of the wilt of cucurbits that Smith established the methods for the critical and faultless study of bacterial plant diseases. He investigated a number of other diseases of this type, such as the brown rot of *Solanaceae*, the black rot of crucifers, the yellow disease of hyacinths, bean blight, mulberry blight, the black spot and canker of peach and plum, the angular leaf spot of cotton, the angular leaf spot of cucumber, the bacterial canker of tomato, the olive tubercle, and others. But his most spectacular studies were those dealing with crown gall. They attracted much attention because of their kinship to cancer. Smith and one of his collaborators, C. O. Townsend, described the causative agent of the tumors under the name *Bacterium tumefaciens* (now in the genus *Agrobacterium*) in *Science* (**25**: 671–673) in 1907, in a paper entitled "A Plant-Tumor of Bacterial Origin." [1]

For two years the writers have been studying a tumor or gall which occurs naturally on the cultivated marguerite, or Paris daisy. It has been difficult to isolate the organism and to demonstrate it unmistakably in stained sections. Recently the bacteria (seen in small numbers in the unstained tissues on the start) have been plated out successfully. With subcultures from poured plate colonies, thus obtained, the galls have been reproduced abundantly and repeatedly during the last few months, the inoculations having been made by needle-pricks. From galls thus produced the organism has been reisolated in pure culture and the disease reproduced, using subcultures from some of the colonies thus obtained and puncturing with the needle as before. More than 300 galls have been produced by puncture inoculations. Under the most favorable conditions (young tissues) the swellings begin to be visible in as short a time as four or five days, and are well developed in a month, but continue to grow for several months, and become an inch or two in diameter.

In some of our experiments one hundred per cent of the inoculations have given positive results (40 punctures out of 40 in one series; 62

[1] Reproduced by permission of the editor of *Science*.

punctures out of 62 in another), while the check plants have remained free from tumors, and also, in nearly every case, the check punctures on the same plant. In the two series just mentioned there were 110 check punctures on the same plants, all of which healed normally and remained free from galls. Old tissues are not very susceptible. The tumors grow rapidly only in young fleshy organs. The organism attacks both roots and shoots. It frequently induces abnormal growths on the wounded parts of young cuttings. Its power to produce hyperplasia is not confined to the marguerite. Well-developed small tumors have been produced in a few weeks on the stems of tobacco, tomato and potato plants and on the roots of sugar beets. More interesting economically is the fact that galls closely resembling the young stages of crown-gall have been produced on the roots of peach trees by needle-pricks, introducing this organism. In eighteen days these growths have reached the size of small peas, the checks remaining unaffected. It is too early, perhaps, to say positively that the cause of the widespread and destructive crown-gall of the peach has been determined by these inoculations, but it looks that way. Of course, the most that can be affirmed absolutely at this writing is that we have found an organism which when inoculated into the peach produces with great regularity galls which in early stages of their growth can not be distinguished from the crown-gall. The matured daisy galls also look astonishingly like the peach gall. Numerous experiments which ought to settle the matter definitely in the course of the next three months are now under way. In the best series of experiments on peach roots (that inoculated from a standard nutrient agar culture five days old) 14 groups of needle-punctures (5 in each group) were made on nine trees, 13 tumors resulting. The fourteenth group was on a weak tree which did not leaf out, and might therefore be left out of the count. In that case we have 100 per cent of infections. On the roots of nine young trees from the same lot, held as checks, 75 punctures were made, using a sterile needle, but no galls resulted. In another series of 9 peach trees inoculated at the same time as the preceding and examined on the twenty-third day, 75 per cent of the punctures had yielded galls (9 tumors on 7 plants). These roots were inoculated by needle-pricks from a culture believed to be rather too old (glycerin agar streak 6 days), but the plants were set out again, and it is not unlikely that galls will finally develop on the roots of the other two plants. The plants, inoculated and uninoculated, were set, immediately after making the needle-punctures, in good greenhouse soil, in new ten-inch pots, and have been subject to the same conditions as to light, heat and water.

That crown-gall of the peach is due to a myxomycete the writers have never been willing to admit, because the inoculation experiments described by Professor Toumey do not clearly establish such fact. He saw often in the tissues of the galls what he interpreted to be the protoplasm of a slime mold mixed in with the protoplasm of the host plant, and he obtained sparingly what he supposed to be the fruiting bodies of this organism on the cut surface of the galls. He made, however, only two series of inoculations with the spores of his *Dendrophagus globosus,* four trees in the first case and six trees in the second, one developing the disease in the first instance and two in the second. Why did not the other seven trees contract the disease when the spores were thrust into the wounded tissue? He did not fully exclude the possibility that the three infections were due to some other cause accidentally introduced on his needle point. The *Dendrophagus* sporangia furnishing spores for the inoculations grew not on culture media but on the cut surface of a gall (an infectious substance). What if a few bacteria had been carried up from the surface of the gall, contaminating the surface, or interior of the sporangia? Then the needle might occasionally have introduced two organisms into the wounds instead of one, as believed, and the unsuspected one might have been the cause of the disease. This supposition is not excluded by any of Professor Toumey's experiments.

The fact remains well established, however, by experiments of various persons: Thaxter, Halsted, Selby, Toumey, Smith, Von Schrenk and Hedgcock, etc., that when minced galls are buried in the earth near the roots of sound trees, the latter develop galls. The disease is therefore a communicable one, but the cause, in spite of much study by many persons, is still in dispute.

For the organism causing these tumors the name *Bacterium tumefaciens* is proposed with the following brief characterization: *B. tumefaciens* n.sp., a schizomycete causing rapid multiplication of the young tissues of *Chrysanthemum frutescens, Prunus persica,* etc., the result being the production of tumors or galls. The organism is motile, especially in young cultures; it is non-gas-forming and aerobic. . . . In young agar streak cultures it is a medium-sized, short rod, with rounded ends, often in pairs with a plain constriction, the elements usually being 1 μ or less in diameter and two to three times as long as broad. The one to three flagella are polar. It is not yellow on any medium, or green fluorescent, nor does it brown the agar. It is rather short-lived on agar. It does not grow in Cohn's solution and does not infect olive shoots. It occurs principally at the bottom of the tumor rather than uniformly distributed in its tissues. It is best isolated from that part of the stem where the tumor joins the healthy tissues. There are slight indications of metastasis. Non-pathogenic yellow organisms are frequently obtained on plates made from older portions of the galls.

Smith continued to broaden his studies on plant cancer. In 1916, then sixty-two, he wrote to a friend: "I am finding my experiments so interesting this summer that I doubt if I shall be able to get in a vacation until some time this fall." However, he was not the only one impressed by his work, for Dr. Charles H. Mayo, elected president of the American Medical Association for 1917, included in his presidential address: "The work of Erwin F. Smith on plant diseases is monumental, especially his discoveries as to the causes of certain plant tumors. . . ."

Erwin Smith was one of the more important links of American bacteriology with the science in Europe. He maintained a prodigious correspondence with plant pathologists on the continent and in England, and he published fairly frequently in the *Zentralblatt für Bakteriologie*. He also made three trips to Europe and was invited to present major papers at a number of international meetings. The first trip was marred by a personal tragedy. Smith had married Charlotte May Buffet in 1893 when he was thirty-nine and she twenty-two. The marriage was, in Smith's words, "deep sweet peace." They shared many common interests. But Charlotte had rheumatic fever. In 1906, while they were in Italy, the disease reactivated, and they were obliged to return with Charlotte in a severely invalided state. She died soon after their return to the United States. Smith's memorial to her was a book entitled *For Her Friends and Mine* that included a biographical sketch and a series of sonnets. He wrote a great many creditable sonnets in his lifetime. Smith remarried in 1914. His second wife was Ruth Warren, highly educated and strongly versed in Latin and Greek.

Smith's second trip, in 1913, was made to present his work on tumors. The last trip, in 1923, was a grand scientific tour that lasted approximately seven months. It included visits to many centers where cancer and plant pathology were studied. He returned with a multitude of ideas for investigation, prepared to extend his studies on cancer to animals, but there was still too much to do on plant studies.

In March, 1927, his heart began to fail, but he was spared prolonged illness by death on April 6 of that year. As he requested, his body was cremated, and his ashes were scattered over the waters at Woods Hole, Massachusetts, where he had enjoyed many summers of work.

References

Austrian, R.: "The Gram Stain and the Etiology of Lobar Pneumonia, an Historical Note," *Bacteriol. Rev.*, **24**: 261–265, 1960.

Bernard, N., P. Hauduroy, and E. Olivier: *Yersin et la peste*, F. Rouge et Cie, Lausanne, 1944.

Cressac, M.: *Le Docteur Roux, mon oncle*, L'Arche, Paris, 1951.

Duclaux, Mme. E.: *La Vie de Emile Duclaux*, L. Barnéoud et Cie, Laval, 1906.

Kervran, R.: *Albert Calmette et le B.C.G.*, Librairie Hachette, Paris, 1962.

Miyajima, M.: *Teacher and Pupil (Koch and Kitasato)*, Tokyo, Japan, 1935.

Rodgers, A. D. III: *Erwin Frink Smith: A Story of North American Plant Pathology*, American Philosophical Society, Philadelphia, 1952.

Smith, E. F.: *Bacteria in Relation to Plant Diseases*, vol. I, *Methods of Work and General Literature of Bacteriology exclusive of Plant Diseases*, 1905; vol. II, *History, General Considerations, Vascular Diseases*, 1911; vol. III, *Vascular Diseases (continued)*, 1914, Carnegie Institution of Washington, Washington, D.C.

Villemin, J. A.: *Études sur la tuberculose, preuves rationnelles et expérimentales de sa spécificité et de son inoculabilité*, J. B. Baillière et Fils, Paris, 1868.

Whetzel, H. H.: *An Outline of the History of Phytopathology*, W. B. Saunders Company, Philadelphia and London, 1918.

Immunology: Cellular

5 As soon as the germ theory of disease was established, the conquest of infectious diseases by vaccination and chemotherapy as well as the study of the mechanisms of immunity began almost simultaneously.

Immunity involves an interplay of cellular and humoral factors. The cellular or phagocytic theory was the first to be developed and was the fruit of Metchnikoff's cogitation. The phagocytic theory was a bold and imaginative concept based upon a deep knowledge of biology. Metchnikoff was to microbiology as Dostoevski to literature and Mendeleev to chemistry. Their lives had much in common.

Metchnikoff's life was marked by crises and suffering replete with two attempts at suicide, yet from his teens to his seventies he found time to contribute consistently to the development first of embryology and zoology and then of immunology.

Elie Metchnikoff came from a typical family of moderately affluent Russian landowners. As befitting their social station, his father, Ilya, had served in the Imperial Guard. Ilya married the charming sister of a fellow officer, and the couple joined in the high life of the St. Petersburg aristocracy. Disdaining gainful employment or management of the sources of the family's income, Ilya squandered his wife's inheritance within a short period of years. With three growing children on their hands, the Metchnikoffs remembered that there still remained in Ilya's possession a domain, Panassovka, in the vicinity of Kharkov. There, they retired to boring provincial life. Two additional boys were born; the last one, Elie, in 1845. With the parents idling away time at card playing and eating, the children were being raised in a casual manner. Elie was capricious and hyperactive; he was nicknamed "quicksilver." He had an endless store of strange questions. Music, however, was one of the charms that could soothe his restlessness. His mother described him as neurotic, but his only sister used the more descriptive term of "little beast."

When Elie was eight, a tutor was hired for one of his older brothers who was an invalid. The tutor was a bright young student called Hodounoff. He developed a strong liking for Elie and started to teach him botany. Soon Elie thought that he knew all the plants growing on his father's domain, and he started to write lectures on scientific subjects. These he read in front of the children of the neighborhood, who, although not overly impressed, graciously accepted to listen and applaud as long as young Elie paid them for attendance.

The family made numerous trips to Kharkov and spent a few winters there in order to marry off their only daughter. While the family was busy with sociability, Elie would be off to the bookshops of the big city, spending his money on natural history books.

He was eleven years old when he entered the *lycée* in Kharkov. At first he applied himself to the curriculum, but he soon started to exercise his individuality by declaring himself an atheist and by neglecting the humanities in favor of botany and geology.

The studies at the *lycée* were insufficient for Elie's active and advanced intellect. He would sneak into the University of Kharkov and sit at some of the scientific lectures. Soon, he timidly approached some of the professors for further instruction. One of them, a young physiologist named Tschelkoff, responded and taught him some histology. At the *lycée* he would read advanced scientific texts during lectures that did not interest him. He was caught in the act during a course on religion. The priest checked the title of the book, returned it to him, and let him go on peacefully with his laic but not immoral reading. As the final examination of the *lycée* approached, Elie shelved his selective love for science, studied all of his subjects diligently, and won a gold medal. In this manner he thought that he might convince his parents to let him go on with the university studies of his choice. He wanted to go to the University of Würzburg, in Germany, where the famous Swiss zoologist Kölliker was teaching. He went all the way to Würzburg only to discover that the summer vacation period in German universities was not the same as in Russian universities. Not mature enough to wait for six weeks by himself, far from his loved ones, he became frightened and returned home. He enrolled at the University of Kharkov even though he knew that it was not an adequate center of learning. Most of the professors were old, as were their methods and their teaching. Elie was prepared to choose medicine, but Tschelkoff was in favor of natural history. To top it all, Elie's mother did not think that her son was rugged enough to stand the sight of human suffering. So natural science it was, and, in 1863, at age eighteen he published his first pages on the nature of the stem of the protozoan *Vorticella*. His interest in fresh water biology led him at this point to create a new order, the *Gastrotricha*, to accommodate a group of organisms first described by Ehrenberg, who considered them rotifers. Other workers thought that they were nematodes. Metchnikoff's proposal to establish the new order was accepted by all, to the great satisfaction of the young student. No matter how much Metchnikoff liked research, he wanted even more to get out of Kharkov. The second year, he did nothing but study, and after only two years he passed brilliantly the final examinations that normally were taken only by fourth-year students.

It was in 1864 that Metchnikoff made a second attempt to get out of Russia. He went to Helgoland to collect marine specimens. Stretching his meager budget, he stayed longer than expected and even managed to attend a congress of naturalists in Giessen, where he presented two communications on the work done in Helgoland. Karl Leuckart, then professor of zoology at Giessen, retained him in his laboratory, and money came as a grant from the Russian Ministry of Education. Metchnikoff studied the life cycles of nematodes. Always excessive in his efforts, he taxed his health and soon left to recuperate in Switzerland, where he met one of his brothers who was dabbling in socialistic endeavors. Happy to see his brother, Elie congratulated himself, however, for having chosen science over politics. During this time, Leuckart published independently some of Metchnikoff's work on nematodes. Disillusioned, Metchnikoff left Giessen to Leuckart and headed for Naples. There he met and made friends with another Russian scientist of his age, Alexander Kowalewsky. Having been stimulated by the concepts of Darwin, the two young friends contributed substantially to the founding of comparative embryology. Their studies were primarily concerned with lower organisms, in whose embryos they demonstrated the evolution of germinal layers, a topic which was made fashionable by Haeckel's studies. In the fall of 1865, overwork and an epidemic of cholera drove Metchnikoff back to Germany. As he went from laboratory to laboratory, it became more and more obvious that he was not a patient experimenter and that he was made to work alone. He went back to Italy, which still showed the scars of the cholera epidemic, and finally he returned to Russia in 1867.

The European trip helped Metchnikoff mature intellectually. First, he was most impressed by the organization of German laboratories, even though he was revolted by some of the crude, bellicose students and the selfish jealousy of some of the German scientists. Second, he had found a line of investigation to follow, namely, the comparative embryology of invertebrates. Third, he had completed a study on the embryology of cephalopods that could be presented as a thesis for a doctorate, and fourth, he had noted that a worm was performing intracellular ingestion, in a protozoan-like fashion. In this, at the time, the Darwinian Metchnikoff only saw a phylogenic link, even though it was the first of his observations on which the phagocytic theory was later to be built.

Upon his return to Russia, Metchnikoff went to St. Petersburg to obtain his doctorate, and he sought a position as docent at the University of Odessa, which by its maritime location seemed proper to carry out his investigations on the embryology of sea invertebrates. In both ventures he succeeded, and the young prodigy found himself a dignified docent at the ripe old age of twenty-two.

Soon, in spite of his success as a teacher and investigator, Metchnikoff got tired of Odessa, where, in his estimation, too much attention was paid to old professors who were far from having a burning love of science. He was transferred to the University of St. Petersburg and went to Italy in the spring of 1868 for another period of study on marine fauna. This trip did not work out too well; his vision began to deteriorate, the weather was debilitating, and Naples was too bustling with noise and outdoor gaiety. He even once tried to quiet the activities in the street, under his window, by throwing a bucket of water on some roving musicians who were fracturing music with too much enthusiasm. Returning to St. Petersburg, he resumed teaching, but he found his research facilities and his salary most inadequate. He started to neglect more than ever all material aspects of life, including eating. During this period, at the home of some friends, Elie met a young lady, Miss Fedorovitch, with whom he became so friendly that when he became sick she nursed him. When she fell ill, he nursed her. From this reciprocal nursing venture evolved plans for their marriage. In a letter, Elie introduced Ludmilla Fedorovitch to his mother: "She is not bad, but that is all. She has beautiful hair; her complexion is not pretty. We are almost the same age; she is little more than 23 years old." Ludmilla's health did not improve with matrimonial plans, and she had to be carried in a chair to the church for the marriage. Then Elie started a losing battle against his lack of funds and the wasting disease of his wife. Having obtained a grant, the young couple left for Italy in the spring of 1869. Here Ludmilla started spitting blood. Following a short period of amelioration in her health, she returned to Russia for one visit but had to leave soon for Madeira, where she died in 1873 in the midst of one of nature's most beautiful and flowered landscapes. Elie's mother had been right; the sight of human suffering was intolerable to him, and even though he had come all the way from Odessa, where he was now a professor, to be at Ludmilla's side, he walked out of the room, defeated, before her demise. The numerous trips necessary to care for Ludmilla's health had been covered by various research grants. When she was well enough, her artistic talent was put to use, and she produced some fine illustrations for his embryological studies. These had been their most happy moments.

During the last days of Ludmilla's life, morphine had been used to reduce her suffering. Elie kept a vial of it, and during his return trip he drank the alkaloid to end his life. The dose was too high, and he vomited most of it. He was blissfully comatose for a few hours, regaining consciousness to find that life still had to be lived.

Back in Russia, Metchnikoff was unable to continue his work in the field of embryology because of poor eyesight, a defect that had plagued him since childhood. As a matter of fact, when he was a child, he used to take advantage of it. The doctor had urged that he should not rub his eyes and that he should not cry. He had soon discovered that he could get his way by threatening to cry. Now he had to adapt his work to this limitation. He rested his eyes by pursuing certain anthropological studies, having found that the measurement of human bodies was not too taxing in this respect. These studies took him on long trips through the steppes, while in Odessa he concentrated on teaching both at the University and in a *lycée*. In Odessa, Elie had an apartment below one that was swarmed by a family of eight children ranging in age from one to sixteen years. Metchnikoff went up to complain about the early morning noises that were emanating from above. Eventually he found that the oldest child, Olga, was interested in natural history. He offered to tutor her in zoology and soon extended the educational relationship to an offer of marriage. Young Olga was attracted by his kind, luminous eyes and his Christ-like appearance. She accepted and wore a long dress for the first time on the day of the ceremony. She arose early, the morning after the nuptial night, in order to do her homework in zoology and thus please her husband.

Olga, influenced by the prevalent ideas of the time, had socialistic notions. Elie did not like to see young people become involved in politics. He thought that politics required mature thinking and that youth should acquire education before attempting to contribute to the *res publica*. He was in favor of social changes through a progressive evolution. In essence his political ideas were those of the Darwinian embryologist that he was. Nevertheless, the intellectual climate at the University of Odessa began to deteriorate. The government was fighting against a liberal trend; academic freedom was in danger. Metchnikoff was completely disheartened in his futile attempts to promote enlightened tolerance. Disheartenment developed to profound depression, and in 1881 he attempted to commit suicide for the second time, by injecting himself with a strain of *Borrelia,* causing a relapsing fever. This, he thought, would also be a method of determining if the disease could be transmitted through the blood. The answer was "yes," but he did not die. He became critically sick and had severe jaundice with cardiac complications. In his fever, he imagined that he had solved all problems concerning human ethics, and this pleased him immensely. He recovered marvelously well, and he felt full of renewed vigor, but a cardiac weakness remained. The Metchnikoffs spent the summer in the country at the home of Olga's parents. There, Elie started experiments to determine if fungal parasites could be used to curb destructive insects. These, together with the second suicide attempt, were his first investigations in the field of infectious diseases.

Olga's parents died, and Elie found himself the responsible head of the family with two domains that had to be administered. He did well and finances were in fair shape when he passed the responsibility over to one of Olga's brothers who had studied agriculture. This was a period of agrarian unrest in the Ukraine. Many farmers were Jewish, and a great tension existed between them and the peasants, on the one hand, and the administration, on the other. The kind Elie tried to find solutions to these problems but was defeated by the lack of good faith and the brutality shown by many of those concerned. The money coming from the inheritance from Olga's parents was enough for existence even though Elie had resigned from the university and no longer held a salaried position. In 1882 the Metchnikoffs left with five of Olga's brothers and sisters for Messina, where the proper biological material for Elie's work was to be found. A small apartment with a beautiful view was rented, and the living room was transformed into a laboratory. A fisherman was found who furnished the necessary transportation for the collection of scientific material. The little family was happy. Later Metchnikoff wrote:

"I was resting from the upheaval which had led to my resignation from the University and I was passionately working in the marvelous setting of the Strait of Messina.

"One day, as the whole family was at the circus to see some trained apes,

I remained home along with my microscope and I was observing the activity of the motile cells of a transparent starfish larva, when a new thought suddenly dawned on me.

"It occurred to me that similar cells must function to protect the organism against harmful intruders. . . .

"I thought that if my guess was correct a splinter introduced into the larva of a starfish should soon be surrounded by motile cells much as can be observed in a man with a splinter in his finger. No sooner said than done.

"In the small garden of our home . . . I took several rose thorns that I immediately introduced under the skin of some beautiful starfish larvae which were as transparent as water.

"Very nervous, I did not sleep during the night, as I was waiting for the results of my experiment. The next morning, very early, I found with joy that it had been most successful.

"This experiment was the basis of the phagocytic theory, to which I devoted the next 25 years of my life."

In Metchnikoff's mind, this simple experiment indicated that in all animals the basis of inflammation was the cellular reaction and that vascular and nervous reactions were only of secondary importance. He postulated further that these migrating cells, which were able to move in order to meet an enemy that had penetrated, were the major guardians of health against bacterial infections. Messina was not an intellectual vacuum, as can be seen by the fact that Metchnikoff was able to discuss there his new ideas with two famous scientists, Kleinenberg and Virchow. They were impressed and encouraged him to publish his theory, which he did in 1883. Back in Russia, Metchnikoff looked for a model system that could be used to demonstrate the validity of his new theory. He found it in a fungal disease of *Daphnia*, as we can see from the following paper that was published in *Virchow's Archiv für pathologische Anatomie und Physiologie* (**96:** 178–193, 1884). It was entitled "A Yeast Disease of *Daphnia*, a Contribution to the Theory of the Struggle of Phagocytes against Pathogens." In this paper Metchnikoff wrote:

> The ubiquitous water flea, *Daphnia*, is a unique animal for the study of certain fundamental pathological processes. It is susceptible to a spontaneously occurring disease and, although small and fragile, it is easily studied because it is transparent. Several investigators have reported that *Daphnia* is host to various parasites belonging to the lowest forms of animals and plants.
>
> The disease which is the subject of this report is caused by a yeast.

To the best of my knowledge it has not been described previously. I first observed it in the fall of the past year, in an aquarium where many *Vallisneria* and *Daphnia magna* comprised almost the entire flora and fauna. On first examination, the white color of the affected *Daphnia* suggested that the disease was pebrine, but microscopic examination revealed that this impression was incorrect. The entire body cavity, extending even into the hindmost antennae was completely filled with fungal cells in various stages of development. I would suggest the name *Monospora bicuspidata* [1] for this fungus. It is a sporulating fungus producing ascospores within asci and forming conidia. The simple conidia are oval or slightly bent, pale cells similar to the corresponding structures in many other yeasts. They multiply by budding. After the buds attain full size they may separate or remain joined to form loose colonies.

The buds frequently are formed at the end of conidia although occasionally budding occurs laterally. Multiplication by fission was not seen. Free conidia increase approximately twofold in length, thereby producing rod forms. With further growth a club shape is attained and this form proceeds to sporulate. The protoplasm, at the wider end, condenses and the condensation spreads to the narrower pole, forming a spore within the cell that has now become an ascus.

These characteristics indicate that this parasite is similar to typical yeasts. Although the definitive position of this organism is not established, it should be noted that various fungi pass through yeast stages. The organism should be recognized as a new genus on the basis of the unique needle shaped ascospore. I have been unable to cultivate this organism in artificial media despite many experiments utilizing acidified meat broth, orange juice, etc. All the developmental stages of *Monospora* are seen in the body cavity of diseased *Daphnia*. In the earlier stages of the disease one finds only sporulating conidia while later, ascospores predominate.

. Healthy *Daphnia* ingest asci originating from *Daphnia* that have died of the disease. Since the asci do not burst in water, and many spores freed of asci are found in the digestive tract, I believe that the ascus walls are destroyed by the stomach juices. Peristaltic movement causes the spores, which are pointed at both ends, as indicated by the specific name *bicuspidata*, to penetrate from the intestinal tract into the body cavity. One often sees spores which have penetrated only partially into the body cavity while a portion of the spore remains in the intestinal wall. Hardly has a portion of a spore appeared in the body cavity when

---

[1] Now placed in the genus *Monosporella*.

one or several blood cells attach themselves to the spore and the battle against the invader begins. The blood cells adhere so tightly to the spore that they are rarely torn loose by the flow of the blood. In such cases they are replaced by other blood cells and finally most of the spores are more or less completely surrounded by them. Often the spores penetrate completely into the body cavity, where they all the more rapidly fall victim to the blood cells. The number of blood cells that surround a spore is quite variable. When several spores group together in the body cavity, an accumulation of blood cells occurs giving as vivid a picture of inflammation as can be conceived in an animal lacking blood vessels. The blood cells collected about the spore do not always maintain an independent existence; they sometimes join to form a plasmodium (giant cell).

Although the spores can penetrate into the body cavity from all parts of the mid-intestine, the horn shaped blind sacs excepted, penetration is more frequent at the bendings of the mid-intestine. Most of the spores in the intestine and feces appear intact, indicating that they are not damaged by the stomach secretions. But the spores behave differently when surrounded by blood cells; they are subjected to a definite series of changes induced by the blood cells. They become thicker, assume a light yellow color and their outlines become irregular. The spores then swell at various points and the swollen areas assume a brownish yellow color, while the remaining portions of the spores retain the light yellow color. Finally the spores disintegrate into brownish yellow or dark brown granules. Blood cells in the meantime have fused to form finely granular plasmodia which retain ameboid motion. These observations on *Daphnia* confirm the reports of others and my own reports on the formation of so-called giant cells or mesodermal plasmodia by fusion of ameboid cells surrounding a foreign body.

The conclusion that the alterations in the spores are produced by the blood cells is justified for the following reasons. When a spore stays for an extended period of time partially in the body cavity, while a portion remains in the intestinal canal, the portion in the body cavity will be destroyed by the blood cells while the part in the intestinal canal will preserve its normal morphology. When a portion of a spore has protruded outside or into the thoracic cavity, while the other portion is in the body cavity, again, the part in the body cavity will be attacked by blood cells and the other will remain unchanged. Occasionally spores may pass from the body cavity to the outside before they have been attacked or during early stages of attack by blood cells. It is evident that spores in the body cavity are attacked by blood cells and probably killed or disintegrated by some of their secretions. Thus, the function of the blood cells is to protect the body against infectious agents.

The disease becomes manifest when too many spores enter the body cavity or when some of the spores in the body cavity are not attacked by blood cells. The free spores germinate by forming buds that develop into conidia. In turn, the conidia multiply and short chains may be formed. In some cases the first spore maintains its original small size but in other cases the spore grows to form a sterigma-type of a structure. In this manner oval conidia arise and bud, intensifying the infection. The newly formed conidia are spread throughout the body cavity by the blood flow. Large numbers of conidia accumulate in such areas as the head and tail cavity where the blood flow is sluggish. The blood cells do not remain passive, however, toward the invasion of the conidia. They devour some of them in order to kill them within their cell bodies.

Since these events can be followed more favorably than an attack of phagocytes on bacteria, I will expand on my observations. To improve the reliability of the results, the same specimen should be observed over a period of hours. One can then see that the blood cells really engulf the conidia at varying rates of speed. The observation of pairs of attacked conidia, of which, one has just been devoured while the other is still free, is instructive since in time the free conidium will also be phagocytized. If the conidium has already assumed a rod form, the blood cells will spread around the rod. If the rod is too long to be handled by one blood cell, two or more blood cells will participate in ingesting the conidium. The number of conidia ingested by a blood cell is variable. Usually a pair of conidia is found in one cell, but in some cases 3, 4 or more conidia can be swallowed. The blood cell which has ingested conidia, retains the power of locomotion. At times several blood cells containing conidia, fuse to form plasmodia.

The phagocytized conidia do not remain intact in the blood cells. They are consistently destroyed as indicated by their shrunken appearance. To be completely certain of this conclusion I have observed for several hours, conidia which were devoured, while in the process of budding. I have never observed completion of the budding process. Although there can be no doubt that the phagocytized conidia are dead, this does not yet prove that they were swallowed while still alive. It is unlikely that budding conidia would die spontaneously, but I have direct evidence showing that blood cells are capable of devouring living fungal cells. The morphology of many conidia in blood cells appears normal and unlike irregularly shaped dead fungal cells. Although a proportion of the conidia must be assumed to be incapable of sporulating, those acquainted with the cultivation of fungi would know that the proportion of dead fungi is much smaller than the proportion seen to be devoured in the *Daphnia* body. Aside from these reasons, I have observed that truly germinating spores are attacked by blood cells. For example, I have observed spores phagocytized with the sterigma and attached young spore still external to the phagocyte.

We have now reached the conclusion that blood cells are able to overcome living fungal cells and spores. While the spores undergo degenerative changes, the phagocytized conidia do not show the same changes, but become thinner and more refractile. Such conidia may be enclosed in vacuoles that are indistinguishable from the vacuoles enclosing food particles in protozoa. The differences in behavior between the conidia and spores may be attributed to differences in the cell membranes surrounding these structures.

While it is certain that the blood cells may attack conidia, there can be no doubt that conversely conidia may damage blood cells. I have seen blood cells engorged with conidia burst before my eyes and the conidia thus escaping. I have also observed the progressive dissolution of blood cells located among large groups of fungal cells, thus indicating that conidia release a component that is toxic for blood cells. This assumption appears plausible in view of the well known observation that ordinary yeast may be toxic to many animals. Massed conidia can also destroy blood cells that have phagocytized spores. The farther the disease progresses the greater is the destruction of blood cells, thus the *Daphnia* body will contain few blood cells when a large number of multiplying and mature spores are present.

In addition to blood cells there are isolated connective tissue cells which may function as phagocytes. The connective tissue cells may also be destroyed by conidia, so that in the later stages of the disease all phagocytic cells will disappear from the *Daphnia* body. Other tissue elements do not suffer such notable losses. Thus conidia may develop in heart muscle, but the functioning of the heart is not disturbed and the delicate fibrils in the sheath also remain intact.

Once the *Daphnia* are diseased, that is once conidia have appeared, the disease progresses to a fatal ending. In the last stages of the disease so many spores are present that the *Daphnia* assume a milky white color, but they retain motility, the heart continues to contract although loaded with spores and feeding continues until the last days before death. The disease lasts somewhat more than two weeks after penetration of conidia. Pebrine may occur simultaneously. In such cases both diseases proceed with their independent courses.

From this study one can see that infection and disease of our *Daphnia* constitute a battle between two types of living forms, the fungus and the phagocyte. The former represents a type of primitively organized unicellular plant, the latter, on the other hand, the most primitive tissue element, strikingly similar to the simplest animals (ameba, rhizopoda, etc.). The phagocyte has preserved the original function of intracellular feeding and, by this means, is the destroyer of the parasite. The phagocyte therefore represents the healing power of nature, which as Virchow first suggested, resides in the tissue elements. The basic idea of cellular pathology, advanced by this master, encompasses the entire course of the *Daphnia* disease, especially here, since the leading role is played by the most independent tissue elements.

It has already been shown above that the outcome of the battle is variable. If it is a question of killing spores, the phagocytes usually gain the upper hand, so that the phagocytes can be considered as the most useful organs of prophylaxis. The situation is quite different when the disease has set in; in this case the parasite dominates the field of battle. We see, as would be anticipated, that the phagocytes are better adapted for the destruction of spores than for the destruction of rapidly proliferating conidia. To obtain a more precise understanding of the quantitative relationships in the struggle, I examined 100 individual *Daphnia* from an aquarium in which the disease had broken out, and classified them under three categories:

1. Apparently healthy: no spores or conidia in body cavity
2. Diseased: conidia in body cavity
3. Infected: phagocytized spores in body cavity

The data were presented in a table and showed:

. . . that of 100 *Daphnia* living for two months in an infected aquarium, 73 were infected but the disease became established only in 14 animals. The 59 *Daphnia* that were protected from the disease by the activity of the phagocytes were isolated, remained healthy and were used successfully as breeding stock. As an example I have at this moment a healthy *Daphnia* which I found 23 days ago infected with a number of phagocytized spores. A few days ago she delivered three healthy youngsters. I have other such isolates. This disease tends to attack young animals. Adults may become infected but generally do not become diseased. Very young *Daphnia* also do not become diseased, probably because they do not take nourishment.

As I noted in the introduction, the yeast disease of *Daphnia* sheds light on certain pathological processes observed in higher animals. The study of this disease supports the hypothesis that the white blood cells and special phagocytes of vertebrates engulf agents of disease, especially bacteria, thus rendering an important service to the body. Prior to the present study a clear example of the ingestion of fungus cells by phagocytes was not available. Now we can criticize the position that phagocytosis by white blood cells does not lead to the destruction of bacteria. Robert Koch claims that the variation in the number of bacteria in the white blood cells of septicemic mice is due to penetration and growth of bacteria in these cells. He could not demonstrate the penetration and multiplication *in situ* but was obliged to use hanging drops or fixed preparations. It appears to me that in the case of mouse septicemia, as in the yeast disease of *Daphnia*, the parasites are engulfed by the blood cells and that the blood cells disintegrate and release bacteria when they have engulfed too many bacteria. Mouse septicemia shows another similarity with the yeast disease; despite the engulfment of parasites by phagocytes in both cases, the parasites gain the upper hand and win, probably because the parasites multiply too rapidly and in addition the parasites may secrete a poisonous substance. All the more interesting is an example cited by Koch where the phagocytes were victorious. I refer to the case of anthrax where this gifted investigator has observed that the frog infected with anthrax remains immune while its body cells are seen to contain bacilli. The nature of these cells has not been studied thoroughly and unfortunately it has not been noted whether or not these cells are capable of ameboid activity. Thus one can only advance as a possibility that it is simply a case of white blood cell activity. These cells contain in addition to the usual bacilli, spirally wound leptothrix-type threads that have developed from bacilli. Koch claims "that the bacilli taken up as short rods, grow in the cells and after filling the cells by bending and turning, the cells burst. In addition to freed bacilli in the form of spirals and bundles, crumpled and empty cell membranes are found." Since this observation is based on the study of drop preparations it is more likely that the engorged phagocytes burst easily under these circumstances. The long thread is probably phagocytized as such and pushed into the spiral form within the cell. I justify this interpretation not only because it is in better agreement with the direct observations on *Daphnia*, but also because it would be otherwise difficult to understand why the frog was not susceptible to anthrax, if the bacterium was capable of multiplying in certain frog cells. I believe that the bacilli are destroyed by the phagocytes, although the influence of other factors that may hinder their development is not eliminated.

The interpretation which Koch offers regarding the tubercle bacilli in giant cells is in closer agreement with my hypothesis. He postulates that the younger giant cells contain living bacilli while the organisms in the older giant cells are dead. This agrees with my observations on *Daphnia*. It may be conceived that the giant cells phagocytize and then kill the bacilli. It is obvious that these events do not necessarily lead to victory for the phagocytes.

When one accepts the concept that phagocytes fight directly against pathogens it becomes understandable that inflammation is a defensive mechanism against bacterial invasion. This interpretation, accepted for some time in medical practice, has just been incorporated into text books. Recently, it has been defended strongly by Buchner who has stated that "the inflammatory process exerts an antagonistic effect only against bacteria in the tissues" and that "the inflammatory change in tissues appears to be the natural useful and healing reaction of the animal body." This hypothesis becomes more tenable if one accepts that the mobilization of the phagocytes is the most important phase of inflammation and that the phagocytes then ingest the irritating agent. In other words, inflammation is a special case of intracellular digestion. The struggle between *Monospora* and phagocytes can be interpreted as a type of diffuse inflammation, a hematitis. When the needle shaped spores accumulate at a specific location, we have a localized mobilization of phagocytes analogous to the events in a local wound. When one examines a large number of *Daphnia*, one finds individuals with small body wounds, probably resulting from bites of other *Daphnia*. Most of these wounds are septic and contain brown detritus in addition to bacteria. On the inner side of the wound, blood cells accumulate in large masses. The wound becomes epithelialized, the detritus resorbed, and the phagocytes then disperse. A similar reaction is observed when the *Daphnia* skin is broken by the pressure of cover slips.

It has emerged that the inflammatory reaction is the expression of a very primitive function in the animal kingdom based on the nutritive apparatus of unicellular animals and lower metazoa (sponges). It can therefore be hoped that these considerations may elucidate the obscure phenomena of immunity and vaccination by analogy with the study of cellular digestion. It may eventually be acceptable to view fever as a convenient device for enhancing the activity of phagocytes. The general conclusion may be drawn that the knowledge of pathological changes in lower animals can lend additional support to the basic concepts of cellular pathology.

Following his studies on the fungal disease of *Daphnia*, Metchnikoff compared the phagocytosis of anthrax bacilli by white blood cells of animals that were sensitive to the disease, with that by cells of vaccinated animals. He reported that phagocytosis was more active in vaccinated animals.

Shortly after, the health of Olga and her sister forced the Metchnikoffs to go again to a warm climate. They spent the winter in Tangier, where Elie studied the embryology of sea urchins. In the spring of 1885 he was in Villefranche, in France, studying the embryology of jellyfish, on which he wrote a monograph. After Pasteur reported on the treatment of his first two rabid patients, the city of Odessa decided to open a bacteriological institute to prepare vaccines. Metchnikoff was named director, and one of his former students, Gamaleia, was sent to France to study the new method. However, Metchnikoff lacked administrative talent, and under his leadership, the operation was a failure. The major source of trouble was that Metchnikoff, not being a physician, was more interested in pure research than in following recipes to make vaccines. He himself said: "Obstacles came from above, from below and from all sides." Disillusioned, he started to prepare for his departure from Russia. Metchnikoff, while he attended a congress in Vienna, sought a position in Western Europe. He went to Paris, the famous Institute was being built, and Pasteur was cordial. He offered a laboratory to Metchnikoff, who, before accepting, felt he should visit other scientific lights of the time. Koch was antagonistic, and a marked contrast was evident between the atmosphere in France and that in Germany. Metchnikoff chose France. He went back to Odessa to transfer the direction of the laboratory to Gamaleia. Shortly thereafter, the first anthrax vaccine produced in the Odessa Institute was tried in the field. It killed many thousands of sheep. It was with a feeling of defeat that Metchnikoff finally left Russia in 1888.

In Paris, Metchnikoff was able to carry out research far from the influence of politics. Young scientists flocked to his laboratory. Pasteur was interested in the progress of phagocytosis; Duclaux and Roux were friendly, and Metchnikoff was happy.

First, he consolidated the defense of his phagocytic theory, which was attacked from many sides. Chemical substances present in the fluid components by the blood could also account for immunity! Metchnikoff sought the link between the cells and the humors.

Second, Metchnikoff started to probe the specificity of disease. He started with cholera. He doubted the available proof of the etiological agent. In Metchnikovian fashion, he tried the test on man. He drank a culture and nothing happened. Another culture went down the gullet of a volunteer, with the same lack of effect. Another volunteer, however, became very sick, much to the great distress and puzzlement of Metchnikoff. He conjectured that differences in intestinal flora might be responsible for the variable reaction he had observed. He developed this idea further in his attempt to combat senescence. For this he came up with an "appropriate" diet from which all raw foods were eliminated, since these might contain harmful microbes, and in which milk products containing "friendly" acid-forming bacteria were to be eaten.

He believed that man and other animals were slowly autointoxicated and that natural death was the result of this poisoning. He saw analogies with microorganisms, such as yeasts, that could kill themselves by producing too much alcohol. In man, the toxins accumulated during the day's activity were detoxified during sleep. With advancing years there developed an unfavorable balance, and the toxins accumulated slowly, producing the changes associated with old age. This intoxication, he felt, might be due to the products of the metabolism of putrefying bacteria located in the large intestine. The ingestion of acid-forming bacteria would be a method of using microbial competition to increase health and longevity.

In 1908, Metchnikoff shared a Nobel Prize with Ehrlich. On the trip occasioned by the award, he and his wife visited Russia after Sweden. They visited Tolstoy, whose interest in science was linked to the role it might play in helping to alleviate the sufferings of humanity. During their discussions on the role of science, Tolstoy said that he did not think that the exploration of other planets was of pressing importance. "What good can it do for man?" This point of view does not seem to prevail any more among the leaders of mankind.

The First World War followed. The Pasteur Institute, dedicated to saving lives, was emptied of its young scientists and technicians and placed under the control of the military authorities. Normal and humanistic life was suspended by the madness of men. Only practical aspects of sera and vaccine production remained.

On the sixteenth of May, 1915, the war did not prevent those remaining at the Institute from celebrating the seventieth birthday of Metchnikoff. On this occasion Roux told him: "Your laboratory is the most lively in the house. . . . Since you read everything, each one of us has depended on you for information . . . this has prevented many errors in translation. . . . Even more than your science, your kindness attracts. . . ."

Toward the end of the year, his cardiac condition worsened. Roux had the Metchnikoffs moved to a small apartment at the Institute, so that the best medical attention could be provided. To his great satisfaction, since he wanted to show the value of his diet, Metchnikoff reached his seventy-first birthday. In June, 1916, he was moved to Pasteur's apartment, where he died, less than a month later, attended by Roux, Martin, and Olga. In death, he had the beauty and serenity of a biblical prophet. He was cremated, and his ashes were deposited at the Pasteur Institute.

Olga, who at the age of sixteen had left her childhood behind to follow a zoologist with a Christ-like face, was left in Paris, far from her native Russia, which was soon to be torn by the socialistic ferment with which she had been imbued many years previously.

Although the theory of phagocytosis was convincing to Virchow, the dean of German pathology, others, such as Baumgarten, Ziegler, and Weigert, were not so impressed. Nevertheless, their objections were not founded on serious experimental studies, such as those eventually begun by Fodor in 1886. He showed that defibrinated rabbit blood destroyed anthrax bacilli in vitro, and he stressed the importance of humoral factors in immunity. Two years later, Nuttall made the same observation but also noted, correctly, the simultaneous occurrence of phagocytosis. Flügge, his superior, chose to infer that the anthrax bacilli had been phagocytized after they had been killed by soluble bactericidal components of blood. The same year, "the humoralists" received further support for their theory when Behring demonstrated that the blood of rat was highly anthracidal, a fact that could be taken to explain the high resistance of this animal to the disease. The clarity of this picture was somewhat clouded, however, when two years later, Behring and Nissen uncovered the fact that the blood of vaccinated animals did not always acquire bactericidal activity. Such observations led Behring to the discovery of the role of antitoxins, substances which combined with toxins but which were not per se bactericidal. As can be seen in the next chapter, the study of the humoral aspects of immunity was most fruitful, and considerable progress was made in that direction while the theory of phagocytosis was not strengthened by any new discoveries. Finally, in 1903, Almroth Wright, in England, with his disciple Stewart Douglas, observed that a humoral component, which they designated opsonin, could render bacteria susceptible to phagocytosis. This was a link between the two theories of immunity. Wright hoped that the knowledge of opsonins would lead to advances in the therapy of disease. Despite long years of devoted effort, however, he did not attain his objective, and a witty observer once referred to him as "Sir Almost Wright." However, as we can see from the following account of his life and work, he was an important figure in the progress of bacteriology and an interesting representative of British virtues and eccentricities.

Almroth Wright was born in the rectory of a Yorkshire village in 1861. His father was an Irish clergyman, his mother, Ebba, the daughter of a

Swedish professor of chemistry. Almroth was the second of five sons. Soon after his birth, his father accepted ministries on the continent, first in Dresden, then in Boulogne. The family returned to the British Isles when Almroth was thirteen to follow Mr. Wright, Sr. in his zealous pursuit of a scholarly and pious career to Belfast, Dublin, and Liverpool. His mother had worked with Florence Nightingale and was described by one of her grandchildren as a "fierce old Lady." It is doubtful that Almroth had much affection for either of his very Christian parents, although he undoubtedly derived at least one positive asset from his austere upbringing, namely, a marked ability for highly scholarly achievements.

At the age of seventeen, he entered Trinity College in Dublin, the traditional Alma Mater of the Wrights, and, in 1882, after having studied modern literature and languages, he received a bachelor's degree with First Class Honors and a gold medal. The next year, he received a degree in medicine. Despite Wright's scholarly abilities, so short a period of training left important gaps that had to be filled. It is not altogether clear why Wright turned to medicine. It is said that in his indecision he requested the advice of Edward Dowden, his professor of literature, who responded with a quotation from Charles Lamb: "Literature is a good stick but a poor crutch," and who terminated the interview counseling him to "go on with medicine." So, like Claude Bernard, Wright started his career with a short trip through "the enchanted streams of literature." As a result, he retained throughout his life a marvelous ability to express himself, and he actually spent the end of his life writing books on philosophy, indeed a scholarly though adventurous way, it might be said, of closing a scientific career.

After having completed his initial medical studies, Wright left for Germany, where he spent some time in the laboratories of some of the great pathologists of the time: Cohnheim, Weigert, and Ludwig. After one year, he returned to England, where he worked in Cambridge, and having won a scholarship, he returned for a few months to Germany. This time he worked with Von Recklinghausen. Upon his return to England in 1889, he accepted a demonstratorship in Australia, where he stayed for two years.

In 1892, Wright had an opportunity that he described later as "the best stroke of luck that ever man had." It was an appointment as professor of pathology at the Army Medical School, which was then housed in the Royal Victoria Hospital at Netley, on the banks of Southampton Water. Indeed, many a man would have also thought it a golden opportunity. The surroundings of Netley were delightful; sailing, tennis, and swimming were all available. The work load was light. It would have been easy to drift into a life of peaceful teaching and pleasant indulgence in sports. But Wright was not such a man. He did not like sports, and to him the light work load suggested an unparalleled opportunity for submerging himself in medical research.

At the Army Medical School there was something of a bacteriological tradition, which was derived from Bruce's discovery of the causative agent of Malta fever. Since Netley was receiving soldiers with fevers of unknown origin, Wright thought of developing a diagnostic test for Malta fever. He did this work with his colleague, Captain F. Smith, and together they discovered a diagnostic agglutination test. The basic observation of agglutination had just been made by a number of independent investigators, which included Gruber in Austria, Pfeiffer in Germany, and Widal in France. These same workers also applied the phenomenon of agglutination to diagnosis at the same time as Wright and Smith. Still fascinated by Malta fever, Wright tried to develop a vaccine for this affection. With a stiff upper lip, he tested it on himself with instructive, though unpleasant, consequences. He learned at one stroke that *brucellae* were not easily amenable to vaccine making, and that some in vitro methods of assaying vaccines would be more desirable and useful.

Undaunted by his attack of Malta fever, he chose typhoid fever as the next hurdle. In the development of a typhoid vaccine, he applied more precise quantitative techniques than had been employed previously in the making of vaccines. He used bacterial counts; he also related dosage to level of antibody response, and he determined the duration of antibody persistence. He made frequent use of a capillary tube to which was attached a rubber teat. This useful little contraption, which he modified as needed, was soon his trademark. His ingenious procedures permitted considerable manipulation of small samples of blood and other reagents; it was, in essence, *microserology*. The vaccine was ready at about the same time as Pfeiffer and Kolle announced the development of a typhoid vaccine (1897). Field trials were to be more of a problem than the laboratory studies had been. Of course, the army was the logical proving ground, but official opposition was strong and spearheaded by one officer who was jealous of Wright's appointment at the Medical School. With the outbreak of the war in South Africa, Wright was permitted to solicit volunteers. But, in good army tradition, few cared to face the needle. Under the supervision of Colonel Leishman, the vaccine was finally tested and shown to be efficacious. Even though the Minister of War, Lord Haldane, had Wright knighted in 1906, typhoid vaccination still was not compulsory in the British army at the outbreak of the First World War.

Pleasant as the surroundings of Netley may have been, they were not a strong enough inducement to keep Wright within the bonds of the army hierarchy. In 1902 he became pathologist at St. Mary's Hospital and Medical School, in London. In spite of a small salary and a meager but "cheerful" two-room laboratory that was located in the cellar directly above the roaring subway, he continued in the battle against infectious diseases with a vigorous mind, heart, and pen. As we see in the following quotations from an article that he published in the Liverpool *Daily Post* (August 30, 1905), *disease* was the world's greatest problem against which scientific *medical research* had to be organized:

> Of all the evils which befall man in his civilized state, the evil of disease is incomparably the greatest. It ought, accordingly, to loom largest in his mind. In comparison with the chance of winning directive control over this evil, every other thing ought to be counted as loss. If the belief is nurtured that the medical art of today can effectually intervene in the course of disease, this ought to be dismissed as illusion. Putting out of consideration the case of one or two infective diseases which can to some extent be controlled by remedies placed in our hands by chance, and, further, the case of diphtheria—a disease for which an effective remedy has been furnished by medical research—it may be affirmed with confidence of the medical art, as at present practised, that it can do practically nothing to avert death from a virulent bacterial invasion or to bring about a cure. . . . If the conclusion is thus forced upon us that the medical art of today cannot cope effectually with disease, let us turn and consider how far the problem of disease can be solved by the resources of sanitation, meaning thereby all those measures of disinfection, isolation, and conservancy, which are adopted for killing off the germs of disease outside the organism or, as the case may be, for holding these off from contact with the healthy.
>
> The modern world prides itself on all these measures. That pride furnishes an object for consideration.
>
> Conservancy must be credited with having lightened the burden of disease in the respect that it has, in towns where it is applied, diminished the incidence of typhoid fever. It has further practically eradicated dysentery. It is doubtful whether much more can be claimed for it. . . .
>
> Consider the life of the busy general practitioner. He has to go from

house to house, from sick bed to sick bed, exploiting in the diagnosis of disease the knowledge which he has acquired in his hospital career, and such personal experience as he has since accumulated. His diagnosis made, he applies to each case the accepted method of treatment, and passes on. He has neither time, nor training, nor opportunity for research, and no one would wish him to desist from the useful work which lies ready to his hand to undertake the task of research. The same conditions, or essentially the same conditions, present themselves in our hospitals. . . .

We are not, in fact, making any effort worthy of the name to solve the problems of disease, and we have not in England any appreciable number of workers engaged upon the task of medical research. This is due to economic reasons. A young man who proposes to take up medical research as his life-work finds himself immediately confronted in his own person with those very fundamental and primitive problems of obtaining subsistence, and clothes, and a shelter over his head. Even if appointed to one of the research scholarships which have been recently founded for the purpose of launching the student upon a career of research, the problem will only be staved off a little. . . .

The great practitioners of Harley Street were not accustomed to having the miseries of medicine exposed to the view of the layman. They countered with articles of their own in which they stressed the recent achievements of medicine and in which they maintained that great investigators were not made by attractive salaries. Nevertheless, Wright had moved to St. Mary's to strike a new path in medicine. Despite the fact that he was one of the discoverers of an important vaccine, Wright was not convinced that prophylaxis was the only possible application of immunology in the control of infections. Enhanced humoral bactericidal activity was an effective mechanism in typhoid fever, but it was not the answer in staphylococcal, streptococcal, or brucellal infections. Metchnikoff's theory was there, and Wright thought that if the phagocytes were not sufficiently active, an immunological stimulus might enhance their action. In this he saw therapeutic rather than prophylactic hopes.

Wright's bright assistant at Netley, Capt. Stewart Douglas, followed him to St. Mary's and worked with him from 1902 to 1920. Douglas was unusually dexterous, an invaluable asset in the elaboration of the delicate techniques necessary to spy on the phagocytes. Together they published in 1903 (*Proc. Roy. Soc. London,* **72:** 357–370) a paper entitled "An Experimental

Investigation of the Role of the Blood Fluids in Connection with Phagocytosis."

It is still a matter of uncertainty whether the blood fluids perform any
role in connection with phagocytosis.

Certain facts suggest that the role of the blood fluids, if it comes into
consideration at all, is very subordinate. The facts we have in view, are,
on the one hand, the facts brought forward by Metchnikoff to show that
bacteria may be ingested in the living condition, and on the other hand
those, brought forward by one of us in conjunction with Captain F.
Windsor which show that the human serum exerts absolutely no bactericidal action on the staphylococcus pyogenes, the micrococcus
melitensis and the plague bacillus.

These facts are, however, not conclusive. They are not inconsistent
with the idea that the blood fluids, apart from actually killing the particular pathogenic bacteria here in question, may in some way cooperate
in their destruction.

What are required for the resolution of the problem are experiments
in which the phagocytes are tested apart from the blood fluids.

The experimental methods which we now pass on to describe enable
these crucial experiments to be made.

We have employed a modification of the method of measuring the
phagocytic power of the blood, which was devised by Major W. B.
Leishman.

In the procedure described by this author equal volumes of a bacterial
suspension of appropriate density and of blood drawn from the finger
are measured off in a capillary tube, mixed on a slide and covered in
with a cover-glass. The blood and bacterial culture are then left in contact for 15 minutes in an incubator standing at blood heat. After this
interval the cover-glass is, if necessary, loosened from the slide by a drop
of physiological salt solution, and the slide and cover-glass are drawn
apart by a sliding movement.

The films thus obtained are stained by Leishman's modification of
Romanovski's stain, and are subjected to examination under an immersion lens. By enumerating the bacteria ingested in a number of poly-
nuclear white blood corpuscles and dividing, an average is obtained.
This average is taken as the measure of the phagocytic power of the
blood. It is compared, when comparative experiments are made, with
the phagocytic power of a normal blood.

We have modified this method for our purposes (a) by conducting
the phagocytosis in capillary tubes, making afterwards film prepara-

tions in the ordinary way; (b) by decalcifying the blood with citrate of soda, thus avoiding the complications introduced by blood coagulation, and making it possible to separate the white corpuscles from the blood fluids by centrifugalisation, decantation and washing.

Three different procedures, varying only in details, were employed in our experiments. . . .

A point which comes up for consideration is the possible effect of the addition of citrate of soda to the blood.

The concentration of the solution in particular comes into consideration . . . phagocytosis is inhibited when the white corpuscles are bathed in a medium containing 3 per cent of citrate of soda. . . . It may be noted that the morphological structure of the white corpuscles is extremely well preserved, and phagocytosis proceeds actively in a medium containing up to 1.5 per cent of citrate of soda.

The last point to be considered relates to the maintenance of the activity of the phagocytes for a sufficient period after they have been withdrawn from the organism and have been subjected to the procedures described above.

A number of experiments undertaken with a view of obtaining information with regard to the point here raised have shown us that the phagocytic power is well maintained under the circumstances of our experiments. Even after the lapse of 3 days (our observations have not extended beyond this limit) the phagocytic power has not declined to less than one-half or one-third of that of the blood freshly drawn. We have found no indication of a variation within the space of a few hours.

These preliminary points having been dealt with, we may pass to the consideration of the problem to which attention was directed in the opening paragraph of this paper.

After having shown that the substitution of citrated serum for citrated plasma had no influence on phagocytosis, the authors demonstrated that heated serum, like a salt solution, acted merely as an inert diluent and had no effect on phagocytosis. Then, they asked the following basic question:

Do the blood fluids cooperate in phagocytosis by exerting a direct "stimulating" effect upon the phagocytes, or by effecting a modification in the bacteria?

The following experiments were instituted with a view to elucidating the problem as to the nature of the activating influence exercised by the blood fluids. It will be seen that a comparison is in each case instituted between serum inactivated (by heating) before it came in contact with either bacteria or white corpuscles, and serum inactivated after it had come in contact with the bacteria, but before it had come in contact with the white corpuscles.

The results showed clearly that the heating of sera ten to fifteen minutes at 60 to 65°C before they were put in contact with suspensions of staphylococci, be they heated or not, reduced the subsequent numbers of phagocytized bacteria from some thirty per white blood cell to about four. The authors continued:

> We have here conclusive proof that the blood fluids modify the bacteria in a manner which renders them a ready prey to the phagocytes.
>
> We may speak of this as an "opsonic" effect (opsono—I cater for; I prepare victuals for), and we may employ the term "opsonins" to designate the elements in the blood fluids which produce this effect.

The next question was:

> Does the unheated serum contain, in addition to elements which render the bacteria more liable to phagocytosis (opsonins), also elements which directly stimulate the phagocytes (stimulins)?
>
> We have sought to elucidate this question by three separate methods.
>
> In a first series of experiments, we experimented with staphylococci which had been exposed to high temperatures (115°C.) with the design of rendering them insusceptible to the opsonic power of the blood fluids. Our expectations from this method—expectations based on the fact that we had noticed that typhoid bacilli acquired, when heated to over 70°C., a resistance to the bacteriolytic effect of the blood fluids— were unrealised. We found that the quantitative differences between the phagocytosis in heated and unheated serum respectively were not less in the case of staphylococci which had been exposed to a temperature of 115°C., than in the case of staphylococci which had not been subjected to high temperatures.
>
> In a second series of experiments we substituted for suspensions of staphylococci suspensions of particles, which we assumed would be uninfluenced by the opsonic power of the blood. The results of these experiments, conducted both with carmine particles and with India ink, were inconclusive by reason of the circumstance that we were not able to obtain any satisfactory enumerations. An impression was, however, left on our minds that phagocytosis was in every case more active in unheated than in the heated serum.
>
> A third method of experimentation was then resorted to. In a first operation we mixed and digested together at blood heat a suspension

of staphylococci and unheated serum. After allowing what we supposed would be a sufficient interval for the exhaustion of the effect of the serum upon the bacteria, we divided the mixture into two portions. While the first of these portions was mixed with the corpuscles without undergoing any further treatment, the other was heated to 60°C., and cooled before it was so mixed. In each case the phagocytic power exerted was greater in the case where the heating was omitted, and the differences were not less marked where the serum had been digested with the bacteria for 50 minutes and 1 hour respectively than in the case where it had been digested with these only for 15 minutes.

These results are ambiguous.

The question as to whether the blood fluids contain, in addition to opsonins, also an element which directly stimulates the phagocytes, remains for the present unsolved. . . .

In conclusion we would briefly refer to the following points:

The opsonic power of the blood fluids disappears gradually on standing, even when the serum is kept in a sealed capsule sheltered from the light.

After 5 or 6 days we have found the opsonic power of the serum kept under these conditions to stand at little more than half of what it was originally.

The opsonic power of the blood fluids is but little impaired by the action of heat until these have been exposed to temperatures above 50°C. The following are the results of a typical experiment: Phagocytic power obtained with the serum before exposure to heat, 12.7; with the same serum heated for 10 minutes to 45°C., 13.1; with the same serum heated for 10 minutes to 50°C., 10.2; with the same serum heated for 10 minutes to 55°C., 5.7. . . .

Lastly, a fact which has a practical importance in connection with the study of immunity may be adverted to. It will be manifest that we have not exhausted the study of a condition of immunity when we have measured the phagocytic power of the white corpuscles, and the agglutinating, bacteriolytic, and bactericidal powers of the blood fluids. We must, in connection with these last, take into consideration also the opsonic effect. . . .

From a practical point of view, what did Wright have in mind? He showed that the blood of a man with recurrent boils had less phagocytic power than that of a healthy man and that the blood of the healthy man contained more opsonins than the blood of the patient chronically affected with boils. His next step was to try to "beef up" the opsonin factory in the body of the affected man. How could this be done? Nobody knew, including Wright.

However, his enthusiasm was infectious, and the number of eager coworkers who surrounded him grew steadily. Of these, Alexander Fleming was to become the most famous. He joined Wright's department in 1906, and he succeeded him as head in 1946. That department was called unofficially the "Inoculation Department." It grew steadily in importance from the "cheerful" two rooms that we described previouly, to the present-day Wright-Fleming Institute.

When the First World War began, Wright transformed most of his Inoculation Department into a factory for typhoid fever vaccine, and he, sporting the uniform of a colonel, moved his own research laboratory into a transformed fencing hall on the roof of the Casino of Boulogne. Surgeons were being faced daily with human bodies badly ripped and shredded by high-power explosives. The wounds were often full of nooks and crannies where septic bacteria would hide effectively from antiseptics. In addition, antiseptics were inactivated by the proteins of body fluids. Wright and his assistants, which included Alexander Fleming, Leonard Colebrook, Georges Dreyer, John Freeman, and Parry Morgan, set about to study the fundamental aspects of wound sepsis and healing. In this regard, Fleming's ingenuity was a tremendous asset. First, it was demonstrated that certain low concentrations of disinfectants not only did not inhibit septic bacteria but even stimulated their growth. Often, to have done some good from a microbiological point of view, the concentration of disinfectant would have had to be so high that it would have killed the patient. Second, these workers observed that many bacteria would not have grown in body fluids at all if the latter had not been hydrolyzed by the proteinases liberated by decomposing leucocytes. Interestingly, the body fluids were found to contain an antiproteinase component referred to as an "anti-tryptic principle," but it was able to counter the destruction of only a modest number of phagocytes. Furthermore, leucocytes were found to elaborate an antibacterial substance that was independent of their phagocytic power. To recapitulate, leucocytes had to be protected from destruction for at least three reasons: (1) their phagocytic ability, (2) their property of releasing unwanted proteinases when they were destroyed, and (3) their substances having antibacterial power

These considerations led Wright to tell the Army surgeons: "The leucocyte is the best antiseptic and you must provide the optimal conditions for its functioning. There is no sense in repeating daily dressings of the clean wound, and you should avoid putting chemicals into it which will destroy leucocytes as well as microbes. . . . Optimal conditions . . . will be obtained if you allow no 'dead spaces' in which fluids can collect and microbes multiply out of reach of the leucocytes. Ensure maximal concentration of these cells—that is very necessary—and protect them from dying." [1] He felt that the wound should be well cleaned by surgery so as to remove dead tissue that would soon putrefy, and to reduce to a minimum the unwanted "dead spaces." After this the wound was to be closed by sutures and grafting, as necessary, and covered with a firm bandage.

The closing of the wound sounded like a great idea, but might it not be a factor helping the growth of the anaerobic bacteria responsible for gas gangrene? Wright did not think so. First he showed that *Clostridium perfringens* would grow even aerobically if the serum did not have its full antiproteolytic properties, or its normal alkalinity or both. This observation was supported by the results of a study of the acid-base balance of patients affected with the gangrene, for the study revealed that the blood of these patients was characterized by a marked acidemia. Treatments with alkalies were tried with inconclusive results.

At the end of the war the tired Wright, who was then close to sixty, served for a short time as the director of the section of bacteriology at the newly organized Medical Research Council, but he soon returned to the Institute in which had been his old Inoculation Department. He took up some of the problems that had arisen during the war, such as the study of the exchange of fluids in damaged tissues. His major concern was still immunity and phagocytosis. He introduced a slide cell technique for the study of the effect of antiseptics and chemotherapeutic agents on leucocytes. This technique was used by Fleming in his studies on penicillin.

The fire of Wright's zeal for research was burning low. In putting down the test tube, he took up the pen, beginning work on the books of philosophy he had always planned to write. One, *Prolegomena to the Logic Which Searches for Truth*, he saw published in 1941. Another, an extension of the first, was in the process of final revision at the time of his death and was published in 1947. He thought these books were his most important contribution to humanity.

Wright was a staunch conservative during his entire life, and he was engaged in controversies up until his last years. In addition to prodding the medical profession continually to take research more seriously, he also considered it his duty to be an active opponent of women's suffrage. His publications include a not-so-thin volume on *The Unexpurgated Case Against*

---

[1] L. Colebrook, *Almroth Wright, Provocative Doctor and Thinker*, William Heinemann, Ltd., London, 1954.

Woman Suffrage. After the public defeat of his ideas the two maids in his household decided to exercise their franchise. One had spoken of voting for the conservative ticket and was driven to the polls; the other had expressed laborite views and had to walk. He had married in 1899 and had had three children. He and his wife gradually became estranged and lived apart for many years. His complete devotion to science and philosophy had left little room for his spouse, whom he remembered only during her last bedridden years. During this period he visited her often, always bringing her flowers for which they both had a great liking.

Wright's mind remained active until the very end. In 1946 he was no longer able to make the trips to St. Mary's, but continued writing in a race against death. In April of 1947 a heart attack required that he be hospitalized. He rallied after a short stay and returned to his country home. When he felt the end was near, he had his bed moved to the dining room, where he could look out on his flowers. He gave his instructions on business and funeral arrangements. Within twenty-four hours he was unconscious, and he passed away peacefully. As he wished, he was cremated and his ashes were spread among the flowers of his garden.

The work of Wright and Douglas stimulated an interest in the effect of humoral components in phagocytosis. It was confirmed that opsonins act on microorganisms and not on phagocytes. Furthermore, it was shown that immunization often enhanced the opsonic power of the serum against the organism used as vaccine. As an added facet, the virulence of pathogens was shown to reside, in part, in the elaboration of antiphagocytic substances called antiopsonins or aggressins.

Since the humoral components and cell components were shown to be interwoven in an intricate pattern, the complex chemicals associated with immunity must be the next subject for consideration.

References

Bordet, J.: *Traité de l'immunité dans les maladies infectieuses*, Masson et Cie, Paris, 1939.

Colebrook, L.: *Almroth Wright, Provocative Doctor and Thinker*, William Heinemann, Ltd., London, 1954.

Metchnikoff, Olga: *Vie d'Elie Metchnikoff*, Hachette et Cie, Paris, 1920.

Immunology: Humoral

6 One of the most important objectives of microbiology, from its beginnings, was the control of diseases of animals and man. Confining ourselves to the immunologic approach, we recall that Pasteur and his coworkers following the principles of Jenner developed attenuated vaccines for anthrax, fowl cholera, and rabies.

In 1886, Salmon and Smith demonstrated that it was not always necessary to have a live, attenuated strain of a bacterial pathogen in order to build up the immunity of animals artificially. They showed that killed bacterial cultures could also be used as vaccines. Their new method seemed to offer definite advantages since such killed vaccines could be stored easily and would certainly be uninfective. Salmon and Smith published "On a New Method of Producing Immunity from Contagious Diseases" in the *Proceedings of the Biological Society of Washington* (**3**: 29–33, 1886). They had just recently isolated a bacterium (*Salmonella choleraesuis*) that they thought was the cause of swine plague. The subcutaneous injection of live cultures of this bacterium could kill pigeons, and the previous injection of heat-killed cultures could induce immunity and permitted the pigeons to survive a fatal challenge. The authors concluded their presentation with the following:

1. Immunity is the result of the exposure of the bioplasm of the animal body to the chemical products of the growth of the specific microbes which constitute the virus of contagious fevers.

2. These particular chemical products are produced by the growth of the microbes in suitable culture liquids in the laboratory, as well as in the liquids and tissues of the body.

3. Immunity may be produced by introducing into the animal body such chemical products that have been produced in the laboratory.

Thus one more way had been found to produce immunity in the animal body. As a result of his work on swine plague, Salmon's name was given to a genus of bacteria. A native of New Jersey who studied veterinary medicine at Cornell and at Maisons-Alfort, the stronghold of Nocard, he first practiced veterinary medicine in Newark, New Jersey. Following this he went to Washington, where he established the veterinary division of the Department of Agriculture. He became the head of the Bureau of Animal Industry, a post that he held up to 1905 when he joined the staff of the University of Montevideo in Uruguay. After five years, he came back to the United States to head a plant which made biological products for veterinary medicine and which was located in Butte, Montana. He died there of pneumonia in 1914. Microbiologically speaking, Theobald Smith, his coworker, led a more important life, the threads of which will be picked up in Chapter 10.

Mere empirical methods of producing or increasing immunity were not sufficient; the whole basis of immunity was now questioned. The first answer, the theory of phagocytosis, was boldly proposed in 1884. Still another explanation was brought forward: that certain diseases were the expression of the action of toxins that could be neutralized by antitoxins, a subject to which we will soon return. A third fact, which could not be ignored, was that the blood of animals contained certain bactericidal agents.

In a paper published in 1888 (*Z. Hyg. Infektionskrankh.*, **4**: 354–394) George Nuttall, a young American-born scientist working in Flügge's laboratories, wrote:

Metchnikoff has sought to establish the principle that the activity of phagocytes is responsible for the protection of the animal body against infectious disease and the development of acquired immunity. The experimental basis for Metchnikoff's theory, excluding the observations on *Daphnia*, rests on experiments with anthrax bacilli in frogs and rabbits. . . . Metchnikoff observed that anthrax bacilli in infected tissue lost viability after several days when such tissues were placed under the skin of frogs. Continuing with rabbits, Metchnikoff introduced capillary tubes containing attenuated or virulent anthrax bacilli under the skin of the ears of test animals. The capillary tubes were then broken. Successive microscopic examination of the pus showed that many of the leucocytes phagocytized attenuated bacilli. Virulent bacilli were not phagocytized in the susceptible animals but were phagocytized in the immune animals. Further, Metchnikoff could demonstrate phagocytosis of anthrax bacilli by leucocytes in preparations which were kept warm on the stage of the microscope. . . .

Summarizing my studies with frogs, I can confirm Metchnikoff's statement that leucocytes accumulated about fragments of anthracic tissues introduced under the skin of frogs. I also confirmed that the phagocytized bacilli were destroyed. However, from then on our findings deviate. I observed that as many, or more, of the extracellular bacilli were destroyed as were phagocytized. . . . The fact that, in the frog, the destruction of bacilli was as extensive extracellularly as intracellularly is in contradiction with Metchnikoff's theory. . . .

Certain phases of the relationship between bacilli and phagocytes are observed best in the living animal, but other phases can be evaluated better by the study of preparations under the microscope where the interaction between leucocytes and bacteria can be observed continually. In this manner it was possible to observe whether phagocytosis occurred immediately or after a lapse of time and whether virulent and attenuated bacilli reacted differently in this regard. One could also determine the degenerative changes occurring in the body fluids outside the cells. This method appeared all the more appropriate because it had been used by Metchnikoff. Experiments in this direction, which were initiated only to check Metchnikoff's observations, gave such significant results that they were extended. This was done by including a variety of animal species in these studies, in which not only blood was used, but also lymph and other tissues. Various bacteria were also included.

A drop of the liquid to be investigated was placed on a cover slip and an inoculum of virulent anthrax bacilli was introduced at the edge of the drop. The cover slip was then applied to a concave slide and sealed with paraffin. . . . When observations were to be made at the temperature of warm blooded animals, the microscope was enclosed in a specially insulated and humidified heated chamber. The slide could be manipulated through an oval opening normally covered with a well-fitted lid. . . .

In experiments with mammals, blood was removed rapidly and aseptically from a small wound. Slides were then prepared as described above and transferred immediately to heated microscope chambers maintained at the body temperature of the donor animal. The drops of blood coagulated shortly after transfer to cover slips, leaving serum at the edge of the droplet. Thus the inoculum of anthrax was surrounded by liquid and not entrapped in the coagulum. Contamination of the blood with other bacteria was rarely observed. We noted that coagulation did not eliminate the antibacterial properties of the blood and a later experiment showed us that defibrinated blood also retained the bactericidal properties of blood. Some of the preparations were observed continually, under the microscope; others were examined from time to time. . . . In general these studies showed that indeed a portion of the bacilli were phagocytized by leucocytes, but the larger number remained free and showed degenerative changes.

The bloods of different species differed in the speed at which phagocytosis occurred and at which degeneration of free bacilli proceeded. In preparations inoculated with large amounts of bacilli, growth occurred after the period of degeneration and the newly developing bacilli overran the entire droplet. In preparations inoculated with small numbers of bacilli, the degenerative changes extended to all bacilli in the preparation. Degeneration occurred most rapidly in human blood, where degeneration was observed after ¾ of an hour. After 4 hours, growth could be observed among surviving bacilli. Although many bacteria were phagocytized, most were free and a higher proportion of the free bacilli than the phagocytized bacilli were degenerated. Anthrax bacilli degenerated in immunized sheep's blood almost as rapidly as in human blood. In the blood of immunized sheep, degeneration was observed after one hour and growth of surviving bacilli could be observed after 26 hours. In the blood of non-immunized sheep, degeneration ensued earlier and growth of surviving bacilli occurred earlier. The blood of birds was less bactericidal than the blood of other animal species. The degenerative changes in the blood of birds, though limited to fewer bacilli, was more frequent among free bacilli than among phagocytized ones. Further, the bactericidal action in bird's blood was influenced by the rapidity of coagulation. When clotting was retarded by prior defibrination, giving a greater yield of liquid component, the bactericidal effect was increased. Rabbit's blood was bactericidal; the action was slow but extensive. A maximal number of involuted forms was observed after 5 hours and growth of surviving bacilli was evident after 28 hours. Phagocytosis was seen 30 minutes after mixing blood with bacilli. This observation is contrary to Metchnikoff's report that rabbit leucocytes in microscope stage preparations do not ingest virulent anthrax bacilli. Mouse blood did not destroy anthrax bacilli; on the contrary, it supported growth of bacilli and phagocytosis was infrequent.

Studies on the action of blood were followed by studies on other tissue fluids. Tissue fluids containing minimal numbers of cells were chosen to demonstrate the presence of a bactericidal component other than leucocytes. Aqueous humor and pericardial fluids were selected for study because they contained few leucocytes. Both liquids caused involution and destruction of anthrax bacilli.

I next determined how long extravasated blood maintained its bactericidal effect. Drops of blood were transferred to slides in the usual manner and kept at 37° for various intervals of time prior to inoculation with anthrax bacilli. The bactericidal effects were lost after 4 hours at 37°. . . .

Nuttall made an observation that was the first indication that opsonins did exist:

Indeed, we have been led to the assumption that the bacilli that were phagocytized were not completely normal and had been subjected to the damaging effect of extra-cellular fluids. The observation that the rapidity of phagocytosis varied directly with the rapidity of extracellular bacterial destruction supports this view. In man, where extracellular degeneration was rapid, phagocytosis was also rapid. On the other hand, in the rabbit, where extra-cellular degeneration was slow, phagocytosis was also slow. In the mouse, where extracellular degeneration was slight, only small numbers of bacilli were phagocytized. The fact that in the frog there was extensive *in vitro* phagocytosis despite slow extra-cellular degeneration is explained by the ability of leucocytes from this cold blooded species to remain more active than leucocytes from warm blooded species when removed from the body and kept at room temperature.

Studies with hanging drops could demonstrate whether the bacilli which appeared morphologically to have degenerated, were no longer viable. Known numbers of anthrax bacilli were mixed with freshly drawn blood and samples were cultured at various intervals after preparation of mixtures. Blood was obtained aseptically and defibrinated by shaking with sterile sand. The containers were maintained at a temperature of 38°. Defibrination was necessary in order to assure intimate contact between blood and bacilli and to facilitate cultivation of aliquot samples of the mixtures. Anthrax bacilli were obtained by suspending spleens of mice just dead from anthrax, in sterile sodium chloride solution. Suspensions of bacilli were introduced into blood with platinum loops. The variations in inoculum with this procedure did not influence the experimental results. Nutrient gelatin was used as culture medium and culture dishes were incubated for 3 to 4 days.

The blood of immunized sheep produced marked reductions in the number of bacilli after 3½ hours; from 4,578 to 185, from 4,872 to 283, from 11,046 to 427, 9,245 to 665. The blood of normal sheep produced a much smaller decline; from 7,938 to 6,664, from 8,330 to 4,728. Considerable variation in bactericidal effect was observed. In one experiment with rabbit blood 7,000 bacilli increased to 90,000 but in another experiment with a similar inoculum the count dropped to approximately 100. Normal sheep's blood also gave variable results. In normal mouse blood, the bacilli multiplied shortly after preparation of mixtures. . . .

. . . the bactericidal power of blood is lost after some time and the blood then becomes a good nutrient medium for the bacilli. Thus the antibacterial property of blood is not due to a disinfecting agent in the usual sense. The antibacterial agent is unusually labile, possibly destroyed by other components of blood, or it may, be of ferment-like nature. . . . If exposed to temperatures to 50–55°, dog's blood lost its antibacterial activity within 30 minutes and rabbit's blood lost its activity after 45 minutes. Ten minutes of exposure to 48–50° did not completely inhibit the disinfecting action of the blood. For rabbit's blood at least, 19 to 38° was without significant effect on the bactericidal activity of blood. . . .

These studies were extended to include organisms other than the anthrax bacillus. Blood samples were infected with 12 hour old spore-free broth cultures. *Bacillus subtilis* was always completely killed within 2 hours. Viable cells of *Bacillus megatherium* decreased in number, but in only one case was complete killing attained. Blood was ineffective against *Staphylococcus pyogenes*.

Nuttall concluded:

These investigations have shown that independently of leucocytes, blood and other tissue fluids may produce morphological degeneration of bacilli. Culture experiments eliminated with certainty that phagocytic action was responsible for the decline in the number of viable organisms. Metchnikoff's position that destruction of bacilli in the living body is due solely to phagocytosis, could not be confirmed. In view of the limited number of experiments in the present study, I cannot formulate a hypothesis regarding the importance of the unique antibacterial mechanisms here observed. Further investigation will establish the importance of these mechanisms.

George Henry Falkiner Nuttall was one of the earliest Americans to have played an important part in the development of bacteriology, but he did not remain an American and did not remain a bacteriologist. He settled in England at the age of thirty-three and lived there until his death at age seventy-five, in 1937. Though he helped found serology, and with Welch discovered the first anaerobic spore-bearing organism causing gas gangrene, the greater part of his professional life was devoted to parasitology.

He received his early education in Europe but returned to the United States to study medicine at the University of California. After he obtained his M.D. degree in 1884, he spent a short period of time at Johns Hopkins and then went to Germany where after four years of study he obtained a Ph.D. His research for the Ph.D. at Göttingen under Flügge was the study of the bactericidal action of blood, as we have seen, an important early step in the development of serology. By 1890 he was back at Johns Hopkins as Welch's assistant, working in bacteriology. Welch, wishing to develop hygiene at Johns Hopkins, sent Nuttall to acquire additional training under Rubner at the Institute of Hygiene at the University of Berlin. Nuttall remained in Germany until 1899, and during this period he and Thierfelder conducted the first study on germ-free maintenance of animals.

Instead of returning to Johns Hopkins, he accepted a lectureship at Cambridge University. He began by teaching bacteriology and preventive medicine, but in 1906 he became the first Quick Professor of Biology. By this time his interests had shifted to parasitology. His combined talents for research and writing permitted him to found two important journals, the *Journal of Hygiene* in 1901, and *Parasitology* in 1908. The parasitology laboratories at Cambridge were rapidly expanded under his direction, and he eventually established the Molteno Institute for Research in this field. As a parasitologist, he made authoritative contributions on tick-borne piroplasmosis and the biology and control of ticks and lice.

Nuttall's paper on the bactericidal action of blood, even though not the first publication on this subject, attracted much attention. It was published, we may recall, the same year as Roux and Yersin announced the production of extracellular toxins by the causative agent of diphtheria. Members of Koch's school, such as Behring and coworkers, investigated the bactericidal power of blood and came to the conclusion that the presence of bactericidal substances in blood did not necessarily indicate protective immunity. From a practical point of view, members of the Koch intellectual family did not want to have recourse to attenuated strains of pathogens to produce immunity, since that would have been in contradiction to the "law" set up by the master himself on the fixity of species. "Once a pathogen, always a pathogen" could be one way of expressing this "principle," which, as important as it was for Koch's disciples, was ignored by bacteria. At that time, the end of the 1880s, Koch was busily preparing for a failure by working on tuberculin. Salvation was to come to the group as a whole in the work of Behring, who sought to neutralize toxins with antitoxins, and he, in so doing, created serotherapy.

Emil Behring was born in 1854 in a village of West Prussia. Like Koch, he was one of thirteen children of poor but purposeful parents. His father was the third generation of a line of school teachers in a position that had begun as a privilege granted by Frederick the Great to an earlier Behring for army service. The school salary was meager, comprising both cash and grain, and the position carried the right to farm a designated plot of ground: Emil assimilated the available learning so rapidly that his father, with true sense of duty, sought more advanced instruction for his son. The pastor of a neighboring town continued Emil's instruction until he entered the gymnasium in 1866. With the aid of a tuition scholarship and free room and board Behring was able to complete this course of study, but university training brought about another financial crisis. Behring wished to study medicine, but he laid this idea aside in favor of preparation for the ministry, a course that would offer free university training. Arrangements were nearly completed for matriculation in the theological division of the University of Königsberg when a friend recommended Behring to the army medical school, the Friedrich-Wilhelm Institute, the *Pépinière,* in Berlin. It was thus, in 1874, that Behring entered medical school. The army physician candidates were subjected to a rigorous yet enlightened training at the University of Berlin, which at that time was favored with a brilliant faculty. The surgeon general, Alwin von Coler, included thorough laboratory courses in the curriculum, in addition to medical studies and courses in mathematics, languages, and the humanities. The medical corps was keenly aware of preventive medicine. For example, during the Franco-Prussian War, the German army lost 297 men from smallpox compared to 23,000 for the French army. Many important bacteriologists emerged among the graduates of the Friedrich-Wilhelm Institute. Among those who have been mentioned previously were Löffler, Gaffky, Pfeiffer, and Kirschner.

Behring was graduated in 1878, an eventful year in the history of bacteriology. Behring later referred to it as the "birth-date of medical bacteriology." Koch had published his anthrax studies one year previously, and in 1878 he published his investigations on the etiology of traumatic infective diseases. Pasteur was turning his attention to animal diseases, and Sédillot had coined the word "microbes."

After two years of internship, Behring was assigned to formal military duty. For the moment he embraced the gay life of a bachelor officer with such energy that at one point he sought a change of posts to avoid the embarrassment of accumulated gambling debts. Nevertheless a more serious interest was maturing at the same time; Behring was thinking of new approaches to the study of infectious diseases. At the army laboratories in Posen, where he was stationed, he injected rabbits and dogs with iodoform with the intention of rendering the host unfavorable for the support of growth of bacteria. He published a series of reports on iodoform and its application to the treatment of wounds. He also prepared for eventual civilian placement by qualifying for the public health service as *Kreisarzt* and participating in a course on bacteriological techniques given by Falk in Wiesbaden in 1886. Falk had received training at Koch's laboratories.

His potential ability was recognized by the command staff of the Army Medical Corps, and in 1887 he was assigned to the Pharmacology Institute of the University of Bonn. He improved his knowledge of organic chemistry and continued the line of research begun at Posen. His papers from this period include one entitled "The Specific Parasiticidal and Specific Antitoxic Properties of Chemical Agents." He used anthrax as an experimental infection. In 1889 he was assigned, at his own request, to the Institute of Hygiene at the University of Berlin, Koch's laboratory. He moved with Koch, in 1891, from the University to the newly formed Institute for Infectious Diseases.

Shortly after this move, Behring reviewed an aspect of serum therapy in a paper entitled "Serum Therapy in Diphtheria and Tetanus" (*Z. Hyg. Infektionskrankh.*, **12**: 1–9, 1892). In this paper, which was not his first publication on the subject, he wrote:

> The relative importance of the cellular and humoral theories of immunity cannot be resolved by theorization. One should take the old and tried point of view, "By its fruits shall you judge." In view of my interest in the humoral concept of acquired immunity, I shall review experimental findings that support this concept.
>
> I was led to consider the blood as responsible for immunity on the basis of the bactericidal effect of the serum of the immune rat on the anthrax bacillus. . . . Nissen and I undertook the painstaking task of investigating a large number of test systems and we found that a correlation was not always present between the bactericidal power of blood and immunity. Thus the body must have mechanisms other than bactericidal components in the blood, for protection against disease-producing agents. To find these other mechanisms was our next problem.
>
> These negative correlations were obtained at the time when Roux and Yersin for diphtheria, and Kitasato for tetanus, found toxins in bacterial cultures that were active enough to account for the death of infected individuals, despite localization of these organisms at the site of infection. Further study of the properties of tetanus and diphtheria toxin showed that they were highly labile to small changes in temperature or to the action of chemicals. . . .
>
> These facts raised the possibility that a more revealing understanding

of diphtheria and tetanus might be reached by emphasizing the toxin produced by these bacteria rather than the organisms themselves. I myself was led to this point of view by my studies on the effects of iodoform. Iodoform appears to act on the paralyzing, inflammatory and pus-producing bacterial products rather than on bacteria themselves. Experimentation showed that a variety of agents were therapeutic for experimental diphtheria or tetanus without a bactericidal effect on these organisms, as for example, iodine trichloride or sodium chloroaurate, when applied to the site of localization of the bacteria very soon after infection. Despite the therapeutic effect the bacteria were not destroyed. The treated animal had acquired immunity that could be demonstrated by subsequent challenge. However, the blood of the immunized animal was not bactericidal for the specific organism. The deduction to be drawn was that the blood had acquired the property of rendering the diphtheria or tetanus toxin harmless. This was a new view of the origin of immunity and the next problem was to determine if this principle could be extended to human therapy. The high level of antitoxic activity that Kitasato and I observed for the blood of rabbits rendered immune to tetanus immediately suggested the use of such blood as a therapeutic agent for tetanus-susceptible rabbits. The results with tetanus exceeded our most optimistic expectations. The therapeutic effect could be demonstrated with serum, thus showing that the active agent was a soluble non-cellular component of blood. Further studies showed that the therapeutic activity in recipient animals was correlated with the level of immunity demonstrable in the donor animal. Satisfactory therapy could be achieved in diphtheria only when the immunity level of the donor animal was raised to a high level. Since the production of a high level of immunity was difficult to attain in some cases, we turned to animal species that were naturally immune to the disease under consideration. It was found that the blood from naturally immune species did not confer immunity to susceptible animals. I therefore infer that natural immunity depends upon the living cell complex or a soluble component of blood that loses activity when the blood is withdrawn from the body. . . .

Behring concluded his presentation as follows:

A generally acceptable explanation for natural immunity is not available. But our knowledge of artificially produced immunity has advanced to the point where we can say that in a series of infections, it is due to an acellular component of blood. With this point of view, specific effective therapy against infectious diseases is clearly indicated. The question as to whether this effect is due to a bactericidal effect or toxin neutralization or a combination of both is not decided. At present, it appears to me more important to produce therapeutic sera that can be used for treatment of disease in man. This is the objective of my research group and we shall now present our studies with diphtheria and tetanus.

Turning the page, we come to a long detailed paper by Behring and Wernicke entitled "Diphtheria Immunization and Therapy in Experimental Animals" (*Z. Hyg. Infektionskrankh.*, **12**: 10–44, 1892). Wernicke was one of Behring's classmates at the Friedrich-Wilhelm Institute. They were to remain lifelong friends. The two army physicians wrote:

> Our experiments on immunization and therapy for diphtheria in animals have led us to the point of considering human therapy possible and feasible. We first developed a method for producing a high level of immunity in rabbits and guinea pigs and have used the immune blood for highly successful therapy in susceptible animals. Seeking a larger animal for adequate amounts of blood, we have succeeded in immunizing sheep. In two sheep we have induced sufficient immunity to anticipate having a supply of blood adequate to treat several children. We have come to the conclusion that the demands of further development of our therapeutic agent for diphtheria exceed our personal means and resources. We have decided, therefore, to admit to a temporary standstill, and to stimulate the interest and efforts of other investigators by publishing the results achieved at this time.

After having noted that there was little chance that the disinfectants available at the time could be used to treat diphtheria successfully, the authors remarked that the most promising method of treatment was the injection of blood from artificially immunized animals. However:

> The next step was to develop a more convenient method for immunizing donor animals. When infected animals were immunized by local treatment with iodine trichloride, we observed that the diphtheria bacillus was not killed, but the toxin produced by the bacillus was weakened. It occurred to us that the toxin could be similarly altered by iodine trichloride outside the body, and the chemically treated culture could be used for immunization with less danger to the animal. We learned subsequently that a germ-free culture could be used as successfully as a bacillus-containing culture. The important factor in immunization is

the level of toxicity. During the past year we have attained high toxicity by incubating cultures for four months. Cultures have been filtered through paper to remove organisms and the filtrates have been preserved without loss of potency by addition of carbolic acid to 0.5% concentration. Such a pooled filtrate is highly toxic; 0.15 cc. kills grown guinea pigs in 4 days and 0.01 cc. kills quite large guinea pigs after several weeks. The toxicity was unaltered after 9 months of storage. . . .

We have observed that effective immunizing doses of toxin treated with iodine trichloride cause local and generalized reactions. In the absence of a reaction, only a slight rise in immunity is obtained and too severe a reaction leads to progressive emaciation of the test animals. The best results are obtained by allowing the iodine trichloride to react with the toxin for at least 2 days and not more than 4 weeks. Detoxification is generally completed in 36 to 48 hours. In our most recent trials we have immunized nine sheep without producing significant toxic effects. We have had less consistent success with rabbits. In one improved procedure with rabbits, toxin is introduced daily for long periods of time into the stomach. The intragastral dose is larger than the lethal subcutaneous dose. The intragastral dosing is continued and raised until there is a decline in body weight. Dosing is then discontinued and resumed when the weight returns to normal. Another procedure has given outstanding results in individual cases. Filtered toxin is dried, treated with calcium chloride solution for one hour at 77° and the toxin-calcium precipitate is used for injection. A small amount of the powder, 0.005 grams, is inserted into an abdominal skin flap. A generalized erysipelas-like inflammation is produced which subsides in 8 days. After 8 more days the animal is reinjected with toxin powder and a mild localized inflammation is now produced. Subsequently, larger amounts of dried toxin, invariably fatal to unprepared animals, can be administered safely to prepared animals. The blood of animals so immunized produces unusually good therapeutic effects. . . .

The potency of immune sera is determined as follows: Graded amounts of serum are injected intraperitoneally or subcutaneously into guinea pigs of known weights establishing a series of 0.5, 1.0, 2.0, et seq., cc. of serum per 100 gm. of body weight. One day later the guinea pigs are injected subcutaneously with 48 hour broth culture of diphtheria known to be lethal to guinea pigs in 3 to 4 days. . . . Untreated animals show a slight edema at the site of the injection which increases on the succeeding day, then changes into a dense area of infiltration and leads to formation of an ulcer. After three to four days, the animals die with symptoms of dyspnea. . . . Successfully treated animals show no symptoms. The potency of the serum is defined by the smallest amount per 100 gm. of body weight that will protect the animal from symptoms of infection. To illustrate, a dose of 0.5 cc. per 100 gm. body weight may protect the animal from death but local and generalized symptoms will still be observed while serum doses of one and two cc. per 100 gm. body weight will prevent local symptoms as well as death. We would thus say that the immune titre of the serum was 1 to 100. For many reasons we have found it desirable to preserve the serum by addition of carbolic acid to 0.5% concentration. . . . The amounts of serum to be injected for successful treatment can be calculated. One first determines the level of immunity of the donor animal. For example, if according to Ehrlich's method a guinea pig has an immunity index of 10, we must inject a guinea pig of equal size with $\frac{1}{10}$ of the entire blood volume of the donor animal to give the recipient an immunity factor of one.

To determine which fraction of the blood contained the immune substance, we tested the activity of whole blood, serum, and clot. Whole blood was less effective than the same volume of serum and the dried clot suspended in saline had only slight immunizing power. Thus the immunizing substance was in the serum.

The relationship of passive prophylaxis and therapy for the same serum was determined. Larger amounts are needed for therapy than for prophylaxis and the amount of serum needed for successful treatment is increased with delay in treatment. In infections which were fatal to guinea pigs in 3 to 4 days, one and one-half to two times as much serum was required for treatment immediately after infection as compared with the prophylactic dose. Eight hours after infection, three-fold the prophylactic dose was required for successful treatment and after 24 to 36 hours, the dose increased 8-fold. We do not know the exact duration of protection following a prophylactic serum injection, but it is at least several weeks. . . .

The intraperitoneal and subcutaneous routes for injection of serum were compared. For treatment before, or immediately after infection, both routes were equally effective, but when treatment was delayed until symptoms had ensued, the intraperitoneal route was more effective than the subcutaneous route.

This paper continued with the presentation of the actual results obtained in the various experimental animals used. The following paragraph sums up the optimistic outlook of the investigators:

We require only an exercise of patience to obtain increasingly effective sera, which finally will permit treatment of larger animals with feasible doses. Continued reinjection of sheep with culture or toxin produces successive increases in serum potency and we have not 'yet attained the maximal limit of potency inherent in this method.

One more paper, this time coauthored by Kitasato, concluded this series. It dealt with the results obtained in the serotherapy of experimental tetanus.

Such publications precipitated a series of priority disputes with Fränkel, a former assistant of Koch's and at this time professor of hygiene at Marburg, Ogata of the University of Tokyo, and Emmerich of the University of Munich. Behring enjoyed such disputes and found them stimulating. In this instance Behring was on the firm ground in impeaching the claims of the disputants. The French referred to this situation as a typical *Querelle d'Allemand*.

The stunning demonstration of the principle of antitoxin therapy was not easily translated into clinical practice. Koch and Althoff, one of the high officials in the ministry for culture and soon to become Behring's friend and protector, suggested collaboration with the German chemical industry that was now developing an interest in pharmaceutical products. At this point in 1892, the Lucius and Bruning firm in Höchst, later to be known as Farbwerke Höchst, approached Behring, and an agreement was reached regarding cooperative effort for development of the antitoxin. One additional investigator, Paul Ehrlich, was drawn into the cooperative effort. He had worked on the development of immunity to ricin, a toxic protein from castor seed, and, working with Wassermann on diphtheria, he had produced antitoxins in goats that were manyfold more potent than Behring and Wernicke's sheep antitoxins. Ehrlich also solved the perplexing problems in developing reliable methods for testing the potency of sera, and, eventually, he became director of the Prussian Institute for Experimental Therapy, which was responsible for certifying the safety and potency of antisera. Ehrlich's procedures and units for evaluation of potency of antitoxin are still the basis for standardization of diphtheria antitoxin.

Ehrlich did not fare well in the arrangement. Behring and Höchst convinced Ehrlich to surrender his financial rights, and in turn Behring was to prevail upon his contacts in the Ministry of Culture to appoint Ehrlich to an appropriate salaried post. At the time, Ehrlich was working in Koch's Institute as an unpaid guest investigator. Behring was unable to keep his promise, and Ehrlich lost a sizable amount in royalties.

The first trials of antitoxin in man were started late in 1892. Among the early clinical investigators, Hermann Kossel, one of Koch's assistants working in Koch's wards at the Charité Hospital, contributed the critical clinical study favoring the antitoxin. Kossel's report on 233 cases in May, 1894, has remained a classic for evaluation of therapeutic agents. He showed that the incidence of survival was high if antitoxin treatment was initiated early. In Kossel's series there were no deaths among cases treated on the first day of the disease, and the incidence of deaths reached 51 per cent for cases treated on the seventh to fourteenth day of the diseases. The first available sera were still so low in potency, apparently 1 to 2 units per milliliter, that doses of 50 milliliter were required. Later antisera contained hundreds of units per milliliter. Diphtheria was the major order of the day at the International Congress of Hygiene held in Budapest in 1894. But Behring was not even present, and the star report on diphtheria was presented by Roux, Martin, and Chaillou. Roux and Martin had succeeded in producing potent antitoxin in the horse, and this serum was submitted to testing by Chaillou in February, 1894. The French group reported a decline in mortality from 52 per cent to 25 per cent. Roux was very careful to give full and proper credit to Behring, but the French newspapers credited the discovery to Roux. Roux quickly attempted to correct this error, but his efforts had little effect on the enthusiasm of some sections of the French press. As long as three years later, *Figaro*, the influential conservative French paper, was gently chiding Roux for his excessive modesty and continuing to recognize Roux as the discoverer of diphtheria antitoxin.

The discovery of diphtheria antitoxin rapidly brought Behring international fame and fired the public at large with a helping enthusiasm for bacteriology. In Germany, public subscriptions and allotment of government funds were made available for purchase of serum from German manufacturers for free distribution. France adopted the antiserum with a display that the reserved *British Medical Journal* referred to as "theatrical." Donations to the Pasteur Institute provided for no less than 136 horses for antitoxin production. The provincial French cities allotted their donations for establishment of local antitoxin production, but such hasty ventures were slowly abandoned, and the task fell eventually to the Pasteur Institute alone. Austria established a government production laboratory in short order. Great Britain and the United States moved more slowly, but funds were provided to make the antiserum available to all.

Students from many countries came to Koch's Institute for Infectious Diseases with the specific intent of working with Behring rather than with Koch. The assistant's fame overshadowed the fame of the director, temporarily blighted by the tuberculin fiasco. Friction developed between Koch and

Behring. Althoff, the Minister of Culture and the chief administrator of the Prussian University System, was faced with the problem of providing a proper position for Behring. He had become acquainted with Behring in 1892 and developed a high regard for his talents, but in placing Behring at a university, he encountered a great deal of resistance from the various faculties. Behring's aggressive style was well known, and opinions on his ability as a teacher, based on his teaching duties at Berlin, were divided. With considerable difficulty Althoff was able to place Behring as acting professor of hygiene at the University of Halle in 1894. Approximately twenty students registered for Behring's course, but before long there were six to eight regular attendants. He lacked knowledge in many fields, and he could not simplify the subject matter for the undergraduate student. Behring went on a deliberately protracted vacation to the Riviera, the Near East, and Italy and asked of Althoff to be transferred to the chair of hygiene at Marburg. Despite Behring's concern with career problems, he had not lost his touch for gambling. A lucky streak at Monte Carlo provided for the purchase of a villa in Capri. After a variety of attempts at a compromise for another position, Althoff appointed Behring to the chair of hygiene at Marburg, over the objections of the faculty. Behring returned to Germany as soon as the appointment was made, and in a letter to Althoff in May, 1895, he wrote that he was preparing diligently for his course, feeling now that he would be able to overcome his inadequacies as a teacher.

Althoff continued to advance Behring's position, and before the end of 1895 Behring was honored with the appointment of *Geheimmedizinalrat*. One year later Behring was admitted to the nobility, and he was now Von Behring. A further official honor raised him to *Wirklicher-Geheimmedizinalrat*, and he was now *Exzellenz* von Behring. The honors from abroad included the French Legion of Honor and the Nobel Prize of 1901.

Once settled at Marburg by 1896 and in comfortable financial circumstances thanks to his contracts with Höchst, Von Behring married twenty-year-old Elsie Spinola, the daughter of the director of the Charité Hospital. On his honeymoon, which could be protracted because faithful Wernicke could carry on for him at the University, he acquired a large house in Capri in place of the small villa he had previously bought. Capri was his favorite retreat. In 1898 he acquired an elaborate villa in Marburg that served frequently as a gathering point for the many foreign scientists in Behring's circle of acquaintances. His hospitality left little room for improvement.

Behring's circle of friends was cosmopolitan; he reserved his fighting moods for his countrymen. Of the German bacteriologists he was the one with the closest ties to foreign circles, especially the Pasteur Institute group. Pasteur was one of his great heroes; in his study, a photograph of Pasteur was hung along with renderings of Frederick the Great, Napoleon, and Bismarck. He visited France often, and was always well received at the Pasteur Institute. With Metchnikoff he maintained a continuing friendship that began in 1888. Both Metchnikoff and Roux stood as godfathers for two of Von Behring's six sons. The friendship with Metchnikoff was unique. The two men were at opposite poles in explaining basic aspects of immunity, defending their positions with equal and effective vehemence but enjoying the stimulus of intellectual debate.

The two men were also opposite in character. Behring was a believer in the hero cult and enjoyed the material symbols of success. The saintly Metchnikoff, as we have seen, had no interest in material success. Behring and Metchnikoff showed great courage and faith, and both shared a belief in science that transcended national boundaries.

Behring built a strong Institute at Marburg. The original quarters assigned to him comprised two floors in an old surgical clinic. He soon established a separate, well-equipped unit for serum therapy that included facilities for large animals. As the scope of his studies extended, he became dissatisfied with the slow and conservative style of Farbwerke Höchst. By 1904 he established an independent commercial laboratory known as the Behring Werke. The Behring Werke was a successful venture and is still an important producer of biological products.

With diphtheria and tetanus antitoxin realized, Behring turned to solving the riddle of tuberculosis therapy, a field in which Koch already had met with an initial failure. In the discussions in scientific circles regarding Koch's early work on tuberculosis therapy, Behring defended Koch against the personal attacks that came especially from Virchow and his followers. Behring believed that Koch's tuberculin was a significant start toward the solution of the problem of tuberculosis. However, Behring and Koch were involved in frequent disputes during the course of their competitive studies on tuberculosis.

Behring's first approach, as announced in 1895, was to produce an antitoxin that might act in the same way as diphtheria or tetanus antitoxin. To produce the antitoxin, the toxin had first to be demonstrated. By 1898 he had described a procedure for toxin isolation. He defatted the fragmented organisms and extracted these preparations with glycerin at 150°C in the absence of air. The resultant toxin was named tuberculin acid.

Von Behring offered a patent, obtained in 1899, on this subject to the

Ministry of Culture with the idea that the Ministry would support the cost of the extensive testing in cows, horses, goats, and small laboratory animals. This offer was declined, and the testing was left as Behring's responsibility. The testing apparently was not favorable, and Behring's interest in tuberculous toxin and antitoxin faded for the time without a formal final evaluation. The second line of attack was discussed in his Nobel Prize address; he suggested immunization in the Jennerian sense as used by Pasteur in anthrax. Von Behring had observed that tubercle bacilli freshly isolated from man were virulent for cows but lost this virulence when maintained in laboratory culture for long periods of time. Passed through goats, the avirulent strains would regain their virulence. Similarly, cultures of bovine origin lost virulence for cows when cultivated in the laboratory. He proposed to inoculate young cows first with avirulent culture of human origin, then with virulent culture of bovine origin, and finally with virulent goat-passed human culture. In 1902 he reported that laboratory-maintained human strains had been injected safely into adult cows without producing tuberculosis, that for young cows doses no greater than 50 milligrams could be used safely, and that such animals had acquired increased resistance to both human and bovine cultures. Such evidence of cross resistance led him to oppose the separation of tubercle bacilli into types. Whereas Koch did not consider bovine tuberculosis important in the human disease, Behring thought the cow the major source of human infection through contamination of meat, milk, and butter. Von Behring considered the intestinal tract of adult humans resistant to such mode of infection, but he believed that infants easily acquired infection by the gastrointestinal route owing to the high permeability of the infant intestinal epithelium.

In 1903, at the Vienna Association for Internal Medicine Von Behring suggested a new approach to prophylaxis in man that utilized artificially immunized cows. He stated, "I am at this moment as convinced of the practical value of my tuberculosis immunization [referring to vaccination of cows] as I was of the value of my diphtheria experiments as I discussed them here in Vienna eight years ago." Based on experimental evidence from cows and autopsy analysis of human infants by the pathologist Heller, Von Behring believed that most primary human infections were acquired at infancy. The organisms passed through the intestinal wall and lodged in the mesenteric lymph glands. At a later time dissemination from this primary focus could lead to the pulmonary form. Rendering the milk supply free of tubercle bacilli would eliminate the disease in man. Extending this point of view, later in the same year, he introduced the concept of immune milk. He believed that immunized lactating cows transferred immunity to calves through milk; might not therefore such milk be used to prevent tuberculosis in infants? "I have every basis for the hope, that we are on the right road, to have in immune milk a weapon against human tuberculosis. . . ." Koch's position was diametrically opposed to Von Behring's and in 1902 Koch stated, "At present we can only say that the harmful effect of milk and milk products from tuberculous cows has not been demonstrated," and, probably referring to Von Behring, "The fight against tuberculosis if it is to be successful, should not be diverted to false paths."

The practical development of "immune milk" became a major objective of Von Behring's group. The bacteria in milk would certainly destroy antibodies, and similarly heat treatment applied to destroy bacteria would also destroy antibodies. Over a period of approximately five years a variety of procedures were devised to achieve a practical "immune milk." They consisted in adding various compounds to the milk, such as low concentrations of formaldehyde. The evaluation of "immune milk" was no simple matter, however, and it drifted into oblivion.

The last category of agents developed by Von Behring for treatment of tuberculosis was vaccine prepared from components of the tubercle bacillus. In the Von Behring manner he made optimistic announcements on the last day of the Third International Conference on Tuberculosis in Paris in 1905. He promptly became lost in a multitude of preparations, none of which emerged as meaningful agents. Nevertheless, as late as 1912 in a letter to Metchnikoff, he wrote of an experiment with sheep wherein he believed he had a serotherapeutic agent that was both preventive and curative for tuberculosis. This appears to have been his last statement on tuberculosis.

From 1907 onwards Von Behring's health began to fail. In late November of that year he entered a sanatorium in Munich to rest and recover from depression and exhaustion. He was also suffering from severe foot pain that could not be attributed to an organic cause. After a period of nine months he was able to return to Marburg. Until 1915 he worked on purifying and improving diphtheria and tetanus antitoxins. Tetanus antitoxin proved invaluable as a preventive treatment in war wounds, and for this agent he was once again decorated by the German government.

After the many years of trial and failure in attempting the conquest of tuberculosis, and now continually ailing, he achieved yet another triumph in the conquest of diphtheria. Through the years Von Behring had tried sporadically to develop a prophylactic agent for diphtheria. Efforts to adequately detoxify the toxin with chemical treatment and preserve immunizing activity had failed. In 1913, he announced before the Congress for Internal Medicine in Wiesbaden that he had developed a prophylactic agent for diphtheria; this was an underneutralized mixture of toxin and antitoxin. It had been proven experimentally and clinically at Marburg and was now ready for controlled clinical testing elsewhere. This time the promise was realized. The immunizing value of the toxin-antitoxin mixture was confirmed. It was a major advance in immunoprophylaxis.

Five months after the presentation at Wiesbaden he suffered a fall and fractured his left femur. The fracture did not heal properly and was followed by arthritis deformans. A further sharp decline in health followed the surgical drainage of an abscess in September, 1916. He then remained bedridden, and pneumonia brought death on March 31, 1917.

Behring, like Koch, failed in his fight against tuberculosis, and in spite of the wishes of the Minister of Culture, the prevention or the effective treatment of this scourge was not to be found by Germans. Instead, it was in France that an effective vaccine for tuberculosis was developed by the methods of attenuation so dear to Pasteur and disdained by Koch.

In the development of this vaccine, called BCG (*bacille de Calmette et Guérin*), Albert Calmette played a leading role. Born in Nice in 1863, he was left motherless at the age of two. His father was an administrator who traveled from city to city following the demands of his civil career. Fortunately, when his father remarried, he chose a sympathetic and educated woman who did all in her power to compensate for the missing mother in the lives of Albert and his two brothers.

When Albert was a teenager and a student at the *lycée*, he decided on a carrer in the navy. He went to Brest, where he could be prepared for naval school. His dream of sailing the high seas was interrupted by an attack of typhoid fever, which left him in such a poor state of health that he failed the medical examination. Nevertheless, the thread of his exotic dream was not altogether broken, and after completing his secondary education, he was able to return to Brest for the first two years of training as a naval physician. At the naval hospital of Brest, one of the most modern of the day, time passed swiftly for the serious and studious Calmette, and it was soon time for him to depart on the sea cruise that he had desired for so long. This was during the year 1882. The medical profession was trying to evaluate the practical value of the principles of Pasteur, the pathology of Koch, and the surgery of Lister. In one camp were the believers in microbes; in the other, the old guard ready to die in its routine. But for Albert Calmette there were other horizons to be explored. He was assigned to the French squadrons engaged in fighting in China, when in 1884 one of the guns on his cruiser exploded, and the ship had to sail to Saigon for repairs. It was his first contact with this enchanting city, which later played an important part in his life.

In 1886, Calmette returned to France to complete his medical studies. His doctoral thesis dealt with human filarial diseases. Soon after graduation, he was again on the seas, this time heading for Gabon, where he was to face all the problems of a colonial physician. Not only were diseases numerous and the climate debilitating, but worst of all, Calmette came to realize that colonial life did not attract the best human material that France had to offer. The problem was both medical and social. He left Gabon without regret in 1887 to return briefly to France.

He was a married man when he left for his next assignment, the islands of Saint-Pierre et Miquelon. These patches of rock, the last remnants of New France, offered respite from the fog and the sea to wooden sailboats that spent months away from the coast of France, reaping a harvest of codfish on the Banks of Newfoundland. Hospital work and private practice kept Calmette busy, but so did salted codfish. Sometimes, as salted cod was prepared, a red color developed. Calmette showed that it was due to the growth of a colorful but harmless *Micrococcus* that was present in some salt batches and not in others. The taint could be avoided by using only uncontaminated salt. This first contact with microbiological research affirmed his vocation and he decided to take "Roux's Course" at the Pasteur Institute. The most expedient method of getting away from the tiny desolated islands was to request a transfer to the newly formed corps of colonial surgeons.

Back in France, in 1890, Calmette found Roux helpful in providing him laboratory space. At this time he met the shy but adventurous Yersin. One day Calmette was asked to see Pasteur. The great man did not have a reputation for charm, and it was with a certain anxiety that Calmette was ushered into his presence. However, Pasteur had only good intentions. He had received a request to provide a director to organize an institute for the manufacture of vaccines in Indochina. Calmette, with his professional skill and colonial experience, appeared to Pasteur to be uniquely fit for the job. Calmette accepted immediately.

In Indochina, Calmette found Yersin in debilitated condition from malaria acquired during an expedition into the interior. In order to marry Yersin's need for adventure to official respectability, Calmette convinced him to join the corps of colonial surgeons.

Diseases were numerous in Indochina and local medicine relied heavily on such remedies as horse urine, amulets, and incantations. Soon Calmette had a modest organization producing vaccines. He prepared smallpox vaccine, using a local species of water buffalo. For rabies, the desiccated section of rabbit medulla were preserved in glycerol. The institute's biological products were adapted to survive tropical conditions, and soon it was supplying other colonies.

In 1891, Calmette started the first of his main lines of research, the study of snake venom, and the institute soon had extensive snake pits. Following the road traced by bacteriologists, he immunized animals against venoms, and he was able to state:

"The serum of an animal immunized against venoms can react with these venoms *in vitro*, and is both preventive and therapeutic, exactly like the serum of animals immunized against diphtheria and tetanus. . . . Such sera are highly effective if used shortly after exposure." From venoms, his interest broadened to include poisons and drugs of natural origin. Often toxicology and microbiology were combined, as when he investigated the fermentation of opium, in which molds play an important role, and the fermentation of local beverages made from rice.

After two years in Indochina, Calmette became affected with dysentery and was sent back to France. For a while he had a dual appointment. Mornings, he worked in Roux's laboratories on problems connected with venoms, and afternoons he was at the Ministry of Colonies in Paris. The bureaucratic duties did not enthuse Calmette. Fortunately, in 1894, Pasteur received a request from the officials of Lille, the city where he himself had started his microbiological career, asking that he establish there a branch of the Pasteur Institute. For the second time, Pasteur called upon Calmette to assume the responsibility of founding an institute.

By the time the cornerstone of the new institute was laid, two months after the death of Pasteur, Calmette, using temporarily rented quarters, had already been producing enough smallpox and rabies vaccine to supply the northern part of France. Soon he felt the need for an assistant competent in veterinary medicine. Nocard was consulted and sent his own personal assistant, Camille Guérin. Thus began, in 1897, a harmonious cooperation, which lasted until Calmette's death.

With the foundation of the new Institute in Lille, Calmette started his second major line of investigation. He did not begin in the laboratory but in the field. At that time, in the northern part of France, 3 persons out of every 1,000 were dying of tuberculosis. Since Calmette did not think that it would be possible to find the necessary money to build an adequate number of sanatoria, he suggested the creation of special dispensaries for detection of tuberculosis and the education of the population on the prevention of the disease and the care of the sick. The idea was received with enthusiasm, and the first few years of the century saw the opening of many such institutions. Laboratory studies were also started, and, in 1924, Calmette and Guérin published a paper entitled "Vaccination of Bovines against Tuberculosis . . ." (*Ann. inst. Pasteur,* **38**: 371–398), in which they thus reviewed their work: [1]

In a series of papers that we have published during the past 20 years, we have established certain facts that can be summarized as follows. . . .

We have shown (1913–1920) that resistance to a tuberculous infection or reinfection depended on the presence in the body of *living* bacilli be they *virulent* or *avirulent,* and that one could produce this resistance artificially in young bovines by inoculating them with our "biled" bovine bacillus, which is *non virulent* and *non tuberculogenic* but which still has a certain toxicity and *induces the formation of antibodies.*

This "biled" bacillus . . . we will recall, emerged from a very virulent bovine strain called *"milk of Nocard"* which had been transferred about every two weeks since January 1908 on potato cooked in bovine bile containing 5% glycerol.

Our aim, in this series of cultures, was to *modify the bacillus* . . . in the hope that the alkaline medium, rich in lipids . . . would perform changes of the bacilli that would make them more easily digestible by phagocytosis. . . .

After four years, our bacillus was no longer virulent for bovines but was still pathogenic for horses.

After thirteen years of successive transfers on "biled" potato (230 transfers as of January 5, 1921), *it was completely avirulent, even at high doses, for all animal species. It did not induce the formation of tubercles by intravenous, intraperitoneal or subcutaneous inoculation or by ingestion.*

We then maintained this organism on ordinary glycerinated potato, in the hope of keeping its properties constant. From the very first transfer we noted the return of the typical appearance and odor of cultures of tubercle bacilli, even though the culture was somewhat less dry and more easy to spread than ordinary tubercle bacilli. The culture also started to produce tuberculin, a property that it seemed to have lost on the bile-containing media. A glycerinated concentrate of this tuberculin

[1] Translated by permission of Professor Jacques Tréfouël, Director of the Pasteur Institute of Paris.

behaved characteristically in tuberculous bovines. . . . *by successive reinoculations, we have never succeeded to restore tuberculogenic properties to this organism.*

We have thus in our hand a true *avirulent strain of bovine bacillus.* This bacillus, even though toxic for the tuberculous animal and a producer of active tuberculin, has lost all ability to induce the formation of tubercles. . . . It is completely harmless to all animal species, including the chimpanzee and man, even after intravenous injection of 40 million viable units (1 milligram of culture). . . .

It is possible to vaccinate young bovines against infection with virulent bacilli. Animals less than two weeks of age were injected subcutaneously with a large dose (50 to 100 milligrams) of our modified bacillus that we will now designate with the initials BCG. . . .

Fifteen months after vaccination, such animals were still resistant to the intravenous injection of a virulent strain capable of killing all the controls of the same age in less than two months. . . .

It was thus that the story of BCG was presented to the world in 1924, the culmination of twenty years of experimental work, which had been interrupted by the First World War. Calmette had remained in occupied Lille. He had certain unpleasant contacts with the Germans that were deplored by Pfeiffer. Most serious, Madame Calmette was taken as a hostage and sent to a reprisal camp in Hanover. In October, 1918, British troops finally broke the German front close to Lille and liberated the city. During the war, Metchnikoff had died, and Calmette had been named, *in absentia,* associate director of the Pasteur Institute of Paris.

Calmette temporarily left Guérin in Lille, to continue to study tuberculosis in bovines, and went to Paris to fulfill his new duties as Roux's associate. The contrast between the two men was striking. Roux was satisfied with a little nook in the hospital as his living quarters and could not understand why any investigator would want to have a decent salary. Calmette was more worldly. He could appreciate gastronomic, artistic, and literary achievements. With diplomacy, Calmette managed to ameliorate the financial situation of the young workers of the Institute, and with ingenuity he found room for his own operation in the Institute, which was already full as an egg. Calmette's administrative duties included the supervision of the Institutes located outside of France. In this capacity, he was able to give much help to his old friend Yersin.

In 1921, the first child was vaccinated with BCG. This was done at the request of a physician who had delivered a baby of a tuberculous mother living in an infected family. Calmette was certain that his BCG strain was harmless, and the child received three doses of 6 milligrams each orally. Encouraged by this unplanned trial, Calmette and his collaborators proceeded more systematically. In 1924, the same year that Calmette and Guérin published the paper that we have quoted, Calmette with numerous medical collaborators reviewed the results obtained from the vaccination of children since 1921. Of the 217 infants that had been vaccinated with the permission of their parents, 9 had died of nontuberculous diseases, 169 were healthy, even when living in highly contaminated environments, and the rest had been lost to survey. There was no indication that the vaccine was dangerous. It was then tried in many countries including Portugal, Uruguay, Argentina, Paraguay, Ecuador, Bolivia, Cuba, Poland, and Czechoslovakia. By 1925, 2,070 infants had been vaccinated in France, Holland, and England. The observed mortality was 0.5 per cent as compared with an expected mortality of 25 per cent. As the years passed, efficacy data emerged from additional countries, and the statistics became more significant. All had gone well; there was no indication of any serious side effects or any tendency for the strain to revert to its original virulence. Thus in 1928 a new building was started at the Pasteur Institute to house the new *Service du BCG*.

In July, 1929, Calmette received a request for a culture of BCG from Dr. Alstaedt, director of the Public Health Laboratories of Lübeck. The culture was sent, and Dr. Deycke, head of the bacteriological laboratory of the city hospital, prepared the vaccine, which was then put in use toward the end of the year. A few weeks later, out of 252 children vaccinated, 71 were dead from a serious tuberculous infection caused by the organism used as vaccine. The reaction of the public was violent; Calmette was assailed in the press and by letters, telegrams, songs, and cartoons. The specialists in tuberculosis who had used his vaccine without any trouble came to his defense. None believed that there was any indication that the BCG strain could revert to a virulent state, not, at least, if the method of culture advocated by Calmette and Guérin was followed. It was in Germany that Calmette found his best supporters. With great fairness, the German experts studied all the available facts and turned their opinion over to the tribunal of Lübeck. The BCG strain had been contaminated with a virulent strain in the laboratories of Lübeck, and, in 1932, Dr. Deycke and Dr. Alstaedt were condemned to prison terms of twenty-four and fifteen months respectively. Calmette was exonerated, but the affair had taken its toll, and he died the next year a few days before Roux.

BCG vaccination had been set back temporarily by the disaster of Lübeck.

Its use continued to give significant protection, and Guérin continued to supervise its preparation at the Pasteur Institute. Trained by Nocard, he was a meticulous experimenter, and the BCG strain could not have been entrusted to better hands.

Returning to the story of serology, we recall that the work of Nuttall, especially, had focussed attention on the bactericidal action of soluble components in blood. Richard Pfeiffer, a military physician working in Koch's laboratories, who had already distinguished himself by isolating and describing *Hemophilus influenzae* in 1892, was further emphasizing the importance of humoral factors in immunity by publishing a paper ". . . On the Nature of Immunity to Cholera and Specific Bactericidal Processes" (*Z. Hyg. Infektionskrankh.*, **18:** 1–16, 1894). We note in the following excerpts that, in the experimental immunization of guinea pigs, he was using dead bacteria, as Salmon and Smith had done previously with the *Salmonella:*

> A guinea pig was injected intraperitoneally with cholera organisms killed by chloroform vapors. Seven days later the guinea pig was injected intraperitoneally with ⅛ of a loopful of living cholera organisms. Samples of peritoneal exudate were then removed periodically and examined in hanging drop and stained smear preparations. After one hour, many of the vibrios were non-motile and there were many granules representing degenerated forms of vibrios. The leucocytes contained large numbers of granules and vibrios. The process of vibrio disintegration and phagocytic activity was complete after 4–5 hours and the animal survived after symptoms of intoxication.
>
> Another guinea pig was injected with a preventive dose of killed cholera organisms and reinjected with two loopfuls of organisms 6 and 15 days later. Samples of peritoneal exudate were examined periodically after the second injection of living organisms. Immediately after injection, all vibrios ceased moving. After 10 minutes there were many granules and swollen vibrios, but almost no leucocytes. After 20 minutes all vibrios had disappeared and there were many granules. Approximately 95% of the granules were extracellular and 5% were in the leucocytes. Before my eyes, the cholera vibrios were lysed without participation of phagocytes. . . .
>
> I obtained the same results by immunizing the guinea pig passively. The following experiment was most striking. One loopful of 20 hour culture was mixed with 1 cc. of 50 to 100 fold dilution of immune serum in broth. The mixture was injected intraperitoneally into a 200 gram guinea pig. Samples of peritoneal exudate were examined at 10

minute intervals. After 20 minutes nearly all vibrios had disintegrated into granules and after 30 minutes the process of disintegration was complete, without participation of leucocytes. As a control, I placed a sample of the same mixture of bacteria and diluted immune serum in the incubator to determine whether the bactericidal action occurs outside the animal body. Contrary to my expectation, the vibrios multiplied rapidly. The animal body plays a significant and active role. The peritoneal cavity is not a mere receptacle, but reacts to the stimulus of the vibrios under influence of the immune substances in serum to elicit the bactericidal action. . . .

After having demonstrated the specificity of the lytic reaction, and having shown that the potency of an immune serum could be evaluated by serial dilution of serum, Pfeiffer concluded:

The reaction wherein cholera vibrios mixed with a small amount of immune serum and injected into the peritoneal cavity of guinea pigs are lysed rapidly, can be used to identify the cholera vibrio in difficult situations as with isolates of vibrios from water. I have already tested many such cultures. The typical cholera culture, regardless of origin, is lysed after 30 minutes of contact with serum in the peritoneal cavity of the guinea pig. Many other types of vibrios so tested, were not lysed. . . .

This method of identifying bacteria was obviously not popular with guinea pigs and those who had to buy them. More economical serological methods were described that could be carried out entirely in vitro. A publication of importance in this field was the fruit of the cogitation of the Austrian bacteriologist Max Gruber and an Englishman, Herbert E. Durham, who worked with him in Vienna. It was entitled "A New Method for the Rapid Identification of the Cholera Vibrio and the Typhoid Bacillus" (*Münch. med. Wochschr.*, **13**: 1–2, 1896) and read as follows:

When cells from agar cultures of the cholera vibrio or the typhoid bacillus are mixed with the corresponding immune serum or peritoneal lymph, the organisms clump together in masses and motility ceases. This effect is closely associated with the protective action of the serum. We have named the specific substance in the immune serum producing this effect "Klumper" or Agglutinin.

Highly active immune sera produce the agglomerative effect in astonishingly high dilution, although high concentrations of serum should be used if a rapid and complete effect is desired. With highly potent sera, less than 1% concentrations of serum will produce complete agglomeration and immobilization within a short time. Microscopic examination is not necessary, since the formation of flakes in the originally uniformly turbid suspension, followed by settling and clarification of the supernatant can be seen with the naked eye.

Extensive experiments have shown that the reaction applies to all cholera vibrios and typhoid bacilli. Twenty strains of typhoid bacilli and twenty strains of cholera vibrios of widely varying origin mixed with various corresponding immune sera, give this reaction.

How do these immune sera react against other types of bacteria? The most effective cholera serum does not agglutinate *V. proteus, V. metchnikovi, V. danubicus,* Brink's vibrio, etc. Similarly, highly potent typhoid serum is ineffective against many types of coli-bacteria. There is a corresponding absence of protection with these immune sera in animals infected with non-sensitive bacteria. However, the activity of the immune sera is not sharply specific. Cholera serum may more or less effectively agglutinate vibrios that are not apparently derived from the cholera vibrio, as the Seine-Versailles vibrios or *V. berolinensis.* In all cases where the serum is capable of agglutinating the organism in question, it is protective in the animal body. . . .

According to more recent studies by one of us (G), the reaction is performed most rapidly as follows:

A guinea pig is immunized by repeated intraperitoneal injection of killed culture; within 4–6 weeks a high level of immunity is achieved. The immunity must be raised to a level where a given volume of peritoneal lymph or blood serum will completely immobilize and agglutinate the specific organisms in an equal volume of bacterial suspension at room temperature within one minute. A loopful of serum is mixed with a loopful of homogenous bacterial suspension on a cover slip and the preparation is inverted on a hollow ground slide. If the reaction is questionable the preparation may be examined after ¼ to ½ hour in the incubator. A negative reaction as with cholera permits the highly certain conclusion that the organism is not a cholera vibrio but this is not the case with a positive reaction. The presumptive positive reaction must be confirmed by further study of pure cultures. . . .

Gruber continued his career in Munich. Durham, after having introduced the use of tubes for the collection of gases to bacterial technology, went to Brazil to study yellow fever and to Malaya to investigate beriberi. During his travels he attracted the attention of entomologists by his studies on the insecticidal properties of *Derris elliptica*. He eventually applied his talents to the growth of apples and their fermentation, an endeavor that kept him busy from 1905 to 1935.

The same year that Gruber and Durham stressed the diagnostic value of the agglutination reaction, Fernand Widal, the Algerian-born son of a French army surgeon, described the agglutination reaction for typhoid fever. This reaction, or test, still bears his name. He was a physician and did not limit himself to bacteriology. In 1907 he discovered that the red blood cells of patients suffering from familial jaundice were unusually fragile, and he demonstrated at about the same time the role of sodium chloride in edema. He was a great teacher, who expressed himself with unusual flair.

Serology was further advanced by another Austrian scientist, Rudolf Kraus, who, using a cell-free system, discovered the phenomenon of precipitation by immune sera. He published his data in a paper entitled "Specific Reactions in Cell-free Filtrates from Cholera, Typhoid and Plague Liquid Cultures in Homologous Serum" (*Wien. klin. Wochschr.*, no. 32, 736–738, 1897). As we can see from the following quotations, Widal and Buchner's work led Kraus to his discovery:

> Agglutination had been considered a function of living bacilli, but Widal demonstrated that dead bacilli could be agglutinated. Typhoid bacilli killed at 56°C. were agglutinated as readily as living bacilli, but bacilli killed at temperatures above 56°C. were not agglutinated.
>
> Widal's observations and E. Buchner's discovery of fermentation with zymase form the starting point for the following observations. The thought occurred to me that solutions of lysed bacterial cells could, like living cultures, react with specific serum. Cholera broth cultures were filtered through bacterial filters (Pukal) and the filtrates tested for sterility by incubation at 37°C. Sterile filtrates were mixed with varying quantities of sterile cholera serum and incubated at 37°C. Within 24 hours the filtrates which were mixed with specific sera became cloudy, floccules were formed which settled to the bottom of the tubes and the liquid became clear. The density of the clouding and the sediment varied. Frequently, large volumes of filtrate mixed with large volumes of serum gave no reaction. The reason for the irregularity of the reaction could not be determined. The settled precipitates were greyish white to brownish; when shaken, the broth became turbid but the precipitates resettled rapidly. A chemical study of the precipitate by Dr. Freund showed that the precipitate consisted of two proteins. . . .
>
> In order to demonstrate the specificity of the precipitates, control tests were performed by mixing filtrates previously shown to precipitate with specific serum, with heterologous serum. To cholera filtrate were added typhoid or coli goat serum, normal horse serum, diphtheria, cholera or

streptococcal horse serum and normal human serum. None of these sera produced a precipitate with cholera filtrate. This reaction was as specific as agglutination with living or dead cultures. To determine whether or not the reacting components in the filtrates were breakdown products of the bacterial cells, cholera cells were mixed with ground glass and subjected to a pressure of 300 atmospheres. The pressed masses were diluted with alkaline broth and passed through bacterial filters. These sterile filtrates were mixed with cholera serum. After 24 hours at 37°C turbidity and sedimentation were observed. The same result could be obtained by extracting dried cells from agar cultures with alkaline broth. The precipitated components may thus be considered as components of the bacterial cells. . . .

Kraus continued his paper with a demonstration that similar results could be obtained with the organisms responsible for typhoid and plague. He also showed that in the case of diphtheria, cell-free toxins were not precipitated by the specific serum.

Kraus was active in the field of serology and directed institutes in this field in Buenos Aires, Sao Paulo, and, finally, back home in Vienna.

All this work was carefully followed by Metchnikoff, who must have resented any advance in the field of immunology that did not involve phagocytes. Ironically, it was from his own laboratory, at the Pasteur Institute, that emanated some of the most brilliant work in the humoral aspects of immunology. It all started in 1894 when Metchnikoff welcomed a bright young Belgian scientist, Jules Bordet, who became a leading immunologist of his time. The son of a secondary school teacher, he was born in a small town located a few miles from Brussels. In 1870, Jules attended the Athenée Royal Secondary School in the capital. His scientific vocation manifested itself at a tender age in the form of chemical experiments performed in the family attic. It was with relief that Jules's parents saw him desert chemistry for the less explosive field of medicine. He entered the University of Brussels at the precocious age of sixteen, and at the age of twenty-two, he had completed his medical studies. The same year, 1892, he published a paper in the *Annals of the Pasteur Institute* in which he demonstrated that serial passages of *Vibrio metchnikovi* (isolated and described by Gamalea) through vaccinated animals led to the selection of cells that were resistant to phagocytosis and were more virulent than the original population. His early interest in bacteriology had certainly been influenced by his brother, Charles, who was at the time investigating chemotaxis in leucocytes. This and a bursary from the Belgian government led Jules Bordet to the very active laboratory of Metchnikoff, at the Pasteur Institute, in Paris. The young star did not wait long to demonstrate again his intellectual greatness, and in 1895 he described in vitro the action of what was to be called "complement" by Ehrlich. Bordet's paper was entitled "The Leucocytes and the Bactericidal Properties of Serum of Vaccinated Hosts" (*Ann. inst. Pasteur,* **9**: 462–505, 1895). Only a few passages of this lengthy paper follow:

Pfeiffer has shown that in experimental cholera, vaccination leads to an efficient method of defense, specific against the vibrios used for vaccination. As observed by Pfeiffer, vibrios injected into the peritoneal cavity of vaccinated animals are killed in the free fluid without having been engulfed by phagocytes. . . .

The bactericidal substance of normal animals [Nuttall's] and the one characteristic for vaccinated animals [Pfeiffer's] are both destroyed by heating at 60°. Sera held at this temperature become excellent culture media. But these two substances are distinguished very clearly from each other in that the one found in normal serum is weakly active and not specific, whereas the one found in vaccinated animals is highly potent and highly specific. Furthermore, the bactericidal substance of vaccinated animals differs from the preventive substance [antibody] found in the same serum. Fränkel and Sobernheim have shown that the preventive substance even resists prolonged exposure to 70°C, while the bactericidal substance is completely destroyed under these conditions. Serum [from immunized animals] which has been heated at 70°C is, as these savants have shown, as protective for animals as fresh serum. Although heated serum does not possess antiseptic action [in vitro] it produces a bactericidal state of the humours in injected animals. It would appear, therefore, that there are in serum at least three active substances capable of playing a role in immunity: two distinct bactericidal substances and a preventive substance. Actually, the situation is not that complicated. The events reveal that the bactericidal substance of vaccinated animals and that of normal animals—the former highly active and specific and the latter not possessing these characteristics— are nevertheless identical. The preventive substance [antibody], when present together with the bactericidal substance, enhances its potency and confers specificity.

Indeed, the addition of only a small amount of anticholera serum either fresh or heated at 60°C and thus devoid of activity against vibrios, to weak normal serum, renders the normal serum highly bactericidal. Thus, two liquids hardly bactericidal when alone, yield a mixture that is highly antiseptic. As we shall see below, the antiseptic activity is specifically directed against the species of microbe which was used to vaccinate the animal providing the preventive serum. The bactericidal activity is obtained either with serum-containing cells or serum that is free of cells. . . .

A freshly drawn sample of normal guinea pig blood is allowed to settle. The clear upper portion is separated from the lower portion which contains red and white blood cells. The following mixtures are prepared in tubes—(1) twelve drops of serum from a sheep hypervaccinated against the cholera vibrio in East Prussia; this serum is heated for one hour at 58°; (2) twelve drops of clear normal guinea pig serum; (3) twelve drops of a mixture consisting of 8 drops of clear normal guinea pig serum and 4 drops of sheep protective serum heated at 58°C; (4) twelve drops of a similar mixture except that the guinea pig serum contains cells.

These tubes are inoculated with a 24-hour culture of East Prussia cholera vibrio, harvested with 10 cc. of 0.6% NaCl. [Samples are removed for plate colony counts at various intervals after seeding]:

Number of Organisms

Duration of contact in hours	Heated preventive serum [antibody]	Clear normal serum	Clear normal serum +	Cellular normal serum +
			Heated preventive serum [antibody]	
0	8,640	9,600	10,200	9,120
1½	4,320	2,160	0	0
4	6,480	3,600	0	0
16	∞	∞	0	0

A very small quantity of preventive serum [antibody] is adequate to render normal serum [containing complement] strongly bactericidal. The potency can be determined by plating or by observing the disintegration of vibrios into granules, when they are added to serum. Therefore the phenomenon of Pfeiffer can be demonstrated *in vitro*. . . .

Bordet continued with a demonstration of the specificity of the preventive substance (antibody) and a discussion of the phenomenon of Pfeiffer, in which he does his best not to step on Metchnikoff's toes:

The transformation of the vibrios [granulation] indicates that they are susceptible to the bactericidal substances of the host. These . . . [substances] are the contribution of the leucocytes. One can presume, contrary to Pfeiffer, that the microbes are altered by the action of substances elaborated by the leucocytes in the presence of vibrios and which diffuse into the extracellular fluid. Metchnikoff has proven that this hypothesis conforms with the observed facts, and has produced the Pfeiffer phenomenon *in vitro* by adding leucocytes from the peritoneal cavity of a normal guinea pig to mixtures of preventive serum and suspensions of culture.

Regardless of the deductions he has drawn from his observations, Pfeiffer has rendered a real service to immunology by observing the phenomenon of vibrio transformation. . . .

Pfeiffer had already stressed the diagnostic value of the specific lytic phenomenon that he had observed in vivo. Since Bordet was able to observe the same granulation of the vibrios in vitro, it was only logical that he would stress the practical advantage of his own method for bacterial diagnosis. In so doing, he noted that some strains of the cholera vibrio resisted the action of sera better than others. This was not surprising, since the same fact had been observed in vivo and complete agreement between the two tests was noted. With great care, Bordet investigated, as best he could, all the aspects of Pfeiffer's phenomenon. Most important, of course, was the effect of temperature on the two main components studied, the antibody and the complement. This, as a sideline, led Bordet to antedate Gruber and Durham in observing agglutination: *"Let us note that preventive serum, in absence of Pfeiffer's phenomenon [when heated] consistently produced immobilization and aggregation of the vibrios . . ."*

Another crucial point was the determination of complement in vaccinated animals as compared with normal animals. Bordet concluded *"that the quantity of bactericidal substance in the serum of the vaccinated animal is not noticeably greater than in normal serum."*

Bordet failed to obtain an increase in resistance to the bactericidal substance by serially passing large inocula of cholera vibrios through bactericidal serum. He concluded his long paper by strongly emphasizing the role of leucocytes in immunity, of which not the least part was their ability to produce the bactericidal substance (complement) that he was to christen "alexine." This name had already been coined by Hans Buchner for bactericidal substances in blood, and it was not the name of a girl but came from a Greek verb meaning "to push back, to defend." It was a pleasant-sounding term, yet it was not universally adopted. Ehrlich's proposed "complement" became the more widely accepted term.

On the basis of this effective performance Bordet was sent to the Transvaal in 1897 to attempt to curb a devastating epidemic of rinderpest. Germany also sent a mission led by the great master, Koch. On this occasion, Bordet fared better than the ill-fated French cholera mission to Egypt; he demonstrated the utility of prophylaxis with injection of immune serum combined with deliberate exposure. Upon his return he resumed his studies on immunology. He observed that normal serum of some species of animals could agglutinate the red blood cells of other species and that this reactivity could be enhanced by using red blood cells as antigens.

Bordet, in a paper entitled "Agglutination and Lysis of Red Blood Cells by the Serum of Animals Injected with Defibrinated Blood" (*Ann. inst. Pasteur*, **12**: 688–695, 1898), wrote:

> Early in 1896, Gruber reported, and so did we, that the ability to immobilize and agglutinate microbes was not restricted to immune sera.
>
> We determined, for example, that normal horse serum could agglutinate the cholera vibrio, *B. coli*, *B. typhi* and *B. tetanus*. Furthermore we have observed that the serum of one species of animal can agglutinate the red blood cells of another species of animal. Thus chicken serum agglutinates the red blood cells of the rat and rabbit. Buchner has shown previously that some normal sera can destroy the red blood cells of another species and that heating serum to 55° destroys the erythrocidal activity. It is easily shown that the phenomena of agglutination and erythrocyte destruction are produced by different substances; the destructive substance is inactivated at 55°, but not the agglutinating substance. . . .
>
> Thus there is an analogy between the changes in the cholera vibrio in contact with cholera serum and those in the red blood cells in contact with serum from heterologous species. As with the cholera vibrio, the serum of animals injected repeatedly with defibrinated blood of another species acquire enhanced agglutinative and erythrocidal activity for the type of erythrocyte injected. . . .

With the knowledge amassed on the immunological behavior of bacterial cells and erythrocytes, with an understanding of the different properties of the antibody and complement, Bordet had the necessary components to develop the now widely used complement fixation test. In this study he had the collaboration of his brother-in-law, Octave Gengou. Their publication was entitled "On the Presence of Sensitizing Substances in Antimicrobial Serum" (*Ann. inst. Pasteur,* **15**: 289–302, 1901). After a review of what was known about alexine (complement), they stated:

> But all microbes . . . are not destroyed or altered visibly when mixed even with potent immune sera. In such cases another method is required. Accordingly we shall present such a method to demonstrate sensitizers [antibodies] in the sera of animals vaccinated against the plague bacillus, the anthrax bacillus, Eberth's bacillus, swine erysipelas bacillus and *Proteus vulgaris.* . . .
>
> Previous experiments established two distinct concepts: (1) red blood cells or microbes, when sensitized, acquire the ability to absorb alexine avidly and to remove it from the suspending medium. (2) The same alexine in a serum can induce either hemolysis or bacteriolysis. In the present report data are presented to establish that absorption of alexine may be used to demonstrate the presence of sensitizer in antimicrobial sera.
>
> Since the procedure based on this principle is the same for the various antimicrobial sera studied, we will select one example for detailed description, *antiplague horse serum:*
>
> This highly protective serum was supplied by Dr. Dujardin-Beaumetz, who prepared and evaluated it at the Pasteur Institute.
>
> This serum and normal horse serum are heated at 56°C. for one-half hour; the heating inactivates the alexine. A 24 hour agar culture of plague bacillus is harvested with a minimal amount of physiological saline, giving a dense suspension of microbes. The alexic source is normal guinea pig serum obtained the previous day and freed of red cells by centrifugation. The following six mixtures are prepared:
>
> **A.** Contains 0.2 cc of alexic serum, 0.4 cc of suspension of plague bacillus and 1.2 cc of plague antiserum (previously heated at 56°C.)
>
> **B.** Also contains 0.2 cc of alexic serum and 0.4 cc of suspension of plague bacillus, but in place of plague antiserum, 1.2 cc of normal horse serum previously heated at 56°C. are used.
>
> **C.** (Same as a) without plague bacilli.
>
> **D.** (Same as b) without plague bacilli. These first four mixtures contain the same dose of alexine.
>
> **E.** Contains 0.4 cc of suspension of bacilli and 1.2 cc of plague antiserum.

F. Contains 0.4 cc of suspension of bacilli and 1.2 cc of normal horse serum.

The last two tubes are comparable to *A* and *B* but do not contain alexine.

These mixtures are kept at room temperature for 5 hours. Into each of the tubes are then introduced 0.2 cc of suspension of highly sensitized red blood cells consisting of (1) 2 cc of serum from a guinea pig previously injected 3–4 times with 4–5 cc of defibrinated rabbit's blood. (The serum heated at 55°C. for one-half hour) and (2) 20 drops of defibrinated rabbit's blood.[1]

Hemolysis occurs very quickly and simultaneously in tubes *B, C, D*. After 5–10 minutes there are no intact red blood cells in these mixtures. There is no hemolysis in tube *A* which contained alexine, bacilli and plague antiserum. The red blood cells in tubes *E* and *F*, which do not contain alexine, also remain intact, as was to be expected. We see then that (1) plague bacilli mixed with normal horse serum does not absorb alexine; (2) these same bacilli in the presence of plague antiserum fix alexine with avidity and remove it from the solution; (3) plague antiserum without added bacilli, does not bind alexine.

Thus the serum of a horse vaccinated against the plague bacillus contains a sensitizer which confers on this microbe the ability to fix alexine. This sensitizer thus behaves like corresponding substances found in cholera serum and hemolytic sera. . . .

Repeating the experiment with significantly smaller amounts of bacilli and plague antiserum but with the same amount of alexine (0.2 cc) the fixation of alexine is not complete! The sensitized erythrocytes that are added undergo hemolysis but at a reduced rate of speed. . . .

Thus Bordet and Gengou had introduced a method for the detection of autibodies that had no known specific visible activity.

Bordet was a prophet recognized and honored in his own country. In 1900, the province of Brabant voted to organize an Institute of Serotherapy and Bacteriology that would include an antirabic division. A distinguished committee of three traveled to Paris to offer the directorship to Jules Bordet, then only thirty years old, but seasoned by seven years of activity at the Pasteur Institute. Bordet accepted and returned to Belgium in 1901. He had by now been married for two years, and a daughter had been born to the family. Eventually the Bordets had another daughter and one son.

[1] In a footnote, Bordet explained that the defibrinated blood was truly washed red blood cells, free of alexine.

The Institute was first housed in borrowed quarters. A building was completed in 1905, and additions were provided in 1909. Despite its productive performance, the Institute remained modest in size; in 1950, there were only ten professional members. The Institute was granted the privilege by Mme. Pasteur in 1903 of being called a Pasteur Institute. Several years later, 1907, Bordet was appointed to the newly created professorship in bacteriology at the Free University of Brussels.

The march of discovery continued, and in 1906, Bordet and Gengou discovered the etiologic agent of whooping cough by exposing plates of their blood medium before the faces of coughing children. Continuing their study of the host-parasite interaction, they demonstrated the pharmacologic effects of the endotoxin from the whooping cough bacterium on the respiratory epithelium. Antigen-antibody reactions in vitro were not forgotten. Bordet and Gengou discovered the coagglutination and conglutination effects that extended the sensitivity of serologic reactions. Bordet's interest in blood components also led him to analyze the factors in blood coagulation and by so doing to clarify the mechanisms of thrombin formation. He described a heat-labile enzyme component and a lipid contributed by platelets. In 1913 Bordet also added anaphylaxis to his array of interests. When the First World War halted activity in Brussels, Bordet used this period to write his classic *Treatise on Immunology in Infectious Diseases* that was published in 1919.

After the war, Bordet became very active in the study of the newly discovered bacteriophages, principally contributing to the study of lysogeny. In 1920, while on a trip in the United States to solicit funds for the rebuilding of Brussels' war-torn university, he learned that he had been awarded the Nobel Prize. Honors were not new to him, and he continued both his work at home and leadership in international circles. He was in turn active in studying phage, bacterial variation, and lysozyme. Yet there was time and interest to serve as the first president of the International Congress of Microbiology in 1930.

With the Second World War the activities of the Pasteur Institute of Brabant were once again curtailed and the staff depleted. Bordet, now seventy years old, remained at the Institute but relinquished the directorship to his son Paul, who had also become a bacteriologist. Failing sight had now obliged him to abandon laboratory work, but his general health remained good, and he continued service on various committees. His eightieth birthday was celebrated by a brilliant international convocation at the Free University of Brussels. Mentally alert to the end, he died peacefully in 1961 at ninety-one years of age.

In 1667, Jean Baptiste Denis, physician to Louis the XIV, performed what is considered to be the first transfusion of blood in man. The experiment was not well received by his fellow physicians, who were accustomed to remove blood from the body rather than to add to it and who considered the idea of reversing this flow too unorthodox. The opposition was strong enough to persuade the prudent Denis not to push the matter any further. Other attempts at blood transfusion made elsewhere at about the same time included those by Purmann in Germany and by Lower in England. Purmann used a lamb as his source of blood, and judging from an illustration of the experiment reproduced in Major's *History of Medicine,* the human patient was not too happy about the whole thing. The practice of transfusion from animals to men was abandoned, and transfusion from man to man remained a risky, unpredictable business until the turn of the century when Carl Landsteiner discovered the blood groups.

Born in Vienna in 1868, Landsteiner studied medicine in his home town and chemistry with Emil Fisher. It was while he was an assistant in the Pathology Institute in Vienna that Landsteiner published the following lines:

> The sera of normal humans agglutinate not only animal red blood cells, but also the red blood cells of other humans. Whether this effect is dependent upon inherent individual differences or is due to damage, possibly from bacterial action, is to be determined. I have observed enhancement of agglutinability of blood from severely diseased people. . . .

This was a footnote in a paper dealing with another subject. The next year, Landsteiner had solved the problem and had laid aside the "red herring" of a possible effect of disease. His paper was entitled "The Agglutinative Properties of Normal Human Blood" (*Wien. klin. Wochnschr.,* **14**: 1132–1134, 1901).

In the introduction Landsteiner established that the reaction that he was reporting was different from one described by Maragliano. He also described studies by Shattock and by Ehrlich and Morgenroth that came close to identifying the isoagglutinins in normal human blood. Then preceeding to his experimental data:

> My results will be presented concisely. The tables that follow are self-explanatory. Equal quantities of serum and 0.5% suspension of blood

cells in 0.6% aqueous NaCl were mixed and examined as hanging drops or in test tubes. "Plus" indicates agglutination.

Only one of Landsteiner's tables is here reproduced in a modified form:

Agglutination of the Red Blood Cells of Six Apparently Healthy Men

	Str.	Lan.	Ple.	Zar.	Stu.	Erd.	
Str.	−	−	+	+	+	+	C
Lan.	−	−	+	+	+	+	
Ple.	−	−	−	−	+	+	B
Zar.	−	−	−	−	+	+	
Stu.	−	−	+	+	−	−	A
Erd.	−	−	+	+	−	−	

On the basis of such data, he concluded:

The sera can be divided into 3 groups. In a number of cases (Group A) the sera agglutinate the red blood cells of another group (B) but not the red blood cells of Group A. Similarly the sera of Group B agglutinate the red blood cells of Group A but not those of Group B. In the third Group (C) the sera agglutinate the red blood cells of A and B while the red blood cells are not agglutinated by sera A or B.

The simplest explanation is that there are at least two types of agglutinins, the one in A, the other in B and both in C. The red blood cells in any group are not sensitive to the agglutinins in the serum of that group. . . .

Landsteiner was surprised to find out that the results could be accounted for by the postulation of the existence of only two agglutinins. After having reexamined his data and dismissed the possibility that the presence of these agglutinins was the result of some pathological process, he stated:

Finally, it should be noted that the observations here presented may explain the variable success of therapeutic human blood transfusions.

This discovery made possible the routine use of blood transfusions. In this domain, an American Army surgeon, Oswald H. Robertson, was most active. By 1918 he had found out how to collect blood under sterile conditions and how to preserve it under refrigeration after mixing with sodium citrate.

Landsteiner's career continued to be filled with interesting discoveries. Before the First World War, he was working on the experimental transmission of poliomyelitis to monkeys. After the war he left Vienna for The Hague and finally made his way to the Rockefeller Institute in New York in 1922. In 1940, with Wiener, he discovered the Rh factor. The designation "Rh" referred to the rhesus monkeys used in the experimental work leading to this important achievement. Landsteiner, who had received the Nobel Prize in 1930, died in 1943.

Blood transfusion is a form of grafting, an operation whereby tissues are passed from one individual to another. Grafting has produced for centuries interesting and practical results in plants. In animals, however, the situation is complicated by immunological factors. Fortunately, in the case of blood, as Landsteiner had observed with relief, the number of agglutinins is apparently low. Actually, it was not as low as Landsteiner thought, but low enough to permit satisfactory results in most cases where transfusion was necessary. Many immunologists are devoting much effort at the moment to fundamental studies that would permit a wider application of grafting, so that not only blood but other tissues as well might be safely transferred from one individual to another.

As serotherapy developed at the turn of the century, it soon became evident that antigen-antibody reactions in vivo could have some harmful effects and even produce death. The most dramatic manifestation of the state of hypersensitivity caused by the injection of foreign substances in the body is called the anaphylactic shock. It was first described and named by two French investigators, P. Portier and Charles Richet. In a paper published in 1902 (*Compt. rend. soc. biol.*, **54**: 170–172) they wrote:

> We call *anaphylaxis*, as opposed to *phylaxis*, the property that a venom may have to reduce rather than to increase immunity when it is injected in non-lethal doses.
>
> It is probable that many venoms (or toxins) are of this type, but, since up to now, emphasis has been on their prophylactic or vaccinal action, little effort has been made to study them from this point of view.
>
> The poison extracted from the tentacles of *Actinia* (sea anemones) produces a striking example of the anaphylactic effect.
>
> In general the manner of poisoning by this actinotoxin is similar to that

caused by the toxin extracted from the tentacles of *Physalia* [Portuguese man-of-war] that we studied during an expedition organized by Prince Albert of Monaco on the Princesse-Alice. It will be enough to mention that the intravenous injection of the poison from the tentacles of *actiniae* in solution in glycerol, is fatal to dogs in four to five days, when the dose exceeds 0.15 cc/kg. . . . When the dose is less than that amount, most of the animals survive after having been sick for 4 to 5 days.

But, if, instead of injecting normal dogs, one injects dogs which have received previously a non-lethal amount, doses ranging from 0.08 to 0.25 cc/kg bring rapid death. This demonstrates the anaphylactic effect of the first injection.

After presenting data on the experimental dogs, indicating the size of the first nonlethal dose and that of the second dose that produced the anaphylactic shock as well as the survival time before death (which ranged from twenty-five minutes to a few hours), Portier and Richet continued:

Our experiments prove another unpredictable fact, that the anaphylactic effect requires a long gestation period. If the second injection is made a few days after the first one, the animal behaves as if normal.

We have, at the moment, several dogs that have received 0.12 cc/kg a first time, then the same dose three to five days after, and that are in good health. If this second injection had taken place 15 to 25 days after the first one, these dogs would probably have died as rapidly as the dogs we discussed previously.

One should note that the anaphylactisized dogs were all in the best of health. . . . Even though a number of hypotheses could be advanced to explain these surprising facts pending further experimentation, we will limit ourselves to calling attention to the similarity between the diminution of immunity following the injection of actinotoxin (anaphylaxis), and the reduction of immunity of tuberculous animals to tuberculin.

In 1913, Charles Richet received the Nobel Prize "in recognition of his work on anaphylaxis." He was born in Paris in 1850, the son of a professor of clinical surgery. Charles followed in his father's footsteps and was appointed professor of physiology at the University of Paris in 1887. That same year, he started to use serotherapy in humans. In 1898 he investigated with Héricourt the effect of eel serum in dogs. A state of hypersensitivity was noted. This first clue was followed by the work with Portier that we have just reviewed. Richet was extremely versatile, evincing a keen interest in psychology, telepathy, and aeronautics. He was an authority in all the fields that he indulged in, and he even designed an airplane and wrote novels and plays. He was also an ardent pacifist. Pneumonia finished his life in 1935 and spared him the anguish of the Second World War.

The use by Portier and Richet of a poison might have tended to obscure the more general nature of anaphylaxis. Indirectly, Theobald Smith was instrumental in bringing into focus the immunological basis of the phenomenon. Smith had been devising procedures for producing antitoxin at the Massachusetts State Board of Health Laboratories. He observed that if guinea pigs that had received injections of toxin and horse antitoxin were reinjected several weeks later with normal horse serum, they died or became severely sick. Smith reported this observation verbally to Paul Ehrlich during Ehrlich's visit to the United States in 1904. Ehrlich assigned R. Otto to investigate this phenomenon.

Otto reported his results in the first volume of *Von Leuthold Gendenkschrift,* published in Berlin in 1906. The title of his paper was "The Theobald Smith Phenomenon of Serum Hypersensitivity." Otto cited the previous work of Héricourt and Richet but apparently had missed the paper that Richet had published with Portier. As the years passed, the name of Theobald Smith was forgotten in this respect, and the phenomenon came to be properly and uniformly referred to as anaphylaxis.

The information accumulating on immunity has raised a number of questions, most fundamental in nature, which have not all been answered as yet. Among those one could list: What is the chemical nature of antigens and antibodies? How do they react with each other? Where are the antigens located, in or on the microbial cell? Where are antibodies formed in the animal body? Are antibodies genetically transmittable?

An answer to this last question was furnished in 1892 by Ehrlich, who showed that male mice artificially immunized against a poison do not beget immunized progeny. Female mice gave birth to offspring carrying some antibodies in their blood, but these antibodies disappeared shortly. The ability to produce some agglutinins, such as those responsible for blood groupings, however, is a genetic property that follows Mendel's laws. An individual receives, as part of its genetic complement, the ability to produce this or that antibody under the prodding of the proper antigen.

The site of antibody formation is not yet a settled question, since not all antibodies may be formed at the same place. Long a serious contender since the work of R. Pfeiffer and Marx in 1898, the lymphoid cells are currently favored as the site of production. Other workers, such as Von Dungern, in

1903, preferred a theory of intravascular formation. Ehrlich, on the other hand, thought that, in the case of antitoxins at least, any part of the body might participate in their formation.

The first three questions that we raised, dealing with the chemical nature of antigens and antibodies, their mode of reaction, and their location, are also not fully answered as yet. Bordet in 1899 noticed that agglutinins were removed from immune serum by the antigen, which was in that case the whole bacterium. This point was further stressed in 1902 by Castellani, who injected rabbits with two bacteria and observed selective removal of the agglutinins by the antigens, which were again the whole bacterial cells. He formulated the following general rule:

> The serum of an animal immunized against two different organisms, A and B, loses after treatment with organisms A its agglutinating activity for organisms A, while activity for organisms B is not affected significantly. Treating the serum with organisms B, activity for organisms B is lost without influencing activity for organisms A. If the serum is treated with both organisms, agglutinating activity for both A and B is lost.

Cell-free systems had to be used to develop the foundation of immunochemistry; thus precipitation reactions came to be favored tools rather than agglutination or lytic phenomena. At first, the conditions of precipitation appeared confusing. Was the antigen present in the precipitate? Was either or both the antigen and the antibody denatured during the process of precipitation? Working in the Naples Zoological Station and supported by a grant from the Ministry of Justice, Culture and Education of the Grand Duchy of Baden, Von Dungern tried to answer these questions.

In a paper entitled "The Nature of the Combination in the Precipitin Reaction" (*Zentr. Bakteriol. Parasitenk. Abt. I*, **34**: 355–380) he wrote in 1903:

> The study . . . revealed some characteristics of the precipitin binding groups of precipitable protein substance (antigen) that were helpful in the understanding of antibody action. I used the blood plasma of cephalops (*Octopus vulgaris, Eledone moschata*) and short-tailed crabs . . . to induce precipitins in rabbits. The combination of precipitin and precipitable substance was studied both in the test-tube and animal. Precise quantitative tests could be performed only with the test-tube. Equal amounts of precipitin serum were mixed with increasing dilutions of precipitable plasma. The precipitates were separated by centrifugation

and the supernatants were tested quantitatively for residual precipitin and precipitable substances. The supernatants were diluted serially and aliquots were mixed with equal amounts of undiluted precipitin serum and 100-fold diluted precipitable plasma. These mixtures were observed for precipitate. The dilution of precipitin or precipitable substance was used as a measure of concentration. . . .

We learn from this series that the absorption of a stated amount of precipitin need not be proportional to the amount of added precipitable substance. If, for example, a given amount of protein removes ¾ of the precipitins, twice this amount of protein added to the supernatant may only remove ⅞ of the remaining precipitin. The volume of liquid does not alter the reaction. Nevertheless in all series there is a middle zone where the relative concentrations of precipitin and precipitable substance are such that both reagents combine completely to form the precipitate.

Ehrlich and Morgenroth suggested that injection of a foreign cell into an animal led to production of a variety of immune bodies specific for different chemical groups in the foreign cell material. Van Dungern demonstrated, by absorption test, cross reactivity between the two types of cephalopod plasma. Following this, he examined the chemical nature of the components in octopus plasma that are precipitated by precipitin.

I wished to determine whether the disappearance of the precipitable substance from octopus plasma corresponded with the precipitation of hemocyanin by ammonium sulfate. Concentrated ammonium sulfate was added in high excess to octopus plasma. The supernatant, separated from the precipitate, did not give a protein reaction and did not yield precipitate when mixed with precipitin serum. Alternatively, concentrated ammonium sulfate was added to 10 cc of octopus plasma so as to produce a slight clouding. Over the next 2 days the hemocyanin separated as a purée of fine crystals. The liquid that was separated from the purée, still contained protein and only a small portion of the precipitable substance. The purée of crystals redissolved in 1% NaCl, contained nearly as much precipitin as the original plasma. When various sera were used for precipitation, significant differences were encountered. Two sera, Numbers 2 and 17 which showed the same precipitin activity with the unaltered octopus plasma, now respectively precipitated the fluid at 4 fold and 64 fold dilution. Those components of octopus plasma that could be precipitated by a partial precipitin present in serum 17 but absent in serum 2, were more soluble in ammonium sulfate than the components precipitated by serum 2.

That the precipitable components in octopus plasma are certainly proteins can also be demonstrated by inspection of the precipitate. The washed precipitate possesses the blue color of the hemocyanin. The color is lost by the addition of carbonic acid and restored by shaking in air. The hemocyanin is not a single protein but is composed of different molecules. These observations show that the precipitin-binding groups of precipitable substance differ chemically. But this does not mean that a single type of precipitable molecule can be bound only by one type of precipitin.

Von Dungern then proved this point by reacting an octopus precipitin serum with *Eledone* plasma and demonstrating that a portion of precipitin remained that combined only with octopus plasma.

The recognition of the inexhaustible variety of antigens led to further studies of the chemical basis of specificity. Obermayer and Pick studied, in 1902, the effects of protein denaturation by heat or enzyme activity on specificity. Denaturation could be controlled so as to maintain immunogenicity, but the changes in specificity were not profound. By 1903 and 1904, they extended their manipulative procedures to controlled oxidation of serum with permanganate. But the relatively extensive chemical treatment was accompanied by splitting the protein molecule. In 1906, however, they produced chemical alterations of the undegraded antigen. That the altered specificity could then be associated with the chemical group that was introduced into the molecule, they reported in a paper entitled "The Chemical Basis of Species Specificity of Proteins.—The Immune Precipitins Produced by Chemically Altered Proteins" (*Wien. Klin. Wochnschr.*, **29**: 327–334, 1906).

If one uses variously modified protein as precipitinogen, the treated animal shows an interesting adaptation to the modified protein, evident in the alteration of the immune precipitin. If a rabbit is injected repeatedly with unaltered cow serum protein, the induced immune serum reacts with the unaltered protein but not with the heated protein. Injecting a rabbit with heated cow serum protein, the immune serum reacts not only with unaltered cow serum protein but with the heated protein and a series of breakdown products. Species specificity was retained and a new specificity was added. . . . We therefore conceive of two types of groups in protein that are concerned with specificity: originative groups that confer species specificity and constitutive groups that confer specificity dependent upon the immediate complete structure of the protein. . . . Heating protein produces changes in some general characteristics, but the changes are not profound. Thus other types of denatured proteins were prepared, alkali albuminate, acid albumin, formaldehyde protein and toluol-treated protein. In all these cases, the degree of alteration was similar to that obtained with heating. Species specificity was not altered but there were changes in constitutive specificity. . . .

We then turned to degradation with ferments such as trypsin . . . and also to oxidative degradation with potassium permanganate in alkaline solution. These procedures did not alter species specificity. . . .

Since the originative groups were resistant, even to severe degradative treatment, we investigated chemical reactions which confer new characteristics on proteins by introduction of certain groups into the molecule. We used iodination, nitration and diazotization. . . . In this manner were obtained characteristic products, chemically and physiologically modified, referable to the groups that were introduced. The diazo proteins were especially interesting; in alkaline solution they may couple with other compounds such as phenols and amines, also naphthol and phenylenediamine. These reactions offered many possibilities because more than one type of group could be introduced into the same molecule. . . .

The results obtained by injecting such substances parenterally were significant. When a rabbit was injected repeatedly with iodinated cow serum protein, immune precipitin was induced that reacted to form specific precipitate not only with iodinated cow serum protein but with iodinated serum of a series of mammals, bird and even some plant proteins. This immune serum was specific for iodinated protein. The same results were obtained with nitro protein and with diazo protein. The group coupled to the diazo protein, did not influence specificity for the diazo group. When this new specificity was acquired, species specificity was lost. The loss of species specificity was so profound . . . that substituted derivatives of protein from the animal producing the immune serum behaved like foreign protein, precipitating the immune serum. . . .

An animal never can be immunized to its own proteins; this phenomenon is basic for the concept of species specificity. To remove any doubt regarding the loss of species specificity, rabbit serum xanthoprotein, as an example, induced immune precipitin in rabbits, that reacted with rabbit xanthoprotein and the xanthoproteins of other species. Thus substitution in the aromatic ring easily induced the loss of species specificity. . . .

The unifying factor in the three chemical reactions used, is substitution in an aromatic ring, and replacing an H or an NH_2. In proteins there are three types of rings: (1) the oxyphenol, (2) the phenyl, and (3) the indol. . . . These rings are highly resistant to acid, alkali or enzymatic action. The proteins are composed of combinations of the rings with their many side chains. . . . The size of the complex is not critical for species specificity. . . . Species specificity is influenced by groups associated with the rings. The aromatic groups as such are not adequate to account for the variations in specificity but they act as a pivot point for the species characteristic groupings of the side chains. By substitutions in the rings it is possible to level these species characteristic differences. . . .

Landsteiner continued the line of investigation started by Obermayer and Pick. Throughout his active career, he maintained an interest in the specificity of serological reactions, a subject on which he published a monograph in 1933. It was a successful text that he had to revise three times during his life.

For years only proteins were recognized as antigens. The meager knowledge of the chemistry of the proteins limited the progress that could be made in immunochemistry. The recognition of the antigenic nature of other chemical entities emerged from investigations of the pneumococci. In 1909, Neufeld and Handel demonstrated the existence of types with the specific protective effects of appropriate antisera. Dochez and Gillespie, in 1913, simplified the recognition of types by agglutination and agglutinin absorption and defined the most important types, I, II, and III.

At that time Avery joined the Rockefeller Institute and participated with Dochez and others in the study of the pneumococci. In 1917, Dochez and Avery observed that a specific soluble substance appeared in culture during active growth and in the infected host. This substance was responsible for the type identity and was demonstrable by a precipitin test. Avery subsequently interested Michael Heidelberger in joining in the study of the chemical identity of this substance. Together they published their results in a paper entitled "The Soluble Specific Substance of Pneumococcus" (*J. Exp. Med.*, **38**: 73–79, 1923) in which they wrote: [1]

In 1917 Dochez and Avery showed that whenever pneumococci are grown in fluid media, there is present in the cultural fluid a substance which precipitates specifically in antipneumococcus serum of the homologous type. . . .

In the earlier studies by Dochez and Avery certain facts were ascertained concerning the chemical characteristics of this substance. It was

[1] Reproduced by permission of Dr. Michael Heidelberger.

found that the specific substance is not destroyed by boiling; that it is readily soluble in water, and precipitable by acetone, alcohol, and ether; that it is precipitated by colloidal iron, and does not dialyze through parchment; and that the serological reactions of the substance are not affected by proteolytic digestion by trypsin. Since the substance is easily soluble, thermostable, and type-specific in the highest degree, it seemed an ideal basis for the beginning of a study of the relation between bacterial specificity and chemical constitution. The present report deals with the work done in this direction.

The organism used in the present work was Pneumococcus Type II. The most abundant source of the soluble specific substance appeared to be an 8 day autolyzed broth culture; hence this material was used as the principal source of supply. . . .

The process for the isolation of the soluble specific substance consisted in concentration of the broth, precipitation with alcohol, repeated resolution and reprecipitation followed by a careful series of fractional precipitations with alcohol or acetone after acidification of the solution with acetic acid, and, finally, repeated fractional precipitation with ammonium sulfate and dialysis of the aqueous solution of the active fractions.

As so obtained the soluble specific substance forms an almost colorless varnish-like mass which may be broken up and dried to constant weight at 100°C *in vacuo*. The yield from 75 liters of broth averages about 1 gm., although it varies within rather wide limits in individual lots.

By the method outlined above all substances precipitable with phosphotungstic acid or capable of giving the biuret reaction were eliminated. The residual material, for which no claim of purity is made, as efforts at its further purification are still under way, contained, on the ash-free basis, 1.2 per cent of nitrogen. It was essentially a polysaccharide, as shown by the formation of 79 per cent of reducing sugars on hydrolysis, and by the isolation and identification of glucosazone from the products of hydrolysis. . . .

The aqueous solution of the substance gave the Molisch reaction out to the limit of delicacy of the test. Reduction of Fehling's solution occurred only after hydrolysis. Phosphorus was present only in traces; sulfur and pentoses were absent. A 1 per cent solution gave no biuret reaction, no precipitate with phosphotungstic acid, mercuric chloride, or neutral lead acetate, precipitated heavily with basic lead acetate, and gave a faint turbidity with tannic acid. Calcium is very tenaciously retained, but does not appear to be an essential part of the molecule, as the specific reaction was also given by calcium-free preparations. No color is given by iodine solution.

The soluble specific substance is remarkably stable to acids. A solution in 0.5 N hydrochloric acid maintained its activity undiminished and failed to reduce Fehling's solution after 36 hours at room temperature, but showed reducing sugars and absence of precipitation with immune serum after transfer to the water bath.

[The most active preparation] . . . gave a specific reaction in a dilution as high as 1:3,000,000. . . .

While it has long been known that the capsular material of many microorganisms consists, at least in part, of carbohydrates, any connection between this carbohydrate material and the specificity relationships of bacteria appears to have remained unsuspected. While it cannot be said that the present work establishes this relationship, it certainly points in this direction. Evidence in favor of the probable carbohydrate nature of the soluble specific substance is the increase in specific activity with reduction of the nitrogen content, the increase in optical rotation with increase in specific activity, the parallelism between the Molisch reaction and specific activity, the high yield of reducing sugars on hydrolysis, and the actual isolation of glucosazone from a small quantity of the material. The small amounts of substance available up to the present have hindered the solution of the problem, and it is hoped that efforts at further purification of the soluble specific substance, now in progress with larger amounts of material, will definitely settle the question.

Polysaccharides, being nitrogen-free, lent themselves more readily than proteins to the development of the quantitative aspects of immunochemistry. Heidelberger and his coworkers were thus able to develop the methods that permit the determination by weight of the amount of antibodies present in a given serum.

Not only did polysaccharides come to be recognized as antigens, but even nucleic acids were added to the list. Thus antigens apparently can be located in the most diverse parts of the microbial cell. The study of the nature of antigens led the immunochemist to use specific immune sera in the determination of the chemical structure of antigens. This approach has been especially fruitful in the elucidation of the structure of polysaccharides. Immunochemistry has thus added biological specificity to the tools of the organic chemist.

References

Bordet, J.: *Traité de l'immunité dans les maladies infectieuses,* Masson et Cie, Paris, 1939.

Heidelberger, M.: *Lectures in Immunochemistry*, Academic Press Inc., New York, 1956.

Himmelweit, F., and M. Marquardt: *The Collected Papers of Paul Ehrlich*, vol. II, *Immunology and Cancer Research*, Pergamon Press, New York, 1957.

Kervran, R.: *Albert Calmette et le B.C.G.*, Librairie Hachette, Paris, 1962.

Marquardt, M.: *Paul Ehrlich*, William Heinemann, Ltd., London, 1949.

Metchnikoff, E.: *Immunity in Infective Diseases*, Translation from the French by F. G. Binnie, Cambridge University Press, London, 1905.

Zeiss, H., and R. Bieling: *Behring, Gestalt und Werk*, Schultz, Berlin, 1940.

Additional Reference

Parish, H. J.: *A History of Immunization*, E. and S. Livingstone, Ltd., Edinburgh, 1965.

From Soil Microbiology to Comparative Biochemistry

Winogradsky, Beijerinck, and Kluyver

7 Soil, the habitat of most microorganisms, is also the arena for the endless chemical transformations that they perform. In the light of the "philosophical studies" of M. Pasteur, the medical profession had come to look upon microorganisms as a possible explanation for certain pathological manifestations. Similarly the soil chemists turned to microorganisms as a possible answer to certain elusive chemical reactions.

It had long been known that nitrates accumulated where organic matter was allowed to decompose. Davy, in 1814, expressed the opinion that nitrates were formed at the expense of both the ammonia present in the soil and the oxygen in the atmosphere. Liebig showed that atmospheric nitrogen played no part in the process of nitrification. Boussingault had stressed the importance of nitrates in the nutrition of plants by showing that those to which a solution of nitrates was given grew better than those receiving only water.

In 1877 Theophile Schloesing, a professor of agricultural chemistry at the French National Institute of Agronomy, and Achille Müntz, a disciple of Boussingault, published the following paper in the *Comptes Rendus de l'académie des sciences* (**84:** 301–303):

It is usually an accepted concept that nitrates formed in soil originate from the combustion of ammonia and of organic nitrogen compounds, although we are far from knowing its mechanism. Is nitrification the result of a direct, purely chemical reaction, between gaseous oxygen and nitrogeneous compounds? Does it proceed through the intervention of organisms acting as ferments? Or is it produced by both processes? These questions have been asked ever since the day when M. Pasteur showed that certain organisms, such as the mycodermae of wine and vinegar, have the capacity to transport atmospheric oxygen to the most diverse organic substances, and are thus the most active agents of destruction of organized matter which has stopped living. On the other hand, M. Pasteur has proved that organic substances, which are at least as subject to transformations as those which are nitrified in soil, are singularly resistant to oxidation, when they are protected from all germs. By analogy, one may suppose that living organisms are at least the agent of rapid combustion and nitrification. This does not exclude the possibility that free oxygen, acting in virtue of physical and chemical forces, may burn and nitrify the nitrogen containing substances at a slower rate. Imbued with these ideas, M. Pasteur declared, as early as 1862, that the study of nitrification had to be undertaken again in the light of the new notions on combustion-causing organisms. The experiment which we are going to report confirms the predictions of the illustrious master. . . .

One of us, in a recent investigation of irrigation with sewage waters, wanted to learn if it was necessary for humic matter to be present in a soil in order to obtain the purification of, or in other words the total combustion of, dissolved matter in this water. For this purpose, a wide glass tube, 1 meter long, was filled with 5 kilograms of quartz sand, which had been heated red-hot, and mixed with 100 grams of powdered limestone. The sand was watered daily with a constant amount of sewage water, calculated in such a way that the liquid took eight days to go down the tube. During the first twenty days, no sign of nitrification was detected, and the concentration of ammonia in the water filtered in this manner did not vary. Then nitrate appeared, and increased rapidly in amount until it was observed that the sewage water at the exit of the apparatus no longer contained traces of ammonia.

If, in this experiment, the organic matter and ammonia had been burned by oxygen acting directly and without intermediate, one might ask why twenty days elapsed before the combustion started. This delay is better explained by the hypothesis of organized ferments which could act, of course, only after due inoculation and adequate development of their germs.

The experiment, started in July, had been going on for four months, when we had the idea to fill the tube with chloroform vapors. One of us had shown, in fact, that this compound stops all activity of organized ferments without in any way hindering soluble ferments. Thus, if the observed nitrification was produced by organisms, this should stop it by paralyzing these agents; if, on the contrary, nitrification was a simple chemical reaction, chloroform should not modify it. We thus put on our sand a small vessel full of chloroform and diffused the chloroform vapors through the tube by a current of air. We have said that the daily dose of sewage water required eight days to travel through the sand; we could not have thus expected nitrate to disappear from one day to the next. However, after ten days during which the sand had been washed by displacement, the spent liquid did not contain any more traces of nitrate and the ammonia of the sewage water was present in its original concentration. After evaporation, the liquid left an appreciably colored and smelly residue, such as that given by filtered but unpurified sewage water.

After having kept the chloroform vapor in our tube for fifteen days (from the 27th of November to December 12th), we removed the small vessel with chloroform. During the following fifteen days, the liquids emerging from the tube still had the characteristic odor of chloroform. This odor disappeared at the end of December; however, during the course of January, when the tube was kept at the mean temperature of 15 degrees, no trace of nitrate was produced. No doubt, our nitrifying organisms were all dead and the sewage water was not bringing back a suitable inoculum, perhaps because it was in a rather advanced state of putrefaction. On the first of February we decided to try an inoculation of these germs. Since they must be present in an organic soil in which nitrification occurs, we suspended in water 10 grams of a soil, which we knew was able to carry out nitrification and this turbid water was poured over the surface of our sand. Nitrate was detected, precisely as expected, on the 9th of February, and its concentration has been increasing ever since. We think that before long the rate of formation which was present before the introduction of chloroform will be reestablished.

Now we still have to discover and isolate the nitrifying organisms. We hope that we will be able to do so by following the procedure so clearly defined by M. Pasteur for this kind of work. . . .

That nitrification was caused by microbial action was confirmed by Warington in England. In addition, Warington discovered that the process of nitrification was a two-step process in which ammonia was transformed first into nitrite and this in turn was transformed to nitrate. Warington, however, was not able to isolate the responsible microorganisms.

This isolation was finally achieved by Winogradsky, a Russian scientist, who was then, in 1890, working in Zurich. At this point we shall let Winogradsky himself give us a summary of his career by reading part of the introduction of Winogradsky's book *Soil Microbiology:* [1]

I started my work in 1885. . . .

Impressed by the incomparable glitter of Pasteur's discoveries, as a young student I entered this field of investigation and I have remained faithful to it to the end.

At that distant time, Pasteurian Science was not widespread. In France, there was Pasteur surrounded by a small group of collaborators, but I could not visualize the possibility of being admitted among them. In Germany, there was Robert Koch, and although many scientists had left his school, their aims and methods which were specifically elaborated for pathological studies, did not attract a non-physician.

This is one of the reasons which led me to start my microbiological investigations in a laboratory of Botany, the one in Strasbourg which was headed by Professor A. deBary, who is mainly known for his mycological investigations. . . .

My first investigations dealt with filamentous bacteria found in sulfur and iron containing springs; these were the first known autotrophs. . . .

In 1889, in Zurich, while I was working in a laboratory of the University, I took up the problem of nitrification. These investigations lead to the discovery of two groups of microbes responsible for the nitrite and the nitrate formation from ammonia nitrogen. Their chemosynthetic power was established quantitatively and the physiological type known as autotroph was firmly established by these investigations.

To this work, I owed the honor of an invitation from Pasteur himself to come and settle at the Institute which bears his name and which was built in Paris in 1890. With much regret I had to decline this great honor as I had to go back after 5 years absence, into my country of origin.

Having returned to Russia in 1891, I was appointed head of the laboratories of general microbiology at the newly founded Imperial Institute of Experimental Medicine in Petersburg. I worked there about 15 years and in 1900 I was promoted to the post of director.

During that time, while continuing experiments on nitrification with

---

[1] Translated by permission of Masson et Cie, Paris, publishers of *Microbiologie du sol, Problèmes et méthodes,* by S. Winogradsky.

the late Omeliansky, and directing investigations on anaerobic decomposition of cellulose, on the retting of flax and other problems, I started to investigate the great problem of the fixation of atmospheric nitrogen. I succeeded without too much difficulty in isolating an anaerobic rod called *Clostridium pasteurianum* that could perform this function. . . .

As early as 1906 various unfavorable circumstances interrupted the normal course of my work. . . .

In 1921, I found myself in Belgrade (Serbia) as professor at the Faculty of Sciences. Unfortunately, due to the devastation of war, the facilities for scientific research were wanting.

It was while in Belgrade that Winogradsky received the following letter:

Paris, the 23rd of February, 1922

Dear Monsieur Winogradsky,

My colleagues and I would be most grateful if you could settle at the Pasteur Institute. You would bring your scientific fame and you could continue, without having any teaching duties, your brilliant investigations. We would be proud of having, after Metschnikoff, Winogradsky among us. You will be for us the master of soil bacteriology. In spite of the difficulties of the present moment, we will give you a laboratory and a position worthy of you.

I hope that you will be able to write to me soon that you are accepting.

Dr. Roux

I accepted with gratitude his invitation to settle at the Pasteur Institute as head of the laboratories of agricultural microbiology.

The small laboratory which was set up for me at Brie-Compte-Robert has been the seat of my investigations for the last period of my studies, 25 well-filled years.

There, in the small village of Brie-Compte-Robert, which is located between Paris and Fontainebleau, the greatest of all soil microbiologists lived out his life, dying in 1953, almost a century after his birth in Kiev. During this last period, Winogradsky studied the methodology of soil microbiology, the decomposition of cellulose, the fixation of nitrogen by the aerobic *Azotobacter,* and the ecology of microorganisms. At the last, he collected all his studies. In 1945, the Academy of Sciences of the U. S. S. R. had offered to publish this work in both a French and a Russian edition. From Poland also came an offer to publish the work of the master in French. It was first published, in 1949, by a French firm, Masson et Cie. Those were hard times; paper was scarce, and Winogradsky's book was almost nine hundred pages long. Help came from America, where the National Academy of Sciences donated the paper, and Dr. Waksman found funds to support the cost of publishing this monumental tome, which certainly was not going to outsell even the most mediocre novel.

Let us go back to 1890. Winogradsky announced from Zurich that he had succeeded in isolating the organisms responsible for nitrification. In the first two papers that we read we note, even though Winogradsky did not state it explicitly, that he isolated only the organism responsible for the oxidation of ammonia to nitrite. These two papers were: (1) a note presented by Duclaux at the French Academy of Sciences (*Compt. rend. acad. sci.*, **111:** 1013–1016, 1890) and (2) a *mémoire* in the *Annals of the Pasteur Institute* (**4:** 211–231, 1890).

In the note, Winogradsky said:

> I would like to go back, before summarizing the investigations that I have been carrying out on nitrification during the past year, to some of my previous studies, which have been their starting point.
>
> They dealt with two groups of organisms which, like those which are the object of this note, have the capacity to oxidize inorganic substances, and to which I had referred as *sulfobacteria* and *ferrobacteria*.
>
> The first inhabit natural waters containing hydrogen sulfide and cannot survive in media so deficient. This gas is absorbed avidly and oxidized by their cells, which store sulfur. The sulfur in turn is excreted as sulfuric acid. The role of the second group of organisms is to oxidize ferrous salts and their existence is closely linked to the presence of such compounds in their nutritive media.
>
> Through my efforts to elucidate the physiological significance of these phenomena, I have been led to the concept that in the existence of these organisms, these inorganic compounds play the role of fermentable matter (in the broad sense of the word) which is, for the great majority of microbes, organic matter. The following logical deduction was then made and confirmed by experimentation: since all the energy necessary for their vital work was being furnished by the combustion of mineral substances, the expenditure of organic matter during their growth is extremely low, and they can utilize, as a source of carbon, compounds which cannot be used as food by other organisms which are lacking chlorophyll.

The remarkable studies of M. M. Schloesing and Müntz have been the first to shed some light on the role of lower organisms in nitrification. Though they showed that the existence of a special agent of nitrification was most probable, they did not teach us how to isolate it from soil, this natural medium so rich in various microbes. The reason is that the isolation and the cultivation in a pure state, which is now essential in any microbiological experimentation, is so difficult in the case of the nitrifying ferment, that attempts by numerous scientists to isolate it were unsuccessful. Thus, the conclusion of M. M. Schloesing and Müntz, regarding the existence of a special nitric ferment, has not as yet been fully confirmed by bacteriologists and botanists.

This question had to be resolved. First I determined that the failures of my predecessors were due to the utilization of gelatinized media, which are so widely employed at present for the isolation and the cultivation of microbes. The nitrifying organism refuses to grow on such media, to the degree that if one seeds them with a mixture of microbes selected from an actively nitrifying soil, all the active organisms die and only ineffective ones are harvested. A careful study of the culture conditions which would be favorable for the first group and unfavorable for the second group of organisms has permitted me, but not without difficulty, to eliminate one by one all foreign species and to get abundant and pure growth of the nitrifying species. This species both showed and retained, under the usual conditions of microbiological experimentation, an activity which was as potent as would have been anticipated from comparisons with recent experiments of M. Schloesing, who was studying nitrification in soil. . . .

Applying to this study my theories regarding the nutrition of organisms which can combust mineral substances, I grew the microbe of nitrification, from the beginning, in a liquid which contained as organic matter only what there is in a very pure natural water. In this liquid medium, which provided the organism with no carbon-containing compounds other than carbonic acid and carbonates, neither the quantity of growth, nor the intensity of its activity seemed to diminish during several months.

It seemed obvious that this organism was able to assimilate the carbon of carbonic acid. This was proved by quantitative determinations that demonstrated that this organism constantly accumulated organic carbon.

The microbe of nitrification, which is colorless, is thus able to carry out a complete synthesis of its own substance at the expense of carbonic acid and ammonia. It carries out this synthesis without light and without any source of energy other than the heat produced by the oxidation of ammonia. . . .

This was, in essence, the announcement of the discovery of one of the autotrophic nitrifying organisms. We will now read part of the *mémoire* in the *Annals of the Pasteur Institute* (4: 211–231, 1890) for further technical details:

It seemed logical by analogy with mycoderma, to look for the ammonia-oxidizing organism, or organisms, in the pellicle, but repeated experiments demonstrated that all the microorganisms which comprised the pellicle were inactive. . . .

My attention was drawn to the stratum of magnesium carbonate on the bottom of the flasks. I noticed that this coat of salt, so finely divided and of such a perfect whiteness, was changing its appearance in the older cultures. It was becoming grayish and somewhat gelatinous. If one shook a flask slightly, after it had been left standing for a long time, this deposit on the bottom was not raised but remained as if bound by a kind of cement which prevented the carbonate from dispersing in the liquid. Stronger agitation would finally break it up into large, gray flakes. Under the microscope these flakes exhibited an interesting structure. They were composed of transparent pellets of salt literally covered by a thick coat of a beautiful oval bacterium. Its perfect resemblance in form and size to the motile organisms previously observed could not be missed. If one would add a small amount of diluted acetic acid, the salt would disappear, but a characteristic zooglea could be observed. It contained holes marking the places where the salt particles had been dissolved. It seemed evident that the oval microorganism was attaching to the particles of salt and was coating them with a gelatinous substance that it secreted. There were very few of these germs on the surface of the liquid or on the side of the glass flasks. They were all accumulating in the stratum of basic carbonate which gradually became invisible having been dissolved by their action. . . .

This was the time to isolate the microorganism in question and to prove by a faultless experiment that it was really the nitric ferment. But how? . . .

After having purified it partially by a series of subcultures in a liquid mineral medium, the idea came to me to use for its final isolation its inability to grow on a gelatine medium. . . .

With a capillary tube a few clumps of the deposit were picked up and placed in a flask containing sterile water. After settling, the particles were picked up again with the same capillary tube and the dilution was deposited as separate drops on the surface of solidified nutrient gelatine. The location of the drops could be seen by the deposit of the particles of carbonate. These plates were kept for 10 days at 18° after which time the residue of the drops were examined microscopically. Some of these which were free of bacterial growth were used to inoculate nitrifiable solutions.

In this manner, I seeded six flasks. In all of them, nitrification started slowly owing to the small inoculum. All but one were pure cultures. . . . I had finally succeeded in isolating a nitrifying microbe.

Let us review the results presented in this memoir:

1. Schloesing and Müntz were correct in attributing the phenomenon of nitrification to a specific organism.

2. This bacterial agent can be isolated and cultured in a mineral medium free of any trace of organic material and in which the phenomenon takes place with intensity until the ammonia present is fully oxidized.

3. The conflicting results of my predecessors are explained by the application of methods which were unsuited to the desired aim.

When speaking about a nitric ferment, I do not mean that there is only one species capable of carrying out this phenomenon. . . . I am referring to a well-characterized physiological type.

The following year, unaware that Warington had already made the same observation ten years earlier, Winogradsky discovered that nitrification is a two-step process with each step involving a different flora. In 1891, Warington had written in the *Journal of the Chemical Society:* "Results obtained in 1880 and 1881 revealed the existence of an organism which energetically converted nitrites to nitrates, but was apparently unable to oxidize ammonia."

Warington was an English soil chemist who worked for many years at the Rothamsted Experiment Station. From 1877 to 1891 he worked on nitrification. He demonstrated the two-step nature of the process, but using the usual organic bacteriological media, he failed where Winogradsky succeeded.

In the following paper, Winogradsky explained that his previously isolated organism carried out the transformation from ammonia to nitrite and that he now had the second organism foreseen by Warington. The following are quotations from a note in the *Comptes Rendus of the French Academy of Sciences* (**113**: 89–92, 1891):

I have already shown (*Ann. inst. Pasteur*, 1890, no. 12) that the principal product of nitrification in pure mineral solution cultures is nitrous acid, nitric acid being formed only in very small quantities. I have also shown (*Ann. inst. Pasteur*, 1891, no. 2) that the conditions of culture do not have anything to do with this incomplete oxidation of ammoniacal nitrogen, since the same results can be obtained by cultivating the nitrifying organism, the nitromonad, on silica gel or in sterile soil. These very clear and very constant results did not help in explaining why, in a nitrifying soil, mainly nitrates are formed. Thus, I have decided that I should extend my observations to a greater number of soil samples of various origins, European and foreign. . . .

In each liquid medium seeded with one of these soils, regardless of origin, nitrification always starts by the formation of nitrous acid, which increases very rapidly. However, when the ammonia has disappeared, the elaborated nitrite is oxidized completely to nitrate.

Repeated microscopic examinations of these cultures showed that all, especially those where the formation of nitrate was in process, contained many varieties of microbes in addition to a predominant one, easily recognizable, which greatly resembled the nitromonad. . . .

All observations showed me that one should look for the cause of the formation of nitrates only in the action of organisms. However, which ones were nitrite oxidizing organisms? . . .

The common species which can cause combustion of organic matter were not associated with the phenomenon and one had to look for a specific microbe, which would be an agent for the oxidation of nitrites. As it had escaped me in my gelatin cultures, I had to use other means, and I used the method which served me so well in my research on the nitrous ferment.

Starting from a culture of the Quito soil series, I made all subsequent transfers to nitrite-containing solutions. Soon I was able to see that the course of the phenomenon was becoming very regular in the new series of cultures. For the definitive isolation I used a silica gel impregnated with a culture of a nitrous ferment. After seeding this medium with a drop of the new Quito series, I saw colonies of two different microbes, one of which was the organism that I was seeking. It was a very small rod, angular and irregularly shaped, which bore no resemblance to the nitrous ferment from the same soil. Seeded into nitrite solutions, it very rapidly transformed the nitrite into nitrate.

Since then, I have sought and found organisms performing the same function in a soil from Java and in a soil from Zurich. Each soil probably harbors a unique species belonging to this group, as is also the case for the group of the nitrous ferment. It is easier to check this point by adding a small amount of fresh soil to the proper liquid. The oxidation is slow to start, but when started, it very rapidly reaches 20 milligrams to 30 milligrams of nitrous acid a day. Using a series of cultures in the same medium one can soon purify the nitric ferment well enough to permit study of its properties. The nitromonad and the organisms having similar properties are not to be found in these cultures.

One of the most curious properties of the exotic nitric ferment that I have isolated, is that it does not oxidize ammonia. Seeded in ammoniacal solutions which would be easily nitrifiable by the nitrous ferment, it gave neither nitrite nor nitrate. I am now studying its action on organic substances.

In 1895, Winogradsky wrote the following in *Archives des sciences biologiques de l'institut impérial de médecine expérimentale à Petersbourg,* vol. 4:

During the last decade much has been done and said about the assimilation of free nitrogen by organisms. . . .

Before starting to describe our investigations, we will try to summarize, as briefly as possible, the results of investigations carried out before ours. We will mainly consider the biological aspect of the problem in order to put in focus the point of departure of our investigations and our aim.

Previous investigations dealt with three different classes of lower organisms.

M. Berthelot, to whom we owe credit for so many investigations on this subject over many years, concluded that soil contains "microbes" which can fix nitrogen.

The rightly famous investigations of M. M. Hellriegel and Wilfarth have focused attention on the fixation of nitrogen by leguminous plants bearing on their roots nodules of bacterial origin. This discovery has inaugurated a series of investigations which have confirmed, by the most diverse methods, the existence of this fixation which is contingent on the symbiosis of the higher plant with specific microbes. However, all efforts to study the obviously complex and delicate mechanism of this fixation have failed. Which one of the two partners in this symbiosis is responsible for the fixing action? One might well have suspected that it was the microbe. But nothing has been demonstrated even though it was isolated by M. Beijerinck and cultured in a pure state by M. Praznowsky and M. Laurent. Although the latter authors tended to consider the microbe the responsible agent of nitrogen assimilation, they were unable to prove this assertion. As far as this aspect was concerned, the leguminous plant and its microbe remained unseparable. . . .

There still remains the green algae, or more exactly the blue-green algae, mainly the *Nostoc*, which fixes nitrogen in soil as Frank, Schloesing and Laurent have wished to demonstrate. There is no doubt that the presence of these algae might be necessary under certain conditions for this fixation. This has been proved mainly by the beautiful experiments of the two latter authors. However, the action of bacteria was not excluded in these experiments, and M. Kossowitch has shown recently that fixation seems impossible in a bacteria-free soil.

One can see from this short summary that many remarkable investigations have simply demonstrated the participation of certain living organisms, legumes, algae and microbes, in this fixation. But what is their mode of action and what are their mutual relationships? Is there a specific species that one could consider as having this function? Or is it always the result of the combined action of two or more different organisms?

While I was following the progress of this problem in general, I kept thinking that it was in the world of the true microbes that one should look for the principal agents of fixation of free nitrogen. No matter how telling is the case for the legumes, it seems obvious that bacterial species *living freely in soil* are likely to be the universally distributed agents of this fixation, which takes place at the expense of organic substances which are found in varying amounts in natural environments. . . .

One finds in nature enormous reserves of carbon-containing substances which are poor in nitrogen. One can ask if all this organic carbon could be utilized and put back in circulation without the existence of organisms able to assimilate free nitrogen. These organisms with the double role of decomposing large amounts of carbon-containing matter and assimilating the small amount of free nitrogen which is necessary for their composition, could only be microbes: bacteria, molds, actinomycetes.

I started my investigations about 1892. Soon I had positive results that I summarized in a note presented at the session of the Academy of Sciences of Paris of June 12, 1893.

I was not the first to announce this result. A note by M. Berthelot on the same subject had been read 6 weeks before mine (session of April 24) in which he announced that he had isolated from soil, bacterial species that were capable of fixing nitrogen.

I regret that I learned about M. Berthelot's note only after having sent my own note to Paris; I was therefore unable to cite M. Berthelot. M. Berthelot remedied this defect by remarking after the reading of my note: "it will not escape the attention of anyone that this report presents a great analogy in method and in result to the memoir that I myself had read to the Academy almost two months ago."

If one considers only the general results, M. Berthelot was correct. He could see in the facts that I presented decisive proof for a general theory of nitrogen fixation by microbes. But there the analogy stopped. . . .

A second note summarizing my results for the year 1893 was published in the same *Comptes Rendus,* session of February 12, 1894.

In this 1895 paper that we have been reading, Winogradsky concluded that the nitrogen fixing bacterium that he had started to isolate in 1893 was a *Clostridium,* and he gave it the name of Pasteur.

We will now go back to his note of 1893 and see how he proceeded.

The method that I used is a methodical culture, in a medium rigorously devoid of combined nitrogen, and containing a fermentable substance, such as a solution of mineral salts with sugar added.

The distilled water and the salts have been specifically prepared for these experiments. The sugar was very pure dextrose, prepared following Soxhlet's method. It did not contain any trace of nitrogen.

Cultivation was performed in flat bottomed flasks that were placed under large glass bell jars that were provided with a ground glass joint and placed on glass plates. Outside air could reach the cultures only after having passed through washing bottles filled with potassium hydroxide and sulfuric acid. I used special culture flasks with a large surface area which was swept with air filtered through cotton. . . .

The cultures made from soil inocula under these conditions soon acquired rather constant characteristics. In every case one could note a liberation of gas and the production of an acid, principally butyric acid. These products were formed, as long as there was still sugar in the solution, by bulging masses of zoogloea, puffed up by gas bubbles. These masses were formed by a large rod which often contained spores. Whereas other organisms present in the liquid showed, without doubt, signs of involution, these rods were striking by their absolutely normal appearance, their ability to be deeply stained by aniline dyes and the relative abundance of their growth.

I have not succeeded up to the present time in isolating this rod which has refused to grow in a state of complete purity on nutritive gelatine. Two other types of rods are still present in the liquid even though their growth is insignificant. . . .

Quantitative determinations of nitrogen by the Kjeldahl method have been performed on the whole dried culture. . . .

In the following table, the differences in nitrogen gains are explained by differences in dextrose concentrations. In several experiments, for example, in the last six, one did not wait for the total utilization of added sugar and what was left was not determined. The decomposition of one gram of sugar in active cultures took only three to five days, at a minimum.

No.	Added dextrose, gm.	Added nitrogen, mg.	Nitrogen found, mg.	Net gain in nitrogen, mg.
1	1	...	3.0	3.0
2	1	...	2.5	2.5
3	1.5	...	4.5	4.5
4	6	...	10.4	10.4
5	3	...	8.9	8.9
6	?	...	7.2	7.2
7	1	...	2.7	2.7
8	1	2.1	4.5	2.4
9	3	...	8.1	8.1
10	6	...	12.8	12.8
11	7	...	14.6	14.6
12	4	2.1	10.5	8.4
13	?	2.1	7.7	6.6
14	4	2.1	16.4	14.3
15	5	3.0	15.5	12.5
16	2	...	3.1	3.1
17	2	...	2.9	2.9
18	2	...	2.5	2.5
19	2	1.8	3.5	1.7
20	2	4.0	4.6	0.6
21	2	3.3	4.1	0.8

Is there a constant ratio between the amount of decomposed sugar and that of assimilated nitrogen? Can this assimilation take place at the expense of organic substances other than sugar, more specifically at the expense of those found in soil? Which are the culture conditions most favorable from the point of view of nitrogen gain? All these questions are being studied.

And the next year (1894), again in the *Comptes rendus de l'académie des sciences* (**118:** 353–355), Winogradsky continued:

In order to isolate the rod that was thought to be the agent of the fixation of gaseous nitrogen, anaerobic culture conditions were used. Even though anaerobiosis seemed to be in contradiction with the fact that we were dealing with a rod that grew for many months in a well aerated liquid, it was suggested by the following two consideration: (1) the rod was a typical butyric ferment, and (2) in the conditions of impure culture, the association with aerobic microbes might have protected it from the oxygen of air. Indeed, cultivation under vacuum in tubes sealed with a flame, with slices of carrot as the solid medium, permitted us to attain our objective.

When seeded as a pure culture into a thin layer of the same sugary liquid, the rod refused to grow. All cultures remained sterile indefinitely. However, if other bacilli, or even a common mold, were seeded at the same time, the fermentation and the growth of the specific rod started immediately.

Those are just the conditions that this rod, which is very common in soil, finds everywhere, and this explains why it is possible for true anaerobes to grow in such a well aerated medium as soil.

Since it is an anaerobe, the rod ferments sugar in the absence of air as long as some ammonium nitrogen is furnished.

From then on, in order to obtain the fixation of gaseous nitrogen by this microbe under conditions of pure culture, one had to use the following culture conditions: sugary liquid free of combined nitrogen in shallow layers which were in contact with an atmosphere of pure nitrogen. Growth and fermentation under these conditions are most energetic.

I shall give here only the results of two experiments of this type with higher levels of glucose than in previous trials:

	1	2
Dextrose in grams	20.0	20.0
Initial nitrogen in mg.	0.0	0.0
Gain	28.0	24.7

The main products of fermentation of glucose are: butyric acid, carbon dioxide and hydrogen.

The chemical mechanism of the nitrogen-fixing phenomenon in this case seems to be due to the effect of nascent hydrogen on gaseous nitrogen in the midst of the living protoplasm. If this hypothesis is correct, ammonia would be the immediate result of this synthesis.

Nitrogen fixation, then, according to Winogradsky, was performed either by the bacteria growing symbiotically on roots of leguminous plants or by specific anaerobic bacteria. The man who had isolated the root-nodule bacteria *Rhizobium* published in 1901 the results of investigations proving that nitrogen fixation can be carried out by numerous microorganisms, including some aerobic bacteria that he called *Azotobacter*. This man, of whom we have already spoken, was the Dutchman Martinus Beijerinck, one of the towering figures in general microbiology. He had been, in his youth, a rather intense and awkward boy with a keen interest in botany. He studied first at the Delft Polytechnical School, where he got a rather good theoretical training in chemistry. There Beijerinck met the future Nobel Prize winner, J. H. van't Hoff, and the two boys supplemented the deficient laboratory training of the Polytechnical School with experiments performed in their own rooms. These were of great benefit to their minds and were the despair of their landlady. After graduating from the technical school, Beijerinck went to the university and by 1877 he received a doctorate. After a few years of teaching and research in botany, his main subject of investigation being plant galls, Beijerinck accepted the position of microbiologist at the Nederlandsche Gist-en-Spiritusfabriek in Delft. He was well paid and was permitted a great latitude in his research. For example, it was during this period of his life that he isolated in pure culture the root-nodule bacteria, even though their relationship to yeasts and alcohol was not obvious. In 1895, Beijerinck was appointed professor at the Delft Polytechnical School, and a laboratory with adjoining living quarters was built for him. He lived there with his two sisters, and the trio was separated only by death. Beijerinck was a difficult man to get along with; he was subject to fits of depression and had the bad habit of berating the students for the smallest errors. Beijerinck did not think that a scientist should marry, and he looked with displeasure on any sign of friendship between students of opposite sex. It is easy to imagine that this unrelaxed and unpleasant man was not a popular teacher, and comparatively speaking, he had few students. Beijerinck was a keen scientific observer with a great background of knowledge. He loved neat experiments performed with simple tools, and few microbiologists ever got more out of a plate culture than he did. He retired in 1921 and died ten years later after having lived in peaceful retirement surrounded by plants, having chosen in the last years of his life to go back to his first love, botany.

We shall complete our historical survey of nitrogen fixation by microorganisms by presenting passages of one of Beijerinck's papers entitled "On Oligonitrophilic Microbes" (*Centralblatt f. Bakteriologie, Abt. II.*, 7: 561–582, 1901).

I call oligonitrophilic microbes those which in free competition with other microbes grow in nutrient solution where one has not willingly introduced nitrogen-containing substances but from which one has not removed the last traces of these compounds. They have the property of fixing free atmospheric nitrogen either alone or in symbiosis with other microbes.

Two types of accumulation experiments can be performed. One can incubate either: (1) in the light, where growth occurs at the expense of carbonic acid of air and one obtains oligonitrophilic organisms which are colored by chromophyll; or (2) in the dark, in presence of carbon-containing food, and one obtains colorless oligonitrophilic organisms. . . .

The selective cultivation of oligonitrophilic organisms in liquid culture media where sugar furnishes organic carbon was first achieved by M. Winogradsky, mainly under conditions which permitted anaerobiosis, with the development of a specific butyric acid ferment that this author called *Clostridium pasteurianum.* . . . His cultures were seeded with garden soil. First a rich flora of aerobic organisms grew which subsequently furnished the necessary anaerobic conditions for the growth of the oligonitrophilic butyric acid ferment. He also conducted experiments in pure culture, in absence of air, by introducing nitrogen into his culture flasks.

While repeating these experiments, I noted that traces of nitrogen-containing compounds were necessary for the growth of the butyric acid ferment. . . . the same is also true for the oligonitrophilic organisms that I have discovered.

The conditions of my own experiments differed from those under which M. Winogradsky was working in that I permitted only aerobiosis. . . . The result was the discovery of an undescribed genus of aerobic oligonitrophilic bacteria. I shall give the name *Azotobacter* to this genus, which is easily recognized by the size of its members. Up to now, I have identified two species. One, *A. chroococcum,* is very common in garden and fertile soils. The other is just as common in the water of the canal of Delft. . . .

The butyric acid ferment cannot exist in complete absence of oxygen, . . . it is a "microaerophilic" organism . . . (a fact that M. Winogradsky did not note), and thus, free access of oxygen is not by itself sufficient insurance against the growth of this ferment in aerobic cultures. This is why I used in my experiments sources of carbon that were easily assimilated by *Azotobacter* but which did not enter easily . . . into butyric acid fermentation. I found mannitol in solutions of 2 to 10% and the propionates of calcium, potassium and sodium in solutions of ½% to be particularly suitable substances. . . .

One should note at this point that aerobic oligonitrophilic organisms do not form spores . . . in contrast to the butyric acid ferment which forms spores that completely resist temperatures of 90 to 100°C. . . .

Beijerinck concluded this paper, from which we have given only a few quotations, with a short description of the new organisms:

Azotobacter chroococcum was found mainly in soil and *A. agilis* was very common in the water of the canal of Delft.

The following recapitulates what was known about various conditions of microbial life:

1. Raulin had shown that certain organisms could grow on mineral media containing only an organic source of carbon (See Chapter 9).

2. Wildiers had shown that for some organisms small traces of certain organic substances were needed (See Chapter 9).

3. Winogradsky had defined chemoautotrophy and showed that some organisms could obtain their energy from inorganic chemical reactions and their carbon from carbon dioxide. Other organisms with photosynthetic pigments were known to obtain their energy from light.

4. Winogradsky and Beijerinck had shown that some anaerobic and some aerobic bacteria could utilize atmospheric nitrogen.

The successor to Beijerinck at the Delft Polytechnical School, Albert Jan Kluyver, stressed that there was biochemical unity in the metabolism of all organisms.

There was nothing in common between the two men. Beijerinck came from a modest background. He was often coarse in his language, and, as we have said previously, he was a rather unpleasant person who was not sought after by students. Kluyver was the son of a distinguished professor of mathematics. He was suave and kind. Beijerinck did not care for such unscientific aspects of life as marriage and tobacco. Kluyver was married and had a great weakness for the host of Beijerinck's *contagium vivum fluidum*. To fill the position that Beijerinck had held was a formidable job, even for a gentle and amiable intellectual. The atmosphere of the laboratory and the house of microbiology at the Delft Technical University was saturated with the greatness of Beijerinck's towering scientific discoveries. Yet Kluyver was only thirty-four years old when he took over, and he did not have the glamour of having accomplished some outstanding scientific feat. He was a former student at the Technical University in Delft. He must have appeared promising, however, since Dr. D. G. van Iterson, a former assistant of Beijerinck and at that time Professor of Microscopical Anatomy, offered him an assistantship. He received his doctorate in 1914. His thesis research was in the field of yeast physiology. Two years after graduation he left for the Dutch East Indies as an employee of the Dutch Government. His task was to develop technological methods for improving the local industries. Later he became director of an industrial laboratory that studied problems dealing with vegetable oils. Conditions were not ideal when the relatively unknown Kluyver was asked to replace the great Beijerinck. In December, 1921, he took possession of the laboratory, the house, the garden, and the books, all of which had been the sanctuary of his predecessor. Kluyver modernized the laboratory and devoted himself to teaching. He directed the research of his associates toward the probing of microbial biochemistry. His laboratory, at first famous because of Beijerinck, gained recognition through his own efforts. Scientists came from all over the world. The laboratories by the canal were, from a biochemical point of view, an international center. Then the war came, and by the time Kluyver's laboratory was ready to function again, American laboratories were, in their turn, biochemical powerhouses. Dr. Kluyver was nominated rector of his university, and he lectured and traveled extensively. He died, while still actively engaged in scientific work, in 1956. During the span of time in which Kluyver was very active in biochemistry, this science made tremendous advances, facilitated by his fundamental concept of comparative biochemistry. Kluyver established that hydrogen transfer was the basic feature of all metabolic processes. He assembled large amounts of scattered information about the "chemism" of living organisms and molded it into a harmonious and coordinated picture. He started essentially with the isolation of an unusual vinegar bacterium that deposited large crystals of the calcium salt of 5-ketogluconic acid in the medium. The new organism was named *Acetobacter suboxydans* to stress its laziness and unwillingness to perform complete oxidations. The isolation of this bacterium demonstrated the possibility of isolating a series of strains that could be arranged in a sequence so that a metabolic product of one organism could be the substrate of the next one in line. In this metabolic diversity, Kluyver saw a unity that we now take for granted. In this attempt to group strains of *Acetobacter* in a metabolic sequence, we see the forerunner of the analysis of biochemical mechanisms with series of mutant strains. We will not review here any further the work of Kluyver's group, but we would like to stress, before closing this chapter, that from his laboratory came a method for the submerged cultivation of molds. The use of shake cultures was not new. In 1908 in Paris, Sartory published a little book on the influence of agitation on lower fungi in which he even cited previous publications on this subject by other workers. However, the paper of Kluyver and Perquin, published in 1933, demonstrated that submerged fungal growth enabled one to obtain cells that were biochemically homogenous. Because of the war, Kluyver did not have a chance to participate in the development of antibiotics. It must have been a consolation for him, however, that the method he developed with Perquin was used so extensively in that field. In this way, we could say that he contributed to the production of antibiotics before they were ever produced.

We can see that few towns have witnessed so many great microbiological achievements as the small town of Delft in Holland. From the secret lenses of Leeuwenhoek to the biochemistry of Kluyver, a span of three hundred years is covered. In this small town, the invisible was first seen by Leeuwenhoek, and Beijerinck understood that there was something living that could multiply only in growing cells, that was still invisible, and that could be handled as a mere chemical solution. There also, Kluyver forgot about the visible and the invisible to think only about the chemistry of the reactions that take place in microbial cells. Thinking of Kluyver, we could quote these lines of Raulin: "If microorganisms . . . excite to a high degree the curiosity of scientists . . . it is because. . . . the simplest of organized beings present striking analogies with more highly developed living organisms, but unlike the latter, they can be used with marvelous facility for the most delicate scientific investigations."

References

Beijerinck, M. W.: *Verzamelde Geschriften van M. W. Beijerinck* (6 vols.), Martinus Nijhoff, Delft, 1921–1940.

Kamp, A. F., J. W. M. La Rivière, and W. Verhoeven (eds.): *Albert Jan Kluyver, His Life and Work*, North-Holland Publishing Co., Amsterdam; Interscience Publishers, Inc., New York, 1959.

Sartory, A.: *Études expérimentales de l'influence de l'agitation sur les champignons inférieurs*, Capiomont et Cie, Paris, 1908.

Waksman, S. A.: *Sergei N. Winogradsky, His Life and Work*, Rutgers University Press, New Brunswick, N.J., 1953.

Winogradsky, S.: *Microbiologie du sol, problèmes et méthodes*, Masson et Cie, Paris, 1949.

Viruses and Rickettsiae

8 We recall that, in 1885, Pasteur had addressed the Academy of Sciences to announce that he and his coworkers had developed a method that would curb rabies. We also recall that in about 1717 Lady Wortley Montagu had been much impressed by a Turkish method of smallpox vaccination, and that in 1798 Edward Jenner had proposed to divulge to the Royal Society the results obtained by inoculating cowpox into human beings to prevent smallpox. These facts clearly demonstrate that it had been possible to curb some viral diseases before the nature of viruses was in any way understood. We can also well imagine the distress of Koch when he saw Pasteur making vaccines of nothing! After all, Pasteur had not isolated the causative agent of rabies, and, yet he was treating it. To the dogmatic Koch, this must have had a smell of charlatanism.

The road to the elucidation of the properties of viruses started in 1884 when a paper entitled "A Filter Permitting to Obtain Physiologically Pure Water" was published by Charles Chamberland in the *Comptes rendus de l'académie des sciences* (**99**: 247–248). The link to modern virology was then not obvious, as we can see from Chamberland's text:

> Ever since the investigations of M. Pasteur and those that he inspired, the doctrine that a living causative agent is involved in contagious diseases has become of great importance. The careful study of the properties of microbes and of the conditions of the propagation of diseases tends to show that pathogenic microbes are not present in air or at least can be found there only in exceptional instances. It is mainly in water that microbes and their germs are present and this is easily understandable if one thinks that the products of all fermentations and of all de-

compositions end up finally either through rain, infiltration or sewage waters in the water of wells and streams. Thus water is considered one of the principal agents in the propagation of diseases. It was thus of the greatest importance, from the point of view of general hygiene, to have a filter capable of removing from water all the microbes that it contains, in order to purify completely drinking waters.

I have been able to do just that by filtration through a porous vase of porcelain, a method of filtration used in the laboratory of M. Pasteur for the separation of microbes from their culture medium. I have noted that even the most impure water, filtered through these vases, was deprived of microbes and germs.

The apparatus that I have the honor of presenting to the Academy can be directly attached to a water tap and water passes through the filter under the pressure of the water system. Under a water pressure of about two atmospheres, as found in M. Pasteur's laboratory, one obtains, with a single porous tube or *filtrating candle* of 0.20 m. of length by 0.025 m. of diameter, about twenty liters of water a day. This seems to me sufficient for the ordinary consumption of a family. By increasing the number of candles, by grouping them in batteries, one can obtain the necessary water flow for satisfying the need of a school, a hospital or barracks. This filter literally permits one to have his own individual spring at home. M. Pasteur has demonstrated that spring waters, collected at their origin are deprived of microbes.

The cleaning of these filters is most easy. Not only can one brush the exterior surface of the candle, which is covered with the substances suspended in water, but one can also dip it in boiling water or heat it directly in a fire in order to destroy the organic matter which deposited on its surface. By so doing the initial porosity is restored. The same candle can thus be used indefinitely.

Charles Chamberland was, as we recall, one of the trusted collaborators of Pasteur. He had participated in the studies on anthrax, swine erysipelas, and rabies. However, his greatest contributions were of a technological nature. Not only did he describe, as we have seen, the bacterial filter, but he also developed the autoclave, without which microbiological laboratories would not operate. In 1885 Chamberland was elected to the French National Assembly as a representative of the Jura. Chamberland, however, did not abandon science for politics. He was one of the section heads of the Pasteur Institute, and at his death in 1908 he was even its associate director. Roux remarked that it was almost impossible not to get along with the practical Chamberland since he had the "happy disposition of a healthy man who is rightfully successful in his enterprises."

A few years after the publication of Chamberland's paper, a young Russian scientist read a paper before the Academy of Science of St. Petersburg on the tobacco-mosaic disease. As we will see from the following passage of his paper, Dmitrii Ivanowski used the filter of Chamberland intelligently but was most reluctant to come to the right conclusion (*St. Petersb. Acad. Imp. Sci. Bull,* **35**: 67–70, Sept., 1892):

> According to my experiments, the filtered extract (filtered through paper) introduced into healthy plants produces the symptoms of the disease just as does the unfiltered sap. . . . However, I have found that the sap of leaves attacked by mosaic disease retains its infectious properties even after filtration through the candle filters of Chamberland. According to the prevailing opinions of today, it seems to me that this can be explained most simply by the presence of a toxin elaborated by bacteria and dissolved in the filtered sap. Of course there is still another equally acceptable explanation, namely, that the bacteria of the tobacco plant pass through the pores of the Chamberland filter candle. Each filter was tested for small breaks or openings by the accepted procedure. . . . As a further check, liquids most favorable for the growth of bacteria remained sterile for many months after filtration through these candles. . . .

Very little is known about Ivanowski, even though his name is found in most textbooks. After the publication of this paper he continued working on tobacco-mosaic disease for a few years. Then his interest turned toward plant physiology. He worked for several years in Warsaw and died in 1924.

The "viral" work of Ivanowski went unnoticed, but the discovery was in the air and was to be made essentially at the same time by two different investigators (or groups of investigators) who were working on different subjects.

In Germany, in 1898, Löffler and his collaborator Frosch wrote a report on foot-and-mouth disease, in which, as we will see, they observed that the causative agent of this disease could pass through a bacteriological filter. Löffler, as we recall, was an associate of Koch, who was mainly famous for having discovered the causative agent of diphtheria. Frosch was a former assistant of Koch. He was a professor of bacteriology in a veterinary school, which explains his interest in a disease of animals. Together they wrote (*Zentr. Bakteriol. Parasitenk. Abt. I, Orig.,* **23**: 371–391, 1898) the reports of a commission that had been set up to investigate the foot-and-mouth disease. They addressed them to the Minister of Education. Altogether,

there were three such reports reproduced in *Zentralblatt*, one after the other.
We quote a few passages:

> Thanks to the telegraphic communication arranged by your Excellency,
> the commission was informed expeditiously of local outbreaks of foot-
> and-mouth disease and was able to obtain ample fresh infectious ma-
> terial for these investigations. The selected sources of material were
> within the distance of one day's round trip to the laboratory.
>
> For etiological studies the contents of fresh vesicles from mouth
> and udder were used; vesicles on the hoof were not used because the
> soiling with manure increased the possibilities of contamination. The
> surfaces of the vesicles were cleansed with absolute alcohol and the
> lymph was removed with sterile glass capillary tubes. Bacteria were
> sought by hanging drop, stained smear and culture. The culture media
> were conventional broth at acid and alkaline reaction, peptone broth,
> glucose broth, liquid and coagulated blood serum, milk, nutrient agar
> and nutrient gelatin. Cultures were grown in atmospheres of air,
> hydrogen, hydrogen sulfide and carbon dioxide.
>
> The results of all these tests were unequivocally negative. . . .
>
> Lymph that was found sterile by bacteriological examination still
> produced typical foot-and-mouth disease three days after injection into
> the mucous surfaces of the upper and lower lips of calves and heifers.
> The fact that the disease was transmitted to stall-mates by animals
> that had been injected with filtered lymph, indicated that the disease
> was not due to a toxin. . . .
>
> A major objective of the commission was to determine whether or
> not animals which survive the disease acquired immunity against further
> infection. An affirmative answer would indicate that prophylaxis is pos-
> sible.
>
> The losses from foot-and-mouth disease are to be measured not only
> in terms of mortality, but more so in terms of morbidity and its conse-
> quent losses in milk, meat, work power and reproductivity. Once the
> disease has appeared, the spread is so rapid that damage cannot be
> reduced by therapy. Only precautions that hinder emergence of the
> epidemic can protect against damage. The measures available at
> present are disinfection procedures and quarantine.
>
> All efforts up to the present, to endow animals with individual pro-
> tection have been unsuccessful. The prevailing opinion of veterinary
> authorities is that recovery from the disease does not confer immunity
> to subsequent exposure. . . . However, the experience of the commis-
> sion is that immunity certainly develops after recovery from the disease.
> Repeated challenge of animals that had survived epidemics showed that

2 to 3 weeks after outbreak of the disease, a majority of calves and steers possess immunity. The immunity lasts at least 5 months. Admittedly, all animals do not acquire immunity; individual animals develop the disease when reinfected one month after recovery from the disease, but acquire immunity after this second attack. This behavior is similar to that observed with diseases like measles and pox which confer high immunity. In the latter diseases too, not all individuals acquire immunity and can experience the disease two or three times. . . .

Aside from practical aspects regarding the development of immunity, the commission has observations to communicate which are of far-reaching importance, not only for further investigations of foot-and-mouth disease, but also for many other infectious diseases of man and animals. We shall spare no pains to report carefully on this phase of our investigations.

In its efforts to develop a practical immunization procedure, the commission investigated, in addition to mixtures of immune blood and lymph, the blood of acutely ill animals with and without the addition of immune blood, and vesicular lymph which by filtration was freed of corpuscular elements. Our investigations have not led to a satisfactory conclusion, but the filtration experiments have led to important observations.

Active lymph was diluted with 39 parts of water and then mixed with a culture of *Bacillus fluorescens,* isolated as a contaminant in lymph. These mixtures were filtered through infusorial earth candles two or three times. The bacteria were added to check the retentiveness of the filters for bacteria. When culture of the filtrates showed that *B. fluorescens* was absent, the filtration was considered successful. Filtrates so tested were always free of bacteria. Measured amounts of filtrate, representing $\frac{1}{10}$ to $\frac{1}{40}$ cc of undiluted lymph were injected intravenously into calves to determine the possible immunizing activity of soluble substances in lymph. The results were striking. The animals injected with filtrate became sick with all the typical symptoms of foot-and-mouth disease at the same time as control animals injected with equivalent amounts of unfiltered lymph. We had the impression that infectivity was not altered by filtration. To confirm this important observation these experiments were repeated many times. The results were always the same: animals injected with filtrates succumb as often as those injected with unfiltered lymph.

How was this striking fact to be explained? There were two possible explanations: Either (1) an unusually effective toxin was dissolved in the lymph or (2) the filtrate contained a previously undiscovered agent of disease, so small as to pass through the pores of a filter capable of retaining the smallest known bacteria. If it were a soluble poison it must have been amazingly active. An amount of filtrate corresponding to $\frac{1}{30}$ cc of lymph, could, in two days, produce disease in calves weighing 200 kg.

Brieger has prepared tetanus cultures yielding filtrates which could kill mice at a dose of 0.00005 cc; that is, one cc could kill 20,000 mice and one liter could kill twenty million mice. One liter of tetanus culture broth yielded one gram of dry solids; thus 0.0000001 grams of substance killed a mouse or one gram was sufficient to kill ten million mice. . . .

Let us assume that the active component in foot-and-mouth lymph was present at 0.2% concentration, consequently $\frac{1}{30}$ cc would have contained $\frac{1}{15,000}$ of a gram of active material. This amount was adequate to produce typical disease in a 200 kg calf. If the toxin diffused uniformly throughout the body, there would be one part of toxin for $15,000 \times 200,000$ grams of calf tissue. Knorr has prepared tetanus toxin with an activity ratio of 1:1,000,000 for the rabbit, 1:150,000,000 for the mouse and 1:1,000,000,000 for the guinea pig. The foot-and-mouth lymph would be more potent than the most active tetanus toxin. We would have an amazing toxin in foot-and-mouth toxin.

With $\frac{1}{50}$ cc. of lymph obtained from the vesicle of an animal that had developed foot-and-mouth disease after injection of filtered lymph, we produced typical disease in a 30 kg. pig. In the course of the second transmission the agent would have undergone a dilution giving a ratio of $1:50 \times 500 \times 3000$, and in relation to the original lymph, a dilution of 1:2½ trillion. Such a toxic effect is simply incredible.

In the case of tetanus, the toxin can be demonstrated easily in the blood or organs of the intoxicated host. A similar demonstration was not possible with foot-and-mouth disease; 50 to 100 cc. amounts of blood were required to produce the disease in healthy animals. It appears likely that all or a large part of the toxin introduced into the circulation becomes localized at the areas where vesicles are found. Let us assume that the test animal becomes sick after injection of $\frac{1}{30}$ cc. of filtered lymph and let us assume further that the contents of all vesicles in the mouth and on the limbs comprise a volume of 5 cc., a fairly accurate estimate. Then the original amount of toxin injected, $\frac{1}{30}$ cc., would have been diluted 150 fold; of this diluted lymph, $\frac{1}{50}$ cc. injected intravenously was adequate to produce disease in a 30 kg. pig within two days. Thus, $1:150 \times 50$, a 1:7,500 dilution of lymph was toxic or if the dilution of toxin in lymph, as in tetanus broth is 1:500, then $1:7,500 \times 500 - 1:3,750,000$ dilution of toxin is active in 30,000 grams of pig tissue or 2,000 grams of pig's blood. Calculated in grams of toxin, one gram of toxin can act against 7,500,000,000 grams of pig's blood.

According to this calculation as well as the previous one, the toxin assumed to be present in foot-and-mouth lymph is excessively active.

Therefore one cannot exclude the possibility that the activity of the lymph is due to an agent that can multiply. The agent must be so small that it passes through pores that retain the smallest bacteria. The smallest known bacterium is Pfeiffer's influenza bacillus which is 0.5 to 1.0 microns long. Were the hypothetical agent of foot-and-mouth disease $\frac{1}{10}$ or even $\frac{1}{5}$ this size, it would be beyond the resolving power of our microscopes, even with the best immersion systems. This simple consideration would explain our failure to demonstrate the agent in the lymph under the microscope.

Thus, Löffler and Frosch were considering most seriously the possibility that there were small living organisms that could pass through bacteriological filters. The same year (1898) Beijerinck, unaware of Ivanowski's timid contribution, ascribed the cause of tobacco-mosaic disease to a *contagium vivum fluidum*. We shall quote from an extensive memoir on the subject that was published in 1900. We can see that, of these early workers, Beijerinck came nearest to understanding the nature of filtrable viruses [*Arch. néerl. sci.*, (2) 3: 164–186]:

> In 1885, M. Adolf Mayer showed that the disease called mosaic, which affects tobacco plant leaves, is contagious. He expressed the sap of diseased plant, filled up capillary tubes with it, and pricked leaves and stems of plants growing outdoors with them. After one or two weeks these individuals were attacked by the same disease. With a microscope the author was able to find neither bacteria nor parasites of other types in diseased leaves. I was, at that time, the colleague of M. Mayer at the School of Agriculture of Wageningen. He showed me his experiments and, no more than he could I detect microbes in the affected plants to which the disease could be attributed. However, my bacteriological knowledge was at that time very incomplete, so that I could not attach too much weight to my own observation.
>
> I have since that time been continuously involved in bacteriological investigations . . . [and], in 1897, I was able to utilize the new bacteriological laboratory of the Polytechnical School of Delft, which included a heated glasshouse that I could use to continue the study of the mosaic disease. I was thus able to perform a faultless series of experimental inoculations, the results of which I shall give here briefly. . . .
>
> 1. *The infection is not caused by microbes, but by a living liquid virus.*

I first observed that the sap of diseased plants remained virulent after having passed through a porcelain filter, which holds back all aerobic organisms. However, I did not limit myself to the search for aerobic organisms alone, but, I made a point, in addition, to perform very delicate experiments aimed at the detection of anaerobic organisms in the sap which had filtered through the candles. The result was always negative and the sap used in the inoculations thus seemed perfectly sterile.

The quantity of filtered sap necessary to inoculate a tobacco plant is extremely small. A droplet introduced with a syringe of Pravaz at the proper location, is enough to infect a large number of leaves and branches. The sap expressed from diseased portions permits one to inoculate an undetermined number of healthy plants and to transmit the disease to them. This shows that the virus, even though liquid, propagates itself in the living host.

Nevertheless, the experiments with sap passed through the candles remain subject to criticism because the corpuscular nature of the virus had not been definitely disproved. This is why I had recourse to the following diffusion experiments which, in my estimation, gave absolutely unassailable results.

I put on the surface of wide and thick agar plates, drops of sap which had been expressed from diseased leaves or fragments of crushed leaves. I thus let the virus spread itself by diffusion in the hope of separating it from the matter of leaves and from all bacteria. Indeed, if the virus was diffusible, it would penetrate into the agar laterally as well as in depth, leaving completely behind all solid particles, be they invisible or bacteria, be they aerobic, anaerobic or spores. The experiment was thus to settle the matter of whether the virus is truly diffusible and thus water soluble, or whether not being diffusible, even though extremely divided, it must be considered as corpuscular, that is to say as a fixed contagium. It was noted that the substance which causes infection can penetrate rather deeply into the agar.

After a length of time that I deemed sufficient for the virus to penetrate to a measurable distance into the agar, in case it was diffusible, I washed the ·surface of the agar first with water and then with mercuric chloride. Finally, at the location where the mashed leaves or the expressed sap had been deposited, I removed with a small sharp spatula a layer of agar which was roughly half a millimeter thick. The agar immediately under was then removed in two successive layers and these two portions were used to inoculate healthy plants. The results left no doubt. In both cases characteristic signs of infection occurred which were more intense with the upper portions of agar than with the deeper layers. . . . Even though this diffusion was demonstrated only over a distance of a few millimeters, it seems at any rate to prove that the virus is truly liquid or dissolved, and not corpuscular. . . .

The sap filtered through the candles acts on plants in a slightly less intense manner than the expressed sap which has not yet been filtered. . . . I would be completely wrong to deduce from that, that the virus is of a corpuscular nature.

Beijerinck's reason for thinking that a certain amount of retention on filters did not mean the presence of corpuscles was based on the observation that certain very water-soluble enzymes were fixed by such filters. This fixation was noted when the solutions were very diluted. Eventually the filters became saturated and the enzymes went through in full strength. Beijerinck also noted that the filterable virus could be fixed on bacterial cells. The bacteria then gave the false impression that they were the cause of the disease.

Let us go back to Beijerinck's text:

2. *Only growing plant organs where cellular division takes place are susceptible to infection. There only does the virus multiply.*

Among the tissues and the organs of tobacco plants, only those which are both rapidly growing and the site of cellular division, are attacked by the virus. All adult tissues are immune to infection but are capable of carrying the virus under determined conditions. Leaves which are still growing, but in which only the elongation of cells take place, are no longer "infectable" even though they are perfectly able to transmit the virus toward the stem. . . .

The virus, even though not able to grow by itself, is carried in the growth of dividing cells where it multiplies enormously, without losing any of its individuality. One thus understands that outside of the plant no multiplication can be observed. I have been able to keep, bacterial free, filtered sap for three months without loss or detectable decrease of virulence. However, I was not able to observe either any increase in contagious properties, even at the beginning of the experiment, even though the sap had been prepared not only by the crushing of diseased portions but also by adding the sap of young healthy buds and leaves. Thus, then, if the virus had been able to grow by nutrition in the usual manner, multiplication would have very probably taken place. . . .

The mode of multiplication of the virus reminds one, in many ways, of that of . . . chromoplasts which also grow only within cellular protoplasm, even though they have an independent existence and function separately. . . .

4. *The virus can be dried without losing its infectious properties.*
Fragments of dried diseased leaves, introduced into healthy plants
were able to infect even after having been kept for two years in my
herbarium. . . . However, the virulence of dried materials is always
lower than that of fresh materials, which, I think, means that this virus
is partially destroyed during desiccation. . . . I shall note, in addition,
that the precipitate formed by adding alcohol to the expressed sap
keeps its virulence after desiccation at 40°C. . . .

5. *The virus can spend the winter in soil, outside of the plant and
in a dry state.*

In the fall of 1897, I let a diseased plant die from desiccation in a
shed in a large flower pot. The plant was pulled out and the soil which
was adhering to the roots was shaken back into the pot which was kept
in a dry state. The next spring I put some of that soil in four pots which
were partially filled with fresh soil. One of these pots was larger than
the others and received three plants, in each one of the smaller pots
only one plant was placed. . . . All these plants were certainly healthy.
After about 4 weeks, one of the three plants in the large pot was diseased and so were all those in the small pot. . . .

7. *The virus is inactivated by boiling temperature.*

. . . Boiling completely destroys the virus; even 90°C is not tolerated
for a short period of time. I have not yet determined the minimum
inactivating temperature, but it seems that we are simply dealing with
ordinary temperatures of pasteurization. In fact, these results are sufficient by themselves to eliminate the possibility of our dealing with
sporulating anaerobic organisms. . . .

Beijerinck completed his paper by emphasizing the likelihood of there
being other diseases of plants caused by such filterable contagious fluids.
He stressed that from the studies of Erwin Smith one might suspect peach
yellows and peach rosette to be caused by such agents. Indeed, Smith had
been unable to find a causative agent for these diseases but had been able
to transmit them by grafting. Beijerinck's interpretation of Smith's observations was correct, and in 1957 another Smith, Kenneth Smith, published a
book on *Plant Virus Diseases* (Little, Brown and Co., Boston) in which over
three hundred separate plant viruses are listed, including the peach yellows
and the peach rosette viruses.

The demonstration that filterable agents could also cause human disease
was made during a study of the etiology of yellow fever carried out in Cuba
by an American commission headed by Walter Reed, an Army surgeon.
Walter Reed was born in Gloucester County, Virginia, in 1851, son of a

Protestant minister. His first medical degree, from the University of Virginia in 1869, was later fortified by a second degree, from Bellevue Medical College in New York City. At the ripe old age of twenty-two he became one of the five inspectors on the Brooklyn Board of Health. In 1875, he entered the Army Medical Corps and during the next eighteen years he served in various parts of the United States. During that time his services were not limited to members of the Army. Often the only physician for miles around, he was frequently called upon to treat members of the civilian population. In this way he gained valuable experience and the habit of making do with very inadequate means.

Reed impressed his superiors with his capabilities, and they felt that the rest of his life should not be spent in the field. In 1890, he was assigned to Baltimore and had the good fortune of having close contact with William H. Welch. In 1893, Reed was appointed curator of the Army Medical Museum in Washington and professor of bacteriology at the Army Medical School.

During the Spanish-American War, Reed was appointed head of a commission investigating the devastating outbreaks of typhoid fever in Army installations. Hundreds of cases were studied, and the commission came to the conclusion that flies, contaminated water, and contact with patients or the articles they had worn or both were the main channels of dissemination of the infection. In 1900, Reed was again asked to head a commission to investigate the infectious diseases of Cuba. Most specifically, the board was concerned with yellow fever. This was a very puzzling disease. The Americans had started to curb most infections in Cuba by introducing sanitary measures. Whereas most diseases showed a decline following the application of these measures, yellow fever was unaffected. If anything, the incidence of this disease was increasing, the increase probably being due to the introduction into the country of numerous foreigners who did not have any immunity to this affection.

Reed suspected, before even starting his study, that an insect of some kind was the distributor of the unknown germ. This hypothesis was based on the analogy to the recently recognized fact that certain mosquitoes played a role in the etiology of malaria. In addition, a certain Dr. Carlos Finlay of Havana had advanced the theory that mosquitoes were the injectors of the unknown causal agent of the disease. This was a bold and accurate guess, but the good doctor was unable to back it up with clinching experiments. There were, of course, more orthodox theories to explain the etiology of the disease. Bacteria were, by then, fashionable etiological agents, and the prevailing theory in 1900 was that a certain *Bacillus icteroides* (probably some *Salmonella*) was the etiological agent. Reed and Carroll had already found this organism in the United States, under conditions having no apparent relation to yellow fever. Their examination of the blood from twenty-four patients suffering from the fever and the thorough examination of the bodies of eleven former patients who had died of the disease completed the discrediting of the bacillar theory by failing to reveal any trace of *B. icteroides*.

Reed obtained the permission to experiment on human volunteers. He also requested money to "compensate" the volunteers for their services. The requests were granted, and many soldiers volunteered for this rather unpleasant duty. Dr. Carroll himself was the first volunteer to receive the bite of a mosquito that had bitten a yellow-fever patient twelve days previously. He became gravely ill. Another member of the commission, Dr. Lazear, performed the same experiment on himself, and was not infected. He repeated the experiment involuntarily, being bitten by a mosquito as he was in a yellow-fever hospital ward. He noted this fact carefully, caught the disease, and died.

A camp was especially constructed to carry out the experiments with human volunteers. Located a few miles from Havana, it was called, appropriately, Camp Lazear. All possible precautions were taken to obtain valid results. Initially the volunteers were protected from the bite of any mosquito for a long enough period of time to insure that they had not been inoculated by natural means before being admitted to the camp. A very complete medical record was kept for each patient in order to eliminate confusion with any nonspecific affection. The mosquitoes to be used in these experiments were the subject of as much care as the patients. They were bred from eggs in a little house of their own before being taken for their human meal in the yellow-fever ward of a hospital and allowed to regale themselves on a patient. As soon as a volunteer would show the first sign of the disease after having been bitten, he was taken to the yellow-fever ward of a hospital. There was not a single case of involuntary infection at the camp. Every possible attempt was made to contaminate the volunteers with objects that had been in intimate contact with yellow-fever patients. These attempts were successful only in demonstrating the strong stomachs and the unlimited devotion of the volunteers. Not only mosquitoes, but also syringes were used to transfer blood from patients to volunteers.

Among the volunteers listed in the report of the Surgeon General one finds: "Miss Clara Louise Maass, of East Orange, N.J., aged 25. Trained nurse in Las Animas Hospital. She served as trained nurse in Medical Department of the Army during the Spanish War. She volunteered and was bitten by infected mosquitoes August 14, and died of yellow fever August 18, 1901."

At the end of the same year, the following paper was presented by Reed and Carroll at the annual meeting of the Society of American Bacteriologists in Chicago:

> In former contributions to this subject, we have shown by observations made on human beings that yellow fever may be produced in the nonimmune individual either by the bite of the mosquito (1) (genus *Stegomyia*) that has previously been permitted to fill itself with the blood of a patient suffering with yellow fever, during the first three days of the attack, or by the subcutaneous injection of a small quantity of blood (2) (0.5 to 2 c. c.) drawn from the general circulation of such a patient during the active stage of this disease. . . .
>
> Although these experiments have demonstrated that the specific agent of yellow fever is present in the blood, we may say that the prolonged microscopic search which has been made by other investigators, as well as by ourselves, both with fresh and stained preparations of blood, taken at various stages of this disease and during early convalescence, has proven thus far entirely negative. We may add that the efforts which we have made with reasonable hope of reward, both in the bodies of infected mosquitoes, dissected in the fresh state, as well as by serial sections of the hardened insect, have likewise given no results which we consider worthy of record at the present time. Leaving out of consideration, therefore, for the time being, the further microscopic search for the specific agent in the blood of the sick and in the bodies of infected mosquitoes, we desire to call attention to some additional observations bearing on the etiology of the disease, which one of us (Carroll) has recently made at Las Animas Hospital, Habana, Cuba, and at Columbia Barracks, near Quemados, Cuba.
>
> We here desire to express our sincere thanks to Dr. William H. Welch, of the Johns Hopkins University, who, during the past summer, kindly called our attention to the important observations which have been carried out in late years by Loeffler and Frosch relative to the etiology and prevention of foot-and-mouth disease in cattle. . . .
>
> It was for the purpose of ascertaining whether observations conducted along the same lines as those above mentioned might throw additional light upon the etiology of yellow fever that the following experiments were undertaken.
>
> Of course it will be thoroughly appreciated that in experimentation on human beings, aside from the grave sense of responsibility, at times

well-nigh insupportable, which the conscientious observer must always feel, even with the full consent of the subjects to be experimented upon, there must be added another factor, viz, the difficulty of finding willing and suitable nonimmune individuals for experimentation just at the proper and urgent moment. It so happened that on the day of Dr. Carroll's arrival at Habana, August 11, 1901, the first patient of the series of seven cases of yellow fever which Dr. Guitéras had produced by bites of infected mosquitoes, was taken sick. The fatal termination of three of these cases produced a somewhat panicky feeling toward experimental yellow fever among the nonimmunes at Habana, which feeling was intensified by the sensational and distorted statements in one of the local Spanish papers. It was, therefore, extremely difficult— in fact, practically impossible—to obtain for inoculation purposes persons who could with reasonable certainty be regarded as nonimmunes.

Further, as it was not practicable to withdraw blood from any case of yellow fever under treatment in the city of Habana, it became necessary to produce cases by means of the bites of infected mosquitoes —*Stegomyia fasciata*—accepting such subjects as were willing to submit to this mode of inoculation. In all, six individuals, supposedly nonimmunes, were bitten by mosquitoes, of whom four gave a negative and two a positive result. . . .

Because of experimental problems, only Case II could be used:

CASE II   J. M. A., Spaniard, recently landed at Habana, was bitten at 4 P.M. October 9, by 8 mosquitoes that had been applied to a severe case of yellow fever (Case I) on the second day of the disease, 18 days previously. The attack which followed was mild. According to his own account, he went to bed on the evening of the 12th feeling in perfect health. He awakened about midnight with frontal headache, but had no chill. October 13, 7 A.M., temperature was 102.2°F., pulse 92; complained of pain in the head and back; later in the day there was marked photophobia; pain in the region of the kidneys, and slight pains in the lower extremities. The eyes were injected moderately and the gums slightly so. On the following day, October 14, the frontal headache was more severe; there was considerable soreness on pressure over the stomach and abdomen, and he complained of sharp lumbar pain. An examination of the fresh blood proved negative for malarial parasites. At 4 P.M. blood was drawn from a vein at the bend of the elbow and 10 drops were inoculated into each of 2 flasks containing 200 c.c. of sterile bouillon. One flask remained sterile, the other developed a growth which proved to be a white staphylococcus.

October 15 the gums were pale, swollen and spongy, their margins distinctly reddened, and blood could be easily pressed out from beneath the lower gums.

October 16 there was free oozing of blood from the gums and margin of the tongue. The case pursued a mild course, the temperature falling to normal at 9 P.M. of the fourth day. A trace of albumin was present in the urine passed on the morning of that day, and for a few days following hyalin and granular casts were found. The patient made a speedy recovery.

On October 15, 11:30 A.M., at the beginning of the third day of illness, the temperature was 101°F.; 65 c.c. of blood were drawn, with antiseptic precautions, from a vein at the bend of the elbow. This was placed in a sterile test tube and set aside in the refrigerator. At 6 A.M., 5 hours later, 19 c.c. of a slightly bloodstained serum were pipetted off into another sterile tube. After the addition of an equal quantity of sterilized distilled water the diluted serum was slowly filtered through a new Berkefeld laboratory filter that had been subjected to previous sterilization in an Arnold's sterilizer. In this way 35 c.c. of a slightly bloodstained filtrate were obtained, a part of which was subsequently used for the inoculation of Cases VII, VIII and IX of this report.

The original level of the blood having been marked upon the tube into which it was drawn, a sufficient quantity of sterilized distilled water was then added to replace the 19 c.c. of serum that had been pipetted off and to make up the original volume of blood. The whole, consisting of clot, remaining serum, and distilled water, was poured into a sterile vessel and whipped up with a sterilized egg beater. The mixture, which approximately represented the partially defibrinated blood, was then divided into two parts, one of which was reserved for the inoculation of a control subject (Case III), while the other part was placed in a double water bath previously heated, and exposed to a temperature of 55°C. for 10 minutes. It was then removed and immediately cooled in ice water. This cooled material was subsequently used for the injection of Cases IV, V, and VI.

It will thus be seen that we have at our disposal, for purposes of inoculation, three kinds of materials, derived from the blood in Case II, viz: (a) The unheated and partially defibrinated blood; (b) the partially defibrinated blood which had been heated to a temperature of 55°C. for 10 minutes, and (c) the diluted blood serum which had been filtered through a Berkefeld filter. Each of these materials was used for the inoculation of one or more nonimmune individuals with the results that follow herewith. .

(a) *The Unheated and Partially Defibrinated Blood*

CASE III   M. G. M., Spaniard, arrived at Habana October 4, 1901. At 4 P.M. October 15 he was given a subcutaneous injection of 0.75 c.c. of the unheated and partially defibrinated blood obtained from Case II, 15 hours previously, which had been kept 5 hours in the refrigerator and 10 hours at room temperature. The earliest symptom, frontal headache, was complained of at 6 P.M., October 20, or at the expiration of 5 days and 2 hours after inoculation. Temperature 100.6°F., pulse 80. At 3 P.M. of the same day the temperature was 98.4°F., pulse 80. At that time the patient did not complain of any discomfort and there was nothing to indicate that he was about to be taken sick. October 21, 5 P.M., nearly 24 hours from the onset, there was flushing of the face, injection of the eyes and gums, and moderately severe headache; pain in the back, and tenderness on pressure in the epigastric region made the picture complete. On the third day the face was deeply flushed, eyes congested and distinctly yellow. There was slight oozing of blood from the gums. The urine passed at 7:30 P.M. contained a distinct trace of albumin. The case was seen by the Habana Board of Yellow Fever Experts and the diagnosis confirmed. The patient passed through a mild but typical attack, the temperature touching normal on the fifth day.

This case, therefore, serves as a "control" for the observations which are to follow, since it demonstrates that the blood drawn from the general circulation of Case II, at the beginning of the third day, contained the specific agent of yellow fever, and, in this respect, confirms the observations which have heretofore been reported by us.

(b) *The Partially Defibrinated Blood Heated for 10 Minutes at 55°C*

CASE IV   A. C., Spaniard, nonimmune, arrived at Habana, October 6, 1901. At 4:35 P.M., October 15, he was given subcutaneously 1.5 c.c. of the partially defibrinated blood which had been subjected to a temperature of 55°C. during 10 minutes. The specimen had been drawn from Case II, 16 hours before. The result of this injection was entirely negative, as the subject remained in perfect health during the 10 days following.

CASE V   B. F. M., Spaniard, nonimmune, arrived at Habana, October 6, 1901. At 4:45 P.M., October 15, he received a subcutaneous injection of 1.5 c.c. of the same material that was used in Case IV. Result negative.

CASE VI   S. O., Spaniard, nonimmune, arrived at Habana, October 7, 1901. At 4:50 P.M., October 15, he was given a subcutaneous injection of 1.5 c.c. of the same material that was used in Cases IV and V. No rise of temperature or other symptoms of ill health followed this injection.

We desire to invite attention to the fact that the four subjects whose protocols have been given above were young Spaniards, who arrived at Habana at a time when yellow fever was not present in the city; that they were carried from the quarantine camp, at Triscornia, across the bay, direct to Columbia Barracks, near Quemados, Cuba, where they were kept for seven full days prior to inoculation; and that after inoculation they were kept under close daily observation for the further period of 10 days, during which time both temperature and pulse were recorded every third hour. Since under these circumstances each of the three nonimmunes (Cases IV, V, and VI) received, without any disturbance to health, double the quantity of heated and partially defibrinated blood that sufficed when unheated to cause an attack of yellow fever in Case III, it follows that the specific agent present in the blood in yellow fever is destroyed, or, at least markedly attenuated, by a temperature of 55°C. maintained for 10 minutes.

(c) *The Diluted and Filtered Serum*

CASE VII   P. H., American soldier, nonimmune, received at 11 A.M. October 15, 1901, a subcutaneous injection of 3 c.c. of the serum filtrate, representing 1.5 c.c. of the undiluted serum 10 hours after the blood had been drawn from Case II. He remained in good health until 3 P.M. October 19, an interval of four days and four hours, when his face appeared flushed and his eyes somewhat injected. His temperature at this time was 101°F., his pulse 80. He did not complain of headache or other pain. From this hour his temperature declined, until at 12 o'clock midnight it registered 98°F., pulse 72. October 20, 9 A.M., temperature 100.8°F., pulse 78. Face more suffused and slight headache complained of. Fever continued on the 21st, with more marked flushing of the face and injection of the eyes. The height of the primary febrile paroxysm was reached at 6 P.M. October 21. Remission occurred at 9 A.M. October 22, when the temperature dropped to 98.8°F., pulse 64. This lasted for 24 hours and was followed by a secondary febrile paroxysm of 42 hours' duration. On the 23rd blood was oozing from the lower gums and the eyeballs were tinged with yellow. Albumin appeared in the urine on the fourth day. The patient was visited by the board of experts and the diagnosis of yellow fever confirmed. Examination of the dried blood for malarial parasites was negative. The patient recovered.

CASE VIII A. W. C., American soldier, nonimmune, was also given at 11:05 A.M. October 15, 1901, a subcutaneous injection of 3 c.c. of the diluted and filtered serum, being the equivalent of 1.5 c.c. of the undiluted serum, 10 hours after the blood had been drawn. He remained in his usual health until about noon, October 19, at which time he felt "out of sorts" and ate but little dinner. This was 4 days and 1 hour after the injection. During the afternoon he lay down and slept until 3 P.M., when he awoke with a severe headache and backache. His face was flushed. Temperature 103.6°F., pulse 102. At this hour his face and eyes were deeply congested, and from this time his symptoms were characteristic of the disease. On the 23rd his eyes were quite yellow and general jaundice followed later. No albumin was found in this patient's urine. He was seen by the board of experts and his illness pronounced a typical case of yellow fever. Careful examination of the dried blood for malarial parasites was negative. The patient made a good recovery.

CASE IX J. R. B., American, nonimmune, at 2:30 P.M. October 15, 1901, was given a subcutaneous injection of 3 c.c. of the diluted and filtered serum, equal to 1.5 c.c. of the undiluted serum. Fourteen hours had elapsed since the blood had been drawn from Case II.

This injection was followed by no symptoms of physical disturbance, until 3 P.M. October 19, an interval of four days and a half hour, when his temperature was 99.4°F., pulse 92. He complained of headache and flashes of heat, with slight pain between the shoulders, symptoms which, the subject stated, were quite unusual to him. At 9 P.M. temperature 98.4°F., pulse 84. There was no further febrile disturbance and the day following the subject was in his usual good health.

We thus observe that of 3 nonimmune individuals who received subcutaneously an injection of filtered blood serum derived from Case II of this report, 2 developed an unmistakable attack of yellow fever, after a period of incubation of 98½ hours and 100 hours respectively, while in 1 case the result must be regarded as negative.

As already stated, the serum used for these inoculations had been slowly filtered through a new Berkefeld laboratory filter. As soon as possible thereafter the filter was resterilized by steam and thoroughly tested as to its effectiveness in preventing the passage of bacteria. For this purpose a recent bouillon culture of Staphylococcus pyogenes aureus was used, of which 50 c.c. were passed through the filter. The filtrate thus obtained was transferred in quantities of 10 c.c. to each of two flasks containing 200 c.c. of sterile bouillon, which were incubated at 37°C. for 4 days and thereafter kept at room temperature for 10 days longer, at the end of which time no growth had occurred. It appears, therefore, that the filter used for the filtration of the blood serum in Case II was to be relied upon for the delivery of a bacteria-free filtrate.

The production of yellow fever by the injection of blood serum that had previously been passed through a filter capable of removing all test of bacteria, is, we think, a matter of extreme interest and importance. The occurrence of the disease under such circumstances, and within the usual period of incubation, might be explained in one of two ways, viz, first, upon the supposition that the serum filtrate contains a toxin of considerable potency; or, secondly, that the specific agent of yellow fever is of such minute size as to pass readily through the pores of a Berkefeld filter. . . .

Against the view that a toxin is present in the serum filtrate, we invite attention to the innocuousness of the partially defibrinated blood when heated to 55°C. for 10 minutes, as shown by the negative results in Cases IV, V, and VI. Here the toxin, which must have been present in just the same quantity as in the serum filtrate obtained from this blood, appears to have been completely destroyed by the temperature above mentioned. Now, although certain bacteria are destroyed by this temperature, as yet we know of no bacterial toxin that is rendered inert by such a low degree of heat continued for so short a time. The tetanus toxin, which has been found to be the most sensitive thus far requires, according to Kitasato, a temperature of 60°C. for 20 minutes, or 55°C. for 1½ hours, in order to destroy its activity.

As a further test and in order to determine whether the serum filtrate contained something more particulate than a soluble toxin, we availed ourselves of the opportunity of observing the effect that would follow the transference to a third individual of blood drawn from one of the patients whose attack had been occasioned by the injection of 1.5 c.c. of serum filtrate (Case VII). If under these circumstances it would be found that the injection of a small quantity of blood was followed by an attack of yellow fever in a third individual, the evidence would point in the strongest manner to the presence of the specific agent of the disease in such blood, since we can hardly believe that a toxin which had undergone so great a dilution in the body of the second individual would still be capable of producing the disease.

CASE X   October 22, 1901, 3 P.M., J. M. B., American, nonimmune, who on October 15, 1901, at 2:30 P.M., had been injected with 1.5 c.c. of serum filtrate with negative result (vide Case IX), and who still desired to have his immunity further tested, was, at the beginning of this, the eighth day after his former inoculation, given a subcutaneous injection of 1.5 c.c. of blood drawn from the venous circulation of Case VII early in the fourth day of the disease. At the time of inoculation the subject's condition was quite normal. October 23, 3 P.M., after an incubation period of just 24 hours, he complained of frontal and slight basal headache and some pain between the shoulders. His temperature was 99.6°F. and pulse 100. At 6 P.M., temperature 100.4°F., pulse 100. Pain in the back quite severe. At 10:15 P.M., he suffered a slight chill. On the following morning the face was flushed and the eyes and gums injected; there was sharp frontal headache and some photophobia. The height of the primary paroxysm was reached at the end of 23 hours. Remission occurred at 9 A.M., October 25, and was followed by a second febrile paroxysm of 45 hours' duration. On the third day, during the secondary fever, the patient presented the typical picture of a mild case of yellow fever; the face was deeply flushed, eyes well injected and slightly yellow; there was sharp headache, and epigastric tenderness and pain in the lower extremities. Heller's test showed albumin in the urine drawn on the fourth day. His fever subsided on the latter day and he made a prompt recovery. The case was seen by the board of experts and the diagnosis confirmed.

In considering this individual's attack, his infection must be attributed either to the injection of the serum filtrate derived from Case II, in which event the onset of his disease was postponed until the commencement of the ninth day after inoculation, or to the injection of blood obtained from Case VII, after a period of incubation of 24 hours.

In our own experience and that of Guitéras of 22 cases of experimental yellow fever, following the bite of the mosquito, in which the period of incubation was definitely and accurately ascertained, the longest period was 6 days and 1 hour, and the shortest period 2 days and 13 hours. If we take the cases produced by the injection of blood, 7 in number, exclusive of the case under consideration, the longest period was 5 days and 2 hours (Case III of this report) and the shortest 41 hours.

In view of these data, we believe we are justified in expressing the opinion that the source of infection in Case X must be attributed to the injection of blood drawn from Case VII, rather than to the injection of the filtered serum derived from the blood in Case II; and further, that the blood in Case VII contained the specific agent of yellow fever, which had, therefore, passed through the filter along with the filtrate with which this latter individual had been inoculated.

The important questions which naturally arise from the foregoing experiments must be left for the future observations to determine.

Following the demonstration that mosquitoes were the peddlers of yellow fever, an eradication campaign was carried out with energy in Havana. By the time Reed was addressing the Society of Bacteriologists in Chicago, Havana, the home city of yellow fever, had seen its last case of the disease, and as a bonus a reduction in malaria was further noted.

Major Reed died of appendicitis, in 1902, in Washington. He had died too soon after having completed his work on yellow fever to receive the honors that his work should normally have brought him. He had discovered the viral nature of yellow fever, the first such disease known to attack man. He had made it possible to save countless lives and suffering by pointing to the potential methods of eradication. His widow received a modest pension. It seemed that the United States of America was not ready to honor handsomely its scientists.

James Carroll had been the first volunteer in this study of yellow fever and had been the main collaborator of Reed. He was a native of England who had emigrated to Canada at the age of fifteen. For a few years he enjoyed the life of a woodsman, and in 1874 he enlisted in the United States Army. He left the Army for the School of Medicine of the University of Maryland and graduated in 1891. He went to the Johns Hopkins Hospital for postgraduate work in bacteriology and pathology, and, as we have seen, he was sent to Cuba as a member of the yellow-fever commission. Born in 1854, he died in 1907.

It was soon found that cell-free filtrates could also transmit certain tumors. The first report came from Copenhagen, where V. Ellerman and O. Bang published a paper entitled "Experimental Leukemia in Hens" (*Zentr. Bakteriolo. Parasitenk., Abt. I, Orig.*, **46**: 595–609, 1908). They said:

> Leukemia was first recognized as a disease entity by Virchow. The types of blood cells affected were described by Ehrlich on the basis of his recognition of granular polymorphonuclear leucocytes and agranular lymphocytes. Accordingly, he differentiated two types of leukemia and showed that the qualitative changes in the leukemic cells were more significant than the quantitative ones. The cause of leukemia is unknown, but an infection theory as well as a non-infection theory has been proposed. . . .
>
> For experimental transmission, portions of liver, spleen and bone marrow from specifically diseased hens were emulsified in 0.9% NaCl and injected intravenously into normal hens. Serial transmission was carried through three passages and 6 of 11 birds in the third passage

acquired the disease. Pathological changes identical with the natural disease were observed three months after inoculation. . . .

Bacteria could not be demonstrated in stained sections, smears or culture. For culture, serum agar was used with aerobic and anaerobic maintenance; serum agar with added chicken hemoglobin was also used. Spirochetes also could not be demonstrated.

The positive transmission in serial passage rules out the possibility of toxin as agent of the disease. In a first trial, passage was tried with a cellular material. Emulsions of infective tissue were centrifuged and the supernatant passed through three layers of filter paper. The filtrate produced leukemia in one of three hens, but the filtrate was not absolutely free of cells.

In a second trial, an organ emulsion was centrifuged and the supernatant was passed through an infusorial earth candle. Five hens were inoculated with the clear filtrate. Two hens developed leukemia.

These experiments indicate that cell-free filtrates can produce leukemia. The transplantation of cells is not involved. It is likely that the disease is caused by an organized virus. . . .

And from the United States in 1911 came a paper by Peyton Rous announcing that not only a leukemia but also a sarcoma could be induced in the chicken by sterile, cell-free fluids (*J. Am. Med. Assoc.*, **56:** 198, 1911). Dr. Rous was working in the laboratories of The Rockefeller Institute for Medical Research in New York.

A tumor of the chicken, histologically a spindle-celled sarcoma, has been propagated in this laboratory since October, 1909, and in the past few months has developed extreme malignancy. From a bit inoculated into the breast muscle of a susceptible fowl there develops rapidly a large, firm growth; metastasis takes place to the viscera; and within four to five weeks often the host dies. The behavior of the new growth has been throughout that of a true neoplasm, for which reason the fact of its transmission by means of a cell-free filtrate assumes exceptional importance.

For the first experiments on the point use was made of ordinary filter-paper and the ground tumor suspended in Ringer's solution. It was supposed that the slight paper barrier, which allows the passage of a few red blood-cells and lymphocytes, would suffice to hold back the tumor and render the filtrate innocuous. Such has been the experience of other workers with mouse and dog tumors. But in the present instance characteristic growths followed the inoculation of small amounts of the watery filtrate, and followed also the inoculation of the fluid supernatant after centrifugalization of a tumor emulsion.

These results led to more critical experiments, which will be here detailed. Tumors of especially rapid growth and young, well-grown, barred Plymouth Rock fowls were used throughout.

EXPERIMENT 1 Tumor material from the breast of a chicken was ground with sterile sand, suspended in a considerable bulk of Ringer's solution, and shaken for twenty minutes in a machine. The sand and tumor fragments were separated out by centrifugalization in large tubes for five minutes at 2,800 revolutions per minute. Of the supernatant fluid a little was pipetted off, and this centrifugalized anew for fifteen minutes at over 3,000 revolutions per minute. From the upper layers sufficient fluid for inoculation was now carefully withdrawn. The pure-bred fowls were injected in one breast with 0.2 cc. of the fluid, in the other with a small bit of tumor tissue. All developed sarcoma at the site of this latter inoculation, and in seven the same growth slowly appeared at the point where the fluid had been injected.

EXPERIMENT 2 Tumor from a chicken was ground, suspended, and shaken as before. But after one centrifugalization the fluid was passed through a Berkefeld filter No. 2 (coarse). Before filtration, it was a pinkish-yellow, cloudy; afterwards, faintly yellow, limpid. Nine fowls were inoculated with 0.2 cc. of the filtrate in each breast, and twenty-two more received filtrate in one breast, a bit of tumor in the other. Of the nine, one slowly developed a sarcoma in each breast, and later microscopic growths were found in its lungs. Of the twenty-two receiving both filtrate and tumor, five developed sarcoma where the filtrate had been injected, and these five showed especially large growths from the tumor bit.

The Berkefeld filter employed was later found slightly pervious to *Bacillus prodigiosus.*

EXPERIMENT 3 The filtrate was similarly prepared except that a small Berkefeld filter (No. 5 medium), impermeable to *Bacillus prodigiosus,* was used. As before, the filtration was done at room temperature. Twenty chickens were inoculated in each breast with the filtrate, but none have developed tumors.

EXPERIMENT 4 In this experiment the material was never allowed to cool. About 15 gm. of tumor from a chicken was ground in a warm mortar with warm sand, mixed with 200 cc. of heated Ringer's solution, shaken for thirty minutes within a thermostat room, centrifugalized, and the fluid passed through a filter similar to that used in Experiment 3. Both before and after the experiment, this filter was found to hold back *Bacillus prodigiosus.* The filtration of the fluid was done at 38.5 C., and its injection immediately followed. In four of ten fowls inoculated with the filtrate only (0.2 to 0.5 cc. in each breast) there has developed a sarcoma in one breast; and though the growths required several weeks for their appearance their enlargement is now fairly rapid. Pieces removed at operation have shown the characteristic tumor structure.

As has been pointed out, the special significance of these results lies in the growth's identity as a tumor. The original sarcoma was found as a unique instance in a flock of healthy fowls; and, though susceptible normal chickens and others with the tumor have since been kept together in close quarters for long periods, no instance suggesting a natural infectivity of the growth has occurred. When inoculated, it is at first a local disease, very dependent on the good health of the host. At this time intercurrent illness of the fowl will check the nodule's growth or even cause it transiently to disappear. For long the sarcoma could be transferred only to fowls of the same pure-bred variety in which it arose, and this only in an occasional individual; but like many tumors, it has gained on repeated transplantation a heightened malignancy, and the power to grow in other varieties of the same animal. Yet in these it does not do well; and it has not been successfully transplanted to other species.

Histologically, the growth has always consisted of one type of cells, namely, spindle-cells in bundles, with a slight, supporting, connective tissue framework. The picture does not in the least suggest a granuloma; and cultures from the growth remain sterile as regards bacteria. At the edge of the invading mass there is often practically no cellular reaction, but lymphocytes in small number may be present, as is common with tumors in general. Metastasis takes place early, through the bloodstream and the secondary nodules have the same character as the primary. Several instances of the sarcoma's direct extension into vessels have been encountered. The secondary growths are distributed especially to the lungs, heart and liver, and in the last organ are sometimes umbilicated. The host becomes emaciated, cold and drowsy, and shortly dies.

Transplantation experiments with the tumors resulting from the filtrate are at present under way. The tumor of Experiment 2, which arose in the fowl that received filtrate alone, has already been successfully transplanted.

By the end of the first decade of the twentieth century it was suspected that there were filterable, invisible agents that could cause diseases, including cancers, of animals and plants. By 1915, the first hint came that even bacteria were susceptible to diseases caused by similar filterable entities. The following are passages of the first communication on the subject, which was published by Frederick William Twort in *Lancet* (**2**: 1241–1243, 1915) and which was entitled "An Investigation on the Nature of Ultra-microscopic Viruses."

> During the past three years a considerable number of experiments have been carried out at the Brown Institution (London). . . .
>
> In the first instance attempts were made to demonstrate the presence of non-pathogenic filter-passing viruses. As is well known, in the case of ordinary bacteria, for every pathogenic microorganism discovered many non-pathogenic varieties of the same type have been found in nature, and it seems highly probable that the same rule will be found to hold good in the case of ultra-microscopic viruses. It is difficult, however, to obtain proof of their existence, as pathogenicity is the only evidence we have at the present time of the presence of an ultra-microscopic virus. On the other hand, it seems probable that if non-pathogenic varieties exist in nature these should be more easily cultivated than the pathogenic varieties; accordingly, attempts to cultivate these from such materials as soil, dung, grass, hay, straw, and water from ponds were made on several hundred media. The material to be tested for viruses was covered with water and incubated at 30°C. or over for varying periods of time, then passed through a Berkefeld filter, and the filtrate inoculated on the different media. In these experiments a few ordinary bacteria, especially sporing types, were often found to pass through the filter; but in no case was it possible to obtain a growth of a true filter passing virus.
>
> Attempts were also made to infect, with the filtered materials, such animals as rabbits and guinea pigs. . . . All the experiments, however, were negative.
>
> Experiments were also conducted with vaccinia and with distemper of dogs, but in neither of these diseases was it found possible to isolate a bacterium that would reproduce the disease in animals. Some interesting results, however, were obtained with cultivations from glycerinated calf vaccinia. Inoculated agar tubes, after 24 hours at 37°C., often showed watery-looking areas, and in cultures that grew micrococci it was found that some of these colonies could not be subcultured, but if kept they became glassy and transparent. On examination of these glassy areas nothing but minute granules, staining reddish with Giemsa, could be seen. . . . Further experiments showed that when a pure culture of the white or the yellow micrococcus isolated from vaccinia is touched with a small portion of one of the glassy colonies, the growth

at the point touched soon starts to become transparent or glassy, and this gradually spreads over the whole growth, sometimes killing out all the micrococci and replacing these by fine granules. . . . The action is more rapid and complete with vigorous-growing young cultures than with old ones, and there is very little action on dead cultures or on young cultures that have been killed by heating at 60°C. . . . The transparent material when diluted (one in a million) with water or saline was found to pass the finest porcelain filters (Pasteur-Chamberland F. and B. and Doulton White) with ease, and one drop of the filtrate pipetted over an agar tube was sufficient to make that tube unsuitable for the growth of the micrococcus. That is, if the micrococcus was inoculated down the tube as a streak, this would start to grow, but would soon become dotted with transparent points which would rapidly extend over the whole growth. The number of points from which this starts depends upon the dilution of the transparent material, and in some cases it is so active that the growth is stopped and turned transparent almost directly after it starts. This condition or disease of the micrococcus when transmitted to pure cultures of the micrococcus can be conveyed to fresh cultures for an indefinite number of generations; but the transparent material will not grow by itself on any medium . . . [but] it will retain its powers of activity for over six months. It also retains its activity when made into an emulsion and heated to 52°C., but when heated to 60°C. for an hour it appears to be destroyed. It has some action, but very much less, on *Staphylococcus aureus* and *albus* isolated from boils of man, and it appears to have no action on members of the coli group or on streptococci, tubercle bacilli, yeasts, etc. The transparent material was inoculated into various animals and was rubbed into the scratched skin of guinea pigs, rabbits, a calf, a monkey, and a man; but all the results were negative. . . .

In the vaccinia experiments described above it is clear that the transparent material contains an enzyme, and it is destroyed at 60°C. It also increases in quantity when placed on an agar tube containing micrococci obtained from vaccinia and this can be carried on indefinitely from generation to generation. If it is part of the micrococcus it must be either a stage in its life history which will not grow on ordinary media but stimulates fresh cultures of the micrococcus to pass into the same stage, or an enzyme secreted by the micrococcus which leads to its own destruction and the production of more enzyme. . . . There is this, however, against the idea of a separate form of life: if the white micrococcus is repeatedly plated out and a pure culture obtained, this may give a good white growth for months when subcultured at intervals on fresh tubes; eventually, however, most pure strains show a transparent spot, and from this the transparent material can be obtained once again. Of course, it may be that the micrococcus was never quite free from the transparent portion, or this may have passed through the cotton-wool plug and contaminated the micrococcus, but it seems much more probable that the material was produced by the micrococcus. Incidentally, this apparent spontaneous production of a self-destroying material which when started increases in quantity might be of interest in connection with cancers. In any case, whatever explanation is accepted, the possibility of its being an ultra-microscopic virus has not been *definitely* disproved. . . . If the transparent portion were a separate virus, it might be vaccinia or it might be some contaminating non-pathogenic ultra-microscopic virus, for it is conceivable that whereas a non-pathogenic variety might grow on micrococci or bacilli, a pathogenic variety might grow only on the animals it infects. . . .

More recently, . . . during the summer and autumn of this year (1915), similar experiments were carried out with material obtained from the intestinal tract. . . . After certain difficulties had been overcome it was found that in the upper third of the intestine, which contained numerous bacilli of the typhoid-coli group, some larger bacilli were also present. . . . These grew only when precautions were taken to eliminate the action of a dissolving substance which infected the colonies so rapidly that they were dissolved before attaining a size visible to the eye. . . . The relation of this bacillus and the dissolving material to infantile diarrhea has not yet been determined, but probably it will be found also in cases of dysentery and allied conditions. . . .

Frederick William Twort was thirty-eight years old when he published this paper. From 1909 to 1944 he was superintendent of the Brown Institution, University of London. Because of the war he was unable to follow up these observations, as his duties took him to Salonica and to Northern Ireland. Previously, Twort had carried out some distinguished work on the nutrition of the *Mycobacterium*, which was then called Johne's bacillus. Twort died in 1950, one year after Félix d'Herelle, who, in the field of bacteriophages, by far eclipsed him in fame. D'Herelle was born in 1873, in Montreal, and during his whole life he was a British subject even though he was "more French than British in temperament and outlook." Young Félix went to the *Lycée* in Paris but went back to Montreal to study medicine. From 1901 to 1909 he practiced bacteriology in Guatemala and Mexico, and from 1909 to 1921 he was at the Pasteur Institute in Paris. There, in 1917, he rediscovered Twort's lytic principle. His discovery was published in the *Comptes rendus de l'académie des sciences* (**165**: 373–375, 1917):

From the stools of various patients recovering from bacillary dysentery and in one case from a urine sample, I have isolated an invisible microbe which is gifted with antagonistic properties against the Shiga bacillus. It is particularly easy to isolate this organism in cases of common enteritis following dysentery. In patients who do not have this complication the disappearance of the antagonistic microbe very closely follows the disappearance of the pathogenic bacillus. In spite of numerous examinations I have never found any antagonistic microbes either in the stools of patients suffering from active dysentery nor in the stools of healthy individuals.

The isolation of the anti-Shiga microbe is simple: a tube of broth is seeded with four or five drops of stool suspension and placed in the incubator at 37°C. for 18 hours. After this time the liquid is filtered using a Chamberland L_2 candle. A small quantity of an active filtrate added either to a broth culture of Shiga bacillus or to an emulsion of these bacilli in broth or even physiological saline, stops the growth of the culture, and brings about the death of the bacilli and subsequently their lysis. This lysis is complete after a period of time varying from a few hours to a few days depending on the number of bacilli present and on the amount of filtrate added.

The invisible microbe grows in the lysed culture of Shiga since a trace of this liquid, transplanted to a new Shiga culture, reproduces the same phenomenon with the same intensity. I have transplanted, up to now, the first strain isolated more than 50 successive times. As a matter of fact, the following experiment gives clear proof that the antagonistic action is due to a living germ. If the filtrate from a lysed culture is added to a fresh Shiga culture in a concentration of 1 to a million and if, immediately after, one drop of this solution is spread on slanted agar, one obtains, after incubation, a layer of dysentery bacilli with a certain number of sterile circles of about 1 mm in diameter where the bacilli did not grow. These loci must only represent colonies of the antagonistic microbe, since a chemical substance could not be concentrated at such definite places. By using quantitative plating methods I was able to determine that a lysed culture of Shiga contains from five to six billion filtrable germs per cubic centimeter. A three billionths of a cubic centimeter of a previously lysed Shiga culture, that is to say a single germ, introduced in a tube of broth, prevents the growth of Shiga even when the culture has been seeded liberally with the bacillus. The same quantity added to 10 cm³ of a Shiga culture sterilizes it and lyses it within five or six days.

The various strains of the antagonistic microbe that I have isolated were originally active only against the Shiga bacillus but after a few transfers in the presence of dysentery bacilli type Hiss or Flexner, I was able to make them antagonistic toward these bacilli. I obtained no results by trying the same experiments with other microbes such as typhic and paratyphic bacilli, staphylococci, etc. The appearance of an antagonistic activity against the Flexner bacillus or the Hiss bacillus was accompanied by a diminution, then the loss of action against the Shiga organism. This activity can be regained with all its previous intensity after a few transfers in presence of the Shiga bacillus. The specificity of the antagonistic action is thus not inherent to the nature of the invisible microbe but is acquired within the system of the diseased person by means of a symbiotic culturing with the pathogenic bacillus.

The antagonistic microbe does not grow in absence of the dysentery bacilli in any medium and it does not attack the bacilli which have been killed by heat, but it grows very well in an emulsion in physiological saline of washed bacilli. The obvious conclusion is that the antidysenteric microbe is an obligate bacteriophage.

The anti-Shiga microbe does not exhibit any pathogenic action on experimental animals. The cultures of Shiga bacilli lysed by the action of the invisible microbe, which in actuality are cultures of the antagonistic microbe, have the capacity to immunize the rabbit against a dose of Shiga bacilli which killed the controls in five days.

I have investigated the possibility of isolating an antagonistic microbe in patients recovering from typhoid fever. In two cases, one from the urine, and another from the stools, I have succeeded in isolating a filterable microbe having lytic properties against paratyphoid A bacillus though not so strongly lytic as the anti-Shiga microbe. These properties have diminished in the subsequent cultures.

To summarize: from certain patients recovering from dysentery, I have observed that the disappearance of the dysentery bacillus was linked with the appearance of an invisible microbe having antagonistic properties against the pathogenic bacillus. This microbe, true microbe of immunity, is an obligate bacteriophage. Its parasitism is strictly specific, but by adaptation it can be trained to lyse various species. It attacks only one species at a time. Thus, it seems that in bacillary dysentery, apart from homologous immunity coming directly from the organism of the patient, there is a heterologous immunity caused by an antagonistic microorganism. It is probable that this phenomenon is not restricted to dysentery but that it is of a more general occurrence since I was able to observe a similar, although less pronounced, situation, in two cases of paratyphoid fever.

Whereas no one paid much attention to Twort, D'Herelle became very famous. His services were requested in various countries, and he, in turn, spent some time in Argentina, Turkey, Indochina, Holland, Egypt, and India. From 1928 to 1933 he was professor of protobiology at Yale University, which he left to found the Laboratoire du Bactériophage in Paris. Before the Second World War he helped Russia to get started in bacteriophage research. D'Herelle was a clear, logical thinker who wrote numerous books and an excellent laboratory technician. He died in Paris in 1949.

The bacteriophages, "true microbes of immunity," did not fulfill all the therapeutic expectations of D'Herelle; however they became marvelous physiological and genetic tools.

It gradually became accepted that viruses were submicroscopic living parasites, and the scientific world patiently waited for the development of high-magnification, high-resolution microscopes in order to have a peek at them. In 1935, Dr. W. M. Stanley, who was then working at The Rockefeller Institute for Medical Research in Princeton, New Jersey, published a disturbing paper on the nature of viruses. It was entitled "Isolation of a Crystalline Protein Possessing the Properties of Tobacco-mosaic Virus" (*Science,* **81**: 644–645, 1935): [1]

A crystalline material, which has the properties of tobacco-mosaic virus, has been isolated from the juice of Turkish tobacco plants infected with this virus. The crystalline material contains 20 per cent nitrogen and 1 per cent ash, and a solution containing 1 milligram per cubic centimeter gives a positive test with Millon's biuret, xanthoproteic, glyoxylic acid and Folin's tyrosine reagents. The Molisch and Fehling's tests are negative, even with concentrated solutions. The material is precipitated by 0.4 saturated ammonium sulfate, by saturated magnesium sulfate, or by safranine, ethyl alcohol, acetone, trichloracetic acid, tannic acid, phosphotungstic acid and lead acetate. The crystalline protein is practically insoluble in water and is soluble in dilute acid, alkali or salt solutions. Solutions containing from 0.1 per cent to 2 per cent of the protein are opalescent. They are fairly clear between pH 6 and 11 and between pH 1 and 4, and take on a dense whitish appearance between pH 4 and 6.

The infectivity, chemical composition and optical rotation of the crystalline protein were unchanged after 10 succcessive crystallizations. In a fractional crystallization experiment the activity of the first small portion of crystals to come out of solution was the same as the activity of the mother liquor. When solutions are made more alkaline than about pH 11.8 the opalescence disappears and they become clear. Such solu-

---

[1] Reproduced by permission of the editor of *Science.*

tions are devoid of activity and it was shown by solubility tests that the protein had been denatured. The material is also denatured and its activity lost when solutions are made more acid than about pH 1. It is completely coagulated and the activity lost on heating to 94°C. Preliminary experiments, in which the amorphous form of the protein was partially digested with pepsin, or partially coagulated by heat, indicate that the loss in activity is about proportional to the loss of native protein. The molecular weight of the protein, as determined by two preliminary experiments on osmotic pressure and diffusion, is of the order of a few millions. That the molecule is quite large is also indicated by the fact that the protein is held back by collodion filters through which proteins such as egg albumin readily pass. Collodion filters which fail to allow the protein to pass also fail to allow the active agent to pass. The material readily passes a Berkefeld "W" filter.

The crystals are over 100 times more active than the suspension made by grinding up diseased Turkish tobacco leaves, and about 1,000 times more active than the twice-frozen juice from diseased plants. One cubic centimeter of a 1 to 1,000,000,000 dilution of the crystals has usually proved infectious. The disease produced by this, as well as more concentrated solutions, has proved to be typical tobacco mosaic. Activity measurements were made by comparing the number of lesions produced on one half of the leaves of plants of Early Golden Cluster bean, *Nicotiana glutinosa* L., or *N. langsdorffii* Schrank after inoculation with dilutions of a solution of the crystals, with the number of lesions produced on the other halves of the same leaves after inoculation with dilutions of a virus preparation used for comparison.

The sera of animals injected with tobacco-mosaic virus give a precipitate when mixed with a solution of the crystals diluted as high as 1 part in 100,000. The sera of animals injected with juice from healthy tobacco plants give no precipitate when mixed with a solution of the crystals. Injection of solutions of the crystals into animals causes the production of a precipitin that is active for solutions of the crystals and juice of plants, containing tobacco-mosaic virus but that is inactive for juice of normal plants.

The material herein described is quite different from the active crystalline material mentioned by Vinson and Petre and by Barton-Wright and McBain, which consisted, as Caldwell has demonstrated, largely of inorganic matter having no connection with the activity. These preparations were less active than ordinary juice from diseased plants, and the activity they possessed diminished on further crystallizations.

The crystalline protein described in this paper was prepared from the juice of Turkish tobacco plants infected with tobacco-mosaic virus. The juice was brought to 0.4 saturation with ammonium sulfate and the precipitated globulin fraction thus obtained was removed by filtration. The dark brown globulin portion was repeatedly fractionated with ammonium sulfate and then most of the remaining color was removed by precipitation with a small amount of lead subacetate at pH 8.7. An inactive protein fraction was removed from the light yellow colored filtrate by adjusting to pH 4.5 and adding 2 per cent by weight of standard celite. The celite was removed, suspended in water at pH 8, and the suspension filtered. The active protein was found in the colorless filtrate. This procedure was repeated twice in order to remove completely the inactive protein. Crystallization was accomplished by adding slowly, with stirring, a solution containing 1 cubic centimeter of glacial acetic acid in 20 cubic centimeters of 0.5 saturated ammonium sulfate to a solution of the protein containing sufficient ammonium sulfate to cause a faint turbidity. Small needles about 0.03 millimeters long appeared immediately and crystallization was completed in an hour. Crystallization may also be caused by the addition of a little saturated ammonium or magnesium sulfate to a solution of the protein in 0.001 N acid. Several attempts to obtain crystals by dialyzing solutions of the protein gave only amorphous material. To date a little more than 10 grams of the active crystalline protein have been obtained.

Although it is difficult, if not impossible, to obtain conclusive positive proof of the purity of a protein, there is strong evidence that the crystalline protein herein described is either pure or is a solid solution of proteins. As yet no evidence for the existence of a mixture of active and inactive material in the crystals has been obtained. Tobacco-mosaic virus is regarded as an autocatalytic protein which, for the present, may be assumed to require the presence of living cells for multiplication.

Stanley's paper was thought-provoking. How could a chemical substance be alive? He received the Nobel Prize in Chemistry, in 1946, for this significant piece of work that had focused much attention on the nature of viruses. He became professor of biochemistry and virology at the University of California. Later he helped develop an influenza vaccine.

However, Stanley's statement that the tobacco mosaic was a protein was not altogether correct, as pointed out in 1937 by F. C. Bawden of the Rothamsted Experimental Station and N. W. Pirie of the Biochemical Laboratory of Cambridge University [*Proc. Roy. Soc. London, B*, **123**: 274–320, 1937]: [1]

> Recently, Stanley has isolated from tobacco and tomato plants suffering from mosaic a protein which he described as crystalline and as possessing the properties of the virus. . . .
>
> Tobacco mosaic virus is known to exist in a number of strains which possess similar properties *in vitro* . . , yet differ from one another in the ease with which they can be transmitted to different hosts and in the type and severity of symptoms they cause in infected plants. We have worked with three such strains and have isolated liquid crystalline proteins from various plants infected with each. . . .
>
> The three strains used were a mild strain of the common tobacco mosaic virus, aucuba mosaic virus and Enation mosaic virus. . . .
>
> Liquid crystalline proteins have been isolated from all the plants [tobacco or tomato] affected with the diseases described, and the kind of protein isolated was found to depend entirely on the virus strain and not at all on the host plant used. The activity of the proteins was tested in two ways: (1) for their infectivity, and (2) for their precipitation end-point with the serum of rabbits immunized by a course of intraperitoneal injection with crude sap from plants infected with the tobacco mosaic virus. . . .

For the preparation of the proteins:

> Infected leaves are minced in a meat mincer and the sap expressed in a hand press. The sap can usually be clarified by centrifuging immediately. . . . One volume of alcohol is added to the clarified sap, and the greyish brown precipitate is centrifuged off immediately. The precipitate contains all the virus. . . . The alcohol precipitate is suspended evenly in from five to six times its volume of water and again centrifuged. The first extract contains little or no virus and is discarded. . . . From six to eight washings are usually necessary to remove all the virus from the chocolate-coloured residue. The collected supernatant fluids are now brought to about pH 3.3 by the addition of N/10 HCl. A white precipitate with a very characteristic satin-like sheen is produced, and this is centrifuged off. This precipitate is now suspended in water and dilute NaOH added to bring the pH to 7, when the fluid is again centrifuged until clear. The darkly coloured precipitate is discarded,

[1] Reproduced by permissions of the editor of the Royal Society of London and Dr. F. C. Bawden.

and a third of a volume of saturated ammonium sulphate solution is added to the supernatant. This again produces a precipitate with a sheen and, after centrifuging, a yellow-coloured supernatant liquid which is discarded. The precipitate is taken up in water and again precipitated by adding a third of a volume of saturated ammonium sulphate solution. . . . [A] salt-free precipitate [is finally obtained and] is then dissolved in enough dilute NaOH to give a solution at about pH 6.5. Such a solution remains active for months when kept cold. . . .

The yield of purified virus obtained has varied from 1 to 2 g/l. of expressed sap. The yield is greater . . . from young actively growing plants that have been infected from about 3 weeks to a month. . . .

Dried neutral solutions of the three virus strains have analytical composition falling within the following range:

	%
Carbon	49.3–50
Hydrogen	7.2–7.4
Nitrogen	14.4–16.6
Sulphur	0.24–0.59
Phosphorus	0.45–0.55
Ash	1.5–3.0
Carbohydrate	2.5

. . . The phosphorus and carbohydrate contents of our preparations are very constant. They are unaffected either by prolonged dialysis in cellophane tubes against dilute acid or alkali, or by reprecipitation ten times with either acid or quarter-saturated ammonium sulphate solution. From precipitates with their antisera . . . the viruses have been recovered with their full activity and with their phosphorus content unaltered. . . . All the treatments that we have tried which in no way inactivate the virus preparations leave the phosphorus content unaltered. Some treatments that do inactivate them, such as heating to 90°C. or exposure to strong acid or alkali, split off a nucleic acid or its breakdown products. . . .

The satin-like sheen produced in solutions of these purified viruses when they are acidified or mixed with quarter of a volume of saturated ammonium sulphate solution suggests that the material precipitated may be in the crystalline state. The individual particles of the precipitate, however, are too small to permit a conclusive microscopical examination. . . .

The process of gel formation can be more easily studied in mixtures containing 0.5–1.0% of the virus, 0.25–0.5% of glycine and 5% of neutralized ethyl formate. The pH is adjusted initially to about 5, and the fluid is immediately put into the vessels in which it is to be observed. After a time . . . sufficient ethyl formate will have undergone hydrolysis to give a mixture at about pH 3.3. During the acidification the optical properties of dilute virus solutions which have been kept quite still undergo no change, although the mixture turns to a fairly rigid jelly. When shaken or stirred this jelly breaks up and changes almost instantly to a suspension with the usual sheen. If now allowed to remain quite still the mixture does not return to the gel state. . . . When more concentrated virus solutions are slowly acidified in this way they behave somewhat differently, for the undisturbed gel breaks up into a mosaic of irregularly disposed birefringent spindles. When stirred, however, this again gives rise to a suspension with the characteristic sheen.

These phenomena suggest that it is more accurate to describe the visible virus precipitates that are obtained with acid and ammonium sulphate as fibrous rather than as crystalline. . . .

When a neutral solution [of a purified virus preparation] in the presence of a trace of salt is heated for 5 min. at 95°C. the pH shifts slightly to the acid side, and the major part of the protein precipitates as a coagulum that can be centrifuged off. This precipitate is free from both phosphorus and carbohydrate. The supernatant fluid still contains some protein, part of which can be precipitated by the addition of from 0.5 to 1.0% of a neutral salt. . . . [It] now contains very little protein but all the phosphorus and carbohydrate content of the virus preparations. When mineral acid is added to the supernatant a curdy precipitate separates immediately, and soon aggregates into resinous masses with the characteristic appearance of nucleic acid. When denatured by heating [or by other means] protein and nucleic acid are the only breakdown products that have been found. . . .

The purified virus nucleic acid resembles yeast nucleic acid closely; it contains a pentose and does not give the reactions with Schiff's reagent characteristic of a desoxy pentose. . . .

It was then established that viruses were nucleoproteins. Of some viruses, about which we are best informed, such as tobacco-mosaic virus, we now believe that the virus particle is formed of a long strand of ribonucleic acid around which protein particles are disposed, forming an outer coat. Of other viruses, such as some bacteriophages, we now believe that they are formed of desoxyribonucleic acid covered with a protein coat. It is obvious that bacteriophages are the most simple viruses to study because of the ease in handling their hosts. For facilitating study of other viruses, virology owes much to the development of tissue culture techniques. With such methods it is possible to study the multiplication of viruses under conditions that permit constant observation of the infected cells. Also one can, in tissue culture, count the number of viral particles present in a given system by a technique similar to the counts of plaques formed by bacteriophages.

Edna Steinhardt, C. Israel, and R. A. Lambert published a paper in 1913 in the *Journal of Infectious Diseases* (13: 294–300) in which they described multiplication of vaccinia virus in tissue culture:

> In previous work with the virus of rabies, done in conjunction with D. W. Poor we applied Harrison's method of growing tissue *in vitro* to brain tissue and were able to produce inclusions in normal ganglion cells closely resembling certain small forms of Negri bodies, but were not able to cause any multiplication of the virus of rabies. After these results with rabies, we decided to continue the work with Harrison's method, applying it in the present experiments to the virus of vaccinia incubated with corneal tissue and plasma, again, with two objects in view: the possible production of vaccine bodies *in vitro*, and the possible cultivation of the virus of vaccinia outside of the body. . . .
>
> Small pieces of rabbit or guinea-pig cornea were placed for a few minutes in [a] weak emulsion of virus. The pieces were then transferred with a small quantity of the virus to cover-glasses to which drops of rabbit or guinea-pig blood plasma were added. The cover-glasses were immediately inverted and sealed over hollow-ground slides, which were incubated at 37°C. To control the microscopic work, similar preparations were put up without virus, and to control the necessity of the cornea in the animal inoculation work, pieces of paraffin were substituted for cornea in a series of preparations. . . .
>
> The results of these studies may be given in a few words. The corneal epithelium shows an active lateral spreading through the clot, forming sheets or groups of cells in the plasma. The cells early show an accumulation of fat in their cytoplasm, but may retain their form for several weeks even when not transferred to fresh plasma. Careful studies have failed to reveal any specific vaccine bodies in the preparations; only smaller, undifferentiated forms have been seen and these have been found in both the controls without virus and the virus preparations after incubation.
>
> We have also studied the corneas of guinea-pigs, inoculated with

virus, which were removed 24 hours later and put up in hanging drops as described above. Only the smaller forms of the vaccine bodies were found in the beginning, and no further developments were seen. These negative findings are of interest on account of the results of the animal inoculations of the incubated preparations as given below.

Altho we have observed numerous granules in the incubated preparations, these have not been sufficiently definite in character with the methods employed thus far in our studies to allow us as yet to make any positive statements in regard to them. . . .

To demonstrate the activity of the virus, we have adopted the method of Calmette and Guérin. They have shown that the virus rubbed on the freshly shaven skin of the rabbit produces a typical vaccinia eruption, and, furthermore, by methods of dilution, that the number of vesicles in the eruption indicates approximately the quantity of virus present. This method is generally used in commercial laboratories for standardizing the virus.

In our experiments, the skin of a rabbit was inoculated at once with a small number of *unincubated* preparations, and similar inoculations were made later with *incubated* preparations to determine whether or not there was an increase in the virus.

The "takes" with unincubated preparations, using eight different viruses, varied from 10–50 vesicles, whereas a similar number of preparations incubated 7–18 days gave in every instance an extensive confluent eruption, showing that a definite increase had taken place. Repetitions of these experiments with the different viruses have in all cases given the same results. As yet we have not found a virus that has not shown increased activity under these conditions of incubation. Preparations containing small pieces of paraffin showed, on the other hand, no increase in activity upon incubation. . . .

Subcultures of the preparations have been made and successful skin inoculations have been obtained from the third transfer. The virus has remained active after 34 days' incubation, and this obviously does not represent the limit of activity under such cultural conditions. In control experiments it was found that the virus alone was greatly weakened after a week's incubation, and that after three weeks' incubation, inoculations on rabbits were negative altho much larger quantities of the virus were used than with the tissue preparations. . . .

And the authors concluded:

From our work thus far on the application of Harrison's method to the cultivation of tissue *in vitro* to corneal tissue plus the virus of vaccinia, we are able to state that there is a multiplication of the virus of vaccinia altho no specific vaccine bodies are found in the preparations.

Nowadays, many tissues from a variety of species, including man, are used as sources of cells. The magnitude of the operation is such that these have even been grown in fermentors, previously used only for the mass multiplication of microbes. The sensitivity of tissue cultures to certain viruses may be higher than that of laboratory animals, and in the case of the ECHO (enteric cytopathogenic human orphans) viruses, in particular, the method has rendered invaluable services.

To the technique of tissue culture, as a means of growing viruses, was added, in 1931, culture in chick embryos. This again was an American innovation and the work of Alice M. Woodruff and Ernest W. Goodpasture from the department of pathology of Vanderbilt University, in Nashville, Tennessee. In their paper entitled "The Susceptibility of the Chorio-allantoic Membrane of Chick Embryos to Infection with the Fowl-pox Virus" (*Am. J. Pathol.*, 7: 209–222, 1931), they wrote:

> For studying a representative of the pox group of virus diseases, fowl-pox has many advantages, among them being the fact that infectious material is readily obtained and easily handled, since the disease is limited to fowls. It is believed that knowledge gained concerning this disease may be serviceable in the study of other members of the pox group. Hence, this virus has been the subject of several problems investigated in this laboratory during the last three years.
>
> Fowl-pox, like the other pox diseases, is characterized by the appearance of eruptive skin lesions. The spontaneous fowl-pox nodules appear especially on the unfeathered parts of chickens, although experimental lesions may be easily induced in specialized epidermal structures such as cornea, feather follicles and oil gland. The lesions consist of a hyperplasia of the epithelial cells, with inclusion bodies in their cytoplasm. It has been shown that these inclusions are composed of groups or colonies of minute (Borrel) bodies. . . .
>
> Heretofore, fowl-pox has been studied only in the grown or newly hatched chicks, or in tissue culture. Tissue culture experiments with this virus have, however, been few and inconclusive. The present paper deals with the inoculation of the virus into embryonic tissues in the incubating egg.
>
> The chorio-allantoic membrane of the chick embryo has been used by a number of investigators for the study of the growth of various implanted tissues. Rous and Murphy were the first to use this technique

for the study of tumors. Danchakoff has used the method to grow embryonic chick tissues. Since the publication of these two papers the technique has been used frequently in experiments with auto- and heteroplastic grafts, as well as in those with auto- and heterogeneous tumors. The production of experimental infection in the chorio-allantoic membrane has, however, been done only in the one instance where Rous and Murphy grew the virus of the Rous sarcoma, and mention is made by Askanazy of the production of tuberculous chicks by the infection of fertile eggs. . . .

After describing four methods of obtaining uncontaminated virus inocula, the authors continued:

The technique for opening the eggs used in our experiments was based on that described by Clark. We omitted the use of a hot box, but kept the air sac immersed in water at 39°C. Keeping the air sac thus immersed prevented sagging of the egg contents when the egg was opened. We found that a piece of plasticene molded to fit the egg was a convenient support. The top surface of the egg was sterilized by bathing in alcohol and flaming. Then, proceeding with sterile precautions, a window, 7 to 10 mm. square was made by cutting or scraping with a sharp point. We found the sharp end of a scissors blade very convenient for this. After the shell was removed, the shell membrane was cut away carefully in order to expose the chorio-allantoic membrane. . . .

Two sorts of inoculation were attempted. The simpler procedure consisted in slightly injuring the chorio-allantoic membrane by pricking with a needle and applying a drop of an uncontaminated virus suspension to the injured area. In the second and more difficult operation, the skin of the embryo itself was inoculated. This involved cutting the chorio-allantoic membrane and amnion and slightly abrading the skin of the embryo, since some injury to epithelial cells favors the invasion of the virus.

In most of the techniques described in the literature the original piece of shell is replaced following the operative procedure and paraffined, so that the egg can be turned daily to continue normal development. Since we desired to watch the effects of the virus and to get sections at once if the chick should die, we substituted a glass cover slip for the original shell, fixing it upon a ring of vaseline, and returned the egg to the incubator immediately after the operation. This technique necessitated keeping the window uppermost during the rest of embryonic growth. The lack of turning caused usually an oval depression and fold in the membrane directly below the opening. No other abnormality due to lack of turning appeared to occur, for a number of chicks hatched normally from eggs which had been subjected to this treatment.

Embryos at various stages of development were used. Since it takes about four days for a well defined fowl-pox lesion to appear after inoculation, it was necessary to inoculate at least that many days before hatching. The most extensive lesions were obtained six to seven days after inoculation, so that 10 to 15 day embryos were used most frequently. Occasionally, a contamination occurred in the inoculation of the egg. Sometimes a mould grew symbiotically with the virus in the embryonic membrane. Such contaminated eggs were discarded. Except as a contaminating organism is introduced upon inoculation, the eggs are relatively free from infection and remain, according to Rettger, a sterile medium, unless subjected to moisture and dirt.

Fowl-pox infection of the chorio-allantoic membrane occurred as the result of inoculation in every case where the embryo survived for at least four days. Infections were first noted when thickened areas on the chorio-allantoic membrane were detected after several days' incubation. That a fowl-pox infection was definitely present was proved by three tests. The tissue, removed with sterile precautions and inoculated onto the scarified epithelium of adult hens, produced a massive fowl-pox lesion. Smears of the lesions, stained by Morosow's method, showed Borrel bodies present in great numbers. Histological sections of the tissue showed the typical picture of the fowl-pox lesion. These lesions are characterized by a marked hyperplasia of the ectodermal layer and an accompanying thickening of mesoderm as well. Frequently hyperplasia of the entoderm occurs also. In the cells of the ectodermal layer many large inclusions are present, while in the entodermal layer, when occasionally a definite infection is present, inclusions are few and small. The lack of an inflammatory exudate, even in an advanced stage of the infection, should be noted. . . .

The inoculation of the embryonic skin caused considerably more trauma than the inoculation of the chorio-allantoic membrane. The percentage of mortality was so great that this operation was soon abandoned, since the membrane inoculations proved very satisfactory. . . .

Viruses were not altogether invisible, but for their examination man had to wait for the invention of the electron microscope. The development of this instrument emerged from the study of the conduction of electricity through gases at reduced pressures. As vacuum was increased in a tube that contained two electrodes, and as a high difference of potential was applied across the two electrodes, electrons moved from the negatively charged cathode to the positively charged anode. Furthermore, the rectilinear flow of electrons could be deflected with magnetic fields. These basic phenomena, which were studied by such pioneers as Sir William Crookes and J. J. Thomson, were rather haphazardly investigated until de Broglie, in 1924, introduced the wave theory of the nature of electrons and laid the foundation of electronic optics. Magnetic coils came to be considered as optical lenses, and a number of experimental microscopes were built that were used more to study the physical principles needed to develop the instrument than to make microscopical observations. These led in turn to the first instruments that could be used in biology. As the Second World War loomed, work was proceeding in Germany, England, Canada, the United States, and Japan.

In Germany, in 1939, Kausche, Pfankuch, and Ruska described the appearance of the tobacco-mosaic virus under the electron microscope: "Using concentrations of 10^{-5} g. of virus protein per cc, dried preparations revealed individual virus particles. . . . The threads were from 10–25 mμ thick. They showed knob-like thickenings that could have been due to folding of the carrier film. . . . Threads were frequently 300 mμ long, but threads of 150 mμ were seen quite often. . . . We must consider the rod or thread structures of the stated dimensions to be the molecules of tobacco mosaic virus" *(Naturwissenschaften,* **27**: 281–299, 1939). Longer threads were observed owing to end-to-end disposition of the short units.

War conditions did not permit the Europeans to develop the electron microscope commercially. The first commercially available instruments were manufactured in the United States.

In 1941, L. Marton, then working at the research laboratories of RCA in the United States, wrote an article in which he introduced the practical electron microscope to microbiologists. It was published in the *Journal of Bacteriology* (**41**: 397–413) and was entitled "The Electron Microscope. A New Tool for Bacteriological Research." [1]

In this Marton wrote:

The science of bacteriology could hardly exist without the microscope, and it is almost providential that pathogenic bacteria are within the limits of visibility of the present-day microscope. However, the limits of microscopical observation have been severely felt since early in the development of bacteriological research, and many attempts have been made to extend the range of observation. These attempts brought the

[1] Reproduced by permission of the secretary of the American Society of Microbiology.

realization that the sizes of micro-organisms extended far beyond the limits of visibility of light microscopes, and therefore the need has been constantly felt for a better instrument which would give more detail. Such an instrument is provided in the electron microscope which, in its present-day development, extends the observation range by a factor of about 50 to 100, with possible further extensions in the future.

Electron microscopy is based on the discovery of geometrical optics for electrons quite similar to the optics of light. To understand the term "geometrical optics" let us first consider the action of an electric or magnetic field on an electron beam. It is well known that an electron beam is deflected by such fields, and we can therefore compare their action on the beam to the action of a refractive medium on a light beam. A lens is nothing but a refractive medium of special symmetry—in this particular case of rotational symmetry. If we create an electric or magnetic field of rotational symmetry, such a field acts on an electron beam as a lens, i.e., the electron beam is concentrated or made divergent in the same way that the light beam is acted upon by a glass lens. It has been proved mathematically that the laws of geometrical optics can be fully applied to such systems, and experimentally that we can obtain electronic images which can be made visible, for instance on a fluorescent screen. An image can be formed of any self-emitting object, as would be the case with light if we observed the image of the source itself, or we can illuminate an object in the same way as we illuminate one in a light microscope, and observe the image of the object with the help of the illuminating beam.

After this discovery, the obvious next step was the building of compound optical systems, and, in particular, a system corresponding to the compound microscope. Such an instrument is built up of different elements for which we can use the same nomenclature as in a light microscope. The light source of the light microscope is replaced by an electron source, the electron beam being concentrated on the specimen by means of the first field in the same way that the light beam is concentrated by means of a condenser lens. The field is created by a coil through which current flows and produces inside the coil the necessary magnetic field of rotational symmetry. Two more similar coils are used in place of the objective and eyepiece lenses of the light microscope. We may call the first one the objective coil and second one the projection coil, because the image is not viewed by applying the eye to an eyepiece, but by projecting it on a fluorescent screen. The function of both lenses is exactly the same as in the familiar light optics. The objective coil produces a first stage magnification which is re-enlarged by the projection lens.

The above described microscope system has some very special properties. In the first place, the electrons travel without hindrance only in vacuum, and therefore the whole microscope must be pumped out to a high degree. This means that the specimens and photographic plates must also be in the vacuum. The operation of an instrument of this kind is necessarily different from the operation of a light microscope. Focusing, for instance, is done in a quite different way. In light optics the optical components of the system are fixed by the construction, and focusing is done by changing the distance between the specimens and the optical system. In electron microscopes the focal length of a lens is given by the strength of the applied field, and therefore the focusing is done by varying the optical constants of the lens similarly to the way in which the focal length of the human eye is varied by changing the curvature of the lens. The distance between the object and the optical system therefore remains constant, and the magnification can be changed continuously from the lowest to the highest magnification, instead of the step-by-step variation obtained in a light microscope by exchanging eyepieces.

The great advantage of an electron optical system lies in its highly increased resolving power. The practical microscopist knows that the resolving power, i.e., the smallest distance separately shown by an optical system, is about one-half of a wave-length for the best light optical systems. Since the fundamental discovery of de Broglie, we know that the electron behaves for some applications as a corpuscle, and for other applications as a wave, the wave-length of which depends on the speed of the electron; and, for the speeds generally used in electron microscopy, it is about one 100,000th of the wave-length of visible light. Under identical conditions, this would mean 100,000 times better resolving power, or 100,000 times higher useful magnification for the electron microscope than for the light microscope. This, however, cannot be realized, the reasons for the limitations being manifold.

In the first place, the lenses of electron optics exhibit all of the optical defects known to light optics, plus a few more, and at the present state of our knowledge we are not able to correct very far the electron optical aberrations. For instance, spherical aberration is about 1,000 times worse than in light microscopy. We do not have any means of correcting the chromatic defect. The result is that the numerical apertures of electron optical systems are much reduced over the numerical apertures of light optical systems, and as the resolving power is not only proportional to the wave-length but also inversely proportional to the numerical aperture, we come to the conclusion that for the present-day practical apertures of about 0.001, we should expect the best resolving power to be about 10 Angstrom Units, or a corresponding useful magnification of about 200,000 times. Incidentally, the use of extremely small apertures has another advantage: as in light microscopy, the smaller the aperture the greater is the depth of focus, and the resulting depth of focus for the herein described electron microscope is at least ten times greater than the depth of focus of the highest power light microscopes.

Another peculiarity of electron microscopy is the mechanism of image formation. In light microscopes we see the image due to differences in absorption, or refraction in the specimen. In electron microscopes the image formation is not due to either of these effects, but to scattering of the electrons. As mentioned before, electrons do not travel in straight lines except in vacuum, and even a few molecules suffice to deflect an electron beam. This means that we do not have any material transparent to electrons in the same sense as to light, and therefore we only have transmission of an electron beam when the substance through which the beam passes is extremely thin. This means that, first, the glass slides of light microscopy must be replaced by extremely thin films as specimen holders, and, second, that what we observe in an electron microscope image are only differences in thickness and in density of a specimen.

The practical application of these principles can be best explained by describing an RCA electron microscope. The electrons are generated by an "electron gun," similar in construction to an X-ray tube. The main tube contains the various optical and mechanical elements of the microscope—three coils which act as lenses, the "object chamber," and the "photographic chamber." The lenses, as mentioned above, are coils consisting of a large number of turns of copper wire, which are housed in iron enclosures. The function of the iron enclosures is to concentrate the field between the gaps in the enclosure, and to give the necessary shape to the field.

The object chamber has two important functions. The first is to introduce and withdraw the specimen to and from the main tube without breaking the vacuum in the microscope, and the second is to permit motion of the specimen in front of the objective, replacing the mechanical stage of a light microscope. . . . At the bottom of the instrument an aperture can be closed by means of a gate which separates the main body from the photographic chamber. This gate can be replaced by a photographic plate, using a carriage-like system. A fluorescent screen is mounted above the aperture so that it can either cover the photographic plate for visual observation of the image, or be swung away to leave the photographic plate open for exposure. Observation of the fluorescent screen is done through two large windows. . . .

Due to the described characteristics of electron microscope image formation, a new kind of technique had to be developed for investigating specimens. The most important conditions are given by the fact that electron scattering is, in a first approximation, proportional to the thickness and density of the substance through which the beam is passing. This means that for very transparent object holders the supporting films must not only be extremely thin, but very homogeneous as well, as the slightest inhomogeneity would give a definite contrast, and might be the source of an artefact. Such films can be prepared by spreading out one or more drops of diluted solution of collodion in amyl acetate (1½ to 2 grams in 100 grams of amyl acetate) over a distilled water surface. After evaporation of the solvent a thin film remains on the water surface and can be taken off on suitable holders. In order to have the film very homogeneous, the water surface should be perfectly quiet and free of dust and gas bubbles. It has been found, also, that good homogeneous films cannot be prepared without previously saturating the water with amyl acetate. As holders for the films, small apertures have been widely used, but we have found it more advisable to replace the small apertures by very fine wire screens, punched out to little disks, which can be easily inserted in the microscope. The advantage of such wire screen holders is that there is more than one aperture free for the inspection of the specimen, and still enough support for the extremely fragile films.

The usual technique of preparing bacteriological specimens for electron microscopy is to suspend them in distilled water and allow a small drop of the suspension to dry on the film surface. . . . The same technique must be applied when working with virus particles. In the latter case sometimes difficulties arise due to low contrast, and the focusing becomes difficult. Different methods can be devised for overcoming this difficulty. One is to take a series of pictures of some specimen at different focuses, and select the best one. Another method, which has been used in Germany, is the adsorption of colloidal gold on the virus particles. Focusing is facilitated in this way due to the high contrast of the gold particles.

Of course the technique just described is not the only bacteriological technique which can be used. Some attempts have been made to incorporate the bacteria into the film material before preparing the film, and spreading them out together on the water surface. Another attempt was made to reproduce, in a way similar to light microscope technique, the so-called "india ink technique" by making a film of higher density than the bacteria itself, which consequently appear black on the electron micrographs. Up to the present time, however, the first-described technique seems to be the best one.

Generally speaking, no staining has yet been applied in electron microscopy. The reason for this is twofold. First the stains of light microscopical technique do not show any advantage in electron microscope work, and second most of the bacteria show sufficient contrast for electron microscope work without staining or even fixing. . . . The foregoing does not exclude a possible need which may arise later for the development of a new staining technique for electron microscopy. Such a technique would need to be a staining in density, which means selective introduction of high density materials into the bacteriological cell. Some knowledge already exists about such materials, and the known technique of fixing with osmic acid might be mentioned in this connection. . . .

Men have always been inclined to battle. The wars that have ceaselessly ravaged the earth have also been a most satisfactory substratum for the development of a number of epidemic diseases that have been more effective in killing human beings than all the weapons of war. Of the numerous diseases that have benefited from wars, typhus has flourished best. At first this disease was considered akin, if not identical, to typhoid fever, and this is why typhoid fever is sometimes called "abdominal typhus" and typhus *stricto sensu,* "exanthematous typhus."

Early in the study of typhus and associated infections two men played a leading role. One of them, Charles Nicolle, lived in France; the other, Howard T. Ricketts, lived in the United States.

Whereas Nicolle had a long life, as did many other Pasteurians, Ricketts died at the age of thirty-nine, the victim of his studies. Fittingly, the group of intracellular parasites that cause typhus and similar infections bears the name of the American who first saw them and who also was the first to demonstrate their transmission by arthropods. Rickettsiae can be considered intracellular bacteria that grow not only in the cells of higher animals but also in the bodies of ticks, mites, lice, or fleas. In their strict intracellular parasitism they are similar to viruses. The rickettsial diseases include not only typhus but also, among others, trench fever, Japanese river valley (Tsutsugamushi) fever, Rocky Mountain spotted fever, and fièvre boutonneuse.

Charles Nicolle, a physician's son, was born in 1866 in Rouen, in the stronghold of Pouchet. It was his older brother Maurice, also a physician, who enticed him to come to the Pasteur Institute to follow Roux's course in microbiology. Charles Nicolle began his studies in microbiology in 1892. Ten years later he was nominated director of the Pasteur Institute in Tunis, a post he held until his death in 1936. At this time he also held a chair at the Collège de France in which had sat such luminaries as Claude Bernard, Magendie, and Laënnec. Typhus was one of the first diseases to attract his attention. In 1910, Nicolle and his coworkers, C. Comte and E. Conseil, published in the *Annals of the Pasteur Institute* (**24**: 261–267, 1910) a paper entitled "Experimental Transmission of Exanthematous Typhus by Body Lice (*Pediculus vestimenti*)."

The authors wrote:

> Physicians who have followed the course of epidemics of typhus have unanimously noted the important role played by famine, poor hygiene, crowding and filth in the dissemination of this disease. Without stating categorically that parasites on body and clothes were the agents of transmission of the disease, many authors inclined toward this mode of transmission.
>
> During the past few years, we have been able to follow personally a number of typhus epidemics . . . especially in Tunis proper. . . . Our observations had led us to consider that an insect was the probable agent of the diffusion of this disease.
>
> Typhus is, in Tunisia as elsewhere, a result of crowding and famine. It attacks the most miserable elements of the population and those who disregard the rules of hygiene. It is not a contagious disease in a clean house or in well-kept hospital wards. Under these conditions, only the parasitic insects found in dwellings, on the body or on the clothes could be suspected. The emergence of typhus epidemics in the spring eliminated the consideration of mosquitoes, ticks and stable flies.
>
> Many observations have led us to limit our study to lice. At the Tunis native hospital, the patients are washed and receive clean clothes upon admission. Not a single case of internal transmission has been observed even during the epidemics of 1902 and 1906 despite the lack of isolation and the presence of numerous bed-bugs in the wards. The only cases of contagion observed have claimed as victims some of the personnel in charge of the collection and the disinfection of the clothes of new patients. In addition, in the Kerkennah Islands which is an endemic focus of typhus, bed-bugs are rare. In Djerid, where typhus is also present, fleas are absent. These insects on the contrary are most numerous in phosphate mines where they attack without prejudice Europeans and natives even though only the latter are affected by typhus. Furthermore we know of two cases where typhus started after the usual incubation period following observed and recorded bites by lice.
>
> These facts were kept in mind when we started our experimental studies in April 1909. We had only few monkeys but our hypothesis

appeared so likely that we consigned 2 of the 6 available bonneted macaques to its verification. . . .

As a source of infective material we used bonneted macaque No. 1 which had been previously infected successfully with the blood of a chimpanzee which had in turn been infected with human blood. The 30th of June, 16 days after inoculation, the 3rd day of infection, and during the first hours of emergence of the rash, we placed on the skin of this animal 29 lice which had just been collected from healthy natives and which had not been fed for 8 hours. . . .

The next day we started to apply these lice to our bonneted macaques A and B. Animal A was bitten on six consecutive days by 15, then 12, 13, 8, 6, and 3 lice, whereas animal B was bitten during 12 days by 14, then by 15, 13, 9, 5, 6, 4, 2 and 1 lice. Each day after biting, the lice were assembled and kept at about 20°. . . .

As can be seen from the clinical picture and from the temperature records, these animals became infected with typhus. . . .

These experiments are highly significant. Human lice transmit typhus from monkey to monkey and there is little doubt that they can produce the same effect in man. Should one conclude that they are the only vehicle of transmission of the disease? Very probably, but further experiments will settle that point.

In any case, the immediate application of this new concept to prophylaxis is imperative. . . .

Further studies in collaboration with A. Conor and G. Blanc permitted Nicolle to determine that blood of typhus patients was virulent from the day before the start of the fever to the second day after its abatement. It was also demonstrated that the pathogen responsible for typhus multiplied in the intestinal tract of the louse for seven days following an infected meal.

The scientific activities of Charles Nicolle were not limited to typhus. He excelled not only in the study of the rickettsiae but also in various aspects of mycology, parasitology, and bacteriology itself. Nicolle also published scientific and philosophical books, and he even rested from his scientific endeavors by writing a few novels. When Hans Zinsser wrote a witty history of typhus entitled *Rats, Lice and History*, he fittingly dedicated it to "Charles Nicolle, scientist, novelist, and philosopher." For his work on typhus, Charles Nicolle received the 1928 Nobel Prize for Medicine.

Even though it was not the only subject studied by the gifted Nicolle, typhus brought him fame and honors. To others, it would bring death. In 1916, Da Rocha-Lima described the causative agent of typhus and called it *Rickettsia prowazeki*, linking forever the names of Howard T. Ricketts and

Stanislaus von Prowazek to the organism that had at the same time martyr-ized and immortalized them.

Ricketts was born in Ohio in 1871. He studied medicine in Chicago, and after serving at the Cook County Hospital, he specialized in pathology and was appointed associate professor of pathology at the University of Chicago in 1902. In 1906, he was able to transmit Rocky Mountain spotted fever to guinea pigs by the bites of ticks. Three years later he summarized his inves-tigations in the *Journal of the American Medical Association* (**52**: 379–380, 1909) in a paper entitled "A Micro-organism Which Apparently Has a Specific Relationship to Rocky Mountain Spotted Fever."

Since the spring and summer of 1906, bodies which I have referred to in my notes as "diplococcoid bodies," and sometimes short bacillary forms, have been found with considerable constancy in the blood of guinea-pigs and monkeys which were infected with Rocky Mountain spotted fever. They have also been seen in the blood of man but not so frequently. Much more time has been spent on the blood of the experimental animals than on that of man in view of the fact that it could always be obtained in fresh-condition.

The form most commonly found is that of two somewhat lanceolate chromatin-staining bodies, separated by a slight amount of eosin-stain-ing substance. The preparation of Giemsa, as furnished by Grübler, has been used almost exclusively, and with variations in the technic the intermediate substance may stain faintly blue.

In spite of the constancy with which these bodies were found, it did not seem justifiable to claim that they represent the microparasite of the disease, for two reasons: (1) the very complex morphology of the blood, especially in febrile states, when various cells and probably their nuclei are subject to unusual disintegration; (2) because of my inability to cultivate a micro-organism of this character from infected blood by the use of ordinary and some unusual culture media, under various con-ditions of cultivation, or by other means to obtain it in satisfactory concentration.

Although infected ticks had been examined previously in a more or less cursory manner, their systematic study was not undertaken until recently. In the pursuit of this work advantage was taken of the fact that the disease is transmitted by the infected female to her young through the eggs, as described in a previous report. A repetition of these experiments in the winter of 1907–8, with the help of Dr. Maria B. Maver, resulted in such transmission in 50 per cent of the ticks used, the fact being determined by allowing the larvae to feed on normal guinea-pigs. This second series has not been published heretofore.

Female tick No. 40, a *Dermacentor*, from Montana, had produced fatal infections of spotted fever in guinea-pigs 1740 and 1764. A number of eggs from the first day's laying were crushed individually on cover glasses, fixed in absolute alcohol, and stained with Giemsa's stain. Each egg was found to be laden with astonishing numbers of an organism which appears typically as a bipolar staining bacillus of minute size, approximating that of the influenza bacillus, although definite measurements have not yet been made. Various forms are seen depending on the stage of development and the arrangement in which two or more may be found. It is very common to find two organisms end to end, with their poles stained deeply and the intermediate substance a faint blue, resembling a chain of four cocci. When the chromatin is not yet sharply limited to the poles, the somewhat lanceolate forms so often recognized in the blood are seen. Not infrequently delicate bacilli with a uniform distribution of the chromatin are found. These are all interpreted as stages in the evolution of a bipolar organism. They are present in varying numbers in different eggs, but as a rule they are surprisingly numerous, and in some instances they would certainly count into the thousands. Many faintly staining, apparently degenerate forms are encountered.

Examination of the eggs of three dermacentors from Idaho (different specifically from the Montana dermacentor), which were infected from the guinea-pig, showed the presence of the same forms (Ticks 5, 7 and 9).

Conference with zoologic scientists who have made a particular study of the structure of eggs brought out the fact that such bodies are not known as a constituent of the egg of any species of animal.

Although it has not yet been possible to examine the eggs of ticks which are known to be free from spotted fever, the equivalent control has been made through a comparison of the visceral organs of infected and uninfected ticks. The salivary glands, alimentary sac and ovaries of infected females are literally swarming with exactly similar microorganisms. On the contrary, they appear to be entirely absent from the viscera of the uninfected tick, both male and female.

The most striking evidence of the probable etiologic relationship of this organism to spotted fever is found in the positive outcome of agglutination tests. Fortunately it is so numerous in the eggs that a bacterial emulsion of reasonable concentration for agglutination tests can be made by crushing forty to fifty eggs in about 0.05 c.c. of salt solution. The material is so scant that only the microscopic method could be used. The preparations were made as hanging drops, incubated for two hours, dried, fixed with absolute alcohol and stained.

The serum of the normal guinea-pig either causes no agglutination at all, or at the most produces only slight agglutination in proportions of 1 to 1 and 1 to 9. Dilutions higher than this cause no agglutination. In testing the agglutinating powers of immune serums, three animals which had been infected from different sources and had recovered were used. One (1751) had been infected with a dermacentor from Idaho; another (1692), with a strain handed down direct from guinea-pig to guinea-pig for nearly three years without the intervention of ticks, the original infection having been obtained from the blood of man; the third (1757) with a strain kept in the same way since last spring. Graded dilutions, beginning with 1 to 1 and going as high as 1 to 200, were used in the different series, with the striking result that a complete agglutinating power was present in the three immune serums in dilutions up to 1 to 160. It was somewhat less in a dilution of 1 to 200. The highest dilution which will cause clumping has not been ascertained.

No fresh immune serum from man is at hand, but tests were made with three specimens which are about five, seven and nine months old, respectively. They have been preserved in the ice chest with the addition of 0.3 per cent of chloroform. The peculiar phenomenon of failure to agglutinate in concentrated solution was noted with all three. With the oldest serum no agglutination occurred until the dilutions of 1 to 160 and 1 to 200 were reached, when incomplete clumping was produced. With the second there was no agglutination in the dilution of 1 to 1, distinct clumping in 1 to 10, 1 to 20 and 1 to 40, with little or none in higher dilutions. In the serum of five months standing, the reaction was absent in the dilutions of 1 to 1, 1 to 10 and 1 to 20, positive but not complete in 1 to 40, 1 to 80 and 1 to 120, with very little clumping in 1 to 160 and 1 to 200. Normal human serum caused clumping in a dilution of 1 to 1, a very slight amount in 1 to 10, and none at all in the higher dilutions. . . .

As a means of concentrating the organisms in the serum of the infected guinea-pig the following experiment was performed: Three cubic centimeters of fresh infected serum were diluted with an equal amount of salt solution, and to this was added 0.3 c.c. of an agglutinating serum from the guinea-pig. The mixture was placed in the incubator for two hours and then centrifugated for about ten hours at a speed of 1800 revolutions. All but the merest drop was then pipetted off and stained preparations were made of the sediment. Examination showed the presence of a moderate number of forms which are identical in appearance and size with those often recognized in ordinary smears of infected blood, and also with the "diplococcoid" forms seen in the egg of the infected tick. No such bodies were found in a control tube of normal serum.

The evidence pointing to this organism as the causative agent in spotted fever, though not complete, is of a striking character. In so far as I know it would be an unheard-of circumstance to obtain such strong agglutination with an immune serum, in the presence of negative controls, unless there were a specific relationship between the organism and the disease. In favor of the specific relationship in this case are also the presence of the organism in large numbers in infected ticks and in their eggs, its absence from uninfected ticks and the presence of similar forms in the blood and serum of the infected guinea-pig.

Morphologically the organism is a bacillus and somewhat pleomorphic as described. Its resemblance to the bacilli of the hemorrhagic septicemias is striking, and in this connection it is important to note that spotted fever is a hemorrhagic septicemia. It has not been cultivated, although work with this end in view is in progress.

I have devised no formal name for the organism discussed, but it may be referred to tentatively as the bacillus of Rocky Mountain spotted fever. A further study of its characteristics may suggest a suitable name.

That a bacillus may be the causative agent of a disease in which an insect carrier plays an obligate role under natural conditions may be looked at with suspicion in some quarters. Yet, even without the evidence in this case, it would seem to be unscientific to be tied to the more or less prevailing belief that all such diseases must, on the basis of several analogies, be caused by parasites which are protozoon in character.

Further study of the relationship of the bacillus to the disease is being carried on and will be reported at a future date, together with illustrations and a more detailed account of its characteristics.

One year after publication of this paper, Ricketts went to Mexico to study Mexican typhus (tabardillo). He found that the disease was transmitted by lice and that the blood of infected monkeys contained organisms similar to those that he had reported in the blood of animals infected with Rocky Mountain spotted fever. The elusive organisms escaped his cultivation methods but found his own body a suitable substratum for reproduction. Ricketts died, the victim of his studies, in 1910. A few years later the same fate awaited Von Prowazek in Germany.

Born in Bohemia in 1875, Stanislaus Von Prowazek studied first in Prague, then in Vienna. He worked with Ehrlich in Frankfurt and with Hertwig in

Munich. In 1907, he succeeded Schaudinn at the Institute of Tropical Medicine in Hamburg, and in 1913 he went to Serbia and Turkey to study typhus. He found the organisms of Ricketts in lice that had fed on typhus patients, but he too succumbed to the disease in 1915.

References

Clark, G. L. (ed.): *The Encyclopedia of Microscopy,* Reinhold Publishing Corporation, New York, 1961.

Kelly, H. A.: *Walter Reed and Yellow Fever,* McClure, Phillips and Co., New York, 1906.

Luria, S. E.: *General Virology,* John Wiley & Sons, Inc., New York, 1953.

Mayer, A.: *Concerning the Mosaic Disease of Tobacco;* Ivanowski, D.: *Concerning the Mosaic Disease of the Tobacco Plant;* Beijerinck, M. W.: *Concerning a Contagium Vivum Fluidum as a Cause of the Spot-Disease of the Tobacco Leaves;* Baur, E.: *On the Etiology of the Infectious Varigation,* Phytopathological Classics no. 7, The American Phytopathological Society, 1942.

Senate Documents, vol. 61, 61st Cong., 3d Sess., Government Printing Office, Washington, D.C., 1911.

Stanley, W. M., and E. G. Valens; *Viruses and the Nature of Life,* E. P. Dutton & Co., Inc., New York, 1961.

Zinsser, H.: *Rats, Lice and History,* Little, Brown and Company, Boston, 1934.

Additional Reference

Lechevalier, H.: "Dmitri Iosifovich Ivanovski (1864–1920)," *Bacteriol. Rev.,* **36:** 135–145, 1972.

Mycology

The Origins and the Development of Fungicides

9

The larger fungi are as conspicuous to the untrained eye as many types of plants. For this reason alone we can assume that they did not escape the notice of primitive men. Their rapid rate of growth, which is often manifested by their sudden appearance overnight, made them a logical choice as an example of living things arising by spontaneous generation. Morphologically, some resemble stars, while others look like bird nests complete with eggs, stalagtites, erected male organs, umbrellas of all shapes, and horns. In addition to imaginative shapes, their colors vary from the purest of whites to the brightest of reds, passing through a large range of delicate shades. Our ancestors must have been impressed by these curiously shaped organisms, some of which could even emit light on a dark night.

Men came to use fungi in a variety of ways, often unknowingly, as when they used yeasts in wine and bread making. Men also started to use fungi as food with variable success. They found that some of the fungi were delicious, while others served simply to fill the stomach without pleasing the palate. Others produced strange psychic effects, and still others put a definitive and lethal end to the gourmet himself.

Toxic fungi, such as certain members of the genus *Amanita,* were eventually used by clever people to dispatch unwanted persons to another world. It is said, for instance that Agrippina II, mother of Nero, used such a method to murder her husband and her uncle, the Emperor Claudius I. Therefore, if we can believe the historians, fungi were responsible for bringing Nero to power. Nero, however, was not sure whether Agrippina would know when to stop and had her murdered as a preventive measure in 59 A.D.

The Greek physicain Dioscorides wrote an herbal in the first century of our era, and he noted that fungi are either good food or poisonous. According to him, their poisonous nature depended on where the fungi grew. He recommended shying away from fungi growing near serpent's holes or on trees known to produce undesirable fruits.

Galen, the wise, advised total abstinence since he thought that few fungi were good to eat and that most of them caused the eater to choke and suffocate.

Before specific fungi could be designated, either as food or as causative agents of diseases, some nomenclature had to be developed. Charles de l'Ecluse, called Clusius for short, was one of the earliest botanists to devote much attention to fungi. Born in northern France, Clusius died in Leiden after having collected plants throughout Europe. Probably tempted by some plant precariously growing on the side of a cliff, he broke an arm in Gibraltar. For fourteen years, he was the director of the Imperial Gardens of Vienna. In a book published in 1611, Clusius described 1,385 plants. They were classified as: (1) trees, (2) bushes, (3) bulbous plants with fragrant flowers, (4) bulbous plants with odorless flowers, etc., and finally the fungi. These were separated in three groups, *esculenti, noxii,* and *perniciosi.* Off hand, it is hard to understand the subtile difference between the two last groups, but obviously the emphasis was still gastronomic.

The next important text on fungi, listing about one hundred species, was published by a Swiss professor of botany and medicine in Basel in 1623. He was Gaspard Bauhin, a former student of Fabricius d'Aquapendente and, like Clusius, a great traveler. This remarkable man had already published in 1596 a book in which he described 2,700 plant species and in which he illustrated the potato plant and named it, as it is today, *Solanum tuberosum.* Bauhin had started to make the important distinction between genera and species. This distinction was improved by another forerunner of Linnaeus, Tournefort, a professor at the Jardin des Plantes in Paris. Bauhin had emphasized the description of the species; Tournefort stressed the description of genera. Some of his generic descriptions were so good that they are still used today. Tournefort was born in southern France, and he was at first intended for the priesthood. His vocation, however, died with his father, and the young Tournefort dedicated himself to the study of plants, which he collected in numerous countries of Europe and the Near East. Traffic was already a menace in Paris in 1708, and a cart hit the famous botanist and killed him in a street that now bears his name. In his *Elements of Botany*, published in 1694, Tournefort recognized six groups of fungi: (1) centrally stalked with cap, (2) centrally stalked without cap, (3) laterally stalked, (4) globose without stalk, (5) corralloid forms, (6) subterranean forms. In group 4, he included some of the myxomycetes that actually do look like puffballs.

Tournefort made no attempt to classify microscopic fungi. Some of these had been seen by Robert Hooke and illustrated in his *Micrographia* in 1665. Hooke recognized the vegetable nature of "a plant growing in the blighted or yellow specks of Damask-rose-leaves and other kind of leaves." From his illustration and description it seems clear that he observed the teleutospores of a rust. He did not recognize the multicellular nature of the spores, but this illustration is probably the first one ever made of the reproductive bodies of a fungus. Hooke also observed and illustrated a "blue mould" that could be found in many putrifying bodies, including green cheese. He stated: "There are a multitude of other shapes of which these microscopical mushrooms are figured, which would have been a long work to have described and would not have suited so well with my design in this treatise, only, among the rest, I must not forget to take notice of one that was a little like to, or resembled, a sponge, consisting of a multitude of little ramifications almost as that body does. . . ." Hooke also remarked that fungi and molds do not need seeds to reproduce; they simply need a convenient substratum and the proper amount of warmth.

This opinion of Hooke is not surprising if we review at this point the various opinions that had been expressed up to the eighteenth century on the origin of fungi. The Greeks and the Romans had held that fungi were "an evil ferment of the earth" and that truffles were the result of lightning striking the ground. It was also felt that fungi, like ferns, did not produce any flowers, fruits, or seeds. The lamellae of agarics were observed, but there is no mention of the spore dust falling from them. Spore dust was certainly observed in puffballs, but its reproductive powers were not suspected.

In an herbal published in Germany by Bock in 1552 one notes that fungi are "the superfluous moisture of earth, trees, rotten wood and other rotting things. This is plain from the fact that fungi and truffles, especially those which are used for eating, grow most commonly in thundery and wet weather." And Cesalpino in 1583 stated: "Some plants have no seeds, they are the most imperfect and spring from decaying substances; they only have to feed themselves and grow, they are unable to produce their like."

The Italian physicist Porta published around 1590 a book called *Phytog-nomonica* in which he had a chapter with the revolutionary title: "Contrary to the Ancients All Plants Are Provided with Seed." Porta did not describe any experiment in which he proved that spores could germinate and give birth to the like of their parents, but he observed the spores, that he called seeds, in a large number of fungi including truffles and the myxomycete commonly called "flowers of tan." He saw the spore dust and essentially came to the logical conclusion that it was formed of a multitude of little seeds.

Malpighi, whom we have discussed in his role in medical science, had not limited his activities to animals. Between 1675 and 1679 he published an anatomy of plants. Some pages were devoted to the lowly fungi, and various molds and agarics were illustrated. Several sketches show the spores of the fungi attached to the sterigmata, which in turn were attached to a common sporophore. Malpighi, as Hooke had previously done, regarded the whole fruiting body as an inflorescence. Malpighi, however, differed from Hooke in that he was not ready to accept the concept of spontaneous generation. He said: "Up to now we do not know how they [the fungi] are multiplied and born. . . ." He guessed that "either they have their own seed, by which their species is propagated, or they sprout from the growth of fragments of themselves." He also suggested that fragments and seeds might be dispersed by the wind.

Malpighi's opinion was also held by Tournefort, who in 1707 gave an account of the methods of mushroom growing used by the growers around Paris. He did not really see germinating spores but came to the correct con-clusion indirectly by remarking that the mycelial threads that give rise to young mushrooms "are nothing but developed seeds or germs of mushrooms." He said this in spite of the fact that he considered the so-called seeds too small to be seen until after they had germinated into filaments.

Of great importance was the work of the Italian botanist Micheli. He was born in 1679 of poor parents, and it was in great part due to his own drive that he managed to acquire an education. He taught himself how to read and write Latin, perhaps the most necessary tool to have at that time if one wanted to become a botanist. His ability was rewarded with the directorship of the public gardens of Florence. In 1729 he published part of the results of his scientific investigations in the first volume of his *New Genera of Plants; Arranged After the Method of Tournefort . . . with Additional Notes and Observations Regarding the Planting, Origin, and Growth of Fungi, Mucors, and Allied Plants*. It was a fine volume with 108 plates the cost of which had been defrayed by a large number of people (197) including some Englishmen, interested in natural history. He died in 1737 without having had the time to publish the rest of his work. He had been a pleasant, modest, and charming man who had contributed much to botany in general but who is mainly remembered as the discoverer of the mode of reproduction of fungi. In the field of mycology, Micheli gave many generic names that are still in use now, such as *Mucor, Polyporus,* and *Aspergillus.* His descriptions were so accurate that specific identification can still be made in most cases. His main interest in mycology was to discover flowers and seeds in the fungi in order to be able to treat them like higher plants in a scheme of classification. Armed with a primitive microscope, he was soon rewarded by finding seeds on the gills of *agarics,* in the tubes of *boleti,* on the spines of *hydna,* on the branches of *clavariae,* inside puffballs, in asci of discomycetes, inside truffles, in the sporangia of some myxomycetes, and finally in the sporangia of Mucor and on the sporophores of *Aspergilli* and *Botrytis.* Furthermore, in the lamellate fungi he observed spores in groups of four but did not see basidia, and he noted the presence of cystidia, which he described as follows: *"In some species of fungi . . . I have observed . . . that the surface of their lamellae is ornamented, not merely with seeds, but also with transparent bodies which in some species are conical and in others pyramidal. They are devised by wise nature in such a way that one lamella does not touch another, so that the seeds produced between them should not be hindered in their development. . . ."*

Most important of all, Micheli cultured certain molds on freshly cut pieces of melon, quince, and pear. In this manner he followed the growth of the three molds previously named and noted that each fungus formed its own seeds and reproduced only its own kind.

We shall complete our sketch of Micheli's work by including an account of one of his mold-culture experiments. The translation was prepared by Buller in 1915.[1]

> On the 30th day of December, I took a piece of melon and shaped it into a triangular pyramid. Then, choosing a piece of a quince and also of an almost ripe pear, commonly called *Spina,* I formed them into truncated pyramids, with their apices removed, giving the piece of quince a pentagonal, and the piece of pear a hexagonal base. On the individual faces of the pyramids, I sowed the seeds of *Mucor, Aspergillus,* and *Botrytis,* keeping each kind separate, so that on the piece of melon I had placed three kinds, on the quince on five sides five kinds, and lastly on

[1] A. H. R. Buller, *Trans. Roy. Soc. Can., IV* (3) **9**: 1–27, 1915.

the pear on six sides, six kinds. . . . All these species of seeds began to germinate from the fourth to the fifth or sixth day of the month, as I observed. They developed into plants according to their seed, of which some attained their maturity on the tenth day, others on the twelfth, others on the thirteenth, and finally others on the fifteenth: and they produced the seeds of their kind. I kept these seeds separate, and again and again planted the seeds produced in like fashion from them; and then I always observed the same mode of growth in them, not in one trial only, but however often and whenever I attempted it, without any difference whatsoever other than in the rate of growth or in the earlier or later ripening.

Micheli also grew mushrooms by placing the appropriate spores on leaves of various trees that he kept under conditions of humidity suitable for the growth of these fungi. The results were gratifying, but the molds were better experimental material for Micheli's rigorous demonstration that seeds always reproduce the organism that formed them.

In 1753, Linnaeus' *Species Plantarum* was published and the fungi that had been described up to that time were assembled in a class of their own. In mycology, Linnaeus' contribution amounted only to the firm establishment of the binomial system of nomenclature, since he did not contribute to a better understanding of the fungi themselves.

M. du Tillet, of Bordeaux, director of the mint of Troye, was not an ordinary bureaucrat. Without agricultural or botanical training of any kind, he nevertheless applied himself assiduously to the development of experimental methods in solving problems in agronomy. He set himself the task of discovering the cause of wheat smut. He showed that only seeds obtained from smutted heads or healthy seeds that had been dusted with the black dust escaping from diseased kernels would produce smutty wheat. He was a laureate from the Academy of Bordeaux, and his work was published in 1755. Tillet did not show that the fatal black powder was a fungus, but he did introduce careful experimental methods in phytopathology and forever established the fact that wheat smut was an infectious disease.

It is not surprising that the nature of such diseases as smut and rust was recognized by people who had been directly influenced by Micheli, the discoverer of the fungal seeds. Two names are associated with such discoveries: Targioni-Tozzetti and Fontana.

Giovanni Targioni-Tozzetti was a disciple of Micheli, and when the latter died in 1737, Tozzetti became his intellectual heir. He considered it his duty to complete the work so excellently started by his master. Like so many men of his time, the gifted Florentine was very active. Not only was he professor of botany, and director of the Botanical Garden, in which capacity he replaced Micheli, but he was also the official physician of the Grand Ducal Court and commissary of the Office of Health. He wrote prolifically and studied many subjects, but our interest is linked to his writing the *Alimurgia,* a book on agriculture published in 1767 "for the relief of the poor." One doubts that the poor gained much from this publication, which has, nevertheless, great scientific value.

In the *Alimurgia,* Targioni-Tozzetti stated that "rust is an entire, very tiny, parasitic plant that grows nowhere except inside the skin of the wheat." And he describes the infection of the cereal plants as follows: ". . . it is necessary to recall that the arterial canals of cereal plants, which carry the nutritive humor to the ear, are all situated in the outer surface of the stems immediately beneath the very delicate cuticle. Whenever the little seeds of the rust come to rest upon the same stalk, finding some open mouths of the exhaling vessels, there they bury their minute radical fibers, and they infiltrate in such a manner that they graft into the tender and delicate arteries. . . ." The "radical fibers" are referred to further along as "hairs" that "steal and suck up for themselves the nutriment prepared and destined for the grain. . . ." From these quotations it seems that Targioni-Tozzetti recognized the parasitic nature of the rust fungi, saw that they had a mycelium, and even described the infection of the host through the stomata. In general, to set the proper perspective, it is now thought that when the rust is in the haplophase, it enters only by direct penetration, and that in the dikaryophase, it is capable of entering the host only through stomata. On the contrary, in the case of the smuts, penetration is always considered to be direct. On this basis, there does not seem to be any reason to doubt that Targioni-Tozzetti did indeed observe rust infection through the stomata.

Also in 1767, another Italian, Fontana, published his *Observations on the Rust of Grain.* Felice Fontana was professor of philosophy at the University of Pisa until he was called to Florence to build the museum of physics and natural history. He built astronomical instruments and made anatomical wax models. He dabbled in all fields, was most prolific, and seems to have been one of these brilliant jack-of-all-trades typical of his time. In the booklet that interests us Fontana remarked: "Because of the disagreement and uncertainty among the most modern philosophers, I believe that the solution of the problem could be found only through correct observation and a more exact and diligent examination of the powder of the rust." And Fontana continued: ". . . I do not see why the rust cannot be considered as a plant that lives on the juice of grain surrounding it. . . ." The effect of Micheli on the thinking of Fontana is obvious from the numerous references to him made in the text of Fontana's work. This text consisted mainly of a long and careful microscopic observation of the rust, which strengthened his feelings about the parasitic nature of the rust, which he thought ". . . might be possibly related to the molds described by Micheli. . . ."

In spite of their intrinsic value, the work of Targioni-Tozzetti and Fontana had little influence and was soon forgotten. It is doubtful that their work was known to Bénédict Prévost, who published in 1807 a work entitled *Memoir on the Immediate Cause of Bunt or Smut of Wheat*. Prévost's study was thorough. After giving a detailed description of the disease, he said that the globules found in bunted kernels were "seeds," "gemmae," or "spores" of some cryptogams. He then spent a considerable amount of time studying the germination and the development of the "bunt plant," which he likened to the rusts. He proved by inoculation that indeed the disease was caused by the bunt plant, which invaded whenever the conditions were proper for its development. To complete his study, Prévost demonstrated that copper sulfate would inhibit the germination of the spores of the bunt plant, and he demonstrated with field experiments how to control the disease on a large scale. He soaked the seeds for half an hour or more in a solution of copper sulfate and removed all materials that would float on the surface of the liquid. The work of Prévost was not widely accepted. A few farmers used his seed treatment with excellent results; however, it was used more extensively in Belgium and England than in France. As is often the case, Prévost was not a prophet in his own country. Born in Switzerland in 1755, he lived most of his life in Montauban, in southern France. He went to Montauban originally to tutor the children of some well-to-do persons. Various positions were opened to him at a later date in various European cities, through the influence of his cousin Pierre Prévost, who was a famous professor of philosophy and physics at Geneva. However, Bénédict Prévost loved Montauban, and like a truly wise man, he preferred the kind friendship that he found in the small learned societies of southern France to the bustling life of the big European cities. He died in 1819, having attained the rank of professor of philosophy in a Protestant school in Montauban.

Prévost's method of treatment was a prelude to the use of the Bordeaux mixture, a discovery made by another scientist in southern France, Pierre Millardet, who published his results in 1885. Like Pasteur, Millardet was born in the Jura. He first studied medicine but gave up his practice to study botany. He studied under De Bary at Freiburg, and after obtaining his doctorate, he returned to France and taught botany in various universities, including Strasbourg, Nancy, and finally Bordeaux. At first, he studied botany as a pure science, but important phytopathological problems loomed large in France, and Millardet's interest shifted to applied botany. His two most important problems dealt with the wine industry. The French grapevine was being attacked by aphids (*Phylloxera vitifoliae*) and by downy mildew (*Plasmopara viticola*). The aphids had been introduced into France around 1860 in rootstocks that had been obtained from the United States. The devastation that these aphids caused in vineyards almost destroyed the wine industry. The disease soon spread to all of Europe. Millardet solved the aphid problem by bringing American grapevine stock to France. These plants, which have the same geographic origin as the aphids, were essentially resistant to the *Phylloxera*. The aphids, in the wingless form, attack the roots of the grapevine. Grafting the desirable French varieties on the resistant American roots solved the problem. Millardet was left facing only the fungal parasite. Here is Millardet's own account of how he solved this problem:

"*Since the appearance of mildew in France* (1878), *I have not ceased to study Peronospora,*" said Millardet in 1885 (*J. agr. prat.,* **2:** 513–516), who used the generic term *Peronospora* to refer to *Plasmopara viticola*. He continued:

I had noted, in the course of my investigations, that the summer spores (conidia) of the *Peronospora* easily lose their ability to germinate. This observation, and the lack of success of all the treatments tested up to then, had lead me to the following conclusion: a practical treatment of mildew must not aim at killing the parasite in the infected leaves, which seems impossible without killing the leaves themselves, but rather at preventing its development by covering the surface of leaves with substances able to, at least, prevent the germination of its spores.

. . . chance put such a substance in my hands.

At the end of October 1882, I had the occasion of passing through the vineyard of Saint-Julien, in Médoc. I was quite surprised to see that all along the road that I was following, the vines still had leaves, whereas, everywhere else, they had long since fallen. It had been a mildew year, and my first reaction was to suspect that some treatment applied along the road had protected the leaves from the disease. Indeed, I noted that the upper surface of the leaves were in great part covered with a thin adhering coat of a pale, bluish powdered substance.

I questioned the manager of the Chateau Beaucaillou, M. Ernest David, who told me that the custom, in Médoc, was to cover the vine leaves with verdigris or copper sulphate mixed with lime, at the time of ripening of grapes, in order to discourage thieves.

I drew M. David's attention on the above-mentioned protection of leaves, and I told him of my hopes. I must say that at first he lacked enthusiasm, but that he eventually helped me so effectively that he is in great part responsible for the final success. . . .

M. N. Johnston, to whom I had communicated my ideas as early as 1882, and who had been following during the past two years the tests made by his manager, M. David, took decisive action in this matter and on his estate alone 150 thousand plants were treated. Everywhere the results have exceeded my expectations.

Today, October 3rd, the treated vines are normal, the leaves are healthy and beautifully green, the grapes are black and perfectly ripened. In contrast, the untreated vines appear miserable. . . .

The value of the experiments is enhanced by the fact that they were performed as methodically as possible. In each treated plot, several rows of vines were left as untreated controls. I will add that the treatment was applied preferentially to varieties which are most sensitive to mildew such as *Malbec, Cabernet franc* and *Petit-Verdot* . . . and that, in addition, this year has been especially bad for mildew.

All these considerations allow me to state most firmly that the treatment is effective. . . . But there are further considerations. Since there are close analogies between the *Peronospora* of grapevine and the disease agents of potato and tomato, I hope that we now have an effective prophylactic method against these last affections.

The formula for treatment consists of . . . [the following:]

In 100 liters of any water one dissolves 8 kg of commercial copper sulphate. Separately, one prepares milk of lime by mixing 30 liters of water and 15 kg of lime. Both solutions are mixed together forming a blueish paste. The mixture is sprinkled on the leaves with a little broom, being careful not to touch the grapes. The procedure is safe. . . .

The Bordeaux mixture was to remain the world's outstanding preventive fungicide for the next sixty years or so. It eventually was rivaled by the lime-sulfur solution that was used first as a livestock dip to control epizoa. This mixture was adapted for use in plant spraying by Cordley in 1906 and became the leading eradicating fungicide.

While the Bordeaux mixture and lime-sulfur solutions were sprayed extensively, Riehm, in 1913, started to develop organic mercurial compounds for the treatment of seeds. About 1940 organic fungicides, such as the carbamates, were introduced into agricultural use. Although organic compounds, and occasionally antibiotics, are now the major fungicides, the old, reliable Bordeaux mixture is still used.

Morphology, histology, and cytology

In order to follow certain of the developments in the field of phytopathology, we have left mycology *stricto sensu* at the point of its development following the publication of Micheli's book.

The next major advance was achieved by a French naturalist, Pierre Bulliard.

In his *History of Fungi of France*, Bulliard followed Micheli's lead and wrote in 1791:

> I hope to demonstrate conclusively that there is not even one mold which is not the product of the seed coming from a representative of the same species. I shall furnish my readers with the technique of repeating these experiments on which this proof rests. These experiments should be given the most serious attention, especially from those who maintain that *putrefaction* gives rise to organized beings. . . . Even though I was convinced that a high level of heat must destroy the vegetative power of *Mucor* seeds, I wanted to assure myself with facts that I was right in order to leave no uncertainty in this matter, I soaked bread in water and prepared a paste that I sprinkled with a large amount of *Mucor* seeds. I even admixed various substances on which mucors of various species had been growing. I put this paste in six *medical vials* and I placed them on a bed of hay in a large pot. Water was added to the level of the necks of the vials and the water was boiled for one hour. I capped two of these vials with tallow coated paper while they were still in boiling water. I removed these two vials and wrapped the necks with additional layers of paper which had been soaked in a very thick gum water. The other four vials, I left open. To two of these, I added *Mucor* seed and nothing to the last two. After five days, the two vials which I had left open and to which I had added the seed of mucors were full of molds . . . in the two vials which I had left open but to which I had not added seeds there were a few "stains" of mold. Those which I had sealed contained no more growth after two months than there was on the first day. I have repeated this truly interesting experiment many times. I have varied it and I have always obtained the same results, with the sole difference that in the vials that I left open and to which I did not add seeds, there

were sometimes only one or two small mold "stains" and sometimes none. This was probably due to the fact that the seeds from outside air became stuck to the sides of the long and narrow necks of the vials and never reached the mixture within.

In spite of these experiments, it does not seem that Bulliard played much of a role in resolving the problem of spontaneous generation. The principal value of Bulliard's flora of fungi was the clarity of his illustrations, which facilitated identification. He introduced many modern concepts in his descriptions. For example, he would collect spores from mushrooms, not only to observe their color, but also to study them under the microscope. In this way he observed the ornamentations on the spore coats. This great morphologist died under circumstances that are not precisely known. It is possible that he was assassinated by error, in 1793, during the upheavals of the Revolution.

Bulliard had been only a describer of species, and the framework of the systematics of fungi had progressed little from Linnaeus' low level. The true founder of the systematics of fungi was Christian Persoon, a Dutch citizen born in South Africa in 1761. He went to school in Germany, where in turn he studied theology, medicine, and botany. He moved to Paris, at the turn of the century, where he was to finish his laborious, solitary, and miserable life. Persoon's herbarium is most important since it contains many type species. It was taken over by the Dutch government before his death (1836) and is still in Leiden; many of the very specimens that he studied are still available. Persoon divided all the fungi known at his time into seventy-one genera, most of which are still valid. He understood that mushrooms, which represented the bulk of the fungi then known, were only the fruiting body of the plant, not the whole plant. He recognized two fundamental types of fruiting bodies: those in which the hymenium is uncovered during the maturation of the spores, as in the common toadstool; and those in which the hymenium is enclosed during the maturation of the spores, as in the puffball.

About this time, in Sweden, Elias Fries, at the age of seventeen, had become conversant with more than 3000 species of fungi, to which he had given his own names. When he went to the University in Lund, he read the works of Persoon and the other mycologists who had already given their own names to his beloved fungi. After writing a thesis on flowering plants, he went back to work with fungi and lichens for the rest of his long life. He taught first at the University of Lund, and from 1834 to his death in 1878, he taught botany at Upsala. His *Systema Mycologicum*, published in three volumes between 1821 and 1832, is, with Persoon's *Synopsis Methodica Fungorum* (1801), the chief starting point of the nomenclature of fungi and has earned its author the title of "Linnaeus of Mycology." As the years passed, Fries became the dean of all Swedish biologists, a position that tempted him to indulge in philosophical discussions. Even though he was a very exacting and precise scientist, he held the opinion that biology was not an exact science but was primarily historical and more closely related to theology than to chemistry and physics. Others, on the contrary, were prepared to demonstrate that biology was far from being relegated to the realm of history.

We may recall that asci had already been observed by Micheli, who had also noted how the spores seen on the gill fungi were often grouped in sets of four. Asci were again observed in the cup fungi by the Hungarian-born German botanist Hedwig, who called them theca. It was then taken for granted that all fungi that had a hymenium also had thecae, but the French physician Léveillé observed hymenia containing basidia. He then proposed to form two groups of fungi with hymenia, the thecospori and the basidiospori, which are, of course, the present-day ascomycetes and basidiomycetes. He introduced the use of the microscope for histological study of the fungi. His studies also led him to the important discovery that sclerotia are mycelial structures, not fruiting bodies.

Mycology benefited by the advances in other fields. It soon became obvious that algae and fungi were in many ways similar, and the knowledge of the sexual processes in algae became a guide and a stimulus to those working with fungi.

The algologists of the first half of the nineteenth century were essentially taxonomists who worked with dry herbarium specimens; they thus failed to discover motile gametes. About 1840, a young amateur scientist in Paris, Gustave Thuret, decided to study *Fucus* since fresh samples were always available in the fish markets of Paris. He soon concluded that a study could be productive only if one could have healthy algae and an endless supply of sea water. Since he was rich, long stays at the seashore were no problem, and after a visit to the coast of Calvados he returned in 1844 to announce to the French Academy of Sciences the discovery of the antherozoids of *Fucus*. Thuret's money not only provided travel, but also the hire of an artist that could illustrate his papers in a truly professional fashion. He established an algological laboratory first in Cherbourg, and then in Antibes. In 1854 he had elucidated the mechanism of fecundation of *Fucus*.

The pioneering work of Thuret was extended by other investigators.

notably Nathanael Pringsheim. In 1855 he discovered the method of fecundation in *Vaucheria*. He did more than just observe fertilization; he gave detailed descriptions of the growth of algae, the formation of sexual organs, and the maturation of the gametes. On this basis, the classification of algae was freed from reference to superficial resemblance and was established along more natural lines based on similarity of sexual types of reproduction. Thus to use Thuret's expression, classification of algae was becoming a summary of the biological knowledge of the time. These important discoveries were also a stimulus to the study of other cryptograms, such as the fungi.

The most important mycologist of the period was Heinrich Anton De Bary, who emphasized developmental morphology. He studied medicine but preferred botany and taught successively at the Universities of Freiburg, Halle, and Strasbourg. He was a great teacher, and many of his students played leading roles in the further development of mycology. We may recall that Millardet and Winogradsky had been among them. By age twenty-two De Bary was graduated from medical school and had published a brilliant paper on the smut fungi. He practiced medicine for two years, and from then on, the only pathology to which he paid any attention was that of plants. Nevertheless, De Bary was not a plant pathologist. He was interested in the development of fungi, particularly in their physiology. Of course, some fungi grew on other plants, and it was in this respect, in spite of himself we might say, that De Bary contributed most heavily to plant pathology. In 1861 he showed that the late blight of potatoes was caused by *Phytophthora infestans*, and in 1865 he published a most remarkable account of the life cycle of wheat rust in which he showed that part of the life of the fungus was spent on the barberry plant. We should note that a relationship between the barberry plant and the rust of wheat had been suspected for a long time. In 1660, in Rouen, an edict was issued that directed that all barberry plants be destroyed because they were held to bear some mysterious relation to the diseases of wheat. De Bary's rigorous demonstration of their relationship cleared up the mystery. De Bary also described sexual processes in some phycomycetes and ascomycetes.

Other workers continued to elucidate the life cycles of organisms that De Bary did not have the time to study himself. One of De Bary's early disciples was Michael Woronin, a wealthy Russian who had been supplied with a private laboratory by his father. The young Michael responded favorably and eventually demonstrated that the myxomycete-like *Plasmodiophora brassicae* was the cause of cabbage hernia (1878).

Others went beyond the work of De Bary, contributing to our understanding of the life cycle of fungi by skilled cytological studies. In this domain, Pierre Dangeard was outstanding. With him and his contemporaries, the most delicate structures that ordinary light microscopy could reveal were observed and interpreted. Dangeard, then professor of botany in Poitiers, observed, in 1893, that nuclear fusion occurs in the teleutospores of a rust. After some hesitation, he interpreted this fusion as a fecundation and predicted that it would be found in other basidiomycetes and also in ascomycetes. He searched and found it the next year in a *Peziza*, as a preliminary step in the formation of asci. A few years later, Dangeard left Poitiers for Paris, a *sine qua non* requirement for complete success in France. He had the unique ability to perceive with revealing clarity, but in essence he created his "own botany" from which his students were not expected to deviate. He could not tolerate the idea that others might have the freedom that was his. With such a personality, it was inevitable that he should have many disputes with others in the field—and indeed Dangeardian cytology and Dangeardian sexuality of fungi were attacked. As time passed, however, it was recognized that basically he had been correct, and he died in 1947 at the age of eighty-five after having received numerous honors from all over the world.

Nutrition

At this point we can leave the studies of the life cycles of fungi and of the peregrinations of their nuclei and turn to the consideration of their physiology. When Pasteur was studying alcoholic fermentation, on what might be considered an approximation of a chemically defined medium, he had no knowledge of the mineral requirements of beer yeast. He circumvented this difficulty by adding yeast ashes to his medium. One of his disciples at the *école normale supérieure*, Jules Raulin, devoted ten years of his life to the elucidation of the mineral requirements of the mold, *Aspergillus niger*. The work was started in Paris, at the *école normale*, and was continued in Brest and Caen, where Raulin was teaching at the local *lycées*. Pasteur regarded Raulin highly and asked his assistance in the silk worm diseases studies. Raulin refused since he wanted to complete his work, which earned him a doctorate in sciences in 1870.

In his classical and lengthy paper, entitled "Chemical Studies on Growth" (*Ann. sci. nat. Botan.*, **11**: 93–299, 1869), Raulin, referring to microorganisms, said:

We know today that the life of these minute organisms is necessarily linked to these infinitely varied chemical transformations to which are subjected in nature substances of dead animals and plants. This is one of the fundamental laws of nature, revealed by M. Pasteur's thorough and philosophical studies. From the chemical point of view the simplest of organized beings present very striking analogies with more highly developed living organisms, but unlike the latter, they can be used with a marvelous facility for the most delicate scientific investigations. The study of these small beings should thus throw a light on the perplexing questions of general physiology. This is a new phase of science opened by M. Pasteur. In the hope of obtaining data to support this concept, I have conducted on a *mucedinae* the chemical investigations which are the principal object of this memoir. . . .

The chemical study of the development of the simplest plants in the most simple artificial media must naturally serve as a basis for the study of the general chemistry of plants. This concept was the driving force behind the work that I present here concerning the growth of a *mucedinae, Aspergillus niger,* in an artificial medium composed of well-defined components.

The problem involved the formulation of the most favorable conditions for the production of this plant; the search for the physical and chemical factors which influence its development; the measurement of the effect of each one of these factors; the determination by analysis of the transformations of the substances which are involved in the nutritional processes of the plant; in short to determine in its entirety the chemical equation which would sum up the phenomenon of life.

One can easily understand that I have not completed this task. After having defined the conditions most essential for the growth of the *Aspergillus,* I have tried to determine the influence of each of these factors on the growth of this *mucedinae.*

One of the most interesting properties of plants is their ability to grow in artificial media composed exclusively of known chemical components. After such a medium, suitable for the nutrition of a specific plant, has been formulated the following method should be used to study the influence of the various chemical and physical factors on development of the plant. A dish of the artificial medium is placed in the environment most favorable for the growth of the plant. The dish is seeded with the germs of the plant and growth allowed to proceed for the necessary time. This procedure, which should always be performed under identical conditions in each series of experiments, is the control against which all the other cultures will be compared. Simultaneously another culture is incubated which differs from the first one *only by one factor, the factor to be studied.* The dried cell material obtained from both cultures is weighed separately *and the numerical ratio of the weight of these two cultures will be a measure of the influence of the factor which was studied.* The analysis of the cellular material can be used, if necessary, to complement the results. By thus varying in turn, all the elements of the problem, the phenomenon of growth can be studied in its most minute detail. . . .

In summary, the value of the method which we shall use depends on three general conditions: (1) An artificial medium must first be devised which is suitable for the growth of the plant which is being studied; (2) the weight of the yields which will be obtained from the growth of the controls, during a given time, on a well-defined medium with a constant weight of nutrient substances, must be, all else equal, as large as possible; (3) the controls kept under the same conditions must furnish yields, whose numerical ratios should be as close as possible to unity. The ratio which will show the maximum deviation from unity will be the maximum relative error of the method. . . .

As we have already stated, M. Pasteur is to be credited with proving that *mucedinae* can grow in a medium composed almost exclusively of chemically defined substances and with having discovered the more essential elements necessary for the growth of these plants. He seeded spores of *mucedinae* in the following mixture: pure water, 1 liter; sugar, 20 g; bitartrate of ammonium, 2 g; ashes of yeast, 0.5 g. He was able to obtain on this medium especially with a *Penicillium*, a defined amount of growth and by suppressing one or the other of these elements, sugar, ammonium salt, or mineral matter, he was able to obtain only traces of growth. M. Pasteur, by this fundamental discovery, had opened a new field of science. But he did not investigate it much further although he understood all its implications. . . .

At the beginning of my study I tried to follow as closely as possible the conditions indicated by M. Pasteur. In a 20° incubator, I placed cultures of the following medium seeded with the *Aspergillus:*

water	2,000 g
sugar	70 g
ammonium nitrate	3 g
tartaric acid	2 g
ammonium phosphate carbonate of potassium carbonate of calcium carbonate of magnesium	traces

After 45 days of incubation I obtained dry yields as follows: in one culture 3.19 g, in the other, 1.77 g. . . .

To obtain in 45 days 3.19 gm of *mucedinae* from more than 70 grams of nutritive substances was an indication of rather sluggish growth. Analysis confirmed this conclusion by showing that seven-eighths of the weight of each one of the nutritive elements which were present in the medium had not been used. . . .

It is mainly by adding to the previously described medium: sulfur, zinc, iron and silicon, in form of salts, by modifying the proportions of essential elements; by raising the temperature to 35°, by using very shallow flasks for the incubation, so as to obtain satisfactory conditions of aeration and humidity, that I was able in six days to obtain 25 g of yield from 80 g of nutrient matter. . . .

Whereas the weights of the yield produced in my controls at the beginning of my study varied in a ratio of 1 to 1.8, it was reduced to a ratio of 1 to 1.05.

I must also add that more than once, during the course of my investigations, the medium, which had been seeded with spores of the *Aspergillus,* was contaminated by foreign vegetation, principally a *Penicillium.* The results of experiments in which there was growth of other kinds of organisms were thus not comparable. I have now overcome this difficulty.

To sum up: the method of experimentation dealing with the growth of mucedinae in medium, composed of chemically defined components, has been known since the investigations of M. Pasteur. I have adapted it to precise experimentation: (1) by augmenting considerably the weight of the yield furnished in a given time by a given weight of nutrient substances; (2) by reducing the variation of the weight of the yield obtained in duplicate experiments; (3) by growing the *Aspergillus niger* in pure culture.

Raulin continues with a description of the constant temperature room, or incubators, that he used for his experiments. He made sure that no gasses that might be toxic to the *Aspergillus* would come in contact with the incubation chamber. To ensure this result he heated the incubator with warm water circulating through heating coils. The temperature was regulated to vary by no more than 1°C. The temperature of incubation in all his studies was 35°C. Raulin installed an interior humidifier, which maintained a high relative humidity in the atmosphere. The de Saussure hygrometer usually registered 70°, in his incubator. The containers that he used to grow the *Aspergillus* were rectangular porcelain dishes, containing about 1,500 cubic centimeters of liquid and measuring about 4 centi-

meters deep, 16 centimeters wide, and 26 centimeters long. These dishes
were incubated without being covered.

Raulin continues:

> After having described the special instruments which were used for
> the investigations described in this memoir, I shall describe in detail
> the operations which I carried out in order to obtain the growth of the
> control cultures. These operations fall into 4 groups: the preparation
> of the culture medium, the seeding of the culture medium with the
> spores of the *Aspergillus,* the incubation of the *mucedinae,* and finally
> the harvest of the material.
>
> First the following compounds were deposited in the dish which was
> to be used for the growth of the control:

water	1,500 g
sugar	70 g
tartaric acid	4 g
ammonium nitrate	4 g
ammonium phosphate	0.6 g
potassium carbonate	0.6 g
magnesium carbonate	0.4 g
ammonium sulfate	0.25 g
zinc sulfate	0.07 g
iron sulfate	0.07 g
potassium silicate	0.07 g

> These substances must be weighed carefully and be as pure as pos-
> sible. The mixture is left standing during a few hours in order to permit
> the solid substances to dissolve in the water. To obtain a homogenous
> solution the liquid is stirred with a porcelain spatula.
>
> Second, the medium is seeded by streaking the entire surface with
> the tip of a paint brush with which one has collected spores of the
> pure culture of the *Aspergillus.* These spores are disseminated and
> float over the surface of the liquid. . . .
>
> Third, after 48 hours at 35°, the membrane that I have mentioned
> is already quite thick, it becomes slightly yellowish, then dark brown.
> Finally after 3 days of growth it becomes completely black on its
> superior portion. This change in color is due to the production of
> spores which pass successively from yellow to brown and finally to
> black. After 3 complete days of incubation the pad of growth can be
> collected.

When this has been collected, the surface of the liquid is seeded anew and the dish is returned to the incubator. Three days later another yield is obtained which is not as abundant as the first one. The liquid under these conditions is about exhausted and not suitable to give an appreciable third yield.

Fourth, the *Aspergillus* is harvested by removing manually the thick membrane which is formed on the third day. The membrane is pressed in order to remove the greater portion of the trapped liquid, and is spread in a saucer where it is dried, first at 50° then at 100° in a Gay-Lussac oven. Weighing is the only remaining task. . . .

When highly precise figures are required, it is very important to compare simultaneously cultures which are kept in the same incubator. The external factors, such as temperature, humidity, length of incubation, etc., are thus similar and the results are more constant than if the two yields had been obtained at different times. . . .

After explaining in detail the experiments that permitted him to select the physical conditions of cultures previously described, Raulin investigated the effect of certain salts that might be toxic to the mold. He continued:

Certain chemical substances seemed to merit study because chemically they are closely related to salts necessary for the growth of the *Aspergillus*, but they contrast physiologically by being highly toxic.

The first inorganic salt studied was silver nitrate, which was found to partially inhibit the growth of the *Aspergillus* at the dilution of 1 to 1.6 million parts by weight. At higher dilutions the toxicity of silver nitrate was negligible.

However, the silver nitrate is reduced slowly under the influence of organic matter in the artificial medium and eventually becomes inactive after a certain number of days. Then, even if the dilution of the salts is less than 1 to 1,600,000, growth proceeds as usual.

While trying to grow the *Aspergillus* in a silver dish, I was at first much surprised to obtain only faint traces of mycelium. This property can be explained by the chemical action of the artificial liquid medium on the metal producing minute amounts of a silver salt which affects the development of the plant. Silver dishes, contrary to expectation, are completely unsatisfactory for experiments of the type which interest us. . . .

Mercuric chloride was similarly studied. It was found to be toxic at the concentration of 1 to 512,000. At higher dilutions the salts had little effect on the growth. Platinum dichloride was also toxic but only up to a concentration of 1 to 8,000. Copper sulfate had a slight toxic action, with some growth produced at the dilution of 1 to 160.

> After having discussed the effect of the factors which are foreign to the culture medium, I shall now discuss the work on the physiological action of the substances which are the components of this medium. . . .

Raulin then lists, in the following order, the twelve elements that he removed sequentially in order to demonstrate the favorable influence of each on the *Aspergillus:* (1) oxygen; (2) water; (3) sugar; (4) tartaric acid; (5) nitrogen-containing substances: nitric acid, ammonia; (6) phosphoric acid; (7) potassium carbonate; (8) magnesium carbonate; (9) sulfuric acid; (10) zinc oxide; (11) iron oxide; (12) silicon oxide.

Raulin continued:

> I shall not reconsider the requirement for oxygen, which I have already discussed, and to which I did not add any new information. . . . It is hardly necessary to mention that the *Aspergillus,* like any other plant, requires a certain amount of water in order to achieve normal growth. The principal role of this oxide is to dissolve the other substances which are necessary for the growth of the plant.
>
> The role of sugar in the development of lower plants has been stressed by M. Pasteur. I shall mention here only one example which will show the great effect of this compound on the growth of the *Aspergillus.*

Its removal from the medium reduced the growth sixty-five-fold.

> The sugar furnishes the plant with carbon, hydrogen, and oxygen, essentially most of its components. . . .
>
> Tartaric acid plays a very remarkable role in the growth of the *Aspergillus,* and in the growth of mucedinae in general. If it is removed, the medium becomes filled with infusoria which prevent the growth of the *mucedinae.* This phenomenon occurs, as a matter of fact, in all neutral or slightly alkaline liquids. Thus the effect of tartaric acid is due to its acidic properties rather than its composing elements. Alcohol and sugar, etc. which contain the same elements cannot replace it, whereas, on the contrary, most organic acids and even mineral acids such as sulfuric acid in small quantity can replace it. I shall not discuss this any further since this has been a well-known fact since the work of M. Pasteur.

Nitrogen is the most important element after carbon, hydrogen and oxygen. I shall study experimentally the effect of the most important nitrogen-containing compounds on the development of the *Aspergillus.*

The ratio of yields obtained with and without ammonium nitrate varied from 24 to 1 and 153 to 1. . . . I tried to replace it with potassium nitrate and ammonium tartrate. . . . In general, one can say that nitric acid and ammonia in the form of salts produce similar effects on the growth. This result can be attributed only to their common property of containing nitrogen. I should add that potassium nitrite and potassium cyanate have not favored the development of the *Aspergillus.* . . .

The apparent effects of phosphoric acid, potassium carbonate, magnesium carbonate, and sulfuric acid on the growth of the *Aspergillus* are quite similar. I have thus grouped together the experiments which were designed to show their requirement for the growth of the organism.

Here is the description of an experiment:

Medium 1:

Water	1,000 g
Sugar	50 g
Tartaric acid	2 g
Ammonium nitrate	2.9 g
Zinc carbonate	0.048 g
Iron citrate	0.048 g
Potassium silicate	0.048 g
Ammonium phosphate	0.48 g
Potassium carbonate	0.48 g
Magnesium carbonate	0.32 g
Ammonium sulfate	0.20 g

Medium 2:

Similar to Medium 1 less ammonium phosphate which has been replaced by an equivalent amount of ammonium tartrate.

Medium 3:

Similar to Medium 1, less potassium carbonate.

Medium 4:

Similar to Medium 1, less magnesium carbonate.

Medium 5:

Similar to Medium 1, less ammonium sulfate, which is replaced by an equivalent quantity of ammonium tartrate.

Ratios of weights of total yield on control medium 1 to the weights of other yields:

$$\frac{No.\ 1}{No.\ 2} \cdots \frac{18.2}{0.10} = 182.0, \text{ effect of } phosphoric\ acid.$$

$$\frac{No.\ 1}{No.\ 3} \cdots \frac{18.2}{0.85} = 21.0, \text{ effect of } potassium\ carbonate.$$

$$\frac{No.\ 1}{No.\ 4} \cdots \frac{18.2}{0.20} = 91.0, \text{ effect of } magnesium\ carbonate.$$

$$\frac{No.\ 1}{No.\ 5} \cdots \frac{18.2}{1.68} = 11.4, \text{ effect of } sulfuric\ acid.$$

This experiment demonstrated the efficacy of phosphoric acid, potassium, magnesium and sulfuric acid in the form of salts (in promoting growth) and at the same time it reveals the quantitative range of the effect. . . .

The action of sulfuric acid on growth is independent of the base with which it is combined, so that all the sulfates in which the base does not have an intrinsic toxic effect, will support the growth in an amount equal to ammonium sulfate. A similar statement can be made for phosphoric acid, for potassium, and magnesium. I have checked this hypothesis by measuring the physiologic effect of phosphoric acid, sulfuric acid, potassium and magnesium which were in the form of various salt combinations. The results have shown that I was correct. . . .

It is in my opinion very remarkable that a certain amount of each one of these simple elements can determine the formation of such a large amount of organic matter. . . .

Thus:

1 gram of phosphorus may permit the development of 157 gm of *Mucedinae*

1 gram of potassium　　 ”　　　 ”　　　 ”　　　　 ”　　　　 ”　 64 ”　 ”
Mucedinae

1 gram of magnesium　 ”　　　 ”　　　 ”　　　　 ”　　　　 ”　 200 ”　 ”
Mucedinae

1 gram of sulfur　　　　 ”　　　 ”　　　 ”　　　　 ”　　　　 ”　 346 ”　 ”
Mucedinae

I would like to mention once that all the quantitative values cited in this study should not be taken in an absolute sense. They can be used only to demonstrate the relative effect of the elements on growth. . . .

We have observed up to now the influence of a certain number of elements on the growth of the *Aspergillus*, atmospheric oxygen, water, sugar, tartaric acid, nitrogen-containing compounds such as nitric acid or ammonia, phosphoric acid, sulfuric acid, and potassium and magnesium carbonates.

A medium consisting of these elements can furnish an abundant yield of the *Aspergillus*, if one does not take too great care to prevent other substances from contaminating the medium naturally. This will occur if the elements that I have just named have been inadequately purified, if the dishes, such as crude pottery, in which these elements have been dissolved furnish traces of their own content. It is under such conditions that my first experiments on the growth of the *Aspergillus* were performed. Later, when I carefully prevented the artificial medium from becoming contaminated by traces of foreign substances I noticed that the yields were becoming strangely smaller. I then realized the importance of contaminating compounds.

In 1862, an artificial medium composed of the substances that I have just enumerated furnished a total yield which was as high as 18.04 grams of dry material for 80 grams of sugar. In 1864, a medium composed of the same elements, almost in the same proportions, produced 6 grams of total yield for a hundred grams of sugar.

Why did the first experiment give a yield which was three times as high as the second? In the first case contaminating substances were present among the components of the artificial medium. This cause of error had been eliminated in the second case.

These facts are self-explanatory; there are certain substances which when added to the artificial medium used in the second experiment will increase the growth from 6 to 18 grams.

To prove his point, Raulin added ground pottery, porcelain, and ashes of wood to his medium. A three- to fivefold increase in growth was noted. Testing of a mixture of mineral salts indicated that the most essential of these mineral salts were part of the following mixture: magnesium carbonate, calcium carbonate, sodium carbonate, iron sulfate, zinc sulfate, and potassium silicate. However, Raulin mused:

> The complex substances that I have added in certain experiments, have brought with them a certain amount of phosphoric acid, potassium, magnesium and sulfuric acid. Is it not then possible, that this increase in elements known to be useful, might have been responsible at least in part for the observed effect?

Carefully, Raulin demonstrated that such was not the case and thereby discovered the necessity for trace elements. He continued:

There are thus mineral oxides which are favorable for growth and which are neither phosphoric acid nor magnesium oxide nor potassium oxide nor sulfuric acid. What are these oxides? Experimentation will reveal it to us. . . .

Having developed a medium containing zinc and iron, Raulin concluded:

The weight of the yields is appreciably reduced by removing either zinc oxide or iron oxide from the medium.

The influence of these oxides on growth manifests itself equally, regardless of the combining acid.

They cannot be substituted for each other. . . .

The ratio of a certain weight of organic matter to the weight of zinc or iron which has contributed to its formation has reached the following maximum figures:

for zinc 953
for iron 857

During his study of iron and zinc, Raulin made this interesting observation about manganese:

If iron and zinc cannot substitute for each other, could other metals replace them? Only experiments can answer these questions. A few years ago I tried to evaluate the action of manganese salts on the development of the *Aspergillus*. These salts produced effects which were similar to those observed with the salts of iron and zinc, but the results were less constant and not so striking. Should we conclude that manganese salts were acting simply because they were contaminated with traces of iron and zinc, or that manganese can replace iron or zinc physiologically as it can very often replace them in chemical reactions? I shall not attempt to answer this question.

Having shown the necessity for zinc and iron, Raulin again added ground earthware, soil, and wood ashes to his medium to see if a further increase in weight would be obtained. This indeed turned out to be the result, and as carefully as before, he determined that the increase in growth was not due to an increase in concentration of the elements already present in the medium. The answer was that in these complex mineral substances there were elements, still unknown, that increased the growth. Silicon, in the form of potassium silicate, produced a further increase in organic substances that was independent of the base with which this element was combined.

And Raulin concluded:

> The *Aspergillus* can grow abundantly in a medium composed of the appropriate mixture of 12 chemically defined substances. The presence of all these substances is essential since removal of a single one produces a yield lowered in a ratio varying from one substance to another. The quantity of each simple compound, which is necessary to yield the same weight of the plant, varies between numbers of the same order of magnitude as the weight of the plant growth, to very small fractions of this weight. . . .

Raulin asked himself "whether or not the medium formed of the mixture of these twelve *supposedly pure* elements is the most perfect realization of a fertile soil." He was unable to increase the growth further by adding his favorite sources of complex mixtures of minerals, and the growth that he obtained in his chemically defined medium was better than that produced in the complex media in use at this time.

Raulin, the slow, careful, thorough, and conservative investigator, rightly felt that he had done as well as could have been expected for his time:

> Apart from the elements which we have found essential for the growth of the *Aspergillus,* it is very possible that there exist others which are no less essential, but their discovery is linked to the methods of purification which have been available for our experiment.

Raulin died in 1896. In 1870, he discontinued the studies that had made him famous. As he had predicted, further advances in this field had to wait for more advanced methods of chemical purification of the media. His work was continued by Gabriel Bertrand and Maurice Javillier in France and Robert A. Steinberg in the United States. As time passed, the role of trace elements in various enzymatic systems was demonstrated. Here, Gabriel Bertrand played a pioneering role.

In 1897, the same year that Edward Buchner published his paper on alcoholic fermentation without yeast cells, Bertrand (*Compt. rend. acad. sci.,* **124:** 1032–1035, 1897) wrote: "From now on, it will be necessary to take into consideration, in the study of soluble ferments, not only this substance, which is both organic and very unstable, to which we have attributed up to now all the properties of soluble ferments, but also these substances that we propose to call co-ferments (which are here mineral, but which could be organic) and that form, with the ferments the truly active system." This statement had been written because Bertrand thought he had demonstrated that manganese was a necessary component of laccase. However, he was wrong since copper was later found to be the metal that is part of the laccase enzymatic system. Wrong in the identity of the metal, Bertrand was, however, right in the basic principle.

From what we have just read, it is clear that at the end of the nineteenth century, it was well established that fungi required only one organic substance for growth, namely, a suitable source of carbon. One can imagine the lack of enthusiasm that greeted a paper written in 1901 by a Belgian scientist, Wildiers, who stated that the much-studied beer yeast required small quantities of an unknown organic substance, which he called "bios."

In order to put these events in the proper perspective, it would be best to review briefly what was known at the time about the compounds we now call vitamins.

For a long time, man had known that it was not enough to fill his stomach to satiety, but that he had to eat a variety of foods and that some foods could prevent or even cure diseases. This became painfully evident during the long voyages of the great geographical discoveries, which kept men at sea for weeks and months. At the beginning of the seventeenth century, oranges and lemons were loaded on ships, and the sailors discovered that the juice of these fruits could do more to prevent or cure scurvy than the best of rum.

At that time, and during previous centuries, many children had deformed limbs, protruding foreheads, concave chests, and protruding abdomens. Often these deformities lasted as long as the individual lived. Around the middle of the nineteenth century some French physicians started to eradicate rickets by dosing grimacing children with nauseating cod liver oil. Indeed only the children were dissatisfied with this treatment, because the results gratified the parents, the doctor, and the fishermen of the Newfoundland banks who caught the cod.

A third disease called beriberi was found primarily in China and Japan. This disease, marked by wasting and paralysis, had been known since ancient times. By 1882, the Japanese navy had curbed the disease by increasing the sailors' rations of vegetables, fish, meat, and barley. The disease was then believed to be due to improper protein nutrition. In the Dutch East Indies, a Dutch physician, Christian Eijkman, noticed that chickens were healthier in some villages than in others. He further noticed that chickens fed whole rice were healthier, as a group, than those fed milled rice, from which the bran had been removed. In fact, the branless chickens exhibited an affliction reminiscent of beriberi. Between 1890 and 1906, Eijkman conducted pertinent experiments with hens. He fed them polished rice and obtained an experimental beriberi that he could cure with the bran of the rice. He attempted to isolate the active factor and showed that it was a substance which was soluble in water and alcohol and which was dialyzable. It was precisely in 1901, as Wildiers' paper was published, that Eijkman's experiments led to the belief that beriberi arose because the diet "lacked certain substances of importance in the metabolism of the central nervous system."

Thus, when people who had been growing yeasts for years and who thought of nutrition in terms of carbohydrates, proteins, and minerals read the title of Wildiers paper, "A New Substance Indispensable to the Development of Yeast" (*La Cellule*, **18**: 313–332, 1901), they must have thought that he was somewhat deranged. However, as can be seen, the paper was lucid:

> The famous experiments of Pasteur on alcoholic fermentation have revealed the true nature of the yeast and have so well established its functions that since then, only details have been added to this fund of knowledge.
>
> When one reads the first long paper by Pasteur on alcoholic fermentation, published in 1860, one is impressed by this scientific monument, which is scarcely one hundred pages long, and which not only definitely establishes the true theory of fermentation, but also creates the science which was later to be called "biochemistry" by Duclaux. . . .
>
> An important part of this first paper dealt with the nature of the nutrients necessary for the life and reproduction of the yeast.
>
> Ever since this milestone, there is not a single living organism that we understand better, from the point of view of chemical mechanism, than the yeast, and the data of Pasteur constitute the most beautiful chapter of all physiological chemistry. One should even add that for a generation no correction has been made of this work of Pasteur.
>
> We must though, in the present paper, describe a correction that should be made of Pasteur's laws. We are certain of the data present, but we cannot yet evaluate their significance.
>
> Pasteur had established a law, valid until now, that the yeast only requires the following elements in order to live and ferment:
>
>
> yeast ashes
> ammonium salt
> fermentable sugar
>

In other terms, *the yeast can transform ammonium salt into living organic matter.* Or: *the yeast can synthesize all its organic components from sugar and mineral matter.*

The importance of this law is great. It places the yeast close to chlorophyll-deficient plants by virtue of its synthetic power, and radically separates the mode of life of yeast cells from that of our own cells.

Although *since* Pasteur's time there has not been any objection to this law, one should note the historical fact that a famous man disagreed with Pasteur on the precise nature of the required elements.

Let us go back to the years 1860 to 1870.

The polemic on the nature of fermentation was open and the whole scientific world was interested. On one side, Pasteur, powerful and marvelously lucid, was building on a firm basis, the new theory of fermentation as we understand it at present. Liebig, on the other hand, was defending the old theory that fermentation was a simple "putrefaction of organic matter."

Neither champion was a rhetorician nor a dreamer. On one side new facts were brought out, on the other side observations of apparently sound controls were opposed.

Pasteur states that the seeding of an exclusively mineral medium, to which sugar had been added, with an amount of yeast cells equal to the head of a pin, was enough to bring about an abundant fermentation with proliferation of yeast cells.

Liebig says that, by exactly repeating the experiment of Pasteur, no fermentation and no yeasts are produced.

Surprising contradiction!

However, Pasteur's data were neat and clear, and Liebig was not a man to make a crude mistake in the constitution of such a simple medium as Pasteur's.

Because of the importance of the matter, Pasteur offered to make the experiment in Liebig's presence, and, as he actually said—"to make in front of Liebig a quantity of yeast as large as he could demand within reason"—with the help of an exclusively mineral medium. This was in 1871, and Liebig, who was getting old, did not accept Pasteur's invitation.

The matter was then definitely settled in favor of Pasteur, and the new theory relegated the old theory and even the recent conflict to obscurity.

No one, as far as we know, has tried to determine the cause of the strange contradiction of facts which caused Liebig to oppose Pasteur.

It is almost by accident, by growing yeast cultures for a special purpose, that we stumbled across the new fact which explains the results of both Liebig and Pasteur and which reveals at the same time that, contrary to Pasteur's law, the yeast is unable to elaborate anything from a mineral medium and sugar and that a third element, which had escaped the attention of Pasteur, is needed . . .

We have not been able to elucidate yet the nature of this chemical element; we can only note certain of its properties which will help in the pursuit of its search.

Using as a basis the classical data on the culture media for yeast, we intended to study the synthesis of phosphorus-containing substances (phosphates, nucleins, lecithins) and their successive appearance in the culture media.

This special aim forced us to eliminate from our culture media the maximum possible amount of phosphorus-containing impurities. Thus the first source of contamination that we tried to eliminate was that carried with the inoculum.

Being extremely careful on this point, we seeded our cultures with a minimum of living cells.

To our surprise, our culture either did not grow or would grow only after a very long time. . . .

However, our yeasts were really alive since when seeded, in the same fashion and in the same proportions, on sterile beer wort we obtained abundant fermentation and multiplication.

On the contrary, none of the mineral media with inverted sugar added, showed any sign of life.

It was soon readily apparent that it was adequate to seed any culture medium a little bit more abundantly in order to observe an immediate and rapid fermentation. A few series of transfers made with particular attention to this detail proved to us that this was the crux of the matter.

We were holding the string which was going to guide us in the search for the unknown.

We must present the experiments we have performed, and the results they gave, in a methodical fashion. . . .

We finally decided to use the following medium:

Water	200 gm.
Sugar (saccharose)	20 gm.
Magnesium sulfate	
Potassium chloride	
Ammonium chloride	at 50 ct.gm.
Dibasic sodium phosphate	
Calcium carbonate	10 ct.gm.

. . . The yeast used was, except in certain control experiments, a top yeast of the *Saccharomyces cerevisiae* type I (Hansen) grown in pure culture on sterile beer wort enclosed in a Pasteur flask. . . .

The culture medium was held for 24 hrs. after sterilization, in order to favor aeration, before being seeded with varying inocula taken from the previously described stock cultures. The culture, seeded in round flasks equipped with Crispo's tubes, were weighed on a balance accurate to a decigram. Crispo's tubes have two enlargements made to receive H_2SO_4, which holds back the water vapors emitted by the culture. Losses in weight were thus exclusively caused by carbonic acid. The culture was then kept in a constant temperature room at 28°C, which is the optimum temperature. Weighings taken daily were showing us the loss in carbonic acid and thus information about the quantity of sugar destroyed by fermentation. The loss in carbonic acid in equavalence, *caeteris paribus*, to about 50% of the destroyed sugar. . . .

The mineral medium of Pasteur and similar media do not permit the development of the yeast if the inoculum is not abundant enough. . . . For each yeast suspension which is used for seeding, one can determine the minimum amount which should be added to mineral media in order to obtain growth . . . (for example) . . . 4 cultures containing 125 grams each of mineral media and thus about 10 grams of sugar were seeded with one of our pure cultures on very rich beer wort.

Two round flasks were seeded with two drops and two others with five drops from the same sterile pipette.

We give the values which represent the weight loss in grams for the various cultures.

Days	2 Drops	2 Drops	5 Drops	5 Drops
2	0	0	0.5	0.4
3	0	0	2.2	1.9
5	0	0	5.0	5.5
15	0	0	finished	finished

. . . These results are clear. We have, of course, repeated this experiment many times.

A determined amount of inoculum is needed. The limit is very low and often very sharp. This limit is not far from a pin head (inoculum measure of Pasteur), when one uses dense yeast.

The following question immediately comes to mind. Does the needed amount act by adding living cells or by the addition of chemical substances?

It acts by adding chemical substances. To prove this point we boiled the yeast suspension and we added increasing quantities to each mineral medium. Then we seeded all the round flasks with two drops of pure living culture. 125 gm of medium per flask.

Days	1 cc of boiled yeast suspension + 2 drops of living yeast	2 cc of boiled yeast + 2 drops of living yeast	3 cc of boiled yeast + 2 drops of living yeast	4 cc of boiled yeast + 2 drops of living yeast	5 cc of boiled yeast + 2 drops of living yeast
2	0	0	0.5	1.2	2.5
3	0	0	1	2.1	4.7
4	0	0	1.2	3	5.6

It is thus evident that the action of these solutions was not due to the addition of living cells.

From now on we shall always seed our flasks with the same amount of living cells, we shall vary only the amount of products from killed yeast cells.

In these emulsions of boiled yeasts, is the activating substance contained in cellular bodies or in the liquid in which they are bathed?

By careful preparation of cell-free filtrates, with the help of Chamberland's filter, and by preparing well washed yeast cells, washed in warm water and separated by settling and filtration, we shall see that only the filtrate is active and that the cellular bodies are absolutely inactive. 125 gm of medium per flask.

Days	Filtrate of 5 cc of boiled yeast suspension + 2 drops of living yeast	Filtrate of 20 cc. of boiled yeast + 2 drops of living yeast	Precipitate from 5 cc of the same boiled yeast + 2 drops of living yeast	Precipitate from 20 cc of boiled yeast + 2 drops of living yeast
2	0.5	0.6	0	0
3	2.0	4.1	0	0
4	3.2	5.2 finished	0	0
5	5.1 finished	....	0	0
15	....	....	0	0

This result is clear: it is the liquid bathing the yeast which contains the indispensable substance. . . .

We are thus, from now on, searching for a water soluble substance, perhaps a group of substances. We know where it is located; it is now a question of determining its physical and chemical properties and to isolate it if possible.

Permit us to offer an apt name for this unknown substance to be used until its chemical nature is determined. Let us call it "bios." May this name soon be replaced by a chemical one!

We are left only with the task of describing the properties that we have been able to ascribe to it up to now. . . .

1. We already know that the substance is *water soluble.*

2. The substance is *insoluble in absolute alcohol and ether,* but it is sufficient to have slightly aqueous alcohol, for example 80% alcohol, in order to obtain a good extraction of "bios." . . . This eliminates the possibility that the enigmatic substance might have been lecithin which, either as traces or in combination might have been carried in the water.

3. The enigmatic substance *is not present in the ash of the yeast.* Thus, it is not an inorganic substance gifted with special properties and which would have escaped the analytical chemists. One knows of such a case, since zinc is absolutely necessary for the growth of *Aspergillus niger,* although traces are sufficient. . . .

4. The "bios" *is not destroyed by boiling for half an hour in a solution containing 5% by volume of* H_2SO_4, which is 10% by weight. To destroy it a concentration of 20% H_2SO_4 at boiling temperature is needed. . . .

5. The "bios" *seems to be altered by boiling* (½ hour) *in solutions of NaOH* at 1% and even more at 5%. . . .

6. The "bios" is *not precipitated by lead acetate.* It is known that lead acetate precipitates many substances in these media of organic origin. . . . We should remark, however, that no conclusion can be drawn as to the nature of bios from these facts. A great number of nitrogenous organic substances are not precipitated by any of the metallic salts. . . .

8. In the course of these experiments we have tried most of the known organic substances: urea, asparagine, alanine, tyrosine, the nucleic bases, adenine and guanine, thymus nucleic acid, creatine (Merck), the products of pepsin and trypsin digestion of chemically pure albumins, such as edestin and ovalbumin. The active substance of "bios" *is none of the above listed substances.*

9. "Bios" *easily dialyzes* through parchment paper.

10. "Bios" is also present in the *meat extract of Liebig, in commercial peptones* and in boiled *wort* of germinated barley before seeding with the yeast. . . .

11. The yeast by multiplication and fermentation *does not produce new bios.* It is thus not a product that the yeast cell is able to manufacture. It is a product that is absolutely needed for its life but which it cannot cause to increase in quantity.

To demonstrate this basic fact we have made cultures with the amount of a "bios" which previous experiments had shown to be necessary for growth. When the development and the fermentation had stopped after a rapid growth . . . we boiled this culture and we made a fresh medium using ⅛ or ¼ of this first culture (plus the necessary amounts of sugar and salts in order to get back the original concentration of these substances). When seeded again as was done the first time, but without the addition of new "bios," the cultures grew very poorly. . . . From such experiments one gets constantly the impression that, far from forming new "bios," the yeast uses it up and does not reconstitute it.

12. Lastly, we have checked whether other yeasts required this same unknown compound in order to grow. We have instituted the same experiments with *various commercial top yeasts* and with *another pure yeast at low temperature,* and always the result was the same.

And Wildiers concluded:

It seems useless to look any further for the reason for the contradiction which existed in the data of the two great scientists. If, at the invitation of Pasteur, Liebig had visited the French scientist in order to see and to perform similar experiments, both would have immediately noted the fact that we are bringing up today.

Pasteur, *in seeding his broths, was introducing, not only living cells, but also an unknown chemical element which was indispensable.*

Since Pasteur, certain scientists have prepared cultures on mineral media which had been seeded with extremely small inocula. In these cases, it seems that the development must have taken place without enough "bios." . . . But if we examine closely the events in these cases, we notice with regularity that the progress of the culture had been extremely slow. It is only after weeks and months that these authors weighed their cultures. . . .

Leaving Pasteur and Liebig, Wildiers made this biologically more basic conclusion.

If "bios" was an organic compound containing nitrogen, as seems very probable, one could no longer state that yeast can synthesize albumins from sugar, ammonia and sulfates.

Thus, if yeast was unable to make albumin, as it is unable to make sugar, its metabolism would then be quite different from the metabolism of green plants and would be closer to the metabolism of animals. . . .

. . . since there is a moiety in albumin that the animal cells do not synthesize and that only plant cells can manufacture from simple elements. But there seems to be also a chemical entity that yeast cells cannot make. It is easy to speculate that it is a very difficult entity to elaborate and that it is the same entity that neither the yeast, nor the animal can synthesize.

In 1924, a Polish biochemist, Casimir Funk, wrote a small booklet on the history of vitamins.[1] There we read:

In 1910, when I was working at the Lister Institute, Dr. Charles J. Martin, director of the Institute, received the vist of Dr. Braddon, who had spent many years in the English colonies, and who gave him much information about the problem of beriberi. Dr. Martin rapidly understood the importance of the subject. First he thought that the proteins of rice were involved and, knowing that I had studied proteins for a few years, he thought that I might be able to throw light on the problem. Dr. Martin thought that polished rice was deficient in an amino-acid present in the bran of rice which was eliminated during the process of polishing rice. This was certainly an interesting suggestion and reflects quite well the state of our knowledge on beriberi in 1910.

I immediately started my study by preparing the protein of polished rice and also that of rice bran. My aim was to produce experimental beriberi by employing the former which, in theory, should have been incomplete. But after a few weeks and after having read the literature dealing with the subject, I decided to tackle the problem in a completely different manner. The book by Dr. Schaumann, of Hamburg (who died recently) had a great influence on my decision. . . . he was a dilettante as far as physiological chemistry, and particularly nutrition, was concerned. In the book that he published in 1910, he discussed a number of diseases including beriberi, that he considered to be caused by a disturbance in the utilization of phosphorus (mainly in organic combinations of the lipoidal type) in the animal body. . . .

After having read the book of Dr. Schaumann and the various papers

[1] Translated by permission of Vigot Frères, Paris, publisher.

published on beriberi, I saw clearly that we were not dealing with a disease linked, so to speak, to the quality of proteins, but that I should rather direct my attention to simple nitrogen-containing organic bases. First, I had to prove that my reasoning was right and show that a simple substance, similar to nitrogen containing bases, was what we were dealing with, rather than a substance with a complicated structure of the protein type. To solve this problem was simple since complex substances, when submitted to the prolonged action of a strong, boiling hot acid are hydrolyzed and disintegrate into simple moieties. . . .

The product that we were dealing with was rather resistant to heat and chemical reagents. The activity of rice bran or of yeast was not much altered even after twenty-four hours of boiling with 20% sulfuric acid. We were thus not dealing with a ferment. . . .

It was easy to demonstrate that ashes obtained from rice bran lacked this activity which was, as a consequence, conclusively linked with organic substances. . . .

Under the influence of phosphotungstic acid, the hydrolysates of yeast or rice bran give an abundant precipitate in which most of the nitrogenous substances are carried down.

The curative substance could be detected there and the filtrate seems to be absolutely without effect. By successive fractionations of the precipitate it was possible to follow the active substance and to observe that it was associated to a well-known class of nitrogen-containing bases: the *pyrimidines*. These are derived from nucleic acids which are found in the nucleus of all living cells. . . .

Except for ferments, few substances are known which are able to act in such small concentrations. It thus seemed strange that such small amounts had so much power and had such an importance in nutrition. Indeed during our investigations we have encountered active fractions a milligram of which (or even less) was amply sufficient to save the life of a pigeon dying of beriberi. . . .

Because of the vital importance of these substances in nutrition, I have called them *vitamins*. During the course of my investigations, I became aware that there were a whole family of similar substances which are indispensable to our normal nutrition, the lack (or *deficiency*) of which causes serious nutritional diseases. These diseases are, strictly speaking, *avitaminoses*. . . .

It was in 1912 that Funk proposed his vitamin theory to explain certain nutritional diseases. Yeasts were recognized not only to require growth factors but also to be a source of them. As time passed, Wildiers' bios was fractionated into Bios I, which is inositol; and Bios II, which contained biotin and pantothenic acid. The role of vitamins, as well as that of trace elements, came to be understood to be associated with the action of enzymes. For example thiamine, the main antiberiberi factor, is part of the molecule of a cocarboxylase that is thiamine diphosphate. Deficiency in this vitamin results in a failure of carbohydrate metabolism with accumulation of pyruvic acid.

Medical Mycology

Robert Remak was, like Henle, a disciple of Johannes Müller. He practiced medicine and carried out investigations in the field of neurology and embryology. He demonstrated conclusively that the egg is a cell that divides into an increasing number of cells and stressed that the division process always starts in the nuclei. In 1837 he observed mold filaments in favus, which he, at first, did not believe were due to a pathogen; but after the studies of Gruby, which we will soon review, he named the causative agent of favus, *Achorion schönleini*. It is now placed in the genus *Trichophyton* but still bears the name of Johann Schönlein, who studied it in 1839. Schönlein was the founder of modern German clinical medicine, and, in contrast to many lesser scientists of his day and ours, his list of publications was amazingly short. He published a doctor's dissertation and two papers, one of which was three pages long and the other, less verbose, only twenty lines. He was a great teacher, and his lectures, clinical demonstrations, and methods of examination made him the leading German clinician of his day. Some of his auditors put together his lectures in a single book that became a standard medical text and went through six editions. For most of his life, Schönlein was located in Germany, but from 1833 to 1839 he was professor of medicine in Zurich. It was there that he duplicated the earlier observation of Remak and came to the conclusion that favus was caused by a parasitic mold.

In medical mycology the focus of attention passed from Germany to France with the publication of the investigations of an Austrian-educated Hungarian Jew, by the name of David Gruby. He concluded his medical studies in Vienna in 1839. He was obviously talented and was offered an academic post in exchange for a certificate of baptism. He was not responsive and left Vienna the next year for London and then Paris. He promptly found opportunity for medical research at the Foundling Asylum in Paris, and from there he presented to the Academy of Sciences between 1841 and 1845 a series of papers that mark the beginning of medical mycology. Around 1850, Gruby started to practice medicine and acquired a fashionable clientele. He became friendly with the leaders of the creative arts, Alexandre Dumas, George Sand, Chopin, and Liszt. His popularity with the elite of the time did not go to his head, however, and he preferred to stay home building watches and clocks rather than to indulge in the gaieties of Paris night life. From his house in Montmartre, he turned his eyes toward the celestial bodies, and he became more and more antisocial. This did not prevent him from contributing to philanthropic endeavors. He died at the age of eighty-eight, a recluse, in his little observatory that none but he had entered for many years.

Let us go back to his first year in Paris and see what he said in his first communication at the Academy of Sciences (*Compt. rend. acad. sci.*, **13:** 72, 1841).

> In order to recognize the true tinea [favus], microscopic examination is adequate. A small amount of crust is collected and mashed in a drop of pure water. The preparation is placed between two glass slides and observed at 300 fold magnification. One will see a large number of round to oblong corpuscles $\frac{1}{300}$ to $\frac{1}{100}$ of a millimeter long and $\frac{1}{300}$ to $\frac{1}{150}$ of a millimeter wide. They are transparent, with a smooth surface, are either without color or slightly yellowish and contain only one substance. One also notes small articulated filaments of a diameter of $\frac{1}{1000}$ to $\frac{1}{250}$ of a millimeter, transparent and colorless. The general shape of these filaments is cylindrical or branched depending on the part of the crust with which they are associated.
>
> The cylindrical filaments are composed of oblong or round corpuscles which often appear like rosary beads. In contrast, branched filaments exhibit every so often vegetable partitions forming oblong cells containing small round and transparent molecules of a diameter of $\frac{1}{10,000}$ to $\frac{1}{1000}$ of a millimeter. Sometimes one finds granules, adhering to the filaments, which are similar to spores of *Torula olivacea* and *T. sachari* presented in the work of M. Corda entitled *Icones fungorum* (Prague, Vol. IV, 1841). The shape of these filaments does not leave any doubt about their plant nature. They belong to the group of the mycodermae. . . .
>
> Since we have not found as yet a single *molecule* of true tinea which was not loaded with a large number of these mycodermae, these are truly characteristic of this disease.
>
> The crust of the tinea has characteristics which are too interesting

not to be mentioned. One must choose an isolated crust, whole, only a few weeks old, with an unbroken external surface. The crust should be taken from a location where the concentration of hairs do not interfere with removal. The crust then looks like a flattened capsule, similar to the nut of *nux-vomica*, that is to say the shape of a disk one surface of which is slightly concave and the other convex. The edge, which is of circular shape is separated by a light groove into two equal parts, the upper one which is exposed to air and the other turned toward the dermis. These two disks are yellow on their external surface and are greyish-white inside. The concave surface is aerial, the convex one is cutaneous.

The crust is enveloped in epidermal cells which are much more numerous on the aerial portion than on the cutaneous one.

There is still a second envelope, surrounding the whole crust, which is formed of an amorphous substance and is placed between the cells of epidermis and the growth of the parasite itself. Toward the inside, the parasitic plant is seen with roots extending into the amorphous substance. The branches of the plant extend toward the center of the crust. A vertical section of the crust reveals a central grey porous tissue which is friable and formed of granules and the branches of the mycoderma. The granules are more numerous than the branches. In the peripheral part, toward the center, one can see the compact tissue of the amorphous substance where the roots of the mycoderma are placed.

The granules seem to be the product of the plant and are probably used for its propagation.

In summary, each isolated crust of the tinea consists of two envelopes with a mass of mycodermae enclosed therein, as are fruits in their pericarps.

The contagiousness of this disease becomes more probable in light of its plant nature, and I shall have the honor to submit to the Academy the results of inoculating man and various classes of animals with this plant.

The tissue of epidermis is the specific seat of the tinea. The cells of the epidermis which, by their structure, approximate the composition of plant tissues, seem very well suited to produce tissue analogous to vegetable tissue. Thus, one sees the mycoderma of tinea growing between epidermal cells by compressing the dermal tissue without destroying it. Nevertheless, one occasionally finds inflamed cutaneous tissue under the crust.

Generally, the crust can be removed with minimal damage to the skin.

After the removal of the crust, a serous liquid is exuded on the surface of the wound. Thus, healing occurs without scarring, which tends to prove that the dermis has not been destroyed by the suppuration or by the ulceration.

The association between the bulbs of hair and these parasites is not so intimate as it had been supposed since it happens very often that the mycoderma is well developed without any noticeable alteration of hair follicule. Sometimes, however, the filaments of the mycoderma extend toward hair follicules and surround the bulbs. This is the cause of the conical shape of the cutaneous part of crust.

Under these conditions the hair softens to such consistency, that placed between two glass slides, light pressure is adequate to split it along its fibers. This aids the study of its filamentous tissue. The follicules of skin are affected to a lesser degree as are the other tissues of skin.

As far as therapy of this disease is concerned, these new facts should encourage new attempts in that direction.

There was only one serious precursor that Gruby had missed, and it was, of course, Schönlein. Even the work of this latter worker was poor when compared with the clear account given by this Hungarian physician who emerged as a great mycologist but a poor clinician. He gave excellent descriptions of fungi and poor descriptions of the diseases they caused. The result was confusion, which was much later resolved by the masterful studies of Sabouraud.

We have already noted that Gruby was associated with a children's asylum. It is thus not surprising that his next paper was a description of the causative agent of thrush that we now call *Candida albicans*. Then he went back to the dermatophytes and described the fungus currently called *Trichophyton mentagrophytes*, and in 1843 he described *Microsporum audouini* (*Compt. rend. acad. sci.*, **17**: 301, 1843):

Porrigo decalvans is characterized by roundish plaques which are covered with a white dust and little grey scales and by the loss of hair.

Careful microscopic examination reveals, surprisingly, that the white dust which covers the skin in porrigo decalvans is composed entirely of cryptogams. Microscopic examination of hairs from individuals having this disease reveals a large quantity of cryptogams which surround the hairs from all sides, and which truly form a plant mantle which covers them upwards from the skin for a distance of 1 to 3 millimeters.

Microscopic examination of the mantle reveals its true plant nature. The cryptogams are marvelously arranged like a felt to form a tube or solid plant mantle around each hair. These cryptogams are formed of branches, stems and spores. The branches have their origin in the tissue of the hair and form the internal layer of the mantle, whereas the spores form the outer layer. (The thickness of the wall of the mantle is $15/1000$ of a millimeter.) The stems have an undulated shape; they follow the direction of the fibers of the hair; they are transparent; their diameter is of $2/1000$ to $3/1000$ of a millimeter.

The branches are differentiated from the stems by carrying spores, and end on the external surface of the mantle with a complete covering of spores. Occasionally one finds spores adhering to branches at the surface of the hair. Spores are usually round but there are a few oval ones. Their diameter is $1/1000$ to $5/1000$ of a millimeter. The oval spores are slightly larger: they are from $2/1000$ to $5/1000$ by $4/1000$ to $8/1000$ of a millimeter in diameter. They are transparent, do not contain molecules and they swell in water.

I shall call these cryptogams *Microsporum*, because the spores are small and in honor of the famous academician who, by his beautiful investigations on muscardine, has done much to focus attention on parasitic plants which destroy the living tissues of animals, I propose the name of *Microsporum audouini*, for the individual plants in porrigo decalvans.

The hair tissue is altered by the quantity of *Microsporum audouini* which is attached to its surface. First the hair becomes opaque where the cryptogams are located and its smooth surface becomes rugose. The epithelium which covers the surface of hair loses its luster and consistency and gradually disintegrates. Such a hair breaks, even by simple bending and accordingly, hair is completely lost where the parasitic plants have invaded. The place where hair has fallen out is white greyish because cryptogams remain in the cells and on the surface of the epidermis. . . .

The microspora of Audouin . . . are rather similar to the cryptogams which cause the disease that I have described under the name of phytomentagra.[1] They differ mainly by their location. The cryptogams of mentagra are located in the hair follicles and even around their roots; the microspora of Audouin on the contrary are located around the aerial part of the hair. The spores of Audouin's microsporum are smaller and its branches are shorter than in the mentagrophytes.

Microsporum audouini begins to grow on the surface of hair, 1 to 2 millimeters from the epidermis. The tissue of the hair becomes less transparent over an area of $30/1000$ to $40/1000$ millimeters. There is a de-

---

[1] Gruby is referring to *Trichophyton mentagrophytes*.

velopment of scarcely measurable molecules which are $\frac{1}{10.000}$ to $\frac{2}{10.000}$ of a millimeter in diameter. The tissue thus altered is associated with fibers or cells which are larger than the fibers of hair, and that are elongated and parallel to the axis of hair. In this region one notes the first traces of *Microsporum audouini* which, by covering the hair completely, and by subsequent intimate contact with several hairs, destroys them and causes alopecia.

The cryptogams grow and multiply with an unbelievable rapidity: within a few days after point infection, a skin area of 3 to 4 centimeters will be covered with parasitic plants. The hair, at the point where it emerges from the skin, becomes greyish and, within eight days, breaks at the very location where it is enrobed by the cryptogam. One can even see around the broken hair accumulation of the cryptogams forming a slight greyish mound of about ¼ millimeter in diameter. These mounds have been considered pustules, vesicles or secretions of the sebaceous follicles.

The plant nature of porrigo decalvans tends to indicate that it should be considered a contagious affection. This being so, the same precautions of isolation should be followed in this case as with favus and mentagrophyta. Thus, the medical practitioners must apply themselves to the task of destroying this plant parasite which has hitherto resisted all empirical treatments.

As mentioned previously, Gruby's lack of clinical knowledge at the time he was writing these papers was a great source of confusion. In the above paper he claims to describe the parasite causing porrigo decalvans, a disease which at that time was described as being characterized by "bald plaques in which the smooth, shiny, white skin was completely devoid of hair." Audouin's organism does not cause this disease but rather an epidemic tinea capitis of children. In the enthusiasm of his mycological description, Gruby completely forgot to use the word child and hopelessly confused everybody. Nevertheless, mycologically it was a great piece of work. This paper was eventually forgotten by dermatologists, and Sabouraud, in 1892, redescribed Audouin's organism under another name. When Sabouraud discovered his error, he remarked with admiration that Gruby's description of 1843 was better than his own!

Audouin's organism, as we have noted, grew as a mantle around the hair. Gruby described another type of tinea in which the dermatophyte was growing within the hair. This basic distinction was stressed later when Sabouraud separated members of the genus *Trichophyton* into two groups.

In one, the organism grew inside the hair and was called an endothrix; in the other it grew on the exterior and was called an ectothrix.

These fundamental papers of Gruby were critically received by the great pathologists of the time. They considered the work to be erroneous even though they did nothing experimentally to test its veracity. Subsequently many scientists came to accept the fact that cryptogamic plants were present in many affections, but it was believed that they were the result and not the cause of the disease. Many clinicians argued sensibly that the overall physical condition of the patient had to be considered and that destruction of the parasitic plants would not produce a durable cure. The true and still unknown cause of the disease might simply have made conditions favorable for the overgrowth of the plant. Gruby's fame as an experimenter had grown, however, and for thirteen years he gave a public course on various aspects of experimental science. Such important people as Magendie and Claude Bernard followed his lectures. Eventually, as we have seen, he became involved in an extensive medical practice, and his interest drifted toward astronomy, clocks and watches, weather forecasting, and the betterment of life of his fellow men.

In 1853, Charles Robin published a 700-page book entitled *Natural History of Parasitic Plants that Grow on Man and on Living Animals*. This was a compendium of the field at the time. The book had fifteen plates, which included illustrations of such pathogens as the causative agents of thrush and favus. Robin remarked that the fungus associated with thrush was not necessarily the cause of the disease since the plant could be found in healthy subjects. He concluded that the fungus grew whenever conditions were favorable for its growth and was not a cause of disease per se. It was a remarkable treatise for a man of thirty-two years of age who had already published in many fields of medicine and biology. Charles Robin was not only a voluminous writer but also a blatant philosopher and, later in his life, a politician. His loudly proclaimed positivism was an obstacle to his election to the Academy of Sciences in 1865. Pasteur was asked to support Robin's election. On this occasion Pasteur remarked: "I am not worried about his philosophical studies, except for the harm that it might do to his work. As a scientist, he must cope with experimental methods; I am rather afraid that if he dabbles in philosophy, it might simply mean that he is a man with fixed, preconceived ideas." Pasteur's reservation seems to have been justified since eventually Robin became an opponent of the biological theory of fermentation and the germ theory of disease. Actually he did admit that wherever there were ferments (microorganisms) there was fermentation, but he said that there could be fermentation without ferments. In the case of bacterial diseases, he did not completely deny the role of bacteria in disease but thought that pathogens would grow only if the environment was ripe and were more the result than the cause of disease: "Just as there is no heat separable from bodies," he wrote in 1885, "but only variations of temperature . . . there are also no viruses which are a kind of matter. . . . Viruses exist as motion, an abnormal and pathologic state of activity of organized substances. . . . the virulent state of organized matter is almost always accompanied, as is often the case where there is any alteration, by the growth and reproduction of schizomycetes."

The first attempts to cultivate dermatophytes had been made by Gruby and Remak, but De Bary cautiously reminded the investigators in the field that it was easier to obtain the growth of saprophytic contaminants than that of the pathogens. Success finally came in 1886 when some dermatophytes were grown both in the laboratory of De Bary and in that of Duclaux. It was shown that these fungi did not grow at an acid reaction and thrived on sugars and peptones.

In 1892 Sabouraud started to study dermatophytes. He came from a family with a long medical tradition, and one of his ancestors had been the surgeon-barber of Louis the XIII. He did his internship at the St. Louis Hospital of Paris, and he took the course of microbiology that was being given at the Pasteur Institute. He elegantly allied dermatology with microbiology to create what essentially was modern medical mycology, as far as superficial mycoses are concerned. He was a dermatologist and published abundantly in all aspects of his field. To the microbiologist, his most important book is *Les Teignes* published in 1910, in which he explained his role in the field as follows: [1]

> It was in 1892 that I started my first investigations on this topic. I was encouraged in this direction by E. Besnier, under whom I was then doing my internship. I knew nothing about the subject. . . . the first investigations that I became familiar with were those of Verujsky [who had cultured some of the dermatophytes in Duclaux' laboratory]. My first investigations on dermatophytes lasted about three years . . . and were summarized in a book: "Human Trichophytoses." . . .
>
> Initially one could not determine in advance which method to use to throw light on the question. One had to examine with great care each patient and record the results, to examine the parasite in the squamae

[1] Translated by permission of Masson et Cie, publishers of Sabouraud's book: *Maladies du Cuir Chevelu III. Maladies Cryptogamiques. Les Teignes.* 1910.

and the hair microscopically, to make permanent preparations for comparative studies and finally one had to make cultures from each case and maintain them in order to compare them with each other.

This method which approached the problem from three sides, clinical, microscopical and cultural, gave immediate and unexpected results. There were two types of children's tinea capitis that could be differentiated by examination with the naked eye, by microscopic examination and by culture. It was the tinea capitis with small spores and the tinea capitis with big spores.

Further this work established that these two parasitic types were not a single species but groups of species. Among the Trichophytons with big spores one could recognize 5 species. . . .

More and more it looked as though the species of Trichophytons were extremely numerous and specifically distinct. The characteristics of each species seemed to be fixed, hereditary and permanent. . . .

These various parasitic species are characterized by their specific cultural aspects on an artificial medium. But, for this, one must have a medium on which dermatophytes grow actively and fully. The best is beer wort, that is to say malt water. I have tried to make a similar medium that would be easy to reproduce. This medium had the general formula:

Pure water	1000	grams
Sugar	37	grams
Peptone	10	grams
Agar to make solid		

It was called the *test medium* and similar ones were made with maltose, glucose, mannitol, lactose. My preliminary experiments led me to conclude that all specific differentiation based on cultural characteristics had to be established as follows:

a. By a series of cultures of various types made at the same time on a differential medium made at the same time. The series permitted one to compare the appearance of cultures which had been subjected to identical physico-chemical conditions.

b. By a series of subcultures of the same species, enabling one to follow during several generations the appearance of the properties on which differentiation was based. This assured the stability of the properties of the cultures. . . .

He cultivated five hundred cultures on his three media (maltose, mannitol, lactose) under the same conditions of temperature and aeration, and, as he admitted himself, was surprised to find the differences among the various species so clearly defined. In all he distinguished nineteen species of *Trichophyton.* He continues:

> It was in 1894, as I was preparing my book on *human trichophytoses,* that I read for the first time the papers of Gruby and that I found out that he had discovered and described, between 1842 and 1845 the three parasitic types that I had found: my small-spored *Trichophyton* was his *Microsporum audouini.* He had also described the endo-ectothrix Trichophyta in the beard and endothrix Trichophyta in tinea capitis. . . .
>
> I must say that new facts that kept cropping up in such large numbers were not described by me without errors. . . . the subject had a mycological aspect that a dermatologist such as I was badly prepared to observe well. . . .
>
> My biggest mistake was mycological. When I saw white downy growth in old cultures of Trichophyta, I thought that these new cultures which could be reisolated from the first ones and which had different properties, were contaminating molds. Instead of understanding that their origin was the result of a pleomorphic transformation of the mother cultures, I thought that there was a commensalism of two species. It took two years to correct this error that a competent mycologist would have elucidated at a glance.
>
> Despite the deficiencies, the errors and the mistakes of interpretation that my papers contained, it seems today that it was with them . . . that the modern era of the field started. They introduced a new method of procedure and new techniques. One could now understand that future studies should be based, not on a few isolated cases, but on hundreds. . . .

Of course, we shall not go into the details of the studies of Sabouraud, nor shall we discuss his classification of dermatophytes. He was a dermatologist; therefore let us review his therapeutic methods. He was in favor of simple methods of therapy. His favorite drugs were Alibour water (solution of zinc and copper sulfates), tars, sulfur, and iodine in alcoholic solution. However, he played an important role in the introduction of X rays in the treatment of tinea capitis. Here is how he describes his role in this field in *Les Teignes* in 1910:

> In 1896, four years after having started the study of dermatophytes I wrote: "Not only is there no curative treatment for tinea capitis but I am even tempted to prophesy that no antiseptic treatment will achieve this goal in the future. . . ."

It is radiotherapy which was to bring the dreamed-of solution to this problem.

As early as 1897 we observed the first case of radiotherapeutic alopecia in a young lady who had participated in a public demonstration. She had remained between an X-ray bulb and a screen so that a necklace could be seen under her clothes. She lost almost all her hair on the lower posterior portion of her head and the regrowth, which I watched, was normal. . . .

As early as 1896, Freund and Schiff had observed the loss of hair in a patient with nevus pilosus treated by X-rays. The idea came to many authors to use this depilation in the treatment of tineae, but the first attempts were made on a great number of diseases under poorly controlled conditions and the results were not satisfactory.

It was then that Noiré and I started to study the problem.

It seemed to us that we should study only tinea capitis and to try to obtain depilation at a given point in one single session since a treatment which would have required numerous applications on the same spot would have had no practical value. Indeed, how could one find again exactly the same spot when one had to treat a hundred children at a time. . . .

We succeeded after a few months, during which we investigated with care the power of the instrument that should be used, the distance from the instrument to the region to be treated and the dose of X-ray which must be administered.

On the fourth of January 1904, as the authors which had pre-empted us in these investigations were still publishing their preliminary experiments and their incomplete results, I was able to announce to the Society of Dermatology the cure of 100 patients who had tinea capitis, and the technical details of the procedure which permitted us to effect with safety and without accident the cure of each plaque by a single application of a measured amount of X-rays.

Thus our contribution was to prove that it was possible, and to show how, one could cure in every case a plaque of tinea capitis in a single dose by a defined amount of X-ray.

The results were impressive. The number of children with tinea capitis in Paris dropped within a few months from 5 per cent of the population to one in a thousand.

Sabouraud was not only a dermatologist but also a man of many talents. He was a truly erudite man whose main interest was history. It seems that often his clinical discussions with his colleagues turned into lectures on history given by Sabouraud. He was also extremely interested in art and had a fine collection of paintings. He himself practiced sculpture and produced almost as many statues as medical papers. A book with forty-four plates published in 1929 was entitled *Raimond Sabouraud Sculptor*. One is quite amazed by the output of such a man, an output so varied in scope and yet of such high quality. Needless to say, he was a hard worker. He no doubt considered that a well-born man should be a productive man. And indeed this is what he was. He died in 1938.

Of course all fungal diseases are not caused by dermatophytes. We have already referred briefly to the yeast *Candida albicans* that had been observed in cases of thrush by Gruby. The agents of the other most important fungal diseases were isolated and described in the last decade of the nineteenth century and the first decade of the twentieth century. The field was not dominated by any single individual, and these discoveries came from various parts of the world. Nevertheless, this area of investigation is not closed. New fungal pathogens and diseases are still described at the present time. Also much of the early work has now been reevaluated and one can say, be it due to the successes of antibacterial chemotherapy or to a better knowledge of the fungi, that medical mycology is a field that is acquiring an increasingly important place in medical study.

References

Davy de Virville, Ad., and Collaborators: *Histoire de la botanique en France,* Soc. d'Edition d'Enseignement'Supérieur, Paris, 1954.

Fontana, F.: *Observations on the Rust of Grain.* Phytopathological Classics no. 2, American Phytopathological Society, 1932.

Funk, C.: *Histoire et conséquences pratiques de la découverte des vitamines,* Vigot Frères, Paris, 1924.

Hooke, R.: *Micrographia,* Dover Publications, Inc., New York, 1961 (republication).

Large, E. C.: *The Advance of the Fungi,* Dover Publications, Inc., New York, 1962 (republication).

Millardet, P. M. A.: *The Discovery of Bordeaux Mixture,* Phytopathological Classics no. 3, American Phytopathological Society, 1933.

Prévost, B.: *Memoir on the Immediate Cause of Bunt or Smut of Wheat,* Phytopathological Classics no. 6, American Phytopathological Society, 1939.

Robin, C.: *Histoire naturelle des végétaux parasites qui croissent sur l'homme et sur les animaux vivants,* J. B. Baillière et Fils, Paris, 1853.

Sabouraud, R.: *Maladies du Cuir Chevelu. III. Maladies Cryptogamiques. Les Teignes,* Masson et Cie, Paris, 1910.

Targioni-Tozzetti, G.: *Alimurgia,* part V, Phytopathological Classics no. 9, American Phytopathological Society, 1952.

Tillet, M.: *Dissertation on the Cause of the Corruption and Smutting of the Kernels of Wheat in the Head,* Phytopathological Classics no. 5, American Phytopathological Society, 1937.

Whetzel, H. H.: *An Outline of the History of Phytopathology,* W. B. Saunders Co., Philadelphia and London, 1918.

Woronin, M. S.: *Plasmodiophora Brassicae, the Cause of Cabbage Hernia,* Phytopathological Classics no. 4, American Phytopathological Society, 1934.

Additional Reference

Le Leu, L.: *Le Dr. Gruby,* P.-V. Stock, Paris, 1908.

Protozoology

10
As we have seen, Bichat, who died in 1802, essentially founded histology without the use of the microscope. He refused to employ the instruments available during his short lifetime because they were so bad that they created more problems than they solved. As a result, Bichat's histology was quite different from that of today. If Bichat could not use microscopes for the study of tissues, we can imagine how inadequate these instruments must have been for the study of protozoa until the development of high-powered achromatic objectives, which must have become readily available only around the fourth decade of the nineteenth century.

That the early microscopes were deficient only increases our admiration for those who managed to use them in the observation of protozoa. The oldest illustration of a protozoön may well be the snaillike creature illustrated in Figure X, Plate V, of the *Micrographia* of Robert Hooke, which, as we have already noted, was first published in 1665. The illustration and the following description by Hooke would suggest most strongly that the brilliant Englishman had observed one of the foraminifera. Describing his observations of sand, Hooke wrote:

> Sand generally seems to be nothing else but exceeding small pebbles. . . . But among many others [small bodies], I met with none more observable than this pretty shell . . . for by it we have a very good instance of the curiosity of nature. . . . I was trying several small and single magnifying glasses, and casually viewing a parcel of white sand, when I perceived one of the grains exactly shaped and wreathed like a shell. . . . With a pin, I separated all the rest of the granules of sand, and found it afterwards to appear to the naked eye an exceeding small white spot, no bigger than the point of a pin. Afterwards

I viewed it every way with a better microscope and found it . . . to resemble the shell of a small water snail with a flat spiral shell: it had twelve wreathings . . . all very proportionably growing one less than another toward the middle or center of the shell. . . . I could not certainly discover whether the shell were hollow or not, . . . and it is probable that it might be petrifyed as other larger shells often are. . . .

The versatile Hooke, a prospector in the world of the infinitesimally small, having viewed the cellular arrangement of corks, the inflorescences of some molds, and the shell of a protozoon, among other oddities of creation, continued his activities in other aspects of biology, as well as physics, architecture, and astronomy.

A variety of protozoa, free-living, and parasitic, were observed by Leeuwenhoek, the draper-chamberlain and amateur microscopist of Delft whom we have met earlier in this book. The importance of his contribution lay in the prodigious number of microscopic forms and structures that he described and sketched. Without formal education, he was aware of his limitations and devoted his time to actual observation with his simple microscopes and refrained from extensive theorizing. He communicated his observations in a series of some two hundred manuscripts of which approximately 120 were letters he sent to the Royal Society of London between the years 1673 and 1723. In a letter dated 1676 (letter eighteen to the Royal Society) he wrote: [1]

In the year 1675, about half-way through September (being busy with studying air, when I had much compressed it by means of water), I discovered living creatures in rain, which had stood but a few days in a new tub, that was painted blue within. This observation provoked me to investigate this water more narrowly; and especially because these little animals were, to my eye, more than ten thousand times smaller than the animalcule which Swammerdam has portrayed, and called by the name of Water-flea, or Water-louse, which you can see alive and moving in water with the bare eye.

Leeuwenhoek went on to describe for the first time what can be recognized as a species of *Vorticella*.

Of the first sort that I discovered in the said water, I saw, after divers observations, that the bodies consisted of 5, 6, 7, or 8 very clear

[1] Translation by Clifford Dobell. Reproduced by permission of Mrs. C. Dobell.

globules, but without being able to discern any membrane or skin that held these globules together, or in which they were inclosed. When these animalcules bestirred 'emselves, they sometimes stuck out two little horns, which were continually moved, after the fashion of a horse's ears. The part between these little horns was flat, their body else being roundish, save only that it ran somewhat to a point at the hind end; at which pointed end it had a tail, near four times as long as the whole body, and looking as thick, when viewed through my microscope, as a spider's web. At the end of this tail there was a pellet, of the bigness of one of the globules of the body; and this tail I could not perceive to be used by them for their movements in very clear water. These little animals were the most wretched creatures that I have ever seen; for when, with the pellet, they did but hit on any particles or little filaments (of which there are many in water, especially if it hath but stood some days), they stuck intangled in them; and then pulled their body out into an oval, and did struggle, by strongly stretching themselves, to get their tail loose; whereby their whole body then sprang back towards the pellet of the tail, and their tails then coiled up serpentwise, after the fashion of a copper or iron wire that, having been wound close about a round stick, and then taken off, kept all its windings. This motion, of stretching out and pulling together the tail, continued; and I have seen several hundred animalcules, caught fast by one another in a few filaments, lying within the compass of a coarse grain of sand.

Leeuwenhoek continued his letter with the description of some sort of ciliate.

I also discovered a second sort of animalcules, whose figure was an oval; and I imagined that their head was placed at the pointed end. These were a little bit bigger than the animalcules first mentioned. Their belly is flat, provided with divers incredibly thin little feet, or little legs, which were moved very nimbly, and which I was able to discover only after sundry great efforts, and wherewith they brought off incredibly quick motions. The upper part of their body was round, and furnished inside with 8, 10, or 12 globules: otherwise these animalcules were very clear. These little animals would change their body into a perfect round, but mostly when they came to lie high and dry. Their body was also very yielding: for if they so much as brushed against a tiny filament, their body bent in, which bend also presently sprang out again; just as if you stuck your finger into a bladder full of water, and then, on removing the finger, the inpitting went away.

Yet the greatest marvel was when I brought any of the animalcules on a dry place, for I then saw them change themselves at last into a round, and then the upper part of the body rose up pyramid-like, with a point jutting out in the middle; and after having thus lain moving with their feet for a little while, they burst asunder, and the globules and a watery humour flowed away on all sides, without my being able to discern even the least sign of any skin wherein these globules and the liquid had, to all appearance, been inclosed; and at such times I could discern more globules than when they were alive. This bursting asunder I figure to myself to happen thus: imagine, for example, that you have a sheep's bladder filled with shot, peas, and water; then, if you were to dash it apieces on the ground, the shot, peas, and water would scatter themselves all over the place.

Leeuwenhoek examined all available materials about him, including his own excrement. He thus discovered parasitic protozoa. In letter thirty-four, dated 1681, addressed to Robert Hooke, who was the secretary of the Royal Society, he wrote of animalcules fitting the description of *Giardia intestinalis* and others that cannot clearly be identified:

I weigh about 160 pound, and have been of very nigh the same weight for some 30 years, and I have ordinarily of a morning a well-formed stool; but now and then hitherto I have had a looseness, at intervals of 2, 3, or 4 weeks, when I went to stool some 2, 3, or 4 times a day. But this summer this befell me very often, and especially when I partook of hot smoked beef, that was a bit fat, or ham, which food I'm very fond of; indeed, it persisted once for three days running, and whatever food I took, I kept in my body not much above 4 hours; and I imagined (for divers reasons) that I could get myself well again by drinking uncommon hot tea, as hath happened many a time before.

My excrement being so thin, I was at divers times persuaded to examine it; and each time I kept in mind what food I had eaten, and what drink I had drunk, and what I found afterwards: but to tell all my observations here would make all too long a story. I will only say that I have generally seen, in my excrement, many irregular particles of sundry sizes, most of them tending to a round figure, which are very clear and of a yellow colour: these were the ones that make the whole material look yellow to our eye. And there were also, besides, suchlike particles that were very bright and clear, without one being able to discern any colour in them.

I have, moreover, at divers times seen globules that were as big as the corpuscles in our blood, and that each of them was made up of 6 separate globules: and further, there lay, among all this material, globules whereof 6 of 'em together would make up the bigness of a blood-corpuscle. These last were in such great plenty, that they seemed to form a good third of the whole material: while there were besides many globules which were so small, that six-and-thirty of 'em would make up the bigness of a blood-globule.

All the particles aforesaid lay in a clear transparent medium, wherein I have sometimes also seen animalcules a-moving very prettily; some of 'em a bit bigger, others a bit less, than a blood-globule, but all of one and the same make. Their bodies were somewhat longer than broad, and their belly, which was flatlike, furnisht with sundry little paws, wherewith they made such a stir in the clear medium and among the globules, that you might e'en fancy you saw a sow bug running up against a wall; and albeit they made a quick motion with their paws, yet for all that they made but slow progress. Of these animalcules I saw at one time only one in a particle of matter as big as a sand-grain; and anon, at other times, some 4, 5, or even 6 or 8. I have also once seen animalcules of the same bigness, but of a different figure. . . .

All told, Leeuwenhoek observed approximately one dozen different types of protozoa. He was much impressed by the number of small moving "animals" that could be seen under his lenses, since he commented: ". . . there are more animals living in the scum on the teeth in a man's mouth than there are men in a whole kingdom. . . ."

There were further developments in protozoology within Leeuwenhoek's lifetime, the most notable of which was the first treatise on protozoa, which was published in 1718 by the French microscopist, Louis Joblot.

Joblot was born in 1645 at Bar-le-Duc and died in 1723, the same year as Leeuwenhoek. Like many scientists of the time, he was active in physics and mathematics as well as biology. He was professor of mathematics, perspective, and geometry at the Royal Academy of Painting and Sculpture in Paris. His interest in optics led him to improve the design of the microscope, and his best products, which incorporated compound lenses, were probably superior to Leeuwenhoek's instruments. Consequently, he was able to observe and draw details of morphology that could not have been seen by Leeuwenhoek. More fanciful than the Delft draper, he designated his animalcules with common names such as the "swan," the "kidney," etc. Joblot's treatise was entitled "Description and Application of Several New Microscopes, Quite Simple as Well as Compound; Along with New Observations on a Multitude of Insects and Other Animals. . . ." Joblot prepared boiled and unboiled infusions from various types of plant materials. The boiled infusions, exposed to air and protected from air, were compared with unboiled infusions. In the boiled infusions exposed to air, he observed the same types of organisms that appeared in the unboiled infusions, while no organism appeared in the boiled infusions that were protected from air. He concluded that the organisms developed from "eggs" in air and that boiling killed the eggs.

As the existence of many protozoa became established, there arose a need for a term to designate them. In 1765 H. A. Wrisberg introduced the term "infusoria" which came into widespread use. The number of forms observed during the latter decades of the eighteenth century increased to the point where O. F. Müller was able to write an extensive monograph, *Animalcula infusoria fluviatilia et marina,* published in 1786, that described approximately 150 valid species and a larger number of forms that were small metazoa and bacteria. Otto Frederik Müller was born in 1730 in Copenhagen. His parents were poor, but his brilliance enabled him to study theology and law. While supporting himself as a tutor, he developed an interest in natural history. As a result, his first publications were on insects. He augmented his knowledge by a trip through Europe as the tutor of a young count. Upon returning to Copenhagen, he accepted a government appointment as an archivist. The next step upward was marriage to a wealthy woman that allowed him to leave government service and pursue his research privately. His monograph was published after his death in 1784.

By this time there was also limited interest in the physiology of the protozoa. In 1778 W. F. von Gleichen described the formation and discharge of food vacuoles by feeding infusoria with carmine.

During the early years of the nineteenth century the pace of morphological study accelerated, and a need for a system of classification was felt by a number of investigators. Ehrenberg was the most prolific writer of this group, and in 1838 he published a monograph entitled *The Infusoria as Complete Organisms.*

Christian Gottfried Ehrenberg was born in 1795, near Leipzig. He took his degree in medicine and had the good fortune of participating in two extensive scientific expeditions in Asia that yielded important collections. The second expedition was under the direction of Humboldt, the great geographer-geologist-biologist. No matter how diverse may have been Ehrenberg's interests, his greatest contribution to science was his study of protozoa. Data in his monograph included the descriptions of more than

500 species, and more than 300 were newly described by Ehrenberg himself. As the very title of the monograph proclaimed, Ehrenberg was fascinated by the concept that protozoa had complex internal structures that were similar to those of higher organisms. He described a digestive system, reproductive organs, and even a musculature. His imagination hesitated only regarding the possible existence of nervous and vascular systems.

Ehrenberg became professor of medical history at Berlin and secretary of the Academy of Sciences of that city. He did little active research work after the publication of his monograph, and he died in 1876.

The impact of Ehrenberg's work was discussed by Felix Dujardin in 1841 in a book entitled *Infusoria, the Physiology and Classification of These Animals, the Method of Studying Them with a Microscope.* In this book published only three years after Ehrenberg's monograph, Dujardin thus appraises the work of the famous Berliner:

> During the present period, which is illustrated by the investigations of M. Ehrenberg and which is characterized by the utilization of the achromatic microscope, one wants at the same time to classify infusoria and to penetrate the mysteries of the organization of these small beings. The results obtained during this period will thus be much more important in all respects than those obtained during the previous periods. . . .
>
> M. Ehrenberg was the first to clearly distinguish the infusoria, that he calls *Polygastrica,* from the systolids that he names *Rotatoria.* But he leaves among true infusoria , . . [among others] the diatoms and the desmids that his love for his own system forces him to consider as animals provided with a mouth and a multitude of stomachs. Thus he was able to bring the number of polygastric infusoria up to 533. His classification, which is based on facts that are completely erroneous as far as the organization of infusoria are concerned, has been accepted by authors and compilers who had no interest in verifying these data. But the true observers, first struck with awe by the announcement of the discoveries of the micrographer from Berlin, did not take long to find out that it was useless to try to verify these facts. When they were well assured that this failure was due neither to the weakness of their eyesight nor to the imperfection of their microscopes, they dared to raise their heads and to negate forcefully the assertions of this man who had been clever enough to obtain the patronage of many illustrious academies and men. . . .
>
> I do not think that the time has come to propose a definitive classi-

fication for the infusoria, but since I have agreed to show what is true in the history of infusoria, I must try to classify them at least temporarily. . . .

Dujardin (1801–1860) had been professor of zoology at the University of Toulouse and was in 1841, as he wrote these lines, dean of the faculty of Sciences at the University of Rennes. He explained his classification system as follows:

> If . . . one tries to establish a system of classification for the infusoria based on reliable characteristics, one soon comes to the conclusion that . . . often one lacks sufficient characteristics to establish species. . . . Indeed, instead of finding precise shapes, and well defined organs, as in other plants and animals, one finds only an unstable shape which is continuously modified by unknown causes. . . . The various exterior appendages, inapparent in the instruments of the older micrographers will without a doubt furnish very valuable characteristics, but these will only be characteristics proper for the establishment of genera or families. . . . The species will have to be distinguished on the basis of size, color, and habitat, which are not true specific characteristics in the Linnean sense. . . .
>
> When studying infusoria, one is struck by their irregular and variable shape. However, a few major types can be seen with variable clarity. . . . One would be inclined to group separately the few *symmetrical infusoria* such as *Coleps* [ciliate whose body is covered with regularly arranged, perforated, ectoplasmic plates], *Chaetonotus* [now member of the class *Gastrotricha,* created by Metchnikoff as we have seen in chapter 5] . . . and to recognize the *asymmetrical infusoria* as the most important group. . . .
>
> The presence or absence of various appendages or organs of locomotion are most valuable characteristics for the classification of infusoria. Indeed, we will note a first order of animalcules in which one does not observe any special organ of locomotion. This may be due to the fact that it does not exist or that its extreme tenuity still permits it to escape our means of observation. These animals which are long and filiform seem to move only by overall body contraction. They constitute a separate family, the *Vibrionians,* which does not have any apparent relationship with other families. A second order, which is more important, is formed by animalcules which do not have any apparent internal organization and which have as external organs only variable extensions of body substance. . . .

This second order contained three families:
Body extensions visibly contractile:

> Naked animals; crawling; of continuously variable shape . . . *Amibians*
> Crawling or fixed animals secreting a shell from which expansions of the body protrude . . . *Rhizopodians*

Body extensions contracting very slowly, animals almost motionless . . . *Actinophryans*

Dujardin continued:

> A third order will be mainly characterized by the presence of one, two or many *flagelliform filaments* which are used as organs of locomotion. These have been considered erroneously as proboscises. This order will be divided according to the presence and the nature of a tegument. A mouth will never be visible in any of these animals.

This third order was divided into six families as follows: Without any tegument: swimming or fixed forms . . . *Monadians*
With a tegument:

> In colonies; floating or fixed:
> Tegument forming a common mass, not attached . . . *Volvocians*
> Tegument forming a branched polypoid structure . . . *Dinobryans*

> Isolated; swimming:
> Tegument noncontractile . . . *Thecamonadians*
> Tegument contractile . . . *Euglenians*
> Noncontractile tegument with a furrow containing vibrating ciliae . . . *Peridinians*

> A fourth order will harbor ciliated infusoria which do not have a contractile tegument. Subdivisions will be made according to the presence or the absence of a row of ciliae, either disposed obliquely or as a moustache, according to the presence of a mouth, appendages or cirri shaped like stilettoes or hooks and finally according to the presence of an armor either apparent or real.

The five families of this fourth order were:
Naked animals:
No mouth; ciliae distributed irregularly . . . *Enchelyans*

Mouth visible or indicated by a row of ciliae; no cirri . . . *Trichodians*
Mouth and ciliae as above; cirri . . . *Keronians*

Armored animals:

Armor soft and subject to decomposition as in the rest of the body . . .
Ploesconians

True armor, persistent; short peduncle present . . . *Ervilians*

A fifth order will include ciliated infusoria with contractile tegument.
Almost all of these have a mouth. . . .

This fifth and last order of the asymmetrical infusoria was divided as follows into five families:

Free animals:

Without a mouth . . . *Leucophryans*
With a mouth; without a row of ciliae arranged "as a moustache". . . .
Paramecians
With a mouth and a row of ciliae arranged "as a moustache". . . .
Bursarians

Fixed animals:

"Able to attach themselves voluntarily". . . *Urceolarians*
Attached by specific organs . . . *Vorticellians*

Dujardin was an accurate observer, and his classification was much
sounder than that of Ehrenberg. As limitations of his work, we may note the
lack of understanding of the nature of nuclei, which Ehrenberg believed to
be sexual organs and which Dujardin simply considered as concretions. Also
a major group of protozoa was not included in Dujardin's scheme, the
sporozoa.

A *sporozoon* had been seen as early as 1826 by L. Dufour, who had
observed it in the intestine of a coleopteron. He noted that individuals were
associated in chains, and he thus called the organism "*gregarina*." He believed that it was a fluke. Its protozoal nature was recognized by Kölliker
in 1848.

Dujardin's lack of recognition of the nature of protozoal nuclei is hard to
understand, since his book on infusoria was published ten years after the
recognition by Robert Brown that every cell contained an "areola" or
nucleus. This fact had recently been given a new impetus, in 1839, by
Theodor Schwann's *Microscopic Investigations on the Similarity in the
Structure and Growth of Animals and Plants.* In this publication, Schwann

expanded the cellular theory of life. It is possible that Schwann's sketchiness with regard to detail did not appeal to Dujardin, who demanded high standards of precision.

The cell theory was first applied to protozoa by Martin Barry, who wrote a paper, "On Fissiparous Generation," which appeared in the *Edinburg New Philosophical Journal* (1843, pp. 205–220) and in which he appraised some of Ehrenberg's publications in the light of the new theory:

> On comparing the figures given by Ehrenberg of successive generations of the *Chlamidomonas*, with the successive groups of cells (two, four, eight, etc.) in the mammiferous ovum I cannot help believing that the process of formation is the same in both; the essential part of this process consisting, as I showed, in the division of a pellucid mass (hyaline) situated in the centre of each cell. And it is deserving of remark, that Ehrenberg describes his *Monas bicolor,* evidently a nucleated cell, as possibly an early state of the *Chlamidomonas* just mentioned.
>
> The curiously symmetrical forms of many of the *Bacillaria* appear to be due to the same division and subdivision of the pellucid nuclei or hyaline of cells.
>
> The delineations of *Gonium, Monas vivipara,* and *Ophrydium,* given by the great naturalist just mentioned, afford satisfactory examples of a pellucid body dividing and subdividing like the nucleus of a cell.

The cellular nature of protozoa was discussed more extensively and accurately by Von Siebold in his *Textbook on the Comparative Anatomy of Invertebrates,* published in 1848, when he was a professor at the University of Freiburg. Von Siebold devoted a twenty-three-page chapter to the "protozoa," a term that he introduced. The scope of his classification did not extend beyond that of Dujardin's, but on the question of structure he stated: "The lowest plants and animals correspond to single cells." And of the nucleus he said: "In almost all infusoria and also in rhizopoda there is . . . within the body a sharply defined structure, a type of nucleus which is distinguishable from the surrounding soft parenchyma. This firm nucleus, which varies in form and number among various infusoria, participates in the process of division. When an infusorium is to divide, changes are seen in the nucleus. . . . The nucleus is constricted at the center before the body is likewise constricted. . . ." He also described species having more than one nucleus, such as *Amphileptus* and *Stylonychia,* and those having nuclei that depart from the round or oval shape, giving *Vorticella, Euplotes,* and *Loxodes* as examples.

Various organelles having been recognized in the cells of the smallest animals, the observers' attention shifted to changes that might occur when there was contact between these cells. Conjugation was observed in *Paramecium* by Balbiani in 1858. He interpreted the process as a sexual exchange among hermaphroditic forms. The micronucleus was viewed as a sperm body, and the macronucleus as an ovarian body. Bütschli, in 1876, corrected this error, showing that the micronuclei divided, the macronuclei were destroyed, and the new micronuclei provided new macronuclei. He believed that this was a process of rejuvenation. A detailed study of the relationships of conjugation to life cycles and rejuvenation in a variety of species of ciliates was reported by Emile Maupas in 1888 and 1889. In recent years, the cytological organization and the process of conjugation have been restudied, and significant advances in the principles of genetics have been made. Of particular importance has been the work of Tracy M. Sonneborn with the so-called "kappa factors" of *Paramecium aurelia* in revealing the importance of cytoplasmic factors in heredity.

Not only the life cycle of the protozoa but also their ability to produce disease began to receive intensive study in the second half of the nineteenth century. Pathogenic protozoa are responsible for a considerable proportion of the disease burden of man and animals, and protozoal diseases have impeded the development of large geographical areas of the world. Malaria is still probably the most prevalent disease of this type, but trypanosomal diseases, such as African sleeping sickness and amebiasis, are also major public health problems.

We recall that the protozoa are subdivided into four main groups: (1) *Sarcodina*—organisms with pseudopods that are used for locomotion and ingestion of food, (2) *Mastigophora*—organisms with flagella that are used for locomotion, (3) *Sporozoa*—organisms without obvious organelles, usually sporulating at some stage in their life cycles, (4) *Ciliophora*—organisms with cilia or sucking tentacles that are used for locomotion and conveying food to the mouth.

Disease-producing agents are found among the members of all groups, and all sporozoa are parasitic. The discovery of the disease-producing role of some amebae, among the *Sarcodina*, was comparatively simple since these organisms do not have life cycles demanding a biological vector. On the other hand, disease-producing *Mastigophora* and *Sporozoa* often have multiple host life cycles. To have observed these organisms in man or other vertebrates was not enough; considerable effort had to be expended to demonstrate the complete chain of transmission.

The first observation of an ameba parasitic in man was made by G. Gros in 1849 in a rambling discourse appropriately entitled "Fragments of Helminthology and Microscopic Physiology" in the *Bulletin de la société impériale des naturalistes de Moscou* (**22**: 549–573). The specific name given by Gros to the organism, *Amoeba gingivalis*, has been kept in contemporary classification (*Entamoeba gingivalis*).

". . . *Amoeba gingivalis*. Within the tartar deposits of teeth are seen vibrios and a hitherto undescribed vesicule that I have drawn. The vesicules move slowly by amoebic extension and contraction; one might be led to say that they assume all forms. Within the body are globules that change position slightly; these globules are similar to structures in certain infusoria, the so-called polygastrica. Their origin and function are not known. . . . Is this a case of spontaneous generation? . . ."

Amebae were observed in India in the stools of choleric cases in 1870 by T. R. Lewis. His description was not sufficiently complete to identify the organism as *Entamoeba histolytica*, and he did not attribute disease production to the organism he observed.

Five years later, F. Lösch described amebae as a probable cause of dysentery in man. His report entitled "Overwhelming Proliferation of Amebae in the Colon" appeared in 1875 (*Arch. pathol. Anat. Physiol. klin. Med.*, **65**: 196–211). The observations were based upon the study of a Russian farmer at a St. Petersburg Hospital. The patient suffered intermittent bouts of severe diarrhea and dysentery, and despite treatment with a series of more than six agents, the patient died. Quinine was the one drug in the series that appeared to have produced a temporary remission of the disease. Lösch's report was thorough, and his description of the organism, limited to the vegetative form, would be acceptable by present standards. He wrote:

At 300–500 fold magnification . . . in the mucous and pussy flecks of the fresh stool . . . there are in addition to the usual components of dysenteric stools . . . a large number of cell-like structures of round, oval, pear or irregular form, in continuous motion and continuously changing shape. The diameter of the round forms is 5–8 times the diameter of the red blood cell. The motion, much more rapid than that of body cells, suggests that these forms are independent organisms. At 500 fold magnification there were 60 to 70 specimens per microscope field. . . .

A transparent knob sharply differentiated from the granular protoplasm, will first be formed at the surface [of the organism]. It may be retracted or enlarged to form a finger-like process (pseudopod), the length of the cell proper. This process can be retracted and another one formed at another part of the cell, or the granular protoplasm can stream into the process. In this manner the form can change, and with several processes the shape will be irregular. The processes are always bluntly rounded. Four to five processes may be formed in one minute.

The size of this parasite varies from 20–30 μ when rounded, to 60 μ when a process is fully extended.

The body consists of coarsely granular and hyaline protoplasm, a round nucleus and several hyaline vacuoles varying in size. A true membrane is not evident. . . . The nucleus is very pale and weakly evident, often obscured by the protoplasmic granules and foreign structures such as white and red blood cells, nuclei of disintegrated intestinal epithelium, bacteria and mycothrix chains that have been taken up by the parasite. . . . The nucleus is 4.86 to 6.95 μ in diameter and contains granules varying in size and refractility. . . . The protoplasmic vacuoles vary in size and are filled with a clear liquid. There may be as few as one or as many as 8. They change in size and occasionally I was able to see them discharge their contents to the exterior and disappear. . . .

From these observations, the parasite described is similar to *Amoeba princeps* in shape and in morphology but differs in size. According to Auerbach, *Amoeba princeps* is 70 to 140 μ in diameter while the organism here described rarely exceeds 35 μ. There is also a difference in the character of motility. *Amoeba princeps* does not form and retract processes irregularly; movement is by wavelike change along the entire body. The organism here described also is different from the ameba described by Lambl which was only 4.5–6.2 μ in diameter and formed pointed processes. Since the ameba that I have described is different from varieties now known I believe it is justified to name it *Amoeba coli* in view of its source.

What is the relationship of this parasite to the severe and persistent intestinal inflammation that resisted all treatment? Was it the primary cause of the entire disease? Did it appear subsequent to inflammation from prior damage? Or is it a chance and indifferent finding? The last possibility was unlikely because so many motile organisms could not be present without causing mechanical irritation, the number varied with the severity of the symptoms, and treatment of the inflammation was ineffective so long as the amebae remained.

To obtain a more certain understanding of the significance of the parasite, the effect of the parasite on dogs was investigated. I injected 3 dogs by mouth and anus with 1–2 ounces of fresh ameba-bearing feces on 3 successive days. I also performed another such test with one dog in which an intestinal inflammation was first induced by croton oil enema. Disease did not develop in two dogs of the first experimental group nor in the dog of the second experiment. The dog that was infected successfully developed an early diarrhea that abated, but subsequently formed stools showed blood and mucus that contained amebae. The dog was sacrificed 18 days after the last injection. At autopsy the mucosa of rectum was spotted with red areas, irregularly swollen and covered with tenacious blood-stained mucus. In addition, there were three superficial ulcers. There were numerous amebae in the mucus covering the rectum and in the bases of the ulcers. Thus, amebae proliferating in large numbers can irritate the mucosa to the extent of ulceration.

There is no doubt that in the human case described, the amebae played a significant role in the disease process. At the least they perpetuated the inflammation and the ulceration. . . . The question remains as to whether the entire disease was due to the amebae or whether the amebae followed other causes. If the course of the disease in the patient and the dog is compared, a significant difference is apparent. In the patient the symptoms developed rapidly to severe dysentery but in the dog the progress of the disease was slow and the symptoms were not severe, although the number of amebae administered to the dog was probably heavier than the infecting inoculum in the patient. It must be inferred that the patient first acquired dysentery and the amebae invaded later to proliferate and perpetuate the inflammation.

Direct transmission of the disease in man by ingestion of *Entamoeba histolytica* was eventually demonstrated by E. L. Walker and A. W. Sellards in 1913 by feeding twenty volunteers with amebic cysts separated from feces. Eighteen became infected; four of these developed clinical symptoms, and fourteen excreted cysts for various periods of time without developing clinical symptoms.

The next advances in the understanding of protozoa as agents of disease required the discovery of a new principle: the participation of a biological vector as a host for an essential stage in the life cycle of the parasite before it could infect a vertebrate host. This basic discovery was made by Theobald Smith, already encountered in the development of bacteriology and immunology.

Theobald Smith was born in Albany, New York, in 1859. He received his primary and secondary education in the local public schools. A scholarship, and organ playing during chapel services, enabled him to attend Cornell University, where his intellectual achievements impressed all who knew him. After receiving his baccalaureate degree, Smith decided to study medicine. The local Albany Medical College offered only a two-year course of training that Smith pursued but which he considered inadequate as a preparation for caring for the health of his fellow men. He felt it more prudent to return to the shadows of the ivy-covered walls of his first alma mater.

Smith returned to Cornell for graduate work under Professor S. H. Gage. However, he was not to remain there long. Dr. Daniel E. Salmon, who was organizing the U.S. Bureau of Animal Industry in Washington, wrote to Gage in the hope of receiving help in the selection of an assistant. The Bureau had been allocated the total generous appropriation of 10,000 dollars. Salmon was faced with the simple problem of doing good work, cheaply. Fortunately, Smith was not looking for a high salary, and Gage recommended him to Salmon as a man who knew French and German almost as well as English. In addition, wrote Gage, Smith was a born investigator whose talent for physics, chemistry, and mathematics seemed inborn. It was a bargain, and Salmon jumped on it. Thus, Theobald Smith found his way to Washington, where he was to approach the study of animal diseases with the eye of a naturalist.

As he arrived in Washington, in 1884, Smith found that the physical setup of the Bureau was compatible with its budget. It was housed in a dingy attic. This did not prevent him from rolling up his sleeves and setting himself to the task of applying the new disciplines from across the ocean to the study of animal disease. The scope of his investigations was broad as it covered swine erysipelas, hog cholera, rabbit septicemia, abortion in cattle, tuberculosis in cows, and the relation of bacteria present in water with the dissemination of infections. We have already discussed the work of Salmon and Smith on killed bacterial vaccines (Chapter 6).

Smith felt that science would best flourish by solving practical problems of economic interest. This feeling was so strong that when Theobald Smith was made a member of the Board of Directors of the newborn Rockefeller Institute, he doubted that it would be possible to perpetuate an organization devoted to pure research. His fears were eventually allayed, and, in 1915, he even became a director of one of the departments of this institute.

In 1889 Smith, together with F. L. Kilborne, began a systematic study of Texas cattle fever. Prior to this, Smith had already strongly implied that he thought the etiologic agent was a pear-shaped parasite living in, and then destroying, the red blood corpuscles. Within the next three years they established its transmission by the cattle tick, and a comprehensive report entitled "Investigations into the Nature, Causation and Prevention of Southern Cattle Fever" was published in 1893, in the 8th and 9th annual reports of the Bureau of Animal Industry for the years 1891 and 1892.

It was a long report, more than 200 pages. After a review of the history of the subject and a thorough description of the disease, the authors stressed the following point:

> The blood . . . grows very thin and watery as the disease progresses. This fact was emphasized by the earliest students of this disease. . . . Its prime significance seems to have escaped them and subsequent ones. In the preliminary pathological examination of four cases in 1888 the destruction of red corpuscles explained, best of all, the conditions observed. Hence the importance of concentrating the attention on the blood and its cellular elements was at once recognized. . . .

This consideration led the investigators to a very detailed study of the blood of affected animals. Before describing the causative agent of the disease and its life cycle, bacteria, the popular hypothetical agents of unknown diseases at the time, were shown not to be involved:

> In our own work the first problem which naturally presented itself was to determine whether bacteria could be regarded as the cause of the disease. Hence the very first and some of the later cases were utilized for this purpose. As to the first postulate necessary to be fulfilled in demonstrating the cause of any infectious disease—to find with the microscope the bacterium or other organism in the body of the diseased animal—this failed utterly in all the cases examined. The thousands of cover-glass preparations of the blood, spleen, liver, kidneys, etc., examined fresh and stained, never showed any bacteria excepting when the animal had been dead for a number of hours. . . .

The true pathogen was then introduced:

The Microorganism of Texas Fever
(*Pyrosoma bigeminum, n.sp.*) [1]

> Although Texas fever is essentially a blood disease, and only secondarily affects the spleen, liver, and kidneys, most observers have failed to recognize this fact. R. C. Stiles in 1867 was the earliest and the only

---

[1] Now called *Babesia bigemina*.

observer who laid any stress upon the changed condition of the blood corpuscles. He says: "The red blood corpuscles when examined immediately after removal from the body were shriveled and crenated without artificial provocation. . . . In one case many of the disks appeared to have lost a portion of their substance, as if a circular piece had been punched out, the addition of water failing to restore the disk to completeness." There can be but little doubt that Stiles saw at that time the microorganism of Texas fever, without, of course, recognizing it, since this description applies very closely to the appearance of red corpuscles infected by this microparasite when the blood and the parenchyma of liver, spleen, and kidneys are examined fresh soon after death. Other observers have examined the blood, but have seen nothing unusual. . . .

After describing the appearance of the parasite in the blood of animals at various stages of the disease, and after comparing these results with what could be observed in healthy animals, Smith and Kilborne pieced the result together and proposed:

The Probable Life-history of the Microorganism in the Body of Cattle

In the early stages of the high fever in a few acute cases, before the destruction of red corpuscles had gone far, very minute bodies were seen in fresh blood. Their form, so far as determinable, appeared as an elongated figure of eight or two short rods attached end to end. They had a very active Brownian motion in addition to a movement which carried them from one place to another in the field. This latter movement may have been due to currents in the liquid. They could not be detected in preparations stained with methylene blue. That this is the free form which precedes the parasitic stage must remain at present a mere conjecture.

The (hypothetical) swarming or motile stage (intraglobular)

We have already referred to certain very minute, well-defined, bright, frequently motile bodies seen within the red corpuscles of healthy cattle at various seasons of the year. As might have been expected, these bodies were found in Texas-fever blood as well. It has also been stated that they vary more or less both in size and form. The question has frequently presented itself, whether some of these bright motile bodies were the progenitors of the coccus-like and the pyriform bodies of the Texas-fever parasite. Inasmuch as they are present both in health and in disease, only a most trying examination of the blood in many cases could decide whether certain forms only appeared in disease or not. These bodies are so minute and so inaccessible that it is by no means certain whether such a prolonged study would bear fruit. . . . Such a motile, swarming stage is one which can readily be conceived of as finding its way into the red corpuscle constantly in motion in the vessels of the body. Why it is not seen in every case may be explained by the same hypothesis which accounts for the presence of the peripheral coccus-like stage in the milder type of Texas fever. This hypothesis assumes a retardation in the intraglobular development of the micro-parasite by which the smaller stages remain long enough in the blood to be detected. If the retardation is still more pronounced, it is easy to conceive of the motile or swarming stage as circulating in the blood long enough to be detected.

The stage of the peripheral coccus-like bodies

After the (hypothetical) swarm-spore has penetrated into the corpuscle it comes to rest, loses its bright, refrangent appearance, and attaches itself near the periphery of the corpuscle as a pale body which is only detected with difficulty in the unstained corpuscle. This body next undergoes division which is probably incomplete, for in the more advanced stages the two resulting bodies are as a rule still attached to each other. These remain close together while the infected corpuscle is circulating in the blood. This stage of the coccus-like body, like the preceding hypothetical stage, must be regarded as recognizable because of a retarded development of the micro-parasite. . . . In acute attacks the enormous multiplication of the parasite in the blood shows how rapid in such cases its development and how ephemeral these intermediate stages must be. The period of retardation may vary in length, but it seems probable that this stage may remain in the circulation at least several days.

The stage of the larger forms (pyriform and spindle-shaped bodies)

The two coccus-like bodies resulting from division begin to grow and assume fusiform outlines. It is probable that they remain attached to each other at least for some time, for in stained preparations a very delicate stained line may occasionally be traced passing from one to the other. In this stage they stain very well in haematoxylin and basic aniline dyes. As they continue to enlarge, the two members of the pair remaining always of the same size, a more elongated, pear-shaped outline is assumed, and in the unstained condition a minute dark particle is observed in the broad end of each body. Under conditions not definable a larger or smaller number of the red corpuscles contain but one body. These unpaired forms are found most abundantly in the circulating blood, where they may manifest amoeboid changes.

The larger forms circulating in the blood do not stain so well as the somewhat smaller bodies found in the capillaries after death. This may be due to degenerative processes or to a transformation into some unknown reproductive state. . . .

Free bodies

These are set free after they have reached the preceding stage by the disintegration of the infected corpuscles. They may be found in capillary blood of the heart muscle in abundance. Their most common location is in the kidneys, however. . . .

After having discussed the action of the parasite in the host and having tried to explain how these blood parasites were responsible for the pathological manifestations of the infection, Smith and Kilborne discussed the experiments in which they demonstrated the transmission of the disease from animal to animal.

The Production of Texas Fever in Cattle by the Inoculation of Blood from Cases of this Disease

The demonstration that Texas fever is caused by a certain microorganism is not absolutely made by showing that it is always associated with this disease and not observed in health. It may be argued, that such bodies are the concomitant rather than the cause of the fever. Nevertheless it may be said that no microorganism constantly associated with a given infectious disease has yet been found which is not demonstrably or presumptively the cause of the disease. Hence the probability that the microparasite described is the cause of Texas fever is very high, although the demonstration can not be made until such organism can be cultivated in some manner outside of the animal body and inoculations made with pure cultures. There is nothing today to encourage us

in the hope that parasites so highly adapted as the one under consideration will ever submit to the crude culture methods successful with many
bacteria.

The high probability that we have the cause of Texas fever before us
is increased by the fact that when blood from cases of this disease is
injected into the circulation of healthy susceptible cattle, the disease
is produced and the microparasite appears in the blood under the same
conditions under which it becomes manifest in the natural disease.
There is still the possibility before us that the microparasite is transmitted in the diseased blood and that some unknown agent has been
transmitted with it which is the true cause of the infection. . . .

In addition, the investigators tried to find experimental animals a bit less
bulky than cows and bulls, but:

The inoculations made by us demonstrate that sheep, pigeons, rabbits
and guinea-pigs are to all appearances insusceptible to this disease,
whereas in cattle the disease may be invariably produced by the injection of infected blood. . . .

Finally the authors came to the most important part of their study, the
demonstration that ticks are the agent of transmission of the protozoon from
animal to animal.

It has been a more or less prevalent theory of cattle-owners in the
districts occasionally invaded by Texas fever from the South that ticks
are the cause of the disease. Mr. J. R. Dodge, in his historical report
of this plague, mentions the fact that in 1869 an outbreak in Chester
County, Pa., was believed to be caused by ticks. Gamgee in 1868
states: "The tick theory has acquired quite a renown during the past
summer, but a little thought should have satisfied anyone of the
absurdity of the idea. . . ."

Nothing positive was thus contributed to the elucidation of the
action of ticks in carrying the disease until the subject was taken up at
the Experiment Station of the Bureau near Washington, in 1889. Here
it was found by experiments to be detailed in the remainder of this
report that the disease can be produced by ticks hatched artificially in
the laboratory, without the presence of Southern cattle. . . .

Space does not permit the reproduction of the details of these experiments, which were briefly outlined as follows:

These experiments were begun in the summer of 1889, and have been continued up to the present. They have been carried on in three different directions:

1. Ticks were carefully picked from Southern animals, so that none could mature and infect the ground. The object of this group of experiments was to find out if the disease could be conveyed from Southern to Northern stock on the same inclosure without the intervention of ticks.

2. Fields were infected by matured ticks and susceptible cattle placed on them to determine whether Texas fever could be produced without the presence of Southern cattle.

3. Susceptible Northern cattle were infected by placing on them young ticks hatched artificially, i.e., in closed dishes in the laboratory.

On the basis of the results obtained, Smith and Kilborne were able to conclude:

The hypothesis which seemed most plausible after the experiments of 1889 was that the tick, while withdrawing the blood from Southern cattle, drew out in it the Texas-fever parasite, which, entering into some more resistant state, perhaps some spore state, was disseminated over the pastures when the body of the mother tick became disintegrated. These spores were then supposed to enter the alimentary tract with the food and infect the body from this direction. The later experiments, however, completely demolished this conception. Neither the feeding of adult ticks and tick eggs nor the feeding of grass from infected pastures gave any positive results. On the other hand, the unmistakable outcome of the experiments was that the young tick introduced the infection into the body. This fact implies two possibilities. Either the tick is a necessary or a merely accidental bearer of the microparasite. If a necessary bearer of the infection, we must assume that the latter undergoes certain migrations and perhaps certain changes of state in the body of the adult tick and finally becomes lodged in the ovum. Subsequently it may become localized in certain glands of the young tick and discharge thence into the blood of cattle. This hypothesis assumes a complex symbiosis between the tick and the parasite on the one hand and the cattle and the tick on the other. According to another simpler hypothesis the tick would be merely an accidental bearer of the infection. The parasite entering the body of the tick with the blood of cattle may be already in the spore state or about to enter upon such a state. The young ticks, as they are hatched near the dead body of the female, may become infected from this. This infection, clinging to their mouth parts, is introduced into the blood of the cattle to which they subsequently attach themselves. Further investigations are necessary before the probable truth of one or the other of these hypotheses can be predicted with any degree of certainty.

The way was prepared for solving the riddles of such diseases as trypanosomiasis, malaria, and yellow fever, in which arthropods serve as vectors.

From this report Theobald Smith also emerges as an uncompromising opponent of charlatans in science. He could cut to the truth sharply:

In a report published in 1888 Dr. Frank S. Billings claimed somewhat pompously, to have discovered the "True Germ" of Texas Fever.

The announcement of this supposed discovery is entitled to quotation, "Hence the germ of the southern cattle plague has been discovered, and I think that I may be pardoned the egotism of claiming this to be the first occasion in American medicine that not only one but two germ diseases of animal life (swine plague) have been traced out and their origin placed upon an impregnable basis."

This germ is said to be like the germ of Billings' Swine Plague. It has been found in the blood, the gall, the urine, the liver, the spleen and kidneys of every diseased animal that was examined. It produces Texas Fever in cattle when inoculated in unquestionably pure cultivations.

This seems to be sufficient proof. In scientific research, however, especially when an important discovery is involved, it is incumbent upon the investigator to give at least some details of his experiments, so that others may form an opinion of their own as to whether the work was properly done and the conclusions or inferences warranted. Instead of a conscientious report of work done, we find in Billings' Bulletin of 138 pages the same padding used in the swine plague report of the same author. Quotations, criticisms, and discussions, mostly foreign to the object of the report.

The germ of Texas Fever as found by Billings stains at the ends. It grows on potato with a delicate straw color, which finally becomes brick red yellow. It does not liquefy gelatin. These meager facts are not sufficient to distinguish this organism from a large group of bacteria living especially in the intestines of all domesticated animals. In fact, the few characters apply very well to *Bacillus coli communis*. . . . This supposition is strengthened by the fact that Billings found in fresh and old manure, bacteria not to be distinguished from the supposed Texas Fever germ, etc.

In 1895, Smith left the Bureau of Animal Industry to join the Massachusetts State Board of Health and the Harvard Medical School. His activities shifted to human infections, but he continued his interest in animal diseases to a limited degree and still chose to combine pure research with practical application. He studied toxin and antitoxin production in diphtheria and tetanus and also studied sanitary bacteriology and tuberculosis.

Another major contribution appeared in his report "The Non-identity of Agglutinins Acting upon the Flagella and upon the Body of Bacteria" (*J. Med. Research*, **10**: 89–100, 1903). This work was done with A. L. Reagh, and its importance was not fully realized until fifteen years later. Smith had returned to his old favorite, the hog cholera bacillus. This organism, as a respectable member of the genus *Salmonella*, is flagellated and motile.

Smith and Reagh, having encountered a nonmotile variant that had been given off by a motile flagellated parent, prepared antisera with both strains. A study of agglutination reactions and the absorption of agglutinins led to the conclusion that the motile form had a flagellar and body antigen, and the nonmotile form had a body antigen only.

This was a basic example of the relation of bacterial variation to antigenic structure. For this phenomenon the terms O and H antigens were introduced by E. Weil and A. Felix in 1917 when they rediscovered the phenomenon with a less important group of organisms, the *Proteus* group.

Smith was also active in serotherapy of diphtheria. He succeeded in establishing antitoxin production in the laboratories of the Massachusetts Department of Health, a considerable accomplishment in 1895. He then studied the problem of prophylaxis. P. J. Nikanaroff in 1898 reported immunization with neutral toxin-antitoxin mixtures. Continuing along this line, Smith, in 1909, showed in guinea pigs that such mixtures with a slight excess of toxin were superior to toxins alone or to neutral toxin-antitoxin mixtures. Smith did not attempt any human therapy, but, as we have seen, Behring filled this gap in 1913.

Smith joined The Rockefeller Institute at Princeton in 1915. He resumed his old interest in animal diseases, emphasizing the natural history and ecological aspects of the interaction between host and parasite. Smith became an inspirational influence for young students, who now began to view disease from the ecological point of view. Some of these, such as Richard E. Shope, have become eminent, in their turn. Smith died quietly at Princeton in 1934.

As the etiologic agents of other protozoal diseases were discovered, it was found that often they too were transmitted by arthropods acting as biological vectors. Such was the case of the African trypanosomal disease of cattle and horses, called nagana, that David Bruce investigated.

David Bruce is another representative of the breed of military physicians, who, cutting across national lines, courageously and with intense dedication solved baffling problems in the etiology and transmission of infectious diseases. Born in Australia in 1855, Bruce went to Scotland with his family when he was five. He attended school until he was fourteen and then was put to work in a Manchester factory. Unusually strong and agile, he first considered becoming a professional athlete as a way of escaping from the drudgery of the factory, but a more compelling intellectual urge led him to medicine.

Bruce entered the University of Edinburg in 1876, where he soon became one of the outstanding students despite his limited formal preparation. Serious at work, he was also serious at love. As soon as he was graduated in 1883, he married Mary Steele and joined the Army Medical Service. Mary became his lifelong helper, often working with him in improvised jungle laboratories. Bruce's first assignment was in Malta, where, in 1886, he isolated the etiologic agent of undulant fever, now known as brucellosis. Subsequently, a commission under Bruce's direction discovered that the goat was the reservoir from which the organism was transmitted through milk. The disease could then be controlled.

Shortly after his isolation of *Micrococcus melitensis,* as he had named the agent of undulant fever, Bruce was called back to England to teach bacteriology at the Army Medical School in Netley. When Bruce subsequently met the Governor of Natal and Zululand, they were soon laying plans to investigate the mysterious diseases that rendered large areas of Africa uninhabitable. The Governor arranged a transfer, and in 1894 Major Bruce and wife, leaving the comfortable life at Netley, journeyed to Natal to study nagana, or literally, "depressing disease," of cattle and horses. Within a year, a severe epidemic in Zululand provided abundant material as well as a great challenge. By ox wagon the Bruces proceeded to the epidemic area and began an intensive study. How the Major discovered the trypanosomal etiology and clarified the natural history of the disease can be learned from extracts of his careful report entitled *Preliminary Report on the Tsetse Fly Disease or Nagana in Zululand* (Bennet and Davis, Durban, 1895). The report started with the following cover letter:

YOUR EXCELLENCY,—I have the honour to inform you that in accordance with instructions received from you, I left Pietermaritzburg on the 21st August, 1895, and arrived at Ubombo, Zululand, on the 8th September, 1895, for the purpose of continuing the investigation of the Tsetse Fly Disease, or Nagana, as it occurs in Zululand.

I have now the honour to forward a Preliminary Report, containing a statement of the results of the investigation up to the present date.

I may here state that on my arrival in Zululand the Tsetse Fly Disease and Nagana were looked upon as separate and distinct diseases, and one of the first results of this investigation was to show that they are undoubtedly one and the same.

This Preliminary Report contains:

1. A description of the Haematozoon discovered by me in 1894, in the blood of animals affected by this disease, a parasite not previously discovered in Africa.

2. A description of the Tsetse Fly, with experiments designed to show the part (if any) this Fly takes in the causation of the disease.

3. The result of experiments having for their object, proof of the connection (if any) which is supposed to exist between the Big Game and the spread of the disease.

4. A description of the disease as it affects domestic animals, with illustrative cases.

5. Inoculation and feeding experiments to show the communicability of the disease from affected to healthy animals. . . .

Following this letter Bruce first defined the disease and discussed its distribution and history. On the etiology he said:

The opinions of the Europeans settled in Zululand and of the Natives themselves are so conflicting that little or no good will be gained by entering fully into their evidence.

Three theories are held:

A. That the disease is caused by the bite of the Tsetse Fly. This is the European theory and as everyone knows has been popularly prevalent ever since white men first landed in South Africa.

B. That the disease is caused by the presence of large game, the wild animals in some way contaminating the grass or drinking water by their saliva or excretions. This may be called the Native theory.

C. That the disease is the product of certain physical conditions obtaining in regions to a certain degree tropical, and that the immediate cause is either Malaria or a vegetable poison.

In this Preliminary Report, and although up to the present I have not been able exactly to prove the part which the *Tsetse Fly* plays in the causation or production of the Fly Disease, I think it well to begin with a consideration of the Fly itself, not only on account of its historical value, but also because I am at present of opinion that the Tsetse Fly does play some part, and perhaps no inconsiderable part, in the propagation of the disease. Be it at once stated that I have not the slightest belief in the notion popularly prevalent up to the present that the Fly causes the disease by the injection of a poison elaborated by itself, after the manner of the leech, which injects a fluid to prevent the coagulation of the blood, or the snake for the purpose of procuring its prey or for defense, but that, at most, the Tsetse acts as a carrier of a living virus, an infinitely small parasite, from one animal to another, which entering into the blood stream of the animal bitten or pricked, there propagates and so gives rise to the disease. . . .

Experiments with the Tsetse Fly

The first series of experiments I instituted with the fly was to find out if flies kept in captivity at Ubombo would cause any local or general disease in susceptible animals. The flies were kept in cages made with muslin sides to give them air, and a glass sliding door for light and to observe them feeding, and sometimes lived for several weeks under these circumstances if given a meal once in two days. They were fed by pressing one of the muslin sides of the cage close against the skin of a suitable animal.

Bruce found that if tsetse flies not previously fed on an infected animal were allowed to bite a dog, the dog remained well. A dog bitten by tsetse flies fed on a diseased dog developed the disease.

From these two series of experiments it is seen in the first place that the fly *per se* does not give rise to any local or general disease, and this is further borne out by other experiments in which I placed minced-up flies under the skin of dogs without any results, and secondly it proves that the fly can act readily as a carrier of the Fly Disease from affected to healthy animals.

Now as one of the objects of this investigation is to establish the part, if any, that the wild animals play in the dissemination of this disease, and as my working hypothesis at present is that some species of animals living in the Fly Country harbour the Nagana parasite in their blood and that the Tsetse Fly carries the infection from affected to healthy animals, much in the same way as the vaccinating needle carries the infection of vaccinia from child to child, I next tried the experiment of taking susceptible animals into the Fly Country for a few hours, there permitting them to be bitten freely by the Fly, but not allowing them to eat or drink until their return to the top of the hill. For this purpose I used three perfectly healthy horses, and the following three cases will show that under the stated conditions, animals become readily affected by the disease. It may be mentioned here that the horses were prevented from even snatching a mouthful of grass by having a stout nose-bag of network fitted over their bits. A nose-bag containing a feed of crushed mealies was also taken in order to give them a feed when off-saddled for the purpose of catching flies.

The case histories of three horses so exposed and infected are given, and Bruce continues:

These three experiments then show abundantly that horses cannot be taken with impunity for a few hours into the Fly Country, even although they are not allowed to eat or drink there. . . .

This last series of experiments . . . does not prove that the disease is carried to them by the Tsetse Fly. There may be other ways of taking the disease, for example by inhalation. The disease called Ague or Malarial Fever in man is of all diseases probably the one most nearly related to the Fly disease in animals. They are both caused by blood parasites belonging to the Protozoa, and they are both found under similar physical conditions. In the case of the much-studied and familiar Malarial Fever, none have, up to the present, had the courage to assert that man would be immune by merely taking care not to eat or drink while in a malarious district, or in other words only to eat cooked food and drink boiled water while there. On the contrary, it is asserted by the latest authorities that merely breathing the air of Malarious Districts is sufficient to set up the disease. In other words that the parasite can obtain entrance to the system from the air. This is a hard thing to understand, and in order to make it possible the parasite of the Fly Disease must be able to exist in some other form than that in which it is found in the blood. The supposition would be that it forms a resting stage or spore form in which it can exist as a dry impalpable dust. For my part I have much difficulty in believing that animals are infected as a rule with Fly Disease by inhaling the *materies morbi,* and until I find animals still susceptible to the disease which are protected in some way or other both from feeding and the fly, I shall continue to be sceptical. . . .

At this point I think it will be convenient to give a definite description of the parasite discovered by me in the blood of animals affected by this disease, and to bring forward my reasons for considering it to be the proximate exciting cause of the disease. For the present I shall call it the Haematozoon or Blood Parasite of Fly disease, although in all probability on further knowledge it will be found to be identical with the haematozoon of Surra, which is called *Trypanosoma evansi* or at least a species belonging to that genus. For the purpose of having this point of identity settled I have sent specimens of blood containing the parasite from the horse, donkey, ox and dog to Dr. Lingard, Bacteriologist to the Indian Government, who has been investigating Surra since 1890. I have received one communication from Dr. Lingard, who doubts the identity of the two diseases, since the Indian disease does not affect cattle, whereas here in Zululand, Nagana according to common experience is almost invariably fatal in these animals. . . .

On looking at an unstained specimen by a microscopical power, say of 500 diameters, the Red Blood Corpuscles are seen as small faintly yellow discs, and among them and causing much commotion among them, can be seen transparent elongated bodies in active movement, wriggling about like tiny snakes and swimming from corpuscle to corpuscle, which they seem to seize upon and worry. They appear to be about a quarter of the diameter of a Red Blood Corpuscle in thickness, and 2 or 3 times the diameter of a corpuscle in length. They are pointed or somewhat blunt at one end, and the other extremity is seen to be prolonged into a very fine lash, which is in constant whip-like motion. Running along the cylindrical body between the two extremities can be seen a transparent delicate longitudinal membrane or fin which is also constantly in wave-like motion. The haematozoon vary among themselves a good deal in size and shape, and seem to take on slightly different forms in different species of animals. . . .

These parasites evidently belong to a very low form of animal life, namely the infusoria, and simply consist of a small mass of protoplasm surrounded by a limiting membrane, and without any differentiation of structure, except in so far as the membrane is prolonged to form the longitudinal fin and flagellum.

A much easier method of demonstrating the presence of the Haematozoon in the blood of animals affected by the Fly disease, is by means of stained preparations of blood. . . .

That this parasite is the true cause of Fly Disease is rendered almost certain from the following considerations:

1. It is found in the blood of every animal suffering from this disease, and is absent from the blood of all healthy horses, cattle or dogs.

2. The onset of the disease is marked by a rise in temperature, and this corresponds with the first appearance of the haematozoa in the blood.

3. As the disease progresses . . . with the destruction of the Red Blood Corpuscles the parasites tend to become more numerous, sometimes reaching the enormous number of 5, 10, or 15 millions in every drop of blood.

4. The transference of the smallest quantity of blood from an affected to a healthy animal sets up the disease in the latter, as I have shown above, even the very small quantity of blood conveyed by the proboscis of a few Tsetse Flies is sufficient to carry the disease from animal to animal. . . .

Bruce then described in detail the disease in the horse, donkey, and dog and then commented on treatment.

Up to the present there has been little opportunity of carrying out any Medicinal treatment in cases of this disease. It appears, however, that Arsenic has a specific action, causing the haematozoa to disappear from the blood to hinder or prevent the blood destruction, and emaciation, and to modify the temperature curve. . . .

This remarkable study was accomplished before the end of 1895. The *Trypanosoma* appropriately was named *T. brucei,* and the biting fly was named *Glossina morsitans.* There remained the problem of the source of the trypanosomes. Within the belts of fly infestation there might not be any sick horses or cattle for periods of months. Bruce found that wild antelopes and buffalos could carry the trypanosomes in their blood without ill effects. The *Glossina* flies could first bite these reservoir animals, acquire the trypanosomes, and then transmit them to horses and cattle.

With the outbreak of the Boer War, Bruce took his place with the troops. Trapped in the siege of Ladysmith, Bruce boldly performed emergency field surgery and saved many of the wounded who might have died. From 1903 to the beginning of the First World War he continued his studies on the trypanosomes. He confirmed the significance of Duxton and Castellani's demonstration of trypanosomes in human African sleeping sickness and then clarified the role of the trypanosome *T. gambiense* and its transmitter, *Glossina palpalis.*

During the First World War he was commandant of the Royal Medical College, by which time he had reached the rank of Major General. He was instrumental in introducing the use of prophylactic tetanus antitoxin and also headed the Trench Fever Commission. After the war, his health began to fail, but he continued participating in scientific activities as a "senior statesman," serving as an officer of the British Association for the Advancement of Science, and on the governing board of the Lister Institute. His wife Mary, who had helped him ably and faithfully throughout his career, died in November, 1931. David Bruce survived her for only four days.

Alphonse Laveran, like other men in our story, was raised to carry on a tradition of military medical service. He was born in Paris in 1845 where his father was a military surgeon and professor at the Val-de-Grâce Army Medical School. After obtaining his medical degree at Strasbourg in 1867, he served at Val-de-Grâce from 1874 to 1878 and then was transferred to a military hospital in Algeria.

Here malaria was prevalent and was the cause of many deaths. Laveran was struck by the constant presence of black pigment in the blood and organs of fatal malaria cases. Investigating the possible origin of the pigment, he discovered the malarial parasite. He reported his observations in a paper entitled "A New Parasite Found in the Blood of Malarial Patients. Parasitic Origin of Malarial Attacks" (*Bull. mém. soc. méd. hôp. Paris,* **17:** 158–164, 1880).

Last October 20, while microscopically examining the blood of a malarial patient, I observed, among the red blood cells, elements which appeared to be parasites. Since then, I have examined the blood of 44 patients and I have observed the same elements, whose parasitic nature now appears demonstrated to me, in 26 of these patients. I was unable to find these elements in the blood of patients suffering from other diseases. I shall describe these elements as bodies No. 1, 2 and 3. The report will demonstrate the utility of these designations, without presumption regarding the nature of the parasite.

Description of the parasitic elements found in the blood

BODIES NO. 1 These are elongated elements more or less tapered at their ends, often crescent-shaped, sometimes oval. Their length is 8–9 thousandths of a millimeter, their width about 3 thousandths of a millimeter. The contours are limited by a fine line; the body is transparent and colorless except at the center where there is a blackish spot consisting of pigmented granules. Exceptionally, this spot can be situated at the extremities. . . . Bodies No. 1 do not seem to be motile. When their shape changes, it is very slowly.

BODIES NO. 2 This form may vary depending on whether it is quiet or moving. In the resting phase there is a transparent spherical body, limited by a faint line, 6 thousandths of a millimeter in diameter. Within, there is a regularly arranged circle of uniform, pigmented granules; one might say a necklace of black pearls.

In the motile phase there are transparent filaments around the pigmented spherical body, which move rapidly in all directions like eels attached inside the sphere at one end. These filaments agitate the neighboring red blood cells. The length of the filaments can be estimated at about 3 to 4 times the diameter of a red blood cell. There seems to be 3 or 4 filaments per body; there may be more, because only the motile ones can be seen. . . . The filaments may be arranged symmetrically or grouped on one side. The free ends of the motile filaments are enlarged. . . . Bodies No. 2 often change shape while under observation. . . . Their motion is reminiscent of that of amebae. . . .

BODIES NO. 3 These are generally spherical and larger than bodies No. 2 (8–10 thousandths of a millimeter and sometimes more) lightly granular, immobile without apparent peripheral filaments. Within, there are pigmented granules, sometimes arranged as in bodies No. 2 but more frequently irregular in arrangement and very variable in number. These bodies can change shape.

Aside from bodies 1, 2 and 3 there are almost always small rounded bodies, bright, motile, and granules with bright red or light blue pigmentation. These pigmented granules are free or enclosed in bodies No. 3, or in leucocytes. The blue pigment appears to be derived from the bright red pigment. . . .

Nature and pathological role of parasitic elements found in the blood

Bodies No. 1, 2 and 3 appear to represent different stages in the development of the same parasite. Bodies No. 3 evidently result from the transformation of bodies No. 2. . . . This conclusion can be proven as follows: In a preparation one locates a specimen of type No. 2 provided with mobile filaments. . . . Periodic observations reveal loss of mobility . . . the appendages disappear, the body enlarges, the granules dissociate . . . the arrangement of the granules becomes irregular, they rejoin or disappear except for one or two that persist.

Proceeding in the same manner with bodies No. 1, one can see that they also change shape even though more slowly than bodies No. 2. They first become ovoid and then spherical or irregular. Intermediary forms are seen. I have never seen a body No. 1 transforming itself into the motile phase of bodies No. 2. . . . Motile bodies No. 2 are mainly seen in the blood of patients, suffering a febrile relapse and who do not follow a regular quinine regimen. . . .

After noting that his observations suggest the presence of an animalcule related to the infusoria, Laveran concluded:

The important role that these parasitic elements play in the pathogenesis of malaria appears to be established by the following propositions:

1. These parasitic elements are found only in the blood of malaria patients . . . even though not always present there. . . .

2. The parasitic elements that are numerous in the blood of untreated patients disappear from those who have taken quinine sulfate for a long time and who can be considered cured. . . .

3. In the blood of individuals who have died of pernicious fever, there are many pigmented elements that are similar to bodies No. 3 or less frequently, similar to bodies No. 1. The presence of these elements in the capillaries of all tissues and organs, especially spleen and liver, is the most characteristic lesion of acute malarial intoxication. . . .

What is the origin of the parasitic elements in the blood of malarial patients? By what route do they gain access to the body? How do they provoke the intermittent fever and the other manifestations of malaria? One can only ask these important questions.

Laveran's discovery was received with skepticism. If it were a valid agent causing infectious diseases, it should meet the requirements of Koch's postulates. Klebs and Tomassi-Crudelli had proposed a *"Bacillus malariae"* as the etiologic agent. Danilewsky and Metchnikoff, who saw a similarity to blood parasites they had seen in birds, gave prompt support to Laveran, and Pasteur himself was also convinced of the correctness of his observations. Laveran returned to a professorship at Val-de-Grâce, where he remained until 1897 when he joined the Pasteur Institute in Paris. In 1907 he was awarded the Nobel Prize for his studies on malaria. He died in 1922.

The answers to the questions posed by Laveran were pieced together by a number of investigators. C. Golgi, from 1886 to 1889, showed that Laveran's bodies were stages in the life cycle of the parasite, wherein the ameboid form subdivides into merozoites, and that liberation of the merozoites by rupture of the host red cells is the point at which the fever occurs.

The difficult problem of transmission was, in the main, solved by Ronald Ross and completed by Battista Grassi. Ross was the Byron of microbiology.

We have his own memoirs to tell us the story of his life. It is recorded with bombast, humor, and serious dedication: "Our books of science are records of results rather than of that sacred passion for discovery which leads to them. Yet many discoveries have really been the climax of an intense drama, full of hopes and despairs, visions seen in darkness, many failures, and a final triumph."

Ronald Ross was born on May 13, 1857, on the fringe of the Indian Himalayan mountains, where his father, Brig. Gen. Sir Campbell Ross, was stationed to pacify fierce hill tribes. Ronald, the first of ten children, remembered that his mother "Like all mothers, . . . was the best in the world."

In 1865 he was sent to England to be educated, and soon he was reading the Bible, Shakespeare, Pope, Milton, and Hume. He was also proficient in mathematics even though the teaching was "atrocious." After attending a succession of schools and living among relatives, Ronald reached seventeen and had to choose a profession. "I wished to be an artist, but my father was opposed to this. I wished also to enter the Army or the Navy; but my father had set his heart upon my joining the medical profession and, finally, the Indian Medical Service. . . . I had no predilection at all for medicine. . . ." Ross entered St. Bartholomew's Hospital Medical School. After diluting his efforts at medicine with writing drama and poetry, and playing and composing music, Ross passed the examinations of the College of Surgeons, in 1879, but failed to qualify in medicine. He could not, that year, compete for the Indian Medical Service and took a position as a ship's surgeon on an Atlantic run until 1880. By then he was reconciled to preparing properly for medical qualification, and he soon was admitted to the Indian Medical Service.

Ross's first assignments in India and Burma left bounteous time for writing epic dramas that were printed privately, but "no one ever bought a copy." Writing was alternated with the study of mathematics and philosophy. But periods of depression became frequent. By 1888 he felt he could no longer endure India, and a threat of resignation brought a year's furlough to England.

By then aware of microbes, Ross took a course in bacteriology given by the Royal College of Physicians and Surgeons in London. His performance was poor; this time it was courting for marriage to Rosa Bloxam that explained his delinquency. After the honeymoon, he applied himself diligently to bacteriology for two months, and in 1889 he became the first member of the Indian Medical Service who could boast a Diploma of Public Health. Literary efforts also were not abandoned. A romance, "The Child of Ocean," was put out by a dubious publishing firm and actually was reviewed by a number of newspapers.

Ross returned to India with a box full of bacterial cultures and took up the study of mosquitoes, but "I could get no books on the subject in India." Also, "I had now determined to labour really at my profession and therefore read the medical books and journals within reach; but they did not help much. . . . At that time Laveran's discovery that malarial fever is caused by a parasite in the blood, began to be talked about in India; but unfortunately those who thought they had found the germ had . . . merely mistaken natural objects in the blood for it. I . . . detected their error at once, and hence became skeptical regarding the whole theory. At the same time I became equally . . . skeptical regarding the theory that so-called malarial fevers are due to aerial miasmas. . . ."

In 1894 the Rosses, then with two daughters, returned to England for another year's furlough. Ross was introduced to Patrick Manson, who demonstrated to him how to observe malarial parasites and acquainted Ross with the literature on this disease. In the course of subsequent discussions, Manson, who had already demonstrated that mosquitoes carry filaria, told Ross that he believed that mosquitoes also carry malaria. This idea intrigued Ross as a great intellectual challenge, and he returned to India with a purposeful objective. Four years of labor began that were marked with many false starts and disappointments. Ross and Manson communicated frequently. Manson, guiding from afar, wrote Ross no less than fifty-five letters during that period. The Indian Medical Service did not deal kindly with Ross while this important work was proceeding. He complained of "Three serious interruptions . . . all of which by some evil fatality occurred just at the moments when rapid advances seemed to be assured." Ross set mosquitoes to biting malarial patients and then examined the mosquitoes for the presence of parasites. At first he used any mosquitoes that could be trapped. His work was interrupted not only by the official indolence of his superiors but also by many diseases, including a severe bout of food poisoning that was actually fatal for one person who had had dinner with him. By late July, 1897, though, he was back at the work of examining "malariated mosquitoes."

"I was up against a very difficult problem indeed . . . there were many species of mosquitoes, and each of these contained many kinds of parasites. How was I to discover which species of mosquito and which kind of parasite were the right ones?" But on August 16, 1897, one of Ross's mosquito collectors supplied a type different from the usual ones. This one was brown rather than brindled and had a tail that stuck outwards. The odd type was an *Anopheles*. Two such mosquitoes, fed on a malarial patient, revealed in their stomachs . . . "the wonderful cells. . . . I saw a clear and almost circular outline before me of about 12 microns in diameter. The outline was much too sharp, the cell too small to be an ordinary stomach cell of a mosquito. . . . In each of these cells there was a cluster of small granules, black as jet and exactly like the black pigment granules of *Plasmodium* crescents." The results were reported formally in a paper entitled "On Some Peculiar Pigmented Cells Found in Two Mosquitoes Fed on Malariated Blood" (*Brit. Med. J.*, Dec. 18, 1897). At this critical point, Ross was ordered to duty at an outpost free of malaria. By his own frantic efforts and with Manson's help back in London, Ross succeeded in being transferred to Calcutta, where he could again continue his investigations.

While Ross was at the malaria-free outpost, in desperation for activity, he examined the small animals indigenous to the area for sporozoa, and he found them with ease in both pigeons and sparrows. In Calcutta he continued his studies of birds along with studies of human malaria. The birds were found to be excellent animals for studying *Proteosoma* species, and the bird parasites behaved much like the human malaria parasite. Ross studied transmission of these sporozoa to birds using the various mosquitoes available. The successful cultivation of *Proteosoma* in grey mosquitoes was published in 1898 in a paper entitled "Report on the Cultivation of Proteosoma, Labbé, in Grey Mosquitoes" (*Indian Med. Gaz.*, **33**: 401–408 and 448–451, 1898).

In a report to the Director-General, Indian Medical Service, dated 19th September 1897, I described some peculiar pigmented cells found by me in August 1897 in two dappled-winged mosquitoes fed on blood containing crescents, and anticipated that they were a stage of the parasite of malaria in those insects. An abstract of the report was published in the *British Medical Journal*, to which were appended some remarks by Dr. Manson [and others] on the original specimens, which they had seen. While Dr. Manson thought that the cells might be the extra-corporeal phase of the parasite of malaria, it was, on the other hand, considered possible for them to be either normal cells of the mosquito, or some parasite in it quite independent of malaria. In September 1897, however, the cells were seen again in two ordinary mosquitoes fed on malarial blood; and the fresh light which these new observations threw on the subject enabled me to show that they must be at least some kind of pathological growth in the insect; indeed, little doubt now remained in my mind that they actually were what I had from the first supposed them to be. . . .

I have lately been able to obtain conclusive evidence to this effect by having succeeded in producing these pigmented cells at will in grey mosquitoes by feeding them on birds infected by proteosoma, Labbé, a parasite very similar and closely related to the haemamoebae of malaria in man. . . .

As it was not the malarious season of the year, and as cases of haemamoebiasis suitable for these experiments were not at all easy to procure, it was thought advisable to commence work on the so-called malaria parasites of birds. Accordingly, a crow and two tame pigeons infected with halteridium were obtained; and on the night of the 11th and 12th March these, with four short-toes larks and six sparrows, whose blood had not yet been carefully examined, were placed in their cages, all within the same mosquito netting, and a number of grey mosquitoes of the species I had lately been experimenting with were released within the net. Next morning, numbers of these insects were found gorged with blood and were caught in test-tubes in which they were kept alive for two or three days.

On the 13th March I commenced to examine them. Out of fourteen of them, pigmented cells were at last found in one.

Believing as I did that these cells are derived from the gymnosporidia, I judged from this experiment that the grey mosquito which now contained them had fed itself on one of the birds which happened to be infected by a parasite capable of transference to the grey species of mosquito. As all the birds had been placed together in the same net, the question now was which of them had the mosquito fed herself upon. This could be easily ascertained. A number of mosquitoes of the same species had meanwhile been fed separately on the crow and two pigeons with halteridium; but out of 34 of these examined not one contained pigmented cells. Hence I came to the conclusion that the mosquito with pigmented cells had not derived them from the crow and the two pigeons. The larks and sparrows remained.

The blood of these had not yet been carefully searched. I now found that three of the larks and one of the sparrows contained *proteosoma*, Labbé, and therefore thought it possible that the mosquito had been infected from one of these. Accordingly, on the night of the 17th and 18th March, a number of grey mosquitoes were released on the three larks with proteosoma, and next morning it was found that nine of these had fed themselves.

On the morning of the 20th March, that is, from 48 to 60 hours after feeding, these nine insects were examined. Pigmented cells were found in no less than five of them. After the long continued negative experiments with this kind of mosquito (and indeed, I may say, after three years' doubtful attempts to cultivate these parasites) this result was almost conclusive. It indicated, as was surmised before, that when a certain species of mosquito is fed on blood containing a certain species of gymnosporidia, pigmented cells are developed. Hence it would follow, as an easy corollary, that the cells are a stage in the life-history of the gymnosporidium in the mosquito. . . .

The youngest coccidia yet seen by me have been found in mosquitoes killed on the second day after feeding on birds containing proteosoma. They are small oval bodies of 8 μ or less in the long diameter; they occupy the muscular or submuscular coat of the insect's stomach; they consist of a colourable *ectoplasm* and an indifferent *entoplasm;* and contain *pigment* like that of proteosoma, and also *vacuoles* and *granules.* They now grow apace, retaining a similar form; and add about 10 μ of diameter for every twenty-four hours of their life. A *capsule* is speedily formed; the contents consist of a larger and larger number of vacuoles and granules; and the pigment, which generally surrounds one or more large vacuoles, gradually diminishes in amount until, on about the fourth day, it vanishes altogether. From this time also the contents are studded with numerous small *oil globules* (?); and the whole cell begins to protrude markedly from the external wall of the stomach into the coelom of the host. Growth continues until on the sixth to the seventh day, the size reached may be 70 $\mu \times$ 60 μ. At this stage the coccidia are often quite separate from the stomach tissues, to which, however, they remain attached by some invisible adhesion. From the seventh to the twelfth day after feeding, no further development has been observed. . . .

From this point, Ross had only to follow the course of the coccidia after they had matured in the stomach wall and by July 1898 Ross had found that the *Proteosoma* move from the stomach to the salivary gland of the mosquito. The proteosomes could now be transmitted to other birds by bite of the mosquito. Ross telegraphed this information to Sir Patrick Manson in time for presentation before the July 1898 Meeting of the British Medical Association. It was published under the title "Mosquitoes and Malaria" (*Brit. Med. J.*, Feb. 18, 1899, 432–433).

Major Ronald Ross, Indian Medical Service, has forwarded to us a preliminary report on the infection of birds with proteosoma by the bites of mosquitoes. The report is dated October, 1898. . . .

Reproduction forms of the proteosoma-coccidia

The examination of a very large number of grey mosquitoes fed on sparrows infected with proteosoma showed clearly that the proteosoma-coccidia after reaching maturity (seventh day) formed two different kinds of reproductive elements, namely (1) a large number of delicate thread-like bodies, or (2) a smaller number of large black spores.

The Thread-like Bodies are from 12 μ to 16 μ in length, about 1 μ in breadth, and flattened in the third dimension. From the middle, which contains vacuoles and chromatin granules, they taper to each extremity; and when viewed under the best conditions they are not unlike small trypanosomes, though Ross has never succeeded in detecting any certain indications of movement in them. They are closely packed in thousands within the capsule of the mother coccidium. . . . On rupturing such a body by pressure on the cover-glass, the thread-like elements can be seen pouring out in vast numbers, and floating away in the surrounding fluid.

The Large Black Spores, when mature, are from 16 μ to 20 μ or more in length, and 2 μ or 3 μ in thickness. They are cylindrical, closed at the ends, and are straight, curved, sigmoid, or variously twisted. Their colour is a dark brown, they have a black contour, and they are evidently capable of resistance to outer forces. . . .

Escape of the reproduction forms

On the eighth or ninth day—that is, shortly after the coccidia have matured and produced their reproductive elements—they burst *in situ* in the living insect, and pour those elements into the general cavity of the body, which contains the so-called blood or circulating juices. The empty capsule of the coccidium remains behind still attached to the wall of the stomach; but the reproductive elements are swept away by the blood current and distributed throughout the tissues of the mosquito. If, for example, a mosquito be killed ten days after it has fed on a sparrow with proteosoma, larger numbers of the thread-like bodies will almost certainly be found in the juices of the head and thorax, and similarly some of the black spores in almost all the muscular and connective tissues of the body.

Further migration of the thread-like elements

While examining the thread-like elements in the blood of mosquitoes, Ross observed that they were frequently to be found collected within the cells of a gland in the thorax, which eventually proved to be the salivary or poison gland of the mosquito. This lies in the neck or the anterior part of the thorax of the mosquito, and consists of a number of separate lobes. The ducts of the several lobes ultimately unite and form a single main effluent. This runs up the under-surface of the head in the middle line, enters the base of one of the stylets or lancets of the proboscis, namely, the central unpaired one, called the tongue or epipharynx, transverses the whole length of it and opens at its extremity in such a manner that the secretion of the gland must be poured into the very bottom of the wound made by the piercing apparatus of the proboscis.

It is in the cells of this gland that the thread-like reproductive elements of the proteosoma-coccidia have the power of accumulating. They can be found in them in large numbers, either floating separately in the grape-like cavity of the cells, or crowded together within them in hundreds. Generally only one or two of the lobes of the gland contain these elements, the other lobes being free from them. Ross estimates that one lobe must contain many thousands.

Interpretation of these facts

The secretion of the gland is obviously meant to be injected into the wound made by the tongue and the other stylets. Its function is, it is contended, to check the spasmodic contraction of the torn capillaries, which would otherwise quickly stay the flow of blood into the wound. The thread-like elements in the secreting cells of the gland remain there, waiting the opportunity when they shall be carried, together with the secretion, directly into the blood vessels of the victim of their mosquito host. There, if that victim happens to be a bird of an amenable species, they doubtless originate an infection of proteosoma, thus completing the cycle of the parasite.

Infection of birds by the bites of mosquitoes

All that was now required to obtain complete proof of this theory was actually to infect healthy birds in the manner suggested. This was accomplished very easily. At the end of June, 1898, four sparrows and one weaver bird, whose blood on several examinations had been found to be entirely free from proteosoma, were subjected nightly to the bites of numerous grey mosquitoes fed more than a week previously on a sparrow containing proteosoma. On July 9th these five birds were found to have become infected with swarms of the parasites. All of them died very soon, and the liver was profusely charged with the characteristic black pigment of malaria. The experiment was next repeated on a number of sparrows and other birds:

1. Out of 28 originally healthy sparrows subjected to the bites of grey mosquitoes previously fed on diseased sparrows, 22, or 79 per cent, became infected, all with a very large number of parasites, in from five to eight days. This excludes a number of birds which died before the end of the incubation period from diarrhoea and other disease to which sparrows in captivity in Calcutta are subject. Out of the 6 birds which failed to become infected after these experiments, one was subjected to a second trial, which gave a successful result.

2. Out of 2 crows and 4 weaver birds, some of which contained halteridium though none contained proteosoma, 1 of the crows and all the weaver birds showed a copious proteosoma infection within nine or ten days of being bitten by grey mosquitoes fed previously on sparrows with the latter parasite.

3. Out of 5 sparrows which originally contained a very few proteosoma, 4 showed a new and much more copious infection a week after being subjected to a similar experiment.

4. The attempt to transmit the proteosoma of sparrows to mainas and some other birds failed.

In all the birds in which the parasites appeared after the experiment, the invasion presented such constant and unmistakable characters that no possible room for doubt was left as to the infection being due to the mosquitoes. The blood of the birds experimented with was examined both before the experiment and on several occasions afterwards. The course of events was always as follows: The blood would remain entirely free from proteosoma (except in the case of the five sparrows which originally contained a few of them) until the fifth, sixth, seventh, or eighth day after the experiment, when one or two parasites only would be found in an entire specimen. Next day it would invariably be seen that the number of parasites had largely increased; and this increase would continue until in a few days, in almost every case, the parasites become so exceedingly numerous that from ten to sixty, and even more, could be counted in almost every field of the oil-immersion lens, while as many as seven distinct parasites in a single corpuscle were often observed. The only exceptions were the case of the crow and that of the five originally infected sparrows, in which the number of parasites did not exceed about one in each field, even this being a high figure for proteosoma. Most of the birds now died, and showed not only the characteristic pigmentation of the liver, but also a distinct inky colour in the blood itself. In the few which recovered, however, the number of parasites rapidly declined. . . .

Ross had been called to Calcutta to work on malaria and kala azar, and again his studies were interrupted by an order to study kala azar. "I left Calcutta with a heavy heart for the kala-azar duty. The great treasure house had been opened, but I was dragged away before I could handle the treasures. . . . I wished to complete the cycle of the human parasites . . . that was left to the Italians. . . ."

The discovery made, Ross believed that the work should be continued and applied to the practical control of this scourge that caused 1.3 million deaths per year in India alone. The British authorities did not respond to Ross's pleas, despite the intervention of such notables as Lord Lister. Ross retired from the Indian Medical Service in 1899 and returned to England.

The first days in England were taken up with priority disputes directed primarily against Grassi and his coworkers, whose work will be reviewed shortly. But the creative impulse prevailed, and soon Ross was busy helping organize and finance the newly conceived Liverpool School of Tropical Medicine. He also accepted a lectureship. Financially, Ross was in poor circumstances despite his great contribution to science. His total income was at best less than £900 per year. His pension was £292 a year, and the university paid him yearly an additional £300 until he was granted a professorship three years later at £600 per year. This position had the advantage of allowing time for scientific expeditions.

In July, 1899, Ross left with an expeditionary group for Sierra Leone. The group stopped first at Freetown, and here Ross confirmed the transmission of malaria to man by the mosquito. The species here were *Anopheles costalis* and *Anopheles funestus*. After a successful and productive stay of approximately two months, the group was back in Liverpool by early October with useful observations on methods of mosquito control based upon the habits of the individual species of mosquitoes. Ross wrote a number of successful didactic and practical pamphlets and reviews on malaria. But he was, in 1900, disenchanted with medical research and prepared to revive his old love for mathematics. Unfortunately, his early work on vector geometry had been discovered independently, and his enthusiasm for mathematics also waned.

The cause of malaria control again impinged on his conscience. Ross attempted to interest the government in undertaking large-scale measures for the control of malaria in the colonies, but he was unsuccessful. Koch was already directing such a campaign in the German colonies. The Americans in Havana, following Reed's demonstrations of transmission of yellow fever by mosquitoes, showed that mosquitoes could be controlled by eliminating breeding places. The incidence of both yellow fever and malaria had been reduced dramatically. Ross "determined on temerarious and quixotic action . . ." and obtained the cooperation and financial support of a group of businessmen, who contributed items varying from ship passage to pickaxes and creosote. And so Ross and staff were again in Freetown in July, 1901. With crews of energetic laborers, pools and puddles were drained, and rubbish containers holding stagnant water were removed and destroyed. House owners were happy to be rid of rubbish that had been collecting for years. A supervisor was left at Freetown and Ross's group proceeded to other colonies along the African coast. An expert from The London School of Tropical Medicine reviewing the work of Ross's group in this expedition stated: ". . . Your efforts have been crowned with a large degree of success. . . ."

The success at Freetown led to an invitation from the president of the Suez Canal Corporation to solve the baffling problem of malaria at Ismailia on the Suez Canal. Drainage and quinine had not solved the problem. Ross immediately found *Anopheles* in innocent looking watercress beds and similar fresh water bodies. *Stegomyiae* and *culices* were found in carefully constructed sewage pits, equipped with vents. Periodic oiling solved this problem.

In 1902 came the Nobel Prize. Ross, with his usual flair, broke the prescription of secrecy before the public announcement. This honor was only seven years removed from the beginning of the research. In his address he named the many contributors, of which there were more than two dozen, all save Grassi, to the solution of the problem.

The British Government continued its policy of inaction on malaria control. "Thus it was that the glory ultimately passed to Gorgas and the Americans."

In 1912 after fourteen years in Liverpool, Ross resigned his professorship and moved to London to take up consulting practice, but endless public obligations took precedence over private consulting. During the First World War, Ross accepted a commission in the Territorial Force and advised on malaria problems. In a military voyage in 1917, the ship carrying Ross was torpedoed not once, but twice. With the luck of the Rosses the ship did not sink. Ross remained on deck to cheer as the escorting destroyer blasted the submarine into surfacing and surrender. With this episode, Ross ended his memoirs. The succeeding years were quiet and rewarding. His literary writings were received more kindly than during his youth, and the government agencies began to support serious efforts to achieve malaria control. He died in London, in 1932, in the very institute that had received his name.

Immediately after Ross had demonstrated the development of *Proteosoma* of bird malaria in the mosquito and the transmission from mosquito to bird, he had planned to complete the demonstration of transmission to man. At this crucial point he was ordered to investigate kala azar. By this time, much to Ross's irritation, Battista Grassi and his coworkers in Italy were at work demonstrating transmission in human malaria. Grassi explained the background in his monograph *Die Malaria* published in 1901.

". . . I must state clearly that I was stimulated to undertake my investigation by the brilliant studies of Manson on the mode of transmission of filaria as well as by the mosquito theory whose validity was being studied by Manson and his pupil Ross; nevertheless I followed a line of my own. I directed my approach from the basic observation that there are many places in Italy where there are enormous numbers of mosquitoes without human malaria and thus concluded that in endemic malarial areas indigenous types of misquitoes must be transmitting malaria. After many thorough comparative experiments I concluded that *Anopheles,* though comprising only a small proportion of such indigenous sucking insects, was the transmitter of malaria to man. On June 22, 1899 I proved that malaria was transmitted only by certain varieties of *Anopheles.* . . ."

The development of human malarial parasites in mosquitoes was carefully followed by Grassi and his coworkers, as can be seen from the following passages from a report by B. Grassi, A. Bignami, and G. Bastianelli entitled "Further Studies on the Developmental Cycle of the Human Malaria Parasite in the Mosquito" [*Atti R. Accad. Lincei Rendic.* (5) VIII, 21–28, 1899]:

> Continuing our study of the developmental cycle of the malaria parasite in *Anopheles claviger,* we have used two approaches: (1) Examination of mosquitoes from the rooms of people suffering from malaria, (2) Examination of mosquitoes that had at known time bitten hospitalized malaria patients. Such mosquitoes were sacrificed and examined daily in order to follow the regular development of the parasite. Parasites were not observed in mosquitoes, similarly maintained, but that had not bitten malaria patients. . . . [In the season

of cold weather] female mosquitoes abiding in the dwellings of people infected with malaria become heavily infected, while those abiding in barns and chicken houses remain uninfected. . . .

Mosquitoes were kept at 30°C after they had bitten. Prior observations had led us to suspect the maintenance at 14–15° during the early hours after biting would not support the development of the hemosporidium. Definite development occurred at 20–22° but was slower than at 30°.

The authors remarked, in the case of estivo-autumnal fever:

We have reported previously that the crescents from man develop further in the mosquito. The crescents must be mature. Positive results were obtained only when the crescents changed rapidly to round flagellate bodies. . . .

After two days, one observes round to ovoid encapsulated structures among the muscle fibres of the mid-intestine. These are developing hemosporidia. The pigment, identical with that of the crescent, is accumulated as a large mass at the periphery or as two parallel threads thickened at the center. The hemosporidium is transparent and vacuolated. After four days the capsules have enlarged, vacuolation is more prominent and the pigment has decreased and dispersed. After six days the capsules have enlarged further and produced herniation . . . toward the coelom. There are many small bright corpuscles. . . . After seven days the parasite contains a large number of filaments radiating from several centers. The filaments are approximately 14 μ long and extremely thin. Pigment is still present. It is readily seen that these are maturing sporozoa and that the process is similar to that seen in many other sporozoa: The parasites increase in volume, become encapsulated, the nuclei subdivide and many sporoblasts are formed. The sporoblasts are then transformed into sporozoites that are filiform. . . .

On the succeeding days one can see broken and collapsed capsules adhering to the intestine and nearby are sporozoites that are also dispersed throughout the coelom. Later they are found accumulating within the tubules of the salivary gland. . . .

For the Common Tertian Fever our observations are less complete. . . . The observations reported up to this point allow us to reconstruct the life cycle of human hemosporidium in the body of *Anopheles claviger*. This process is for the most part analogous to that observed by Ross for the proteosoma of birds in the grey mosquito.

[Comparing] the cycle of development for parasites of Estivo-Autumnal fever with that of Common Tertian Fever, we have observed that not all forms of these two parasites are capable of development in the new host. For Estivo-Autumnal fever this occurs when the blood of the patient contains mature crescents capable of transformation into flagellated forms, and for Tertian fever when there are large pigmented sterile bodies which can flagellate. Thus, development can proceed when there are forms in the blood which Grassi and Dionisi have considered to be gametes. The early stages in the intestine of the mosquito appear different, dependent upon whether they are derived from crescents of Estivo-Autumnal fever or from the pigmented bodies of Tertian fever. On this basis a differential diagnosis can be made between the two forms. . . .

Battista Grassi was one of the great masters of zoology and parasitology of his generation; regrettably he has not been accorded the recognition his accomplishments merit. Grassi was born in northern Italy in 1854. After completing medical training at the University of Pavia, he turned to zoology. As a first step in this new direction he studied with Kleinenberg at Messina, with Bütschli and Gegenbaur at Heidelberg, and with Semper at Würzburg. He served as professor of zoology at the University of Catania in Sicily from 1883 to 1895 and then accepted the chair in comparative anatomy in Rome. Despite chronically delicate health, he worked prodigiously during all his years until death in 1925. The list of important contributions of Grassi and his coworkers is impressive. By the early 1880s he had demonstrated the mode of transmission of various parasitic amebae and the life cycles of *Ascaris lumbricoides* and *Hymenolepis nana*. During the next decade he elucidated the social organization and behavior of termites and had made a start on malaria. In 1898 he concentrated his attention on malaria, and before the end of 1899 he and his coworkers had demonstrated the role of anopheline mosquitoes in the transmission of malaria. Subsequently, he turned his attention to economically important insects including *Phylloxera*, which attacks the grapevine; *Phlebotomus*, a transmitter of human disease, and his old forte, mosquitoes.

He was awarded the Darwin medal of the Royal Society in 1896, before his dispute with Ross on priority regarding the mosquito transmission of malaria in man. Grassi was not blessed with a kindly temperament and was involved in more than one priority dispute. Now that the years have passed, all this seems to matter very little; the important thing was that the riddle of malaria had been elucidated. We find that there is enough credit to honor the memory of a Frenchman, an Englishman, and an Italian.

References

Cole, F. J.: *The History of Protozoology,* University of London Press, Ltd., London, 1926.

Dobell, C.: *Antony van Leeuwenhoek and His "Little Animals,"* 1932, Dover Publications, Inc., New York, 1960 (republication).

Dujardin, F.: *Histoire naturelles des zoophytes. Infusoires, comprenant la physiologie et la classification de ces animaux et la manière de les étudier à l'aide du microscope,* Librairie Encyclopédique de Roret, Paris, 1841.

Grassi, B.: *Die Malaria,* Gustav Fisher, Jena, 1901.

Hooke, R.: *Micrographia, or Some Physiological Descriptions of Minute Bodies Made by Magnifying Glasses with Observations and Inquiries thereupon,* 1665, Dover Publications, Inc., New York, 1961 (republication).

Kruif, Paul de.: *Microbe Hunters,* Pocket Books, Inc., New York, 1940 (republication).

Kudo, R. R.: *Protozoology,* 4th ed., Charles C Thomas, Publisher, Springfield, Ill., 1954.

Manson-Bahr, Sir Philip: "The Story of Malaria: The Drama and Actors," *International Review of Tropical Medicine;* vol. 2, Academic Press, Inc., New York and London, 1963.

Ross, R.: *Memoirs, with a full account of the great malaria problem and its solution,* E. P. Dutton & Co., Inc., New York, 1923.

Smith, T.: "Reproduction of Some of the Papers of Theobald Smith together with Biographical and Bibliographical Information," *Med. Classics,* **1** (5): 341–669, 1937.

Additional Reference

Foster, W. D.: *A History of Parasitology,* E. and S. Livingstone, Ltd., Edinburgh, 1965.

Chemotherapy

11 Chemotherapy is the treatment of infectious diseases with chemical substances. Its development has followed three major paths, which commingled into spectacular therapeutic successes. When the men of the Middle Ages left the shores of Europe for the long exploratory trips of the Renaissance, they discovered not only lands of spices and gold but also new medicaments such as cinchona bark, the source of quinine, and ipecac root, the source of emetine. The exploitation of natural products was thus the first of the roads that led to modern chemotherapy. The second approach was the search for effective therapeutic chemicals that could be synthesized in the laboratory. The last path, and maybe the most fruitful, was the exploitation of the synthetic powers of microorganisms.

The basic phenomenon that makes chemotherapy possible is that certain chemicals are selectively more toxic for pathogenic microorganisms than for the cells of the infected host. The full understanding of this selective toxicity requires a basic knowledge of the organization and function of the various forms of life that we are far from having acquired.

Natural Products in Chemotherapy

The first sources of effective chemotherapeutic agents were the plants of nature. The first physicians to use them were medicine men who dispensed, together with incantations and amulets, some of the drugs that are still used in a more purified form today. Among these one can include coca, emetine, curare, ephedrine, and reserpine. The therapeutic effect, if any, was accomplished not only by the power of the drug but also by the psychotherapeutic effect of the magico-religious methods of these first healers. Such methods of treatment are not unique to one country or region and must have been practiced in similar forms at equivalent stages of human civilization all over the world.

The successor to Aristotle as the head of the Lyceum was Theophrastus, who died some 285 years before Christ. He wrote on a number of subjects, some of which were quite literary. His collection of *Characters* gave an idea of the ethical environment of his time and made him a forerunner of La Bruyère. He is most famous, however, for his writings on plants, and these have earned him the title of "father of botany." The ninth book of his *History of Plants* discusses their medicinal properties. In this domain, Theophrastus acknowledged the oriental sources of most such remedies. Among other observations reported in this book was the use of fern rootstock against intestinal worms.

The herbal of antiquity reached the height of perfection with Dioscorides's *Materia Medica*. Pedanios Dioscorides was a Roman army surgeon who lived during the first century after Christ. He took advantage of his many travels in Asia Minor, Italy, Gaul, Spain, and Greece to collect information about plants and their medicinal utilization. He was not a Roman, but, like Galen who lived a century later, a Greek from Asia Minor. Galen was so full of praises for the work of his predecessor that Dioscorides's book became the standard text for centuries to come. Dioscorides was familiar with the soporific effect of opium and mandragora. He had an understanding of anesthesia, a word that he even used.

With the coming of the Middle Ages, all this was not forgotten. Rare manuscripts were copied in monasteries as medicinal plants bloomed in gardens attended by monk pharmacists who might even have been monk physicians. Yet there must have been some conflict of interest in treating both the body and the soul since in 1131 the Council of Rheims forbade clerics to practice medicine. Fortunately, however, the medicinal garden remained.

With the exploratory trips of the Renaissance the knowledge of medicinal plants was enriched by the American and Oceanian plants. The most spectacular of the chemotherapeutic products introduced from the new countries was the bark of *Cinchona* trees, already discussed in Chapter 1.

The medieval medicinal-plant garden was revived in more recent times by another group who favored a monastical society, the Shakers. Fleeing prosecution in England, these believers in the second coming of Christ established in 1774 their first settlement in the United States, near Albany,

New York. They believed in equal rights and duties for men and women but not in the reproduction of the human species. Thus, the Shaker brothers and sisters lived saintly and celibate lives. Their interest in agriculture led them first to enter the seed business and subsequently the medicinal plant business around the turn of the nineteenth century "for the supply and convenience of apothecaries and druggists." Here there seemed to be no conflict between treating both body and soul, but both their spiritual and herbal remedies fell in disfavor before the end of the century.

The modern physician believes in the curative capabilities of certain plants as much as the physician of the past, but he now often has the purified active principles at his disposal. This has the obvious advantage of permitting the doctor to prescribe the proper dosages. The study of purified products has allowed assaying of their specific therapeutic activity and toxicity. It has also permitted the study of their chemistry, allowing the preparation of improved derivatives.

The isolation of morphine from opium by Friedrich Sertürner, in 1806, opened up a new chapter of pharmacy and medicine. For many centuries, opium had dispensed relief from suffering, but the active principle was unknown and the dosage uncertain, so that the results of administering this medication were, at times, disastrous. Sertürner, a young German clerk working in a Hanoverian *Apotheke,* was twenty-three years old when he announced the isolation of morphine. The word morphine was derived from the name of the Greek god of dreams, Morpheus. Sertürner had not hesitated to test the effect of morphine on himself and on some science-loving friends of his. He had noted that it was alkaline in nature and that it formed salts both with inorganic and organic acids. Later, the name alkaloid was proposed for such alkali-like substances. In spite of his outstanding discovery, which was important enough to stimulate jealous reactions throughout the scientific world, as well as honors and recognition, Sertürner died disillusioned in 1841.

At about this same time in France Pierre-Joseph Pelletier and J–B. Caventou dedicated their little research laboratory in the back of a Parisian drugstore to the isolation of natural products. They studied in order, as sources of interesting materials, the ipecac root, *nux vomica,* and the cinchona bark. From the first source they isolated emetine (1817), from the second, strychnine and brucine (1818), and from the last, quinine (1820).

Emetine, in the form of the ipecac root, was used by the natives of Brazil for the treatment of dysentery before the discovery of America. It was introduced into European medicine during the middle of the seventeenth century, but its killing action against amebae was demonstrated only about a century after its isolation. It is used in the treatment of acute amebiasis, amebic hepatitis, and amebic abscesses, but it does not eliminate amebae in chronic carriers.

In September, 1820, Pelletier and Caventou were then thirty-two and twenty-five years old, respectively. They wrote a long paper entitled "Chemical Investigations on Cinchonas," and Pelletier read it at two different sessions of the Academy of Sciences. The text has been reproduced in the *Annales de chimie et de physique* (**15**: 289–318; 337–365, 1820).

Cinchonas are at the head of the list of materials periodically subjected to new investigations. It would be difficult indeed to list all the investigations that have been carried out on these barks. . . . We take only the liberty of recalling the great dissertation of Fourcroy which has long remained a model for plant analysis. Let us also note the examination of 18 species of cinchona which was started by M. Vauquelin. This led to a remarkable memoir in which M. Vauquelin has enriched plant chemistry with a new acid and a reliable procedure for the identification of truly febrifuge cinchonas. . . . The investigations of M. Gomès, from Lisbon, also require careful consideration. . . .

The discovery of plant alkalis is a milestone of science. It explains a number of abnormalities that were encountered in the analyses of plants. Since salifiable organic bases have unique properties and since they are so often the active principle of plants containing them, it was only logical to look for them in the cinchonas. Cinchonine was discovered by M. Gomès in grey cinchona. He believes that it is the active principle of the bark and that it is neither an acid nor a base. It was necessary to make certain that there had been no error in the last stages of his preparation. . . . Once our work on cinchonas was started, we carried it on much farther than expected. The properties of cinchonine are so closely linked to those of the other substances which accompany it, that we have not been able to refrain from investigating them all. . . .

Since we have studied various types of cinchonas, we have divided our text according to the species studied. . . .

We have treated 2 kilograms of ground *grey cinchona* with 6 kilograms of strong hot alcohol. We have repeated this operation four times. The alcoholic extracts were combined and distilled in order to remove all the alcohol. We were careful to add 2 kilograms of distilled water in order that the substance dissolved in alcohol would be protected from heat as the alcohol was removed. The substance that remained on a filter that let the water solution pass, was reddish and resinous in appearance. It was washed on the filter itself with water which had been alkalinized with potassium hydroxide. . . . After a few days of such washing, the alkaline wash became limpid and colorless. The substance that remained on the filter after washing with a considerable amount of distilled water, was greenish-white, very fusible, soluble in alcohol and yielded crystals. It was the cinchonine of Doctor Gomès. In this state it had some resinous properties, but, if it was dissolved in a very dilute aqueous acid solution, it left behind a considerable amount of a greenish fatty substance. . . .

The acid solution was golden yellow. Upon evaporation it gave crystals that were soluble in alcohol and in water. . . . Their solution was treated with very pure magnesia and heated slightly. After cooling, the mixture was filtered and the magnesic precipitate was washed with water. The first washings were yellow, but they eventually became colorless. The magnesic precipitate, having been properly washed and dried, was treated three times with 40° alcohol. The alcoholic solutions were very bitter and slightly yellowish. Upon evaporation, dirty white, needle-shaped crystals were formed. These crystals were redissolved in alcohol and recrystallized. The crystalline matter was very white and shiny. . . . These crystals were very pure cinchonine. . . .

Sulphuric acid unites with cinchonine to form a very soluble neutral salt which can be easily crystallized. . . . We have subjected this salt to the analysis that we have described in our memoir on strychnine. . . . According to this analysis we have calculated the molecular weight of cinchonine to be 38.488, the molecular weight of oxygen being taken as one. We will recall that M. Thomson has calculated the molecular weights of the known plant alkalis and has found the following numbers: Morphine 40.250, Picrotoxine 45.000, Strychnine 47.625, Brucine 51.500.

After having described the properties of a number of salts of cinchonine, Pelletier and Caventou turned to the consideration of all the fractions that they had been able to recognize in grey cinchona. They listed these as follows: "(1) A green fatty substance, (2) Cinchonine, (3) Kinic acid (described by Vauquelin), (4) A small amount of gum, (5) A red soluble dye (tannin), (6) A red insoluble dye (cinchonic red), (7) A yellow dye."

Pelletier and Caventou described the chemical properties of these various fractions and concluded their work on grey cinchona with the description of a new method of extraction for cinchonine that consisted of: ". . . extracting the alcoholic extract of grey cinchona with very dilute aqueous hydro-

chloric acid. The acid dissolved the cinchonine and left behind cinchonic red and the fatty substance. . . ." Further purification was essentially carried out as it had been previously.

The authors then passed to the following:

Chemical examination of yellow cinchona (Cinchona cordifolia)

We thought that we should first assure ourselves that cinchonine was present in yellow cinchona. . . . Thus we have prepared extracts from yellow cinchona which, treated with potassium hydroxide, left behind a yellowish substance that was in great part dissolved in diluted aqueous hydrochloric acid, and the fatty substance that was left behind, differed only by its yellow color from that obtained from grey cinchona. The acid solution was yellow colored and had a bitter taste. It was very similar to an equivalent solution of cinchonine. Thus we added an excess of magnesia in order to remove the hydrochloric acid. The solution became less colored. The magnesic precipitate was washed, dried and treated with alcohol.

The alcoholic solutions were first distilled, then left to evaporate slowly. At this point we expected a beautiful crystallization of cinchonine, but to our great surprise we obtained only a transparent yellowish substance which was not at all crystalline!

This substance should have been, according to us, cinchonine mixed with some foreign substance particular to yellow cinchona. We then attempted to separate this presumed foreign substance from the so-called cinchonine. . . . Since we knew that ether was a poor solvent for cinchonine, we used it in the hope of removing the foreign substance. But all of our substance dissolved readily in ether, and even by slowly evaporating the ether we were not able to obtain any sign of crystallization. Finally we dissolved our substance in acetic acid. Upon addition of ammonium oxalate, a bright white precipitate was instantly formed that could have been mistaken for calcium oxalate if it had not been soluble in alcohol. This precipitate, treated with magnesia and then by alcohol, furnished again a non-crystalline substance. Finally, very remarkable! this substance, thus treated, was soluble in all acids . . . forming very white salts which were easier to crystallize than cinchonine salts from which they also differed by their shape and their appearance. It is thus that this turn of events led us to consider the bitter substance from yellow cinchona to be a particular salifiable base differing from cinchonine. . . . We have been completely convinced of the value of this distinction by the presence of these two substances in some species of cinchona. They can be separated from each other in such natural mixtures. . . . We thought that we should name the bitter substance from yellow cinchona *quinine*, in order to differentiate it from cinchonine by a name which is also an indication of its origin. . . .

Quinine never crystallizes. Dehydrated, and completely deprived of humidity, it appears as a dirty white porous mass. It is only very sparingly soluble in water. . . . It is very bitter. One cannot deny that it has a certain affinity for water, since if one evaporates a solution of quinine in non-absolute alcohol, it strongly retains water. The result is a kind of a transparent hydrate fusible at 90°. . . .

Alcohol dissolves quinine very easily. It is much more soluble in sulfuric ether than cinchonine. It is also soluble to a limited extent in fixed and volatile oils.

Exposed to air, quinine is not altered. . . . It is decomposed by the action of heat, and like cinchonine, it gives off the products of non-nitrogenous plant substances. . . .

Quinine restores the blue color of litmus which has been reddened by an acid. It unites with acids and forms salts that are in general less soluble and more easily crystallizable than those of cinchonine. . . .

Pelletier and Caventou then proceeded with a review of the properties of the various salts of quinine. Using the method they had previously employed for cinchonine, they came to the conclusion that quinine had a molecular weight of 45.906.

After having extracted and examined quinine, we have analyzed yellow cinchona essentially as we had done with grey cinchona. We obtained: Kinate of quinine, cinchonic red, soluble red dye (tannin), fatty substance, kinate of calcium, yellow dye. . . .

The authors analyzed one more species of cinchona, red cinchona (*Cinchona oblongifolia*), in which they recognized the following chemical principles: *Kinate of cinchonine, kinate of quinine, kinate of calcium, cinchonic red, soluble red substance (tannin), fatty matter, yellow dye.* . . .

Pelletier and Caventou had found one species of these therapeutic trees that contained together with the cinchonine of Gomès the new, bitter base, quinine. They concluded their presentation with the following discussion:

What is the active principle of cinchonas? What is there in these barks that is effective in the treatment of fevers and that combats intermittent fevers so energetically? Perhaps we should not attempt to answer this question. However, we are convinced that the active principle is the salifiable base, cinchonine in grey cinchona, quinine in yellow cinchona and both in red cinchona. . . . We feel that we should present the reasons for our conclusion.

One differentiates the cinchonas of good quality from inactive barks not only by their exterior appearance but also by the combination of several physical and chemical properties. One knows that the good cinchonas have a bitter taste, a styptic quality and a particular aroma that cannot be confused with those of other indigenous or exotic barks. It should be noted, if we take grey cinchona as an example, that among all the products it contains, only cinchonine has this bitter taste. Once the cinchonine has been extracted, it is almost tasteless. . . .

We are not yet ready to maintain that one should no longer use natural cinchona. . . . But when there is an active principle in a drug, it seems to us that it should be purified and its properties should be established. . . .

We hope that some able and cautious practitioner will investigate the therapeutic action of the alkalis of cinchona, and in so doing will demonstrate the medical utility of our work.

Pelletier and Caventou exploited their discovery by the preparation of quinine on an industrial scale. Despite the financial potential of their discovery, they published their method of preparation in order to encourage the free competition that would extend the availability of quinine. As early as 1822, quinine was manufactured in the United States by a Philadelphia firm. Pelletier and Caventou remained active in pharmaceutical research, but the isolation of quinine was their masterpiece.

The Peruvian forests supplied the world with the therapeutic bark until an Englishman named Ledger exported some of the seeds to England. Eventually, the seeds made their way to Java, where the enterprising Dutch eventually became the world's major suppliers by 1933.

Quinine was expensive, acted slowly, and was not free of toxicity, and relapses were numerous. All these factors, to which one could add the Japanese invasion of Java in 1942, acted as incentives in the search for substitutes that could be synthesized. And indeed, numerous substitutes were discovered by the chemists. But the final downfall of quinine has not been complete; as a matter of fact it is now more popular than ever, as it is harmoniously combined in alcoholic and carbonated beverages under the optimistic designation of "tonic."

Synthetic Chemicals in Chemotherapy

". . . We know of a number of infectious diseases, especially those which are caused by protozoa, where serum therapy either does not work at all, or only with much loss of time. . . . In these cases . . . chemotherapy must be used. . . ." These lines were written by a cigar-smoking, gesticulating, and muttering genius who had an addiction for mystery stories and mineral water. His name was Paul Ehrlich. He was born in 1854 into a well-to-do family of innkeepers in Upper Silesia.

Paul was the first and only son after a long succession of daughters and was raised with a freedom that favored his scientific turn of mind. The bent for chemotherapy started early; as a boy he had the town pharmacist prepare cough drops according to his own formula. The family servants, like the parents, were inclined to indulgence even though their patience was at times taxed by the occasional invasion of the laundry room by scores of frogs, salamanders, and snakes that Paul had carefully collected.

During his schooling, Paul Ehrlich was very brilliant in certain subjects, such as Latin, and hopelessly without talent in others, such as German composition. His overall intelligence and excellent performance in selected subjects led the examiners to blink indulgently at his shortcomings. Thus as a schoolboy, Ehrlich had already given indications of some of the personality traits he would have as an adult. During the rest of his life his love of Latin would manifest itself by the abundance of Latin clauses with which he would lard his German. His dislike for German composition persisted, and he would write letters, papers, and reports only with the utmost reluctance, often after having overshot the deadlines for the completion of manuscripts. His liking for Latin might also explain why French was the only foreign tongue that he mastered.

Paul Ehrlich was interested in medicine, and he started his studies in this field at Breslau. Not finding there the realization of his high scientific goals, he left Breslau for Strasbourg, where he attracted the attention of certain members of the faculty by his interest in histology. He would pass hours staining tissues and performing more than the required exercises. He was rewarded by the discovery of the "mast cells," a term which he himself proposed in his doctoral dissertation defended in 1878. Always restless, he left the University at Strasbourg, to graduate from Leipzig, after having returned first to the academic halls of Breslau and having passed through the University of Freiburg.

Paul Ehrlich's doctoral thesis was entitled "Contributions to the Theory and Practice of Histological Staining." In this thesis lay the whole seed of his life work. He had been struck by the affinity dyes have for specific cells or tissues due to chemical substances present there. Similarly, at a later date, Ehrlich was to be intrigued by the affinity of antitoxins for toxins. Therapeutically, the active components of immune sera were "magic bullets" that would seek their target in the chemical labyrinth that is the animal body. Not so specific, but still a directed chemical missile, was the synthetic chemotherapeutic agent Salvarsan with which Ehrlich climaxed his scintillating career.

Back in 1878, Ehrlich had passed the state medical examination and had been appointed first as assistant and then as senior house physician in Professor von Frerichs's service at the Charité Hospital in Berlin. Ehrlich's laboratory was not very impressive. As he described it later: "One went to the so-called laboratory through a dark entrance, stumbling over brooms, water pails and boards. . . . There was a long laboratory bench with a Bunsen burner and a water-tap . . . and that was the entire equipment." Nevertheless, this did not interfere with his research, and all his life he showed a marvelous ability for performing outstanding experiments with the very minimum of laboratory paraphernalia.

A large part of his slim salary went to cigars that he smoked continuously. Good, strong cigars seemed a necessary stimulus for his brilliant mind. One might even wonder what would have happened to his intellect had he lived before the introduction of tobacco from America. For years, Ehrlich was to walk around with a box of twenty-five cigars under his arm in order not to run out of the precious stogies at a critical moment of mental activity.

At the Charité Hospital Ehrlich was noted for the quality of his laboratory techniques, techniques that many a famous man came to admire, and for his gentle dexterity at collecting the specimens necessary to diagnose and analyze the course of the patients' diseases. Professor Frerichs was a sympathetic chief who, having recognized the potential of Ehrlich, did all in his power to let him work in peace at his dyes and his chemical reactions.

Ehrlich was a clinician by training, but he had a special gift for chemistry. He could visualize formulae and their interactions with a facility that bordered on instinct. When Frerichs died, Ehrlich realized that he had lost both a friend and a position, since the new chief of the service had no appreciation for his work. To complicate matters further, Ehrlich, inspired by Koch's work on tuberculosis, had applied his skill with stains to perfecting the methods of detecting tubercle bacilli. Possibly as a result of handling this infectious material he became tuberculous. In 1887 he left Germany for Egypt, where, fortunately, he completely recovered his health in two years. Just prior to this he had married Hedwig Pinkus, the daughter of an industrialist of Upper Silesia, so that his trip was like a protracted honeymoon. When Ehrlich returned to Berlin, he established a small private laboratory of his own. At this point Robert Koch offered Ehrlich more suitable quarters at the Institute that he was then directing. There Ehrlich met Behring and helped him with the development of effective therapeutic sera. Ehrlich, as we have seen in Chapter 6, was left under the impression that he had been exploited. Years later, Von Behring, by then an old man, went to Ehrlich's funeral and said: ". . . If we have hurt you . . . forgive us!" The enthusiastic Paul Ehrlich had long since forgiven his former friend for the monetary wrong that might have been done him, but he probably never forgave him for what he considered a betrayal of his friendship.

In 1896, Dr. Althoff, the director of the Prussian Ministry of Education and Medical Affairs, asked Ehrlich to become the director of the Institute for the Investigation and Control of Sera that had been opened in Steglitz, a suburb of Berlin. Ehrlich was the logical man for the job since he knew more about the assay and control of sera than any other man of his time. He accepted this position. Despite the imposing title, the Institute was housed in a former bakery, and an old stable served as animal quarters. Ehrlich moved in with his cigars, his dog, and his feverish good humor. The state budget for research was slim, but the new director had the ability of doing much with modest facilities. As he said: "As long as I have a water tap, a flame and some blotting paper, I can work just as well in a barn." But Althoff was not to leave Ehrlich in these quarters long. Taking advantage of the fact that the Lord Mayor of Frankfurt am Main was interested in developing scientific research in that city, the Institute was transferred in 1899 to suitable new quarters in Frankfurt. A second Institute was built next to the Serum Institute. It was called the Georg Speyer-House. It was privately endowed and was dedicated entirely to chemotherapy. When its doors were opened in 1906, Ehrlich became director of two institutes. His prodigious ability for work, matched by his dedication to the welfare of his fellow men, permitted him to direct all the activities of the various laboratories with great success.

The vivid image of Paul Ehrlich, the director of one, then of two institutes, has been preserved through the devoted writings of Martha Marquardt, his former secretary. She depicts him as a kind man who could boil into a rage when he thought that one of his important directives had been ignored. He was a man lost in his scientific thought, who paid little attention to his clothes, which were chronically covered with cigar ashes. He wore large shirt-sleeve cuffs that he would cover with illegible notes when he had nothing better at hand. The notes might even spill over to the rest of the shirt when the cuffs were filled. Cuffs and shirts having only a limited surface, and being hard to distribute, Ehrlich's note writing was mainly confined to bundles of multicolored cards on which he would daily write instructions for his assistants. He might have had lapses of memory on most of the pedestrian aspects of day-to-day living, but he would never forget the directions that he would have written on any of his sacred cards.

Like one of his favorite heroes, Sherlock Holmes, Ehrlich did not believe in encyclopedic knowledge. He was interested in the disciplines that could contribute to the advancement of his line of work. He considered harmful the cluttering of his mind with information that would be irrelevant. His liking for music was limited to some of the less intellectual expressions of the art, such as popular songs and opera tunes. He enjoyed the sound coming from street organs. His appreciation of painting and literature was on par with his appreciation of music. Asked once why he worked so hard, he answered: "One simply has to, one is urged by some force from within." Few forces in nature have produced so much for the treatment of diseases.

The following is a translation of part of Ehrlich's speech as the Georg Speyer-House was dedicated.[1] Ehrlich explains his work:

> Since, throughout my whole life, my thoughts have been most intimately interwoven with the same fundamental ideas as are now due to give a lead to the enterprises of the Speyer-Haus, I ask for your indulgence, and hope that you will not impute it to me as lack of modesty, if now, somewhat unconventionally, I retrace some of my personal memories and unfold to you the story of the origin of my own ideas. I would not have done this, had I not thought that such a developmental presentation would be the simplest way of making clear to you the nature of the problems which give to this institute its special character.
>
> It was some 33 years ago—I was then still quite a young student— when I came across a publication on lead poisoning, by Heubel. In order to elucidate the nature of this poisoning, the author had estimated quantitatively the lead content of the liver, the kidney, and the heart, and had discovered that there were remarkable differences in the

[1] Reproduced by permission of Pergamon Press, Inc., publishers of *The Collected Papers of Paul Ehrlich.*

amount of lead to be found in the various organs. When he immersed organs of normal animals in dilute lead-solutions and subsequently subjected the organs to chemical analysis, he believed that he obtained exactly the same differences. This experiment seemed to me, at that time, a revelation. The possibility emerged that this technique might be used also to ascertain the sites of action of poisons. That lead was found to be present in certain organs, e.g. the brain, provided merely the starting point for an investigation. The brain is a large structure and is made up of many constituents—cells, fibres, etc. The real problem was to determine in which of these cells the poison was stored. The immediate effect of this idea became almost a disaster for me, since it disrupted, more than a little, the normal course of my studies, without bringing me any nearer to the desired goal. I had nothing but failure from any of my attempts to detect, with the aid of the microscope, the presence of metals applied in high dilutions, and I was not a step further forward.

It became necessary, therefore, to approach the task from a more general standpoint and, first of all, to obtain some insight into the manner and the method of the distribution of substances within the body and its cells. When we see that certain poisons, e.g. strychnine, produce spasms which originate from the nerve cells of the spinal cord, and when we see that the American arrow-poison, curare, causes a paralysis of the extreme nerve-endings, which extend to the muscle, the probability becomes clear that these effects can be caused only through strychnine making a direct connection with the cells of the spinal cord and curare with the ultimate finest nerve-endings. Conclusions such as these seem at once to be self-evident—to be, as I might express it, part of man's inborn inheritance. They can be traced back into antiquity and have assumed importance in several of the by-paths of medicine. They appear quite clearly, for example, in a statement of a mediaeval physician, who thought that drugs must possess spicules, by the aid of which they are able to anchor themselves in the various organs. But these ideas, like so many axioms, were as easy to express as they were difficult to prove; and this may well be the reason why they were completely ignored in the practical study of drugs and, despite their fundamental importance, have played no role in the development of pharmacology. If one had wished to put the "storage-axiom" to the test, one would have had to demonstrate, by the use of the microscope, that the poison under consideration was, in fact, present at these minute sites. This, however, was found to be impracticable.

It was, therefore, necessary to begin the investigation by an entirely different line of approach, and to utilize substances which, like dyes, are easily detected, even by the naked eye; it would then suffice to remove and examine a small piece of an organ of an animal after it had been killed. A glance through the microscope would then give evidence immediately, and in the finest detail, concerning the distribution of the substance.

The fact that a large number of dyes exists which differ widely in their constitution, and the fact, too, that some of these dyes exhibit a high degree of toxicity, made such endeavours all the more feasible, and thus the method originated which, nowadays, is known briefly as "vital staining." I do not intend to dwell on the wealth of results which this technique has yielded, especially in microscopic anatomy, but will just refer to the fact that the various dyes show quite characteristic differences in their distribution and localization. Thus, for example, methylene blue causes a really wonderful staining of the peripheral nervous system.

If a small quantity of methylene blue is injected into a frog, and a small piece of the tongue is excised and examined, one sees the finest twigs of the nerves beautifully stained, a magnificent dark blue, against a colourless background.

With many vital stains it is therefore extremely easy, almost at a glance, to ascertain their distribution in the different parts of the body —which parts they favour, which organs they avoid.

Of course, staining of the dead organs and tissues has for a long time been one of the most important tools of histological research. But staining of this kind can only give information concerning the purely anatomical structure of the tissues. If, however, one wishes to acquire an understanding of the properties and functions of the living cell, then the staining reaction must be made to take place in the body itself, i.e. one must stain the living substance. In this way one can gain an insight into the relationship between the individual tissues and certain dye-stuffs. I have denominated this affinity of the stains and other foreign substances by adjectives with the ending "tropic," and, for example, describe a dye which stains only a single specific tissue as "monotropic," and speak thus of "neurotropic" and "myotropic" substances, etc., while substances which have the capacity to stain several tissues should be called "polytropic." In 1886, in my early study "On the methylene-blue reaction of the living nerve-tissue," I had already indicated the lines along which a further analysis of the process should proceed. Two questions had first of all to be answered:

1. Why does methylene blue stain nerves?
2. Why are nerves stained by methylene blue?

As to the first question, the answer, by virtue of the nature of the problem, had, of necessity, to be in terms of pure chemistry; and I was able to prove that the nerve-staining property of methylene blue is conditioned by the presence of sulphur in the methylene-blue molecule. Synthetic chemistry has, in fact, given us a dye which, apart from the absence of sulphur, corresponds exactly in its chemical constitution to methylene blue. This is Bindschedler's green. With the absence of the sulphur, there is associated the inability to stain living nerves. The interest of the second question was heightened by the circumstance that, in higher animals, not all the nerve endings are stained by methylene blue. I have shown it to be probable that these differences between the individual nerve endings are not due to different degrees of avidity for methylene blue, but rather to certain associated environmental conditions; for bluing of the nerves is intimately associated with the degree of oxygen saturation, inasmuch as it is precisely at those places which are best supplied with oxygen that staining of the nerve endings by methylene blue also occurs. Further, one can easily ascertain that the nerve fibres that stain have also an alkaline reaction; and thus oxygen saturation and alkaline reaction provide the conditions which make possible the staining of nerve endings by methylene blue. Just as methylene blue accumulates only in alkaline fibres, so one must suppose, in the light of the investigations of Lieberkühn and Edinger, that certain other stains, such as alizarin blue, would stain the acid regions. One is, therefore, compelled to differentiate between alkaline, acid and also neutral fibres, and it is evidently to such differences, in conjunction with the degree of oxygen saturation, that the determinant role must be ascribed, in regulating the distribution and action of injected substances among the particular regions of the nervous system. It seemed, moreover, according to my earlier investigations on the methylene-blue staining of the nerve fibres, that an irreversible combination very soon takes place between dye and certain constituents of the nerve substance, since one can see that intensely blue granules appear in the axis cylinder, a phenomenon which may well be intimately associated with the "fibrillic acid" recently described by Bethe.

Very special conditions govern the uptake of dyes by the brain. The early anatomists, themselves, noticed that, even in the most severe jaundice, the brain remains snow-white, while all the other tissues are of a deep orange tint. I obtained this same effect on introducing into animals a large number of the synthetic dyes, all of which, however, were alike in containing an acid group, such as the sulphonic acid radicle. In contrast to this, a large number of the basic dyes which, like the alkaloids, form salts with acids, stain the brain very effectively. I have assumed that the reason for this behaviour is that the alkalinity of the blood plays a decisive role. On this basis, it is now possible to look at all these phenomena from a common point of view. The difference in the behaviour of acid and alkaline dyes may accordingly be ascribed to the fact that the former are firmly bound in the blood in the form of salts, whereas the latter remain free. Thus, the brain plays the same role as does the ether in the method of recognizing poisons devised by Stas-Otto; this method, as is well known, depends on the fact that basic substances, such as alkaloids, are bound in acid solution and are, on that account, difficult to extract, whereas they can easily be shaken out with ether from an alkaline solution. I was, in fact, able to confirm this idea experimentally; for if one introduces acid groups, e.g. the sulphonic acid radicle, into neurotropic stains, then the neurotropic property of the resulting derivatives is immediately abolished. We can similarly explain the fact that toxic substances are so often weakened in their toxic properties by the introduction of a sulphonic acid radicle into the molecule. The accumulation of the toxic agent in the central nervous system is made impossible, merely by the introduction of this sulphonic acid radicle.

The analogy between the roles played, respectively, by the brain and by ether, in the storage of dyes in the one case and in the method of identifying poisons in the other, is further emphasized by the fact, discovered by me, that neurotropic and lipotropic properties, as a rule, go together, i.e. that those dyes which are taken up by the brain are also deposited in adipose tissue. This similarity in the behaviour of the adipose and nervous tissues finds a simple explanation in the fact that the brain contains an abundance of fat-like substances—myelin and lecithin. Fat and brain, therefore, behave in the body exactly as ether does in the extraction of alkaloids. This theory, moreover, many years after the publications dealing with my work on it, was taken up again by Hans Meyer and Overton, and, as the "lipoid-theory," it plays to-day an important role in medicine.

These observations, based on my work with dyes, I developed also, during the years 1886 and 1887, in relation to a series of remedies. I demonstrated that thallin, like a large number of dyestuffs, is lipotropic In order to detect thallin in the tissues, I made use of the property that, even in very dilute solutions, it is transformed by oxidizing agents (ferric chloride) into a dark-green dyestuff. If this is immediately fixed, *in statu nascendi*, in an insoluble form, it is then easy to obtain an insight into the distribution of the thallin. The pronounced affinity of thallin for adipose tissue causes the thallin to be held for a long time in the body; and this long retention of so readily oxidizable a substance is explained by my discovery that there is no free oxygen in adipose tissue, but that this, on the contrary, has a maximal reducing power. All these observations led me to the view that the hitherto dualistic approach to the problem of the connection between chemical constitution and pharmacological action has been much too narrowly conceived, and that there exists another and, indeed, decisive element which must be considered—the distribution in the body. My investigations have shown that the distribution, i.e. the selective affinity for certain organs and systems, is a function of the chemical constitution.

But the establishment of the existence of this relationship between constitution, distribution and action fulfilled only one part of the programme which I had planned for myself. For an action to be produced on an organ, the first requirement is the fixation of chemical substances; but the simple storage of substances of any kind is not of itself enough to cause specific toxic effects. This requires a second determinant factor to be present in the chemical substance. With the alkaloids, which we will consider first, the conditions are very similar to those which had already been observed with the simpler dyestuffs. In these there are two different chemical constituents which are responsible for the dyeing property, a so-called chromophore group and, further, the auxochrome groups. In exactly the same way, one must postulate, when considering the alkaloids, that in the constitution of these powerfully acting substances two different factors must be distinguished:

1. a selective group which governs distribution, and
2. a pharmacophore group which evokes the specific activity.

You will allow me, perhaps, to clarify this with an example. As is well known to you, cocaine, which, in medicine, plays so important a role as a local anaesthetic, is the benzoyl derivative of an ecgoine ester. A large number of chemical homologues of cocaine can be synthesized by replacing the benzoyl radicle by those of other acids, e.g. of acetic acid or formic acid. All the substances obtained in this way are, by virtue of their chemical nature, homologues of cocaine, and all these different cocaines follow the same pattern of distribution in the body, because they belong, chemically, to the same class. But of all these substances, only one, the benzoyl-cocaine obtained from the coca plant, acts as a local anaesthetic; and from this it is immediately obvious that the benzoic acid residue is the source of the specific action, and thus, that the benzoyl radicle functions here as the anaesthesiophore group.

The science of immunology provides the most striking examples of the relation between distribution and action. It has been found in this connection that the group which is responsible for the distribution of bacterial poisons, which I have called the haptophore group, is a quite separate complex, and that the toxic action is attributable to the presence of a second group, the usually very unstable toxophore group. A still further differentiation between the factors of distribution and of activity is to be found if we take into consideration such cell-poisons (haemolysins and bacteriolysins) as are present even in a normal blood-serum, or are produced by immunization. For, in the case of these, each of the two properties is connected with a special molecule, one of which, called the amboceptor, is the carrier of the haptophore group, and the other, called the complement, that of the toxophore group. The haemolytic action of snake venom follows the same pattern. The factor contained in the snake venom represents solely the distributive component, whilst the appropriate pharmacophore group is present in lecithin. Neither snake venom nor lecithin is able, of its own accord, to destroy the blood cells. On the other hand, the product formed from them, the snake lecithide, corresponds, as was shown by Kyes, to a toxin, the haptophore group of which comes from the snake venom and the toxophore complex from the lecithin.

The concepts which have thus been developed indicate the direction which must be followed in the construction of new organotropic medicaments. It will, therefore, be one of the main tasks of the new institute to persevere along this path; and this entails, in the first instance, the discovery of substances and chemical groups which have an affinity for particular organs. The organotropic substances must then be furnished with pharmacophore groups which will bring about a therapeutic and pharmacological activity. We intend, as it were, to use certain chemical complexes as vehicles to carry appropriate pharmacophore groups to the desired types of cell. To begin with, however, the main emphasis will be put on the haptophore group, the distributive factor. For this represents the *conditio sine qua non* for any therapeutic action.

But this theoretical problem represents only one aspect of our objective; on the other hand, our main effort will be directed to the discovery of new, rational, curative remedies. As you are aware, the study of medicaments, or pharmacology, is a long-established field of research, which has been cultivated and developed in numerous institutes. But most pharmacological research is directed merely to determining the effects, on animals in good health, of the substances which are used as remedies, and to fixing the limits for their safe administration in the clinic, by observations on their side-effects and their toxicity. Prior attention is, therefore, given to those substances (especially alkaloids) which produce interesting and important toxic effects. These highly toxic substances, however (if one excludes a few. alkaloids such as morphine, cocaine, atropine, etc.), are frequently quite useless in clinical practice, and thus a large part of the research work is carried out in realms which are remote from any practical application in medicine. Admittedly, one obtains information about the risks which excessive dosage and otherwise unsuitable methods of administering particular substances involve, and this, of course, is of the greatest importance. Application of drugs in practice must be based on toxicological examination, and it need not be emphasized that toxicology, as such, represents a thoroughly justified and necessary type of science, which may even be of the greatest importance in biology and physiology—the very foundation of our medical knowledge. But it seems to me that, with the line of approach so predominantly in this one direction, some of the most important tasks of pharmacology are pushed all too readily aside, and practical medicine does not receive sufficient benefit. We must certainly be grateful to those who have safeguarded our departure [on the voyage of discovery] with beacons against toxic action—but they have not charted our course into the open sea of curative medicine.

The scientific study, even of the medicaments on which we have learned most to rely (e.g. potassium iodide, mercury), has for long not been practised with the required degree of activity. One learns, of course, how toxic the substances in question are, and of what the animals die—but the question of chief importance, how it is that a particular substance can cure a particular illness, remains shrouded in darkness, and must remain so while the present-day methods of investigation prevail. If we observe that preparations of iodine or mercury, for example, in doses which are practically harmless to the organism and its cells, cure diseases of known aetiology in a specific way, then the simplest reflection will show that this *specific*, curative effect can only be scientifically analyzed if one can infect animals with the diseases in question, and then make experiments on them. Investigations on the *normal* animal—as used almost exclusively in pharmacology—are devoid of value for the solution of this preeminently important problem.

Thus it is also intelligible that we are indebted to empiricism for the best remedies which we possess—quinine, mercury, opium, digitalis. A whole host of new synthetic drugs has, furthermore, had its origin in the factories of the chemical industry, which has, indeed, devoted its best resources to the service of medicine. Anybody who has had the satisfaction of observing, at the bedside, the beneficial action and the rich variety of the available medicaments, will most gratefully acknowledge their value and efficacy, and know how to appreciate them; but he will, nevertheless, not overlook the fact that the *materia medica* with which he is working is essentially a collection of mere palliatives which, though they are qualified, indeed, to modify favourably certain sequelae and symptoms produced by the disease, will leave untouched its cause or its focus. The cause and the seat of disease are, however, the most important problems of medicine; *de causis et sedibus morborum* was, in fact, the title given by Morgagni to his famous treatise. Correspondingly, then, the motto for the therapeutics of the future will have to be *de sedibus et causis pharmacorum*.

And if today we have already been given a truly amazing insight into aetiology and pathology, thanks to the tremendous progress made in the past century, it appears today also to be the foremost task of medicine to lead therapeutics into similar paths, and to achieve remedies which are curative in the true sense of the word, as quinine is in malaria, remedies which will sterilize the infected organism and thus break the neck of the disease.

There cannot be the least doubt how this is to be achieved. A sick man is, for various reasons, very little suited to be the subject of experiments for the discovery of curative remedies. The patient comes into the picture only when the drug has been recognized by an extended series of experiments on animals. A prerequisite, therefore, for such chemotherapeutic investigations is that one should be able to reproduce specified diseases and then to undertake the therapeutic experiments on them. This method has, of itself, a general applicability, and it has already given the best evidence of its value in the study of certain metabolic disorders. I need only mention here the treatment of myxoedema and cretinism, diseases which are caused by disordered or deficient activity of the thyroid gland. The greatest triumph, however, of experimental therapeutics has been achieved in the field of the infectious diseases, and we owe it, indeed, to the rapid opening of this line of research, that modern medicine has been raised to an entirely new level. In this connection I may content myself with a reference to Koch's researches on tuberculin and the remarkable discovery of the local tuberculin reaction—but then also, above all, to Behring's discovery of the antitoxins, with its widespread and almost unpredictable extensions in the field of the infectious diseases.

From what has been said, it is easy to see why it is just in this field of the infectious diseases that immuno-therapy has yielded such splendid results. If we picture an organism as infected by a certain species of bacterium, it will obviously be easy to effect a cure if substances have been discovered which have an exclusive affinity for these bacteria and act deleteriously or lethally on these and on these alone, while, at the same time, they possess no affinity whatever for the normal constituents of the body, and cannot therefore have the least harmful, or other, effect on that body. Such substances would then be able to exert their full action exclusively on the parasite harboured within the organism, and would represent, so to speak, *magic bullets* which seek their target of their own accord. By means of an astonishing natural process such substances are, indeed, formed during the process of immunization against a variety of bacteria, substances which may be quite different in nature from one another, but which, as I have shown, are all characterized by the fact that they are very firmly anchored, thanks to their haptophore groups, by the bacterial cell and its contents. Within the framework of the postulated laws of distribution, these substances are all monotropic and, moreover, bacteriotropic, and thus aetiotropic and not organotropic.

The investigation of these substances was, and still is, the task of the Königliches Institut für experimentelle Therapie, which I direct. The projects of the new Speyer-Haus are planned on different, although parallel, lines. Here we shall still be concerned with the problem of curing organisms infected by certain parasites in such a way that the parasites are exterminated within the living organism, so that the organism is disinfected, but in this case, not by the use of protective substances produced by the organism itself through a process of immunization, but by the use of substances which have had their origin in the chemist's retort. Thus, the task of the new institute will be a *specific chemotherapy of infectious diseases*. It is easy to see that this line of approach, by its very nature, must be a much more difficult one than that of serum-therapy. Magic substances like the antibodies, which affect exclusively the harmful agent, will not be so easily found in the series of the artificially produced chemical substances. It must be regarded as in the highest degree probable that substances of this kind, foreign to the body, will be attracted also by the organs, and that, since we shall be dealing with a range of different substances, all with pronounced activities, these are not unlikely to injure the organism as a whole, or some part of it. This point of view has been specially justified when a chemical therapy has been tried on the common infections by pathogenic bacteria. Sublimate in aqueous solution, even in high dilutions, will kill anthrax bacilli. If, however, as was shown by our veteran master Koch, one injects a considerably greater quantity of sublimate into an anthrax-infected animal, its death is in no way retarded, but is, in fact, accelerated; the bacteria themselves suffer no harm whatsoever.

But even if the substances in question are free from toxicity, they may still have no curative action in the body of the host if the affinity of the tissues for these substances is greater than that of the parasites. In collaboration with Dr. Bechhold I found, among the highly brominated phenols, substances which have an extraordinarily powerful disinfectant action on certain pathogenic bacteria in the test tube, and which, at the same time, are of so low a toxicity that it is possible to inject into the animal body, without harming it, doses of which less than one-hundredth part would have been sufficient to suppress (*in vitro*) the continued development of bacteria. But, in spite of these apparently favourable conditions, these substances produced no curative effects in experimental animals, obviously because the affinity of the body tissues for the disinfectant was much greater than that of the parasites; thus, expressed in the suggested terminology, the organotropism, in consequence of its avidity and its strength, had made the bacteriotropism completely ineffective.

Fortunately, not all diseases present such difficulties, and it is an especially happy coincidence that in certain infections, in which immunization is extremely difficult to achieve and always incomplete, chemotherapy has a much better chance of success. For we know that the syphilitic infections, and the various forms of intermittent fever caused by the plasmodia of malaria, are strongly affected, or cured, by mercury and by quinine respectively. But we must not forget that, in fact, these very substances, with the greatest powers of healing, were originally discovered by the detective instinct of primitive peoples—a faculty of which the nature and the keenness can first be fully recognized when we remember that in Africa, Asia and America the indigenous races, by natural preference had introduced the use of just those plant-products, such as coffee, cocoa and tea, which are desirable luxuries on account of their contents of caffeine, or of the related theobromine. In former times, nobody could have suggested any scientific grounds for supposing that a common bond unites these different stimulants; yet the instinct of the primitive peoples had long ago recognized it. I do not think it likely, however, that new and efficacious plant-products will be easily found; we shall therefore be compelled, where nature fails us, to rely upon artifice, and to set our hopes upon the wealth of substances which chemistry can afford us.

Even the first experiment, which I made many years ago in this modern direction, gave an encouraging result. The striking affinity of methylene blue for the nervous system stimulated me to investigate in association with Dr. Leppmann, whether methylene blue has any effect on pain. The therapeutic results confirmed our expectations. Methylene blue did, indeed, markedly relieve the pain in all neuritic and rheumatic afflictions. The effect first makes itself felt several hours after the administration and increases gradually, a characteristic sequence of events which finds an explanation in the fact, also discovered by me, that, after a passage of time, an insoluble compound in the form of blue granules is produced by the interaction of the dye and certain constituent substances of the nerves; this alters the chemical nature of the nerves and thus the pain is temporarily reduced.

In my further experiments, however, I started from the supposition that dyes with a maximal tinctorial activity might also have a special affinity for parasites within the host-organism, and would thus be parasitotropic—hence aetiotropic. I chose the malaria parasites and was able, in association with Professor Guttmann, to show that methylene blue can cure malaria. It was found to be, in fact, a trustworthy curative agent for the form of the disease which is indigenous in this country, although in the treatment of the severe tropical forms it must give place to quinine, which has shown its worth throughout the centuries. A second type of disease, on which I worked more extensively, comprises the group due to the trypanosomes. The causative agents of these diseases are peculiar flagellates which are found in blood-serum. They are relatively large organisms, much larger than the red blood corpuscles. . . .

In the tropics, these parasites are responsible for disastrous effects. It was previously thought that a large number of completely different animal-diseases were concerned, as is shown by a list of over 80 names. In fact, however, only a few trypanosomiases need be differentiated— the Indian Surra, the American Mal de Caderas, Dourine, and the dreaded South African tsetse-fly disease, or Ngana. You will be able to appreciate the importance of these parasites as pathogens from the statement of Bruce that, in Africa, wide areas have been emptied of almost all mammals by the tsetse-fly disease, and also from the geographical distribution of the trypanosomiases. . . . The pathogenic significance of the trypanosomes became infinitely greater when, a few years ago, both Forbes and Dutton were able to show that the mysterious sleeping sickness is caused by *Trypanosoma gambiense*. Sleeping sickness plays a fateful role in Central Africa, especially in the Congo and in Uganda. You will appreciate the shocking nature of this pestilence from the fact that, in Berghe Sainte Marie in the Congo, the mortality from this disease rose from a figure of 13% to 73% in the period 1896– 1900. The disease continues, moreover, its steady and rapid advance.

Sleeping sickness, like other trypanosomiases, is spread by an inter-mediate host, in this case by *Glossina palpalis,* an inconspicuous little creature. . . .

You can see from all this that it is the imperative duty of anyone who is in a position to do so to combat this most destructive of all pesti-lences, which, in its relentless advance, threatens to depopulate the whole of Central Africa.

The first step must be to ascertain, by experiments on animals, whether it is possible, in any case, to cure infected animals with chemi-cal preparations. Laveran, one of our most distinguished scientists, has already shown that arsenious acid on the one hand, and human serum on the other, cause a temporary disappearance of the parasites from the blood of infected animals, but that, generally, the parasites reappear in the blood after several days, with the result that the animals, with few exceptions, finally perish. I, too, used mice, a type of animal which is very susceptible to infection with trypanosomes. If blood from a mouse which is suffering from trypanosomiasis is injected into another mouse, the presence of parasites will, according to the species under investi-gation, become demonstrable on the first or, at the latest, on the second day; the parasites then multiply with such an extraordinary rapidity, during the following days, that death occurs on the fourth or fifth day. Thus, in mice, the trypanosomes have an extremely acute action, whereas the disease in man, acquired naturally, frequently drags on for months, and even years. In association with Shiga, I tested a large number of dyes and finally found one, a dye related to the benzo-purpurin series, which had a small effect, though this amounted only to a retardation of the fatal outcome by one or two days.

Thereafter, with the aid of my esteemed friend Dr. Arthur Weinberg, who for three years has untiringly put at my disposal his whole knowl-edge and ability, and to whom not only I, but also science itself is indebted for this help, I gave the benzopurpurin dye, which I had used previously, an increased solubility by the introduction of a sulphonic acid group. It was found that the dye obtained in this way, trypan red, was able to cure, in a very large number of cases, mice which had been infected with the parasite of the South American Mal de Caderas; in other words, to disinfect the body of the living mouse completely. Further cures were achieved of mice which had been infected with the parasite of Mbori, a disease of the camel, and of mice harbouring the parasite of the horse-disease, Dourine.

The red dye failed, however, in mice infected with other types of trypanosome, e.g. *Trypanosoma brucei* and *gambiense,* and failed altogether in rats. Laveran confirmed these results and demonstrated, moreover, that certain infections can be cured by the combined use of arsenic and trypan red, a result which could not be attained with either substance alone.

Meanwhile further progress has been achieved through the discovery that Atoxyl, an arsenical preparation which was first prepared in the *Charlottenburger* Werke under the direction of Dr. Darmstädter and which is about 20 times less toxic than the arsenious acid used hitherto, is by far the best of all the arsenicals in that, even with mice infected with the trypanosome of African Ngana, against which trypan red is ineffective, at least in some of the experimental animals cures have been obtained. I have also tested, with Dr. Weinberg's support, a series of trypan red derivatives, such as hydroxyamino-trypan-red, of which some have proved to be somewhat more effective than trypan red itself.

All my findings have recently received full confirmation in a publication of the Institute Pasteur, France, by F. Mesnil and M. Nicolle. With the help of the *Bayer Farbwerke,* they have had an extremely large number of dyes prepared for them, more than several hundreds in fact, and, apart from the derivatives of trypan red discovered by Dr. Weinberg and myself, have produced some blue trypan dyes which were successful against the more refractory kind of trypanosome infection, particularly Ngana, although in only 25%–30% of the cases. The trypan blue dyes, on the other hand, had no curative action against infections by *Trypanosoma gambiense,* the cause of the sleeping sickness of man.

At the same time I was continuing and expanding my own experiments, in so many direction and with such success that, with the passage of time, I have now become able, by different methods, to cure practically every mouse which is infected with the very resistant trypanosome of Ngana. I have even succeeded in curing two Ngana-infected animals some 36 hours before their expected death, although the blood, at this time, was already swarming with millions of trypanosomes.

I think that these results should encourage us to continue, without deviation, along this path. Our highest aim must always be the conquest of sleeping sickness; but we must not close our eyes to the fact that this is an especially difficult task, since the disease is caused by a very resistant kind of trypanosome. For this reason, I did not place great hopes even on the treatment with trypan red. A few cautious trials in Uganda, by the English Sleeping-Sickness Commission, were in fact unsuccessful. Of course, at that time, the doses given to the human patients were not very large. The fundamental difficulty in the treatment of human beings is that the quantity of the drug required to destroy the trypanosomes approximates closely to an amount which is highly dangerous to life. Whereas in the case of animals the loss of one or two individuals is of no great consequence, in comparison with the advantage of curing so many others, in the treatment of man, on the contrary, the position is entirely different, and the predominant policy of the institute will have to include a wide conception of the different ways of experimental approach to this difficult problem. I should not wish, however, to give the impression that the experiments which I have been discussing have been confined exclusively to small laboratory animals. At my request, Professor Lingard, at Muktesar, in India, has investigated the efficacy of arsenious acid in combination with trypan red, using large animals which had been infected with the trypanosome of the Indian Surra, a disease which has been causing great harm to the stock of horses in that country. Arsenious acid by itself, as he had previously confirmed, has no adequate curative effect. Consequently, a preliminary dose of arsenious acid was given and then, two days later, trypan red was administered, both of them being given in large doses. As a result four horses and one donkey were completely cured, but with a camel, for which animal it had not been possible to determine the toxic dose, a cure was not obtained. In these experiments, treatment commenced only after the trypanosomes had made their appearance in the blood.

Thus did Ehrlich recount the story of the orientation of his research toward chemotherapy. The date was, as we recall, 1906. Two years later, the magnitude of Ehrlich's work in the field of immunity had so impressed the Nobel Prize Committee that the prize for medicine was split between Metchnikoff and Ehrlich. A year later, in 1909, a young Japanese scientist, Sahachiro Hata, came to work with Ehrlich at the Speyer-House. Hata was a specialist on experimental syphilis. After the discovery of the spirochete of syphilis by Schaudinn and Hoffmann in 1905, he had started to produce syphilitic infections in rabbits in his laboratory in Japan. Ehrlich and his staff of chemists, among whom one should not forget to name Dr. Alfred Bertheim, had synthesized an extensive series of organic arsenical compounds. Ehrlich wanted to test some of these compounds and turned over to Hata, as a start, substances 418 and 606. Number 418 had already shown some action on some spirochaetal infections; number 606 had been tested and had been found inactive. Hata probably did not know about the previous tests, but at any rate he was not influenced by previous results and concentrated on his own experiments that were long and tedious. When he reported that 606 was very effective, Ehrlich did not know whether to trust his new assistant or his old one. As the experiment was repeated, it became clear that Hata's technique was flawless. Ehrlich's reaction was to swear liberally at "the incompetent nincompoop" who had led him to believe that 606—to be known as Salvarsan, or arsphenamine—was inactive.

In 1910, the discovery of arsphenamine was announced to the world. Ehrlich's staff manufactured this compound on a large enough scale to permit the distribution of the drugs to many physicians for clinical evaluation. Ehrlich himself supervised the distribution of the drug. In keeping an account of the number of doses manufactured and distributed, he found his shirt cuffs were inadequate, so he scribbled the necessary data on the door of a bookcase. With the distribution of arsphenamine, Ehrlich was faced with a number of unpleasant problems. First of all, necrosis was observed at the site of intramuscular or subcutaneous injection of the drug. This was remedied by giving intravenous injections, a hazardous procedure in 1910! Some cautious clinicians were giving doses that were too small, leading to severe relapses. As an added source of worry, there were the jealousy of all those who begrudge success and the usual attacks from mentally disturbed people.

Chemical research was continuing as the clinical testing of Salvarsan was proceeding; in 1912, Ehrlich introduced Neosalvarsan, No. 914, that had the advantage over Salvarsan of being readily water soluble.

Ehrlich's dream in chemotherapy was to discover the *therapia sterilisans magna,* the drug that would cure in one single dose. Penicillin has displaced Salvarsan in the treatment of syphilis, but Ehrlich's methods established the direction that has been followed in the development of other chemotherapeutic agents.

Ehrlich's recipe for success was "Patience, ability, money and luck." As the years passed, he did not run out of any of these ingredients, but his health began to decline. The years of hard work, poorly energized by little food and much cigar smoking and mineral-water drinking, led to a rapid deterioration of his health. In March of 1914 Ehrlich's sixtieth birthday was celebrated by his many faithful friends and students. A few months later he experienced a slight stroke and then admitted that he "did not feel as fresh as usual."

The war had come "that was not going to bring any thing good," according to Ehrlich. His prediction became reality with the death of his able and devoted chemist, Bertheim. This death, in essence, was an illustration of the useless wastefulness of war. Bertheim had been summoned to join his cavalry unit. He caught his spurs in a carpet, fell down a staircase, and fractured his skull.

Ehrlich's health continued to decline, and he died in August, 1915. He was buried in the Jewish Cemetery of Frankfurt, even though he had attached little importance to religion during his lifetime.

Ehrlich's ceaseless work in science left only a small number of carefully rationed hours for his family activities. But a loving and devoted wife, fully aware of her husband's importance, helped make these hours happy. The two Ehrlich children, both daughters, expressed their confidence in the peculiar ways of science by marrying, respectively, a physician and a professor of mathematics.

Germany was an active center of research in the field of chemotherapy because the outstanding activities of Paul Ehrlich were complemented by industrial research. In essence, the German dye industry might be considered the mother of chemotherapy since it was the major supplier of stains, some of which, like trypan red, played such an important role in the development of this discipline. The German industry followed the road blazed by Ehrlich and started screening programs that led to a number of significant advances, which included the development of sulfa drugs.

However, as was the rule in all the aspects of microbiology that we have reviewed, France was never far behind Germany and vice versa. The development of chemotherapy in France was in great part due to Fourneau.

Ernest Fourneau was born in 1872, in Biarritz. As a youngster, he helped one of the pharmacists of the town who encouraged him to continue his studies in this field. After graduating from Paris in 1898, he worked first in a hospital, but finding the study of chemistry fascinating, he went from one famous chemical laboratory to another. In this fashion three years of study went by, two of which were spent in Germany, where he was for a while in the laboratory of Emil Fischer.

Fourneau recognized that France was hopelessly lagging in the development of drugs. Upon his return home in 1901 he convinced the director of the Poulenc Laboratories to support him in the development of this field. His first success was the discovery of a substance that could be used as a local anaesthetic. He called it "stovaïne," humorously utilizing the English translation of his own name.

In 1911, Fourneau left the Poulenc Laboratories, which were to become the present-day firm of Rhone-Poulenc, to found a laboratory of chemotherapy at the Pasteur Institute. The creation of this laboratory was fostered by M. Roux, who believed that the house of Pasteur should not neglect aspects of microbiology that might have escaped attention at the time of Pasteur.

Fourneau was an excellent organizer who programmed the testing of his chemicals in a very systematic fashion in order to insure their evaluation against a variety of microorganisms together with a proper pharmacological study. The Pasteur Institute with its numerous colonial branches was ideally located for the testing of new drugs, many of which were designed to cure such exotic diseases as yaws and sleeping sickness.

Fourneau and his collaborators extended Ehrlich's work by introducing acetarsone, the first arsenical compound that could be used orally. It was effective in treponemal diseases and in amebic infections. Chemotherapy was also advanced by Fourneau's synthesis of compound No. 309 in 1925. It was identical to a product discovered in Germany by the scientists working for the Bayer firm. This compound, called Germanin, or Bayer 205, was effective in the treatment of sleeping sickness. The German firm had not disclosed the structure of the compound and thus had maintained exclusive control up to that time.

In contrast to Ehrlich, Fourneau was a man of broad culture to whom all aspects of arts and letters were of prime interest. Fourneau also was a linguist with an excellent knowledge of English and German. After his productive career, Fourneau retired in the Basque countries, where he died in 1949.

One of the greatest accomplishments of Fourneau's laboratory was the discovery of the antibacterial action of sulfanilamide. This discovery was announced in a paper entitled "Activity of *p*-Aminophenylsulfamide [sulfanilamide] on Experimental Streptococcal Infections of the Mouse and the Rabbit," which appeared in *Comptes rendus de la société de biologie* (**120**: 756–758, 1935). The authors of this paper were Mr. and Mrs. Jacques Tréfouël, F. Nitti, and D. Bovet. They wrote: [1]

> The recent investigations of Domagk in the field of sulfamido-chrysoidines, which have yielded the Prontosils . . . have reintroduced, as the order of the day, the chemotherapy of bacterial infections. Having prepared and tested various products resulting from the coupling of the diazo derivative of aminosulfamide with mono- and poly-phenols,

[1] Translated by permission of Professor Jacques Tréfouël, Director of the Pasteur Institute, Paris.

alkylated and unalkylated, we came to realize that derivatives that were very different from Prontosil, from the point of view of their chemical and physical properties, had however, similar antistreptococcal activity. . . . In contrast, the derivatives in which para-aminosulfamide is replaced by aminobenzamide, aminophenylacetonitrile, aminobenzonitrile and phenetidine are deprived of therapeutic activity.

Already Heidelberger has noted the bactericidal action of sulfamidophenylazo-hydrocupreine. How could one explain that such different compounds act in an almost identical manner? Could it be that, in the organism, they are subjected to a series of modifications, the first of which would be the breaking of the double bond with the formation of aminophenylsulfamide? This hypothesis has led us to study the activity of the hydrochloride of p-aminophenylsulfamide (1162 F.).

In these tests mice and rabbits were used.

1° The protocol of an experiment in mice is as follows: The animals were inoculated with a strain of hemolytic streptococci, Dig. 7, that had been isolated from a fatal puerperal septicemia. Inoculation was made intraperitoneally with $\frac{1}{20.000}$ of a cc of a 24 hour culture in ascitic broth. The controls died in less than 48 hours with streptococci in the blood. Of the six mice inoculated the same way but having twice received on two consecutive days, 1.25 and 2.5 mg. of 1162 F. per 20 gr. orally, three died between the fourth and the sixth day and three survived more than ten days. These results are equivalent to those obtained under the same conditions with Prontosil.

2° Following this first experiment, we have investigated the therapeutic action in rabbits. We obtained a strain of hemolytic streptococci of animal origin (strain Pion) that was very virulent for rabbits. . . . The animals received $\frac{1}{500}$ of a cc or twenty lethal doses one half hour before treatment. Compound 1162 F. was administered orally on two successive days in doses varying from 0.05 gr. to 1 gr. per kg. In one case, with the lowest dose, survival was prolonged by two days. At higher doses, survival was longer . . . but it never exceeded a week. Treated animals, in general, give negative hemocultures up to the day before their death.

In parallel with the testing of 1162 F., a series of experiments were carried out with Prontosil. Doses varying between 0.01 gr. and 0.50 gr. per kg. have prolonged the survival of treated animals over that of the controls.

Compound 1162 F. shows very little toxicity for rabbit. An intravenous injection of 0.5 gr. per kg. is tolerated but for reactions that can be ascribed to the acidity of the product. Orally and subcutaneously, doses of 1 gr. per kg. are also tolerated.

Numerous derivatives of aminophenylsulfamide are also active and their study is in progress. . . .

The authors concluded with a comment that indicated that they had fully recognized the importance of their discovery: *"The therapeutic activity of such a simple molecule, which is not itself a dye, opens a road to the systematic chemotherapeutic studies comparable to that offered by the pentavalent arsenicals."*

In this paper, Tréfouël and his collaborators referred to the discovery made by Michael Heidelberger of the bactericidal action of sulfamidophenylazo-hydrocupreine. Neither the authors of this book nor Michael Heidelberger are aware of any paper by Heidelberger in which this activity is reported. Heidelberger and Jacobs (*J. Am. Chem. Soc.*, **41**: 2132, 1919), in a paper in which they reported on the synthesis of many azo dyes derived from hydrocupreine and hydrocupreidine, said: "Many of the substances described in this paper were highly bactericidal *in vitro*, a property which will be discussed in the appropriate place by our colleague, Dr. Martha Wolstein."

Apparently Doctor Wolstein never published these data, and it is felt that only a direct personal communication could have informed the French workers of the near discovery of the sulfa drugs, in New York, in 1919.

The investigators of Fourneau's laboratory continued the study of derivatives of sulfanilamide. They concluded that sulfur was an important functional part of the molecule. They prepared a number of compounds such as dinitrodiphenyl sulfone (1937) containing sulfur in various states of oxidation. At the same time similar investigations were going on in England, where Dr. G. A. H. Buttle of the Wellcome Physiological Research Laboratories prepared diaminodiphenyl sulfone. The sulfones were active against bacteria, but more important they were also active against tubercle bacilli. They were not effective in the treatment of tuberculosis, but they have turned out to be effective in the treatment of leprosy.

In these studies on sulfanilamide and the sulfones, a husband-and-wife team of chemists played a leading role. Jacques Tréfouël, who became director of the Pasteur Institute of Paris in 1940, was born not far from Paris, in 1897. His father was a businessman trading in silk fabrics. After having served as an artillery officer during the First World War, and after having completed his chemical studies at the University of Paris, he entered the Pasteur Institute in 1920, where he distinguished himself both scientifically and administratively. A man of many interests, he is able to apply his talents not only to the building of new molecules but also to the more mundane manufacture of pieces of furniture and objects of wrought iron. In 1921, a young chemical engineer from Bordeaux, Thérèse Boyer, also joined the staff of the Pasteur Institute. Soon thereafter she married Jacques Tréfouël and became his charming, able, and devoted collaborator. She eventually became the head of a department of chemotherapy at the Pasteur Institute.

In October, 1939, Gerhard Domagk was notified that he had been awarded the Nobel Prize for the discovery of the antibacterial activity of Prontosil, a discovery that had led the Tréfouëls to that of the antibacterial activity of sulfanilamide. He was then working for the I. G. Farbenindustrie in Nazi Germany, and he had to refuse the prize under the pressure of the authorities.

Gerhard Domagk was born in 1895. He was a student at the University of Kiel when the First World War exploded. With the patriotism of youth, he volunteered for the army and was wounded in 1915. After the war, he returned to Kiel and completed his medical studies in 1921. After teaching at Griefswald and Münster, he became director of an industrial laboratory for experimental pathology and bacteriology located at Wuppertal. Two of his colleagues, Drs. Fritz Mietzsch and Joseph Klarer, synthesized a number of azo compounds that Domagk tested for activity against microorganisms.

It was in 1932 that the two chemists synthesized a group of compounds that did not look very promising since they had little action against bacteria in vitro. To this group belonged the Prontosil that Domagk found low in toxicity and very active against streptococcal infections in animals. It was left to the French workers to explain the lack of in vitro activity of Prontosil by the discovery of the antibacterial activity of part of its molecule, sulfanilamine.

In 1947, Domagk, a free man again, was able to visit Stockholm, where he received his gold medal and his diploma and presented his address. He did not, however, receive the money that accompanies a Nobel Prize; according to the rules, money not claimed within a year's time reverts to the Nobel Foundation funds. Domagk died in 1964.

With the discovery of Prontosil, infections caused by true bacteria had been shown to be susceptible to chemotherapy. This was the great contribution of the work of Domagk, whose career, like that of many others, had been so tragically influenced by the folly of the "leaders of men."

Antibiotics

Antibiotics form a group of "unnatural" natural products. Produced by microorganisms, they are chemical substances that have the property of inhibiting the growth of other microorganisms, even in very minute concentrations. In nature, such as in the microworld of the soil, microorganisms do not produce sufficient amounts of antibiotics for adequate study in the laboratory. In practice, these chemical substances are the products of the growth of microorganisms under laboratory or factory conditions.

With the knowledge that we now have we may read into some of the folklore of the past the empirical use of antibiotics. This subject has been reviewed in the first volume of Florey et al.'s book on antibiotics. We would like simply to illustrate this aspect of history with a quotation from Jules Brunel (*Rev. can. biol.*, **3**: 333–343, 1944): [1] ". . . The antiseptic properties of Penicillia have been recognized for a long time as is known to those who have studied central European folklore. Doctor A. E. Cliffe, a biochemist from Montreal made the following statement . . . 'It was during a visit through Central Europe in 1908 that I came across the fact that almost every farmhouse followed the practice of keeping a moldy loaf on one of the beams in the kitchen. When asked the reason for this I was told that this was an old custom and that when any member of the family received an injury such as a cut or bruise, a thin slice from the outside of the loaf was cut off, mixed into a paste with water and applied to the wound with a bandage. I was assured that no infection would then result from such a cut.'"

The first scientific observations of antagonisms among microorganisms are more pertinent. As we may recall from Chapter 2, in 1874, Roberts of Manchester had noted that a medium in which *Penicillium glaucum* was growing could not be contaminated easily by bacteria. Tyndall, in 1876, observed the struggle for life between bacteria and molds and concluded that as a rule, each one of the organisms seems to have an even chance of winning the fight.

The first therapeutic attempts were modest. Pasteur and Joubert in 1877 noted that the simultaneous inoculation of harmless bacteria and *Bacillus anthracis* into animals repressed the symptoms of anthrax. Pasteur wrote with his usual flair for accurate prediction: "These facts might permit the highest hopes for therapy." Cantani, in 1885, tried to treat tuberculosis by spraying a fine suspension of *Bacterium termo* into the lungs of patients; and Gas-

[1] Reproduced by permission of the editors of the *Revue Canadienne de Biologie*.

perini, in 1890, was probably the first one to report the antagonistic properties of some actinomycetes.

The various aspects of these heroic, if not too effective, attempts at "antibiotic" therapy have been reviewed in detail in a book written in 1928 by Papacostas and Gaté. The two most important events of the early period were the utilization of "pyocyanase" and of "mycolysates of Streptothrix" in human therapy. In these early applications, one can read the promise of successes to come.

In 1899, Emmerich and Löw introduced an antibacterial agent from aged cultures of *Pseudomonas*. They named their concoction pyocyanase, thinking that it was an enzyme. It was reported useful in the treatment of diphtheria. However, those working on pyocyanase did not think in terms of chemotherapy. Medicine was then enjoying the great achievements of immunology, so that pyocyanase, instead of being purified in order to extract the active principle, was mixed with serum proteins in order to prepare "immuno-proteids." It finally fell into an oblivion that only the great success of antibiotics has disturbed.

Toward the end of the second decade of this century, Belgian investigators developed therapeutically effective preparations of actinomycetic origin. Gratia and Dath, as early as 1924, had recognized the antimicrobial action of a certain actinomycete, then referred to as a *Streptothrix*. More precisely, they recognized the bacteriolytic power of this actinomycete. The injection of pathogenic bacteria that had been lysed by the *Streptothrix* was reported effective in the treatment of bacterial infections, particularly of streptococcal origin. For good measure, bacteriophages were at times added to these "mycolysates." Welsch, a coworker of Gratia, used the word "actinomycetine" in 1937 to refer to the sterile filtrate of the actinomycete that had bacteriolytic properties.

Early investigations, such as these, had little influence on the course of therapy because here again the investigators were not thinking in terms of chemotherapy but in terms of immunity.

One had to wait for the beginning of the Second World War before the importance of antibiotics was recognized. The very word "antibiotic" was not even in fashion at that time. The discovery of penicillin by Fleming and its subsequent development by the Oxford group was the greatest achievement in this domain.

Alexander Fleming's father farmed a leasehold in the remote hill area of Ayrshire, Scotland, that was suitable only for raising sheep and cattle, and growing oats and hay for the stock. From a first marriage there were four children. The widower remarried at age sixty and fathered four more children. Alexander, born in 1881, was next to the youngest of the second group. By the time Alec was a young boy, his father was a kindly but ailing old man, passing his days by the fireside. After his death, his widow and oldest son continued to manage the farm. All the children could not hope to earn an adequate living from the sparse economy of the area, and education was a necessary solution. One of the elder sons, Tom, was already off to study medicine in Glasgow. Eventually he settled in London to develop a specialty of ophthalmology. In turn, the younger brothers John, Robert, and Alec also moved to London to develop trades under Tom's sponsorship, and Mary, one of the sisters, came to keep house for the boys. When she married, sister Grace moved down to take her place, and at a later point, when Grace married, their mother moved down to keep house for the small clan. Alexander became a clerk for a shipping company. When the Boer War broke out in 1900, John, Robert, and Alec joined a Scottish regiment, which turned out to have an excess of volunteers. The Flemings were kept home, but Alec, a good shot from boyhood days on the farm, helped Company H win a shooting trophy. He continued membership in the regiment and maintained his shooting skill as a sport. Fleming had a life-long passion for athletics.

An unexpected legacy of £250, representing an eighth of an eighth share from a proverbially prudent Scottish bachelor uncle, led Alec to enter the study of medicine at age twenty. Lacking a qualifying diploma, Alec had to seek admission by examination. A considerable memory, keen intelligence, and gift of expression helped him pass at the head of the list, and he had the choice of one of the twelve medical schools in London. He chose St. Mary's because it had a good water-polo team. The scholastic pace set with his entrance examination was maintained, and he placed first in all subjects. In the same way he passed the examination for fellowship in the Royal College of Surgeons without experience in practical surgery.

Interest in sports led him to St. Mary's for medical education and also led him to join Almroth Wright's bacteriology group, the famous "Inoculation Service" at St. Mary's. The St. Mary's rifle club had come on bad days. Freeman, of Wright's staff, was one of the leaders of the rifle club. Learning of little Alec Fleming's skill as a marksman, he approached Wright, emphasizing Fleming's prowess with the rifle and incidentally suggesting that Fleming had a scientific mind. Wright was taken with Freeman's approach, and Fleming joined the laboratory in which he remained until his death, fifty years later. This feeling for the importance of sportsmanship is corroborated by Fleming's words to a class of medical students years later: "You should know even at this stage of your career that there is far more in medicine than mere book work. You have to know men and you have to know human nature. There is no better way to learn about human nature than by indulging in sports, more especially in team sports. When you are one of a team, you have to play for the side and not simply for yourself, and this is marvellous training for a man who hopes to become a doctor. . . ."

There were eight persons in Wright's group sharing the meager space described in Chapter 5. In a small room adjoining the laboratories, Wright received his distinguished visitors, which included Ehrlich, Metchnikoff, Balfour, and Shaw.

Fleming, as quiet as Wright was talkative, rapidly acquired the skills of the laboratory and attempted to develop a firmer basis for the opsonic index as a measure of immunity. Very soon, Fleming was spending long nights as well as days in the laboratory, improving the methods for the diagnosis of infectious disease. By 1908 he had extended his interests to include treatment, and when Salvarsan was announced by Ehrlich, Fleming was among the first in England to administer the therapy to syphilitic patients. Fortunately too, this skill provided a limited private practice that augmented his meager salary from research.

Gradually, he acquired friends outside his immediate medical circle. Of these, artists and writers were his favorites, and he became one of the few honorary members of the Chelsea Arts Club. The hours spent at the club became one of his lifelong pleasures. For the members, Fleming provided free medical care and even hospitalization at St. Mary's. At one point one of his artist friends decided that Fleming should paint a picture to be truly qualified for membership. Under protest, Fleming painted a farm scene, which was hung in a commercial gallery. To complete the ritual, one of the Chelsea group was designated to ask the gallery owner for the price of Fleming's picture. The price was quoted and declared too high, but Fleming was now an official Chelsea Arts Club member.

When the First World War broke out, the St. Mary's laboratory group went off into laboratory service for the army. Wright, as we saw in Chapter 5, organized a research laboratory at Boulogne-sur-Mer with a staff consisting of Douglass, Morgan, Fleming, Colebrook, and Freeman from the St. Mary's group. The improvisation of services and equipment was left to Fleming, and he soon contrived the necessary apparatus and services for a functional laboratory. Thanks to Wright's vaccine, typhoid fever was not a problem as in previous wars. But the increased effectiveness of explosives had resulted in severe gangrenous infections of wounds. Listerian antiseptic techniques were of little avail.

Fleming demonstrated the ineffectiveness of antiseptics by use of models of wounds made with glass tubing. The tubes were filled with contaminated serum, emptied, and refilled with bactericidal concentrations of antiseptic. The tubes were emptied again and refilled with uninfected serum. Bacteria that had escaped the action of the antiseptics would multiply from the deep interstices where they were hidden, and the serum putrified. A partial answer was the use of washes, such as highly concentrated saline, that would stimulate exudation of lymph and phagocytes. Fleming went further in stating: "What we are looking for is some chemical substance which can be injected without danger into the blood stream for the purpose of destroying the bacilli of infection, as Salvarsan destroys the spirochetes."

Despite long hours of work, Fleming and his colleagues maintained the British way of life. A golf course was not far away, and the games were enlivened occasionally by Fleming's lack of respect for the rules. When the urge for exercise became strong, and the time short, an impromptu wrestling match might be organized. Once a group of high-ranking French medical officers, appearing for an unannounced visit, found such a wrestling match in progress. The wrestlers merely stopped and phlegmatically proceeded with the conference desired by the French officers.

The war ended without a successful conquest of gas gangrene.

During the war, Fleming had acquired a wife in his characteristic underplayed way; while on leave in England during December of 1915, he had married Sarah (Sareen) Marion McElroy, a nurse whom he had known for some time. He had not announced any plans for marriage before he went on leave. When he returned to Boulogne-sur-Mer, he would make reference to his wife, and it was a time before his friends were convinced that Alec was indeed married and not perpetrating one of his straight-faced jokes.

Sareen had established a private nursing home, which she sold after her marriage. Part of the funds so obtained were used to buy a small but charming country house, the Dhoon, at Barton Mills, in Suffolk. The Dhoon remained their weekend and summer retreat all their lives. In London they leased a house in Chelsea, and Sareen ran the household with an economical turn that allowed Fleming to abandon part-time practice. His appointment as assistant director of the Inoculation Service in 1921 helped further ease the financial situation. Their only child, a son Robert, was born in 1924. Like his father he also attended St. Mary's medical school and became a physician. Life at home was quiet and happy.

After the war Fleming returned to his bench at St. Mary's, and for a period of three years he made modest contributions to methodology. When he came down with a cold, sometime in 1921, he was far from imagining that he was on the road to greatness. Curious as to the organisms associated with the nasal exudate, he placed his own nasal mucus on a blood agar dish. After leaving the plate on the bench for more than a week, he observed that yellow colonies had appeared on the plate. Their distribution, however, was not uniform. There were none in the zone bordering the mucus; the colonies in the next zone were translucent and glassy, and, finally, proliferating yellow colonies were present in the outer zone. Fleming concluded that the mucus contained a substance that was capable of lysing this organism and that its concentration was graded by diffusion. He found that the sensitive organism was a large nonpathogenic Gram-positive coccus. Eventually he named the organism *Micrococcus lysodeikticus,* and the lytic substance, lysozyme. Repeating the observation with a broth culture of the organism, he found the turbid suspension had turned "clear as gin" within a few minutes. Testing other body secretions, he found tears to be particularly rich in lysozyme. Adequate supplies of tears were obtained by subjecting laboratory personnel to an "ordeal by lemon" and collecting the tears with capillary pipettes. Fleming presented his first report before the Medical Research Club in London in December, 1921. It did not provoke a single question. A second paper on lysozyme was presented before the Royal Society in February, 1922, and again, the interest was minimal. But Fleming considered lysozyme important as a possible therapeutic agent, and continued to study it with Allison during the next five years. Lysozyme was shown to be distributed in additional tissues of the body and in other natural materials. Egg white was an especially rich source. Unlike the available chemical antiseptics, which had a toxic effect on leucocytes, a lysozyme-rich substance, such as egg white, was harmless to tissues and blood cells. Going a step further in this direction, he found that leucocytes also contained lysozyme and that it was a component in the bactericidal activity of these cells. The logical step forward would have been to isolate lysozyme in pure form, but there were no chemists among the staff of the Inoculation Service. Lysozyme eventually was purified by Roberts and Abraham in 1937, and Chain and Epstein showed that lysozyme was indeed an enzyme and that it acted on a constituent of bacterial cell walls.

An important aspect of the studies on lysozyme was the development of reliable and efficient in vitro methods for testing inhibitory action on bacteria and evaluating toxicity for tissues. Fleming's culture plates with wells for lysozyme and culture streaks radiating from the well were one of the techniques used in the studies of penicillin.

The discovery of penicillin was a by-product of a commitment to contribute an article on staphylococci to the *System of Bacteriology*. As was his custom, Fleming allowed the culture dishes from the study to accumulate rather than discarding them promptly. As he examined the plates for one last time before discarding them, he noted that in one of the dishes there was a contaminating mold and a zone about the mold free of staphylococci. Fleming immediately removed a fragment of the mold to a culture tube. He also preserved the culture dish. The broth culture of the mold, a *Penicillium*, was mixed with agar placed in a gutter dug in the agar medium of a Petri dish. Cultures of a variety of pathogens were cross-streaked. The culture filtrate of the *Penicillium* inhibited the pathogens and was active even when diluted five hundred fold.

Fleming, with Stuart Craddock as his assistant, found that the active substance would pass through asbestos filters, and though unstable, its activity could be maintained longest at neutral pH. Active broths were free of toxicity for animals and not irritating to human conjunctiva. Frederick Ridley, one of the physicians on the St. Mary's staff who knew somewhat more chemistry than Fleming, and had tried previously to help with lysozyme, also joined in the effort. Craddock and Ridley achieved a ten- to fifty-fold purification but ended up with a syrupy mass that was unstable. In 1928, Fleming presented a report on his substance that he named penicillin, before the Medical Research Club, and he wrote a paper for the *British Journal of Experimental Pathology*. He described the spectrum of activity, its chemical properties as best he could, its lack of toxicity, its use to favor the isolation of the Gram-negative *Hemophilus influenzae* in throat cultures by inhibiting the growth of Gram-positive organisms, and he suggested that it might be used for application to, or injection into, areas infected with penicillin-sensitive organisms. The last statement was included in the paper over the protest of Sir Almroth Wright, who still believed that the immunity mechanism was the only way in which the challenge of infection could be met. This aspect of Fleming's interest was contrary to the philosophy and faith of the St. Mary's Inoculation Service.

Penicillin had not gone completely unnoticed. In 1931 Harold Raistrick, the professor of biochemistry at the London School of Tropical Medicine and Hygiene, who was interested in mold metabolites, took up the problem of its purification. With the collaboration of Lovell and Clutterbuck, he made good progress. By the use of a synthetic medium, some of the difficulties encountered by Fleming's group were overcome. A pigment and proteins were separated as inactive components, leaving the still impure penicillin as a third fraction. A successful solution was close, but Clutterbuck and Lovell developed other interests. In the interim, Fleming used his crude preparations of penicillin for local application, but such efforts could not lead to a conclusive evaluation.

The time was now 1935, and the sulfa drugs had arrived on the scene full of promise. Fleming was still discussing the value of penicillin and still seeking a chemist who would extract it. It looked like the end of a dream. But unknown to Fleming, penicillin was finally on the verge of being developed by the now famous Oxford group led by Howard Florey. In 1935 Florey was appointed professor of pathology at the Sir William Dunn School at Oxford. It was a well-organized and well-equipped institute that dealt with bacteriology, pathology, and biochemistry. Florey, stimulated by the success with lysozyme, encouraged Chain to continue with investigation of antibacterial substances of microbial origin. The Rockefeller Foundation provided the funds to initiate the investigation, and the program was started in 1938. After a careful survey of the literature, penicillin was selected as the most promising agent, and the study went on in spite of the war that started in September, 1939.

In May, 1940, there was sufficient purified material for trials in mice, and activity was tested against infections with a staphylococcus, a beta-hemolytic streptococcus, and *Clostridium septicum*. Penicillin was strikingly effective.

The first report of the Oxford group, in August, 1940, took Fleming by surprise. The paper was entitled "Penicillin as a Chemotherapeutic Agent" (*Lancet*, **239**: 226–228, 1940). The authors of this paper were Chain, Florey, Gardner, Heatley, Jennings, Orr-Ewing, and Sanders.[1] They wrote:

> In recent years interest in chemotherapeutic effects has been almost exclusively focused on the sulphonamides and their derivatives. There are, however, other possibilities, notably those connected with naturally occurring substances. It has been known for a long time that a number of bacteria and moulds inhibit the growth of pathogenic micro-organisms. Little, however, has been done to purify or to determine the properties of any of these substances. The antibacterial substances

[1] Reproduced by permissions of the editor of *Lancet* and Professor E. B. Chain.

produced by *Pseudomonas pyocyanea* have been investigated in some detail, but without the isolation of any purified product of therapeutic value.

Recently, Dubos and collaborators (1939, 1940) have published interesting studies on the acquired bacterial antagonism of a soil bacterium which have led to the isolation from its culture medium of bactericidal substances active against a number of gram-positive microorganisms. Pneumococcal infections in mice were successfully treated with one of these substances, which, however, proved to be highly toxic to mice (Hotchkiss and Dubos, 1940) and dogs (McLeod *et al.*, 1940).

Following the work on lysozyme in this laboratory it occurred to two of us (E. C. and H. W. F.) that it would be profitable to conduct a systematic investigation of the chemical and biological properties of the antibacterial substances produced by bacteria and moulds. This investigation was begun with a study of a substance with promising antibacterial properties, produced by a mould and described by Fleming (1929). The present preliminary report is the result of a cooperative investigation on the chemical, pharmacological and chemotherapeutic properties of this substance.

Fleming noted that a mould produced a substance which inhibited the growth, in particular, of staphylococci, streptococci, gonococci, meningococci and *Corynebacterium diphtheriae,* but not of *Bacillus coli, Hemophilus influenzae, Salmonella typhi, P. pyocyanea, Bacillus proteus* or *Vibrio cholerae.* He suggested its use as an inhibitor in the isolation of certain types of bacteria, especially *H. influenzae.* He also noted that the injection in the animals of broth containing the substance, which he called "penicillin," was no more toxic than plain broth, and he suggested that the substance might be a useful antiseptic for application to infected wounds. The mould is believed to be closely related to *Penicillium notatum.* Clutterbuck, Lovell and Raistrick (1932) grew the mould in a medium containing inorganic salts only and isolated a pigment—chrysogenin—which had no antibacterial action. Their culture media contained penicillin but this was not isolated. Reid (1935) reported work on the inhibitory substance produced by Fleming's mould. He did not isolate it but noted some of its properties.

During the last year methods have been devised here for obtaining a considerable yield of penicillin, and for rapid assay of its inhibitory power. From the culture medium a brown powder has been obtained which is freely soluble in water. It and its solution are stable for a considerable time and though it is not a pure substance, its anti-bacterial activity is very great. Full details will, it is hoped, be published later.

Effects on Normal Animals

Various tests were done on mice, rats and cats. There is some edema at the site of subcutaneous injection of strong solutions (e.g. 10 mg. in 0.3 c.cm.). This may well be due to the hypertonicity of the solution. No sloughing of skin or suggestion of serious damage has ever been encountered even with the strongest solutions or after repeated injections into the same area.

Intravenous injections showed that the penicillin . . . was only slightly, if at all, toxic for mice. An intravenous injection of as much as 10 mg. (dissolved in 0.3 c.cm. distilled water) of the preparation we have used for the curative experiments did not produce any observable toxic reactions in a 23 g. mouse. It was subsequently found that 10 mg. of a preparation having twice the penicillin content of the above was apparently innocuous to a 20 g. mouse.

Subcutaneous injections of 10 mg. into two rats at 3-hourly intervals for 56 hours did not cause any obvious change in their behaviour. They were perhaps slightly less lively than normal rats but they continued to eat their food. Their blood showed a fall of total leucocytes after 24 hours, but after 48 hours the count had risen again to about the original total. There was, however, a relative decrease in the number of polymorphs, but the normal number was restored 24 hours after stopping the administration of the substance. One of these two rats was killed for histological examination; there was some evidence that the tubule cells of the kidney were damaged. The other has remained perfectly well, and its weight increased from 76 to 110 g. in 23 days. It is to be noted that these rats received, weight for weight, about five times the dose of penicillin used in the curative experiments in mice. No evidence of toxic effects was obtained from the treated mice, which received penicillin for many days.

Other pharmacological effects

On the blood-pressure, heart-beat and respiration of cats no effects have been observed after intravenous injection of 40 mg.—enough to bring up the concentration in the blood just after injection to $\frac{1}{5000}$. Perfusion of the isolated cat's heart, with Ringer-Locke solution containing $\frac{1}{5000}$ penicillin, produced progressive slowing during 15 minutes and at the end of that time the heart looked as though it would stop beating; however, it was quickly revived by perfusing with Ringer-Locke solution alone. The same depressant action was seen at $\frac{1}{10.000}$

dilution but the effect was less than at 1/5000. Solutions are absorbed from the intestine in the rat without causing any observable damage to the mucosa. They are also readily absorbed after subcutaneous injection and the substance can be detected in the blood. It is excreted by the kidneys, the urine becoming bright yellow. At least 40–50% appears in the urine in a still active form. Human leucocytes remain active in a 1/1000 solution for at least 3 hours.

It must be emphasized that the results of these preliminary tests have been obtained with an impure substance and such slight toxic effects as have been noted may possibly be due, in part at least, to these impurities.

Effects on Bacteria In Vitro

In view of this slight evidence of tissue toxicity it is all the more striking that the substance in a dilution of one in several hundred thousand inhibits *in vitro* the growth of many micro-organisms, including anaerobes. Of those so far tested in this laboratory the following are sensitive to the inhibiting action of our preparation: *Clostridium welchii* (2 strains); *Cl. septique* (1 strain, Nat. coll. type-cultures No. 458); *Cl. oedematiens* (1 strain, N.C.T.C. No. 277); *C. diphtheriae* (1 strain, mitis type); *Streptococcus pyogenes* (Lancefield group A); *Str. viridans* (1 strain from tooth); *Str. pneumoniae* (type 8); staphylococci (3 strains). Penicillin is not immediately bactericidal but seems to interfere with multiplication.

Therapeutic Effects

From all the above tests it was clear that this substance possessed qualities which made it suitable for trial as a chemotherapeutic agent. Therapeutic tests were therefore done on mice infected with streptococci, staphylococci and *Cl. septique;* the results are summarized in the accompanying table, the preliminary trials on small numbers of mice being omitted. . . .

The general principle has been to keep up an inhibitory concentration of the substance in the tissues of the body throughout the period of treatment by repeated subcutaneous injections. No extended tests have been made to determine the minimum effective quantities or the longest intervals possible between injections. The doses employed have been effective and not toxic; they may have been excessive. The solution contained 10 mg. per c.cm. of substance.

TABLE 1 *Results of Therapeutic Tests on Mice Infected with Strep. pyogenes, Staph. aureus and Cl. septique*

Experiment		Duration of treatment	Single dose (mg.)	Total dose (mg.)	No. of mice	Survivors at end of 10 days
Strep. pyogenes—Lancefield, Gp. A.						
1	Controls	. . . . .	. . .	. . .	25	4
	Treated	12 hrs.	2	10.0	50	25
2	Controls	. . . . .	. . .	. . .	25	0
	Treated	45 hrs.	0.5	7.5	25	24
Staph. aureus						
1	Controls	. . . . .	. . .	. . .	24	0
	Treated	55 hrs.	0.5	9.0	25	8
2	Controls	. . . . .	. . .	. . .	24	0
	Treated	4 days	0.5	11.5	24	21
Cl. septique						
1	Controls	. . . . .	. . .	. .	25	0
	Treated	10 days	0.5	19	25	18
		10 days	1.0	38	25	24

In the first streptococcal experiment the treatment was continued for 12 hours only. That this was inadequate was shown by deaths occurring during the arbitrarily chosen 10-day period. In the second experiment the time of treatment was lengthened, with improved results. In this and the two staphylococcal experiments injections were given 3-hourly for the first 32–37 hours, then at longer intervals.

In preliminary experiments with *Cl. septique* it was found that the infection could be satisfactorily held in check so long as penicillin was being given (i.e., for 2 days) but when the administration was stopped the infection developed. In the experiment quoted, therefore, the injections were given for 10 days, 3-hourly for 41 hours, then at longer intervals, and twice daily for the last 2 days of the period. No deaths have subsequently occurred (22 days after beginning of experiment).

The behaviour of the mice infected with streptococci and staphylococci was interesting. For some hours after the start of treatment they looked sick—some even appeared to be dying—but as the experiment went on they progressively improved till at the end of 24 hours in the case of the streptococci and about 36 to 48 hours with the staphylococci it was difficult or impossible to distinguish them from normal mice. The survivors of the *Cl. septique* infection on the other hand remained well throughout, except for a few in which leg lesions appeared near the site of injection and cleared up in a few hours.

Summarizing the data given in the table we see that in the final streptococcus experiment (no. 2) whereas $^{25}/_{25}$ controls died, $^{24}/_{25}$ treated animals survived. With *Staphylococcus aureus* the final experiment (no. 2) shows $^{24}/_{24}$ deaths of the controls and $^{21}/_{24}$ survivals among the treated. Lastly, with *Cl. septique* when the larger doses of penicillin were given (bottom line) the figures are $^{25}/_{25}$ control deaths and $^{24}/_{25}$ treatment survivals.

Conclusions

The results are clear cut, and show that penicillin is active *in vivo* against at least three of the organisms inhibited *in vitro*. It would seem a reasonable hope that all organisms inhibited in high dilution *in vitro* will be found to be dealt with *in vivo*. Penicillin does not appear to be related to any chemotherapeutic substance at present in use and is particularly remarkable for its activity against the anaerobic organisms associated with gas gangrene.

Although not directly involved in the studies of the Oxford group, Fleming maintained a paternal interest in the project.

The first clear test in a human being administered penicillin by a parenteral route took place in 1941. It was a case of septicemia. Death was delayed but not prevented because there was insufficient material for continuation of therapy. The Oxford group then enlisted the cooperation of scientists in the United States, both in government and commercial laboratories, while continuing with their own efforts to produce adequate supplies for clinical testing.

Fleming was appointed as a regional pathologist to assist in the war effort. The Flemings came to have intimate knowledge of the hazards of war. Their London house was hit by incendiary bombs and later suffered severe damage from an explosion. Fleming spent most of his nights at the hospital, using the dark room as a bedroom, taking his turn with the others on the roof watching for incendiary bombs.

By 1943, penicillin was in large-scale production and radically changing the treatment of many infectious diseases. Fleming's part in the penicillin story was now well known, and honors were beginning to be conferred from all directions. Of special meaning to Fleming was his election as a Fellow of the Royal Society. A year later, he was knighted, and in 1945 he made a triumphal tour in the United States. A number of universities granted him honorary degrees. The visit to the United States was followed by similar trips to France, Italy, Denmark, and Sweden.

In December, 1945, he was back again at Stockholm with Florey and Chain to receive the Nobel Prize. In 1946, Sir Almroth Wright retired as principal of the Inoculation Department, now an elaborate institute, and Sir Alexander Fleming was appointed in his place.

As principal, his habits did not change. He would not have an office; his room was still to be a laboratory. With the death of Wright, antifeminism came to a halt at St. Mary's, and women came to fill professional positions. The first was a Greek physician, Amalia Voureka, whom Fleming admitted to his own laboratory. Much as he might have wanted to work at the laboratory bench, he was now obliged to act as an unofficial ambassador of science for England. On one of his trips to Spain, which he made with his wife in 1948, Sareen became seriously ill, and she died in October of the next year. The loss was a severe shock to Fleming, who sought refuge in the laboratory, and, as in his younger days, he began to spend long hours at the microscope. The demands for foreign appearances and lectures soon pressed upon him again and helped overcome the void left by Sareen's death. When in London, he began to depend upon Dr. Voureka, who was starting to fill the void in his life. After having spent five years in England, Dr. Voureka had acquired both a superior training and a scientific reputation. She was offered an attractive appointment as laboratory chief in the leading hospital of Athens. The call could not be refused. Fleming was again without the feminine companionship he enjoyed, but the international obligations and trips continued to fill much of his time. World Health Organization commitments gave him the opportunity to make a trip to Greece in October, 1952. Once again he could spend time with Amalia Voureka; she was designated to organize Fleming's tour and act as guide and interpreter. Before he left, he asked Amalia to marry him. Fleming's diary for November 9 has the word "yes" on a line by itself. Scheduled trips delayed the marriage until April, 1953. The religious marriage was at a Greek church in London, an odd environment for a dour Scotsman.

In January, 1955, Fleming retired from his administrative post but planned to continue laboratory work. By February he developed a persistent gastric upset, but this did not keep him from the laboratory. On March 11, an attack of nausea so alarmed Amalia that she called their physician. Fleming spoke to him over the telephone and assured him that a visit was not urgent. Fleming then lay down, and a few minutes later, as his wife came to take his pulse, she was frightened by his coldness. "Are you sure that it is not your heart?" she asked. He was calm and serious, concentrating on the proper analysis of his symptoms. No, he did not think that it was his heart. Within a few minutes, he was dead of coronary thrombosis.

He died as he wished; quietly, without a gradual decline in physical or mental capacity, and even without inconveniencing his physician.

Fleming's acute powers of observation had indeed proven the veracity of one of Pasteur's laws, namely, that "chance favors the prepared mind." In the development of the systematic screening programs that yielded so many important antibiotics, Waksman played a leading role.

Selman A. Waksman was born in 1888, in a small town of Ukraine that was a mere dot in the endless expanse of the steppes. Since Russia did not have much to offer to a young man of Jewish extraction, he left his native land in 1910 to come to the United States of America, where some of his cousins were already established on a farm not far from New Brunswick, New Jersey. There, Selman was able to discuss the prospect of furthering his education with Dr. Jacob G. Lipman, a distinguished soil microbiologist who was also a former immigrant from Russia. Dr. Lipman was teaching at the College of Agriculture of Rutgers University, an institution of which he eventually became the Dean. Dr. Lipman, who was mainly famous for his work on nitrogen-fixing bacteria, thought that Selman should study agriculture at Rutgers rather than medicine somewhere else. The choice of a school and the selection of a field of study were further facilitated by the fact that Selman received a scholarship from Rutgers.

The young man found himself surrounded by much younger students and obliged to speak a language of which he was not yet a master. He was a studious person, and his maturity helped him profit greatly from the instruction he received. Unlike young Fleming, young Waksman had no interest in sports. He had been trained from childhood to exercise his mind, not his body. The various ball games that were the delight of his schoolmates were meaningless to him, and he felt that they amounted to nothing more than a waste of time.

The College Farm of Rutgers was a very convenient place for a young man not only to get an education but also to make a few pennies doing odd jobs. In addition, inexpensive food was available. Waksman even found lodging in an old farmhouse located on the grounds of the College. Thus he started a long life of association with the College of Agriculture that was severed only when the Institute of Microbiology of Rutgers University opened in 1954. An additional source of income for Waksman was tutoring freshly immigrated foreigners in English. From this industrious and frugal beginning, he remembered the financial problems of students, and later, in his more affluent days, he always offered his foreign graduate students the opportunity for odd jobs.

During his student days, Waksman organized at Rutgers a chapter of the Menorah Society, dedicated to the study of Jewish culture. He always showed great interest not only in Jewish religion but in religious problems in general. One of his favorite conversational gambits, later in life, consisted of asking point-blank questions about the religious belief of his interlocutor. He would follow that with a learned discourse on some aspects of Jewish culture.

Waksman showed high aptitude, and during his last year of college studies he was permitted to begin research work under the supervision of Dr. Lipman. Starting with a basic and fundamental aspect of microbiology, he investigated the distribution of soil bacteria at different levels of soil profiles. He dug trenches on the College Farm and collected samples of soil at different depths. In cultures from these soil samples he noted that a large percentage of the microbial colonies were strange to him. At first glance the young colonies might have been confused with those of bacteria, but when the colonies became older and especially when examined under the microscope, they looked more like colonies of molds than colonies of bacteria. If they were molds, they were of a very special type, since their hyphae were extremely fine. Waksman asked Dr. Lipman and his other teachers about these organisms that they had never mentioned in class. To his surprise, he discovered that very little was known about them except a name: *Actinomyces*. Thus Waksman started his study of actinomycetes while still an undergraduate student, and he maintained an active interest in them all his life. He was eventually well rewarded for his dedication since the actinomycetes turned out to be the most important producers of antibiotics.

After graduating, Waksman stayed one more year at the College of Agriculture of Rutgers. He prepared a master's thesis under the kind but distracted supervision of Dr. Lipman, who assigned him a problem that he ignored completely. Instead, Waksman worked on actinomycetes. Dr. Lipman did not attach much importance to this change of topic and acted as if he had not noticed it. Lipman's example did not serve to guide Waksman to intellectual tolerance in this domain. Later, he himself was always careful that his graduate students should not deviate from the problem that he had given them. He would leave them latitude in the methods they employed in their investigations as long as they stuck to the problem that he had assigned them.

Following his master's studies, Waksman was awarded a predoctoral fellowship at the University of California. The day before his departure, he married a vivacious young lady who was affectionately known by the nickname of Bobili. As the train that was to take them across the continent started, he sat on Bobili's elegant new hat, and the romance that had started a few years previously in Russia almost came to an abrupt end.

At Berkeley, Waksman studied the proteolytic activity of fungi and actinomycetes. As the work toward the Ph.D. progressed, money became scarce, and he found a part-time position at Cutter Laboratories, a firm specializing in sera and vaccines.

Dr. Lipman continued his interest in Waksman, and after the completion of his doctorate, Waksman accepted a position at Rutgers that was strong in promise but weak in salary. For the first two years, he fortified the family income by working part time at the nearby Takamine Laboratories. This provided him with experience in the field of chemotherapy as Takamine produced Salvarsan. Eventually, in 1921, the University salary was high enough to permit the support of the little family that included a son, Byron, born in 1919.

As this point, Waksman added sulfur bacteria to his interest. Together with Jacob Joffe, who later became a distinguished pedologist, he isolated and named *Thiobacillus thiooxidans,* an autotrophic organism that was able to oxidize sulfur to sulfuric acid and that could grow at very low pH values. Even though not the first organism of this general type to be isolated from nature, it was such an unusual organism that it brought to Waksman's laboratory a fame that further studies in other fields only increased.

It was at the time of this work on sulfur that Robert L. Starkey came to work with Waksman. After obtaining a doctor's degree at Rutgers, Starkey went to the University of Minnesota. In 1927 he came back to Rutgers, where he worked mainly on nitrogen and sulfur transformations by microorganisms. When, many years later, Waksman and his assistants isolated numerous antibiotics, Starkey remained faithful to his own studies.

Students and visiting investigators started to flock to Waksman's laboratory. For many years the basic problem studied in the laboratory was the decomposition of organic matter in soil and the mechanism of formation of humus. Waksman reviewed the field of soil microbiology with care and published a very comprehensive book on the subject in 1927.

As the years went by, Professor Waksman became one of the outstanding microbiologists in the world. His role in furthering the study of soil micro-biology was second only to that of such pioneers as Winogradsky and Beijerinck.

It was about the time of the beginning of the Second World War that the work of Waksman was influenced by that of one of his former students, René Dubos. Dubos was French, a graduate of the Institute of Agronomy of Paris. He came to the United States to study with Waksman, and he received a Ph.D. degree from Rutgers in 1927. His dissertation dealt with the decomposition of hydrogen peroxide in soil. He left Rutgers to work with Dr. O. T. Avery at the Rockefeller Institute, in New York. At first interested in isolating from soil microorganisms that were able to lyse the capsules of pneumococci, he wound up, in 1939, by finding a bacterium that produced an antibiotic, tyrothricin, that was a mixture of polypeptides too toxic for clinical parenteral administration. It found a use when applied topically, but most important, Dubos' work influenced Waksman and incited him to start a thorough survey of soil microorganisms in order to detect potential antibiotic producers.

The first of the "Waksman antibiotics" was reported in 1940, the year of the publication of the paper on penicillin that we have quoted previously. This antibiotic, isolated by Waksman and his assistant, H. Boyd Woodruff, was called actinomycin. It was produced by an actinomycete; it had a beautiful red color, and it was highly toxic. Years later, other workers found that there was more than one actinomycin, that these actinomycins were polypeptides with a phenoxazine chromophore, and that they were thera-peutically active against certain tumors. But at the time of its discovery, actinomycin was an interesting failure. After investigating a few products which were elaborated by fungi and which did not appear adequate for chemotherapy, Waksman and Woodruff reported in 1942 on the isolation of streptothricin. It was the product of the metabolism of an actinomycete and was active not only against Gram-positive bacteria but also against Gram-negative organisms and mycobacteria, organisms not very susceptible to the action of penicillin. The major drawback of streptothricin was its delayed toxicity. Later it was found that there was more than one strep-tothricin and that they represented a family of unusual polypeptides.

The general properties of streptothricin, that is, broad antibacterial action that included activity against mycobacteria, basicity, and water solubility, were also found to be characteristic of streptomycin, reported, in 1944, by Albert Schatz, Elizabeth Bugie, and Waksman.

The following paper written in December, 1944, by W. H. Feldman and H. C. Hinshaw of the Mayo Clinic of Rochester, Minnesota, was the first report of the activity of streptomycin on experimental tuberculosis (*Proc. Staff Meetings Mayo Clinic,* **19:** 593–599, 1944): [1]

In April 1944, Dr. Waksman kindly furnished us a small quantity of streptomycin prepared in his laboratory. This material had a potency of 37 units per mg. The amount available was sufficient to permit treatment of four guinea pigs only.

Twelve male guinea pigs weighing approximately 500 gm. each were inoculated subcutaneously in the sternal region with 0.1 mg. of a sixteen day old culture of tubercle bacilli, strain H37RV. Eight of the infected animals received no medicament; these served as controls both for this and for additional chemotherapeutic studies being done concurrently. Administration of streptomycin to two of the animals was started on the day of inoculation with *Mycobacterium tuberculosis* while treatment of the other two animals was delayed for two weeks.

The dose of streptomycin for the guinea pig was set arbitrarily at 75 mg. (2,775 units) for each twenty-four hour period. One of the two animals in which treatment was started immediately after infection, received only half of this amount. The substance was given subcutaneously five times daily at three hour intervals. The first injection of each day was at 9:00 A.M. and the last injection at 9:00 P.M.

Since a daily dose of 75 mg. (2,775 units) of streptomycin was tolerated satisfactorily by the one animal in which treatment with this amount was started at the time of inoculation with *Mycobacterium tuberculosis,* a similar dose was used to treat each of the two remaining animals, starting on the fifteenth day after infection.

Treatment was continued until the fifty-fourth day of infection, when the supply of streptomycin was exhausted. At the termination of treatment the four animals receiving streptomycin were killed for necropsy. During the preceding two weeks, two of the untreated controls had died. To provide additional material for comparison two more controls were killed. The remaining four animals in the control group were killed sixty days after infection.

[1] Reproduced by permission of Dr. H. C. Hinshaw.

At the time of necropsy the spleen of each of the four animals that had received streptomycin was removed aseptically. The spleens were divided approximately into two equal portions; one portion was preserved for subsequent histologic study and the other was ground and suspended in sterile physiologic solution of sodium chloride. Each splenic suspension was used to make cultures and to inoculate subcutaneously two normal guinea pigs. The recipients were killed for necropsy fifty-six days after inoculation.

Results of First Experiment

Toxicity

In the dosage used, streptomycin appeared to be well tolerated by each of the four animals, since all remained in apparently good health. Each of the four guinea pigs increased in weight. However, the greatest gain occurred in the two animals that had received streptomycin for the shorter period.

The hemoglobin values of the treated animals, determined from blood specimens removed by cardiac puncture at the time of necropsy, were but slightly below normal and in no sense critical. The values expressed in grams per 100 c.c. of blood were 12.8, 12.8, 13.2 and 12 respectively. The average value for the four treated animals was 12.7 gm., a figure that was in excess of the average hemoglobin value for the six untreated controls killed for necropsy, which was 11.

A further indication of the lack of toxicity of streptomycin in the doses given was the absence of recognizable tissue changes in the kidneys, adrenals, liver, lungs, urinary bladder, lymph nodes and bone marrow of each of the treated guinea pigs. Unlike most of the sulfone compounds which we have studied, streptomycin did not induce large, dark spleens.

Antituberculosis effects

The rather severe and widely distributed tuberculosis that occurred in the eight guinea pigs constituting the untreated control group provided satisfactory material for judging the effect which was exerted by streptomycin. The disease was strikingly evident grossly in each of the controls and microscopically the lethal destructiveness of the advancing tuberculous process was characteristic and at marked variance with the situation in the animals treated with streptomycin. The amount of tuberculosis in the subcutis at the site of injection and in the contiguous lymph nodes was minimal in three of the four animals treated with streptomycin. In one guinea pig, lesions were not found in either of these situations. Among the treated animals the amount of visceral tuberculosis was likewise minimal. Among the four spleens there was only one in which tuberculosis was recognizable grossly and in this instance the involvement consisted of a single nodule 0.1 cm. in diameter.

On microscopic examination lesions were found which could not be seen grossly but the changes in the spleens were quantitatively of minor degree and when present were definitely nonprogressive or even retrogressive. The same was true of the lungs and livers of the animals treated with streptomycin. The disease either was not present or when found was minimal and arrested or nonprogressive. The average numerical index of infection for the four guinea pigs treated with streptomycin was 2.8. For the eight controls the average index of infection was 81.9.* The relative amounts of tuberculosis in the treated and untreated groups of animals are shown in Table 2.

Mycobacterium tuberculosis was recovered in cultures made from the spleen of only one of the treated guinea pigs. This animal had received a daily dose of 1,387 units of streptomycin for fifty-four days. However, that viable and virulent tubercle bacilli were present in each of the spleens from the four treated animals was demonstrated by the fact that tuberculosis was present at necropsy in one or both of the two normal recipients that had been inoculated with the splenic suspensions from each.

In this preliminary and admittedly inadequate experiment, the differences in the amount and character of the tuberculosis in the two groups of guinea pigs—the controls and those treated with streptomycin—were sufficiently impressive to warrant a second experiment to determine if the deterrent effects observed in the first experiment could be confirmed.

Second Experiment

Each of twenty male guinea pigs weighing approximately 500 gm. each was inoculated subcutaneously in the sternal region with 0.1 mg. of a culture of *Mycobacterium tuberculosis*. Ten were inoculated with

---

* Based on the arbitrary selection of the numeral 100 as representing the theoretical maximal amount of tuberculosis possible.

TABLE 2   *Summary of Results Obtained in Experiment 1 and Experiment 2*

Experiment	Animals	Duration of infection, days		Duration of treatment, days	Organs showing macroscopic tuberculosis			Index of infection determined microscopically*
		Died	Killed		Spleen	Liver	Lungs	
1 Controls (8 animals)	1	43	...	0	0	1	1	81.9
	1	53	...	0	1	1	1	
	2	...	54	0	2	2	2	
	4	...	60	0	4	3	2	
1 Treated (4 animals)	2	...	54	39	1	0	0	2.8
	2	...	54	54	0	0	0	
2 Controls (9 animals †)	9	...	61	0	8	7	7	67
2 Treated (9 animals †)	4	...	61	47	1	0	0	5.8
	5	...	61	61	0	0	0	

* Tissues examined included spleen, liver, lungs, tracheobronchial lymph nodes, subcutis at the site of injection and the axillary lymph nodes (100 units represents theoretical maximal amounts of tuberculosis possible).

† Of the ten animals in the group originally, one died prematurely.

strain H37RV and ten with a strain (3728) isolated within the past year from the sputum of a patient suffering from pulmonary tuberculosis.* The animals were divided equally into two groups, each group containing five animals inoculated with strain H37RV and five inoculated with strain 3728. One group of ten animals was to be treated with streptomycin and the other ten were to serve as controls.

Streptomycin was administered to six animals—three inoculated with H37RV and three with strain 3728—beginning the same day as the infective inoculum was introduced. Treatment of the other four animals —two inoculated with H37RV and two with strain 3728—was delayed until the fifteenth day after inoculation.

Dosage of streptomycin

The streptomycin available when the second experiment was started was supplied by Dr. Waksman and represented a different lot of the substance from that furnished by him for the first experiment. Each of the six guinea pigs in which treatment was started on the day of infection received a daily dose of streptomycin of 3,500 units divided into four daily doses at six hour intervals. In all instances the streptomycin was dissolved in sterile physiologic solution of sodium chloride and injected subcutaneously. At the end of one week of treatment the weight of each of the animals either had remained stationary (one guinea pig) or had decreased (five guinea pigs); consequently the daily dose of streptomycin was temporarily reduced to 1,750 units. This dose schedule was continued for the ensuing nineteen days, when a more purified lot of streptomycin was received from Dr. Waksman. The dose was then increased to 3,000 units daily for a period of fifteen days. A further refined product was then received † and the dose was increased to 6,000 units. This dose was maintained for the next twenty days, at the end of which the experiment was terminated. The dose schedule for streptomycin for the four animals in which treatment was delayed until the fifteenth day of infection was the same as that being currently employed for the other animals.

Treatment was continued until sixty-one days after the animals had been infected. The treated animals and the untreated controls were then killed for necropsy. At the time of necropsy blood was taken by

---

* Previous observations had revealed this latter strain to be fully virulent for guinea pigs.

† Kindly supplied by Merck & Co., Inc., Rahway, New Jersey, through the courtesy of Dr. R. T. Major, Dr. Hans Molitor, Dr. J. M. Carlisle and Dr. D. F. Robertson.

cardiac puncture for the determination of hemoglobin values and a study of the morphologic aspects of the blood. Approximately half of each spleen was removed aseptically and, after appropriate preparation was cultured for *Mycobacterium tuberculosis.* Tissues from the spleen, liver, lungs and kidneys, the site of inoculation and the axillary space, including the axillary lymph nodes, of each animal were preserved for subsequent microscopic examination.

Results of Second Experiment

Toxicity

The preparation of streptomycin used in the earlier phase of the second experiment was tolerated less satisfactorily than was true of the streptomycin used in the first experiment. As mentioned previously, most of the animals lost weight after administration of the substance had been started. In one case edema and congestion of the external genitalia developed and the animal finally died after having received streptomycin for only nine days. At necropsy the important lesions were those of severe acute peritonitis. One other animal experienced considerable loss of weight, which continued even after the daily dose had been reduced to 1,750 units. Treatment was withheld intermittently in order to permit the animal to recover its loss of weight. After the more refined product was obtained, no further difficulty was experienced. At the end of the experimental period each of the nine surviving guinea pigs weighed more than at the beginning. The gain in body weights varied from 85 gm. to 300 gm. with an average gain of 177 gm. for the nine animals.

The concentrations of hemoglobin for the nine surviving animals that were treated with streptomycin varied from 11.6 gm. to 13.6 gm. per 100 c.c. of blood, the average being 12.8 gm. The hemoglobin values for eight of the nine untreated controls that survived the sixty-one day period of observation varied from 5 gm. to 13.6 gm., the average being 11 gm. Studies of the morphologic aspects of the blood picture revealed nothing of significance.

Microscopic examination of the tissue of the animals that had received streptomycin failed to disclose in the parenchymal organs recognizable changes that could be interpreted as evidence of toxic effects. In some instances the tissue from the axillary space, which was the site of frequent injections of streptomycin, showed some cellular reaction, edema and hemorrhage of minor degree.

Antituberculosis effects

There was a marked and striking difference in the results of the tuberculous infection between the controls and the treated animals. This was evident from the gross appearance of the respective animals at necropsy. The disease in the untreated controls was widely disseminated and in most instances destructive. Among the animals that had received streptomycin the reverse was true, evidence of the disease being absent or barely detectable. The disease had in most instances remained localized at the site of inoculation and the contiguous lymph nodes of the axillary region. In only two of the treated animals was tuberculosis observed grossly in any of the organs. In one instance the spleen contained two minute foci while in the other one tiny nodule was present. In three of the nine treated guinea pigs tuberculous involvement of the tracheobronchial lymph nodes was noted by careful palpation of these structures. The differences in the amounts of tuberculosis in the treated and untreated groups of guinea pigs are given in Table 2.

Microscopically, the disease in the untreated controls revealed the usual advancing morbid processes in the spleen, lungs and liver that one expects in guinea pigs after inoculation with virulent *Mycobacterium tuberculosis*. Microscopic examination of the spleen, liver and lungs of each of the treated animals showed the situation to be entirely different. No lesions were found in the lungs of any of the guinea pigs while the livers of only two were involved. These lesions were too small to be seen grossly, being exceedingly few and atrophic. No lesions were found microscopically in the spleens of four of the treated animals. In one, the tuberculosis was limited to a single microscopic focus composed largely of so-called foam cells and in each of the other four an occasional solitary hard tubercle could be seen, which had not been noted grossly. Incidentally, the latter spleens were from the four animals in which treatment had been delayed for fifteen days after the animals had been infected.

Expressed numerically, the index of infection for the untreated controls was 67; for the animals that had received streptomycin the figure was 5.8. There was no recognizable difference in the ability of the streptomycin to affect favorably infection induced by strain H37RV and strain 3728.

The attempts to recover *Mycobacterium tuberculosis* by culture from the splenic tissue of treated guinea pigs resulted in positive results from the spleens of three animals and negative results in six. Two of the three positive results were obtained from the spleens of animals in which the beginning of treatment had been delayed. Each of the animals in which *Mycobacterium tuberculosis* was recovered had been inoculated with strain H37RV.

Sensitivity to tuberculin was demonstrated in seven of the treated animals at the termination of the experiment. In two guinea pigs the results of the tuberculin test were indefinite.

Conclusions

One may conclude that streptomycin is an antibiotic substance well tolerated by guinea pigs, which is capable under the conditions imposed of exerting a striking suppressive effect on the pathogenic proclivities in guinea pigs of the human variety of *Mycobacterium tuberculosis*.° The results with streptomycin are comparable to those observed previously with certain drugs of the sulfone series.

In these preliminary tests, as the authors stated, streptomycin compared favorably with the sulfones. In practice, it was superior to the sulfones for the treatment of tuberculosis. It became the standard drug for the treatment of this disease and other infections. Streptomycin, however, did not signal the passing of synthetic drugs from the field of chemotherapy of tuberculosis, since *p*-aminosalicylic acid (PAS) and isoniazid, both compounds with simple chemical structures, have been found effective in the treatment of tuberculosis and have often been used in combination with streptomycin.

Chemically, streptomycin was not a polypeptide. Its molecule was composed of three moieties: methylglucosamine, a new pentose called streptose, and streptidine, which is a sugarlike moiety with two guanido substituents. Biologically, it was a very effective chemotherapeutic agent but with certain drawbacks. Upon prolonged administration, such as that needed in the treatment of tuberculous patients, otic and vestibular disturbances were observed; also streptomycin was an efficient agent for the selection of microbial mutants resistant to its action. Streptomycin was even a growth factor for certain bacterial mutants. All these drawbacks were circumvented mostly by the use of streptomycin in combination with other drugs, mainly PAS and isoniazid.

Streptomycin was a great scientific, medical, and financial success. This last aspect of success often brings trouble. Schatz did not feel that Waks-

---

° A third experiment with streptomycin in which treatment was begun forty-nine days after the animals were inoculated with *Mycobacterium tuberculosis* was started some months ago. Treatment will continue for an indefinite period.

man and Rutgers University were treating him fairly in this matter. He sued, with the usual unpleasant side effects of such legal actions. Eventually, the whole matter was settled out of court with a distribution of credit and royalties that was considered proper by those concerned.

The search continued in Waksman's laboratories for the isolation of better antibiotics. In all, about twenty antibiotics have been isolated in the laboratories of microbiology of Rutgers University. One, neomycin, reported by Waksman and Lechevalier in 1949, has found practical application. It was very active against a number of bacteria, and bacterial resistance was less of a problem with it than with streptomycin. However, it was more toxic than streptomycin, and its medical use has been mainly limited to oral and topical applications. Chemically, it represented the first of a series of new antibiotics that could be called amino-polyosides. The molecules of such compounds are formed by the glycosidic linkage of a number of amino sugar residues (or sugarlike aminated moieties) to which may be attached some nonaminated sugar residues.

The royalties coming from the commercialization of streptomycin and neomycin were used primarily for the building and the support of an Institute of Microbiology at Rutgers University. For the discovery of streptomycin, Dr. Waksman received a Nobel Prize in 1952.

As important as streptomycin might have been, the greatest contribution of Dr. Waksman to chemotherapy was to pioneer the systematic search for antagonistic soil microorganisms and to stress the tremendous potentialities of actinomycetes in this domain. Huge industrial screening programs have exploited this mine, and the unknown, mysterious filamentous organisms, which a young Russian immigrant was plating out in 1915, have become, in great part through his efforts, the producers of the most important antibiotics of today, except for penicillin.

Space would not permit to review all the useful antibiotics that are elaborated by actinomycetes. This subject has been reviewed in a book published in 1962 by Waksman and Lechevalier. One should not fail to mention, however, the discoveries of chloramphenicol and the tetracyclines. These compounds have extended the range of activity of antibiotics to the therapy of diseases caused by rickettsiae and the so-called "larger viruses." In addition, these antibiotics are superb antibacterial agents, active against both Gram-negative and Gram-positive organisms.

The discovery of chloramphenicol was the result of the cooperation of scientists from Parke, Davis and Company, headed by Dr. John Ehrlich, Dr. Paul R. Burkholder, of Yale University, and Dr. David Gottlieb from the University of Illinois. The discovery of this important drug was reported in 1947. Chloramphenicol has a rather simple molecule, containing nitrobenzene and two atoms of nonionic chlorine. Four isomers of chloramphenicol exist, but the D-threo form of the molecule is vastly superior to the others and is the only one used clinically. It is mainly synthesized chemically rather than produced by fermentation.

The first of the tetracyclines to be isolated was chlortetracycline, long known as Aureomycin. Its discovery was reported in 1948 by a group of scientists from the Lederle Laboratories. A key role in the isolation of this compound had been played by a distinguished mycologist, Dr. B. M. Duggar. In 1950, scientists from the Charles Pfizer Company reported the isolation of a similar compound, oxytetracycline, first named Terramycin until the elucidation of its chemical structure. Later another similar antibiotic was isolated. Since it is considered as the parent compound of these substances, it is called tetracycline.

Chloramphenicol and the tetracyclines, together with penicillin and streptomycin, form the backbone of the antibiotic arsenal at the disposal of the clinician. All but penicillin are produced by actinomycetes.

The methods used in the study of antibiotics and the methods of synthetic chemistry were harmoniously married in the more recent developments in penicillin chemistry. Antibiotics are usually formed by microorganisms as a family of closely related substances. We have already mentioned this for the actinomycins and the streptothricins. The same was true for the penicillins. The most useful of the early penicillins, and the one that was used at first, was penicillin G, or benzyl penicillin. This substance could be produced selectively by adding specific precursors such as phenylacetic acid to the culture medium in which the producing organism was growing. However, benzyl penicillin, like all drugs, had certain limitations. It was allergenic; it did not provide prolonged blood levels, and certain bacteria, mainly Gram-negative ones, were not affected by this form of penicillin. Modifications of the molecule of penicillin have yielded products better than penicillin G in some respects. All penicillins can be considered as derivatives of the same basic nucleus, 6-aminopenicillanic acid. In 1959, F. R. Batchelor and coworkers, from Beecham Research Laboratories, in England reported (*Nature*, **183:** 257–258, 1959) that a strain of *Penicillium*, in a medium free of substances that could be used as precursors for the formation of one penicillin or another, produced free 6-aminopenicillanic acid. This basic nucleus could also be obtained by the enzymatic cleavage of benzyl penicillin.

The availability of 6-animopenicillanic acid was a boon to chemotherapy. It opened a road to the synthesis of series of compounds, just as atoxyl and sulfanilamide had done in the past, and optimistic reports started to refer to "tailor made" penicillins that would be predicted from the "blueprints" to have a given property.

Mode of Action of Chemotherapeutic Agents

In order to design drugs able to perform a specific chemotherapeutic task, it is helpful to have some idea about the mode of action of the clinically successful drugs. In addition, a knowledge of the mode of action of drugs permits their utilization not only as chemotherapeutic agents but as a tool in physiological studies. The knowledge of the mode of action of drugs has developed concomitantly with that of the structure and physiology of microorganisms.

The basic questions to answer in the elucidation of the mode of action of a chemotherapeutic agent are: "Why is this substance toxic for one organism and not for another?" "Why is this substance toxic for one kind of cell and not for another?" That there are differences from one organism to another and from one cell to another is obvious since all cells and all organisms are not alike. This is true in spite of a certain unity in some of the basic biochemical processes. One of the fundamental differences in the "chemism" of organisms is that some of them require certain growth factors and others do not. We recall in this domain the Pasteur-Liebig controversy and the role of Wildiers in solving the mysterious cause of the disagreement between the two men. This subject was discussed in Chapter 9. Differences in vitamin requirements also explained the mode of action of sulfa drugs. D. D. Woods published a paper in 1940 (*Brit. J. Exp. Pathol.*, 21: 74–90) in which he reported that p-aminobenzoic acid could inhibit the action of sulfanilamide. Woods, who was working in the biochemistry laboratories of the Middlesex Hospital in London, concluded his paper as follows: [1]

> The present investigation was based on a general working hypothesis that anti-bacterial substances act by interfering with some substance essential to the bacterial cell ("essential metabolite," Fildes, *Brit. J. Exp. Pathol.*, 21: 67, 1940). The experiments . . . provide strong evidence that the inactivation is due to competition for an enzyme between

[1] Reproduced by permission of the director of the *British Journal of Experimental Pathology*.

the essential metabolite and the inhibitor. A clearer hypothesis of the possible mode of action of sulphanilamide may now be built up and may prove useful as a basis for further work. . . .

In the first place it is suggested that p-aminobenzoic acid is essential for the growth of the organism. It is, however, normally synthesized in sufficient quantity by the present strain of streptococcus (and by *coli*), since it is not necessary to add it to a medium containing only known substances or preparations known to be free from anti-sulphanilamide activity. It can also be extracted from the streptococcal cell. On the basis of the experimental work it is next suggested that the enzyme reaction involved in the further utilization of p-aminobenzoic acid is subject to competitive inhibition by sulphanilamide, and that this inhibition is due to a structural relationship between sulphanilamide and p-aminobenzoic acid (which is the substrate of the enzyme reaction in question). . . . It was found that the concentration of p-aminobenzoic acid required to overcome this inhibition is $\frac{1}{5000}$–$\frac{1}{25,000}$ of the concentration of sulphanilamide used. The further course of events in a culture may now be considered as follows:

1. *p-aminobenzoic acid is present preformed in the medium.* Growth occurs. As the anti-sulphanilamide factor appears to be widely distributed in small amount, this may account in part for the difficulty in getting complete inhibition on more complex media.

2. *p-aminobenzoic acid is absent from the medium or present in insufficient concentration.* This would be the position under the test conditions used here. There are now two possibilities:

> 2a. *The organism is unable to synthesize enough p-aminobenzoic acid.* Growth is therefore inhibited. This would normally be the case with the streptococcus under the conditions used in the present experiments.

> 2b. *The organism is able to make sufficient p-aminobenzoic acid.* In this case the competitive inhibition is overcome and growth occurs. This is presumed to be the case with organisms that are insensitive to sulphanilamide.

The conditions determining whether (2a) or (2b) shall take place are delicately balanced, and this may explain why different organisms, and even the same organism under differing growth conditions, exhibit many degrees of sensitivity to sulphanilamide. It is here suggested that such differences in sensitivity are correlated with quantitative differences in ability to synthesize p-aminobenzoic acid. On meat infusion broth (in which its growth rate approaches optimum) *Bact. coli* is almost indifferent to sulphanilamide, whilst on the ammonium lactate medium (where the growth rate is sub-optimal) inhibition is well marked. Similarly, complete inhibition of streptococcal growth is not obtained (a) with rich media on which growth is very rapid, or (b) with poorer media if the inoculum is large or consists of young actively dividing cells. To account for such variability it is suggested that the original inoculum contains sufficient *p*-aminobenzoic acid to reverse the inhibition and permit some (non-visible) growth to take place; such growth is known to occur in the early stages of inhibition. When this supply of p-aminobenzoic acid becomes exhausted by the further enzyme reaction under discussion, subsequent growth will depend on the rate at which more can be synthesized, and this in turn on the number of organisms present and thus on the initial growth rate. Presence of precursors of *p*-aminobenzoic acid in the medium may also influence the rate of synthesis.

Another possible interpretation of the experimental results is that sulphanilamide inhibits the enzyme reaction involved in the *synthesis* (and not, as above, the further utilization) of *p*-aminobenzoic acid, and that it does so this time by virtue of its chemical similarity to the product of the reaction. The balance of evidence is against this view, as in this case it would be expected, on any simple interpretation, that the addition of just sufficient *p*-aminobenzoic acid for the needs of the organism should cause growth; the amount of *p*-aminobenzoic acid needed should thus be independent of sulphanilamide concentration.

p-Aminobenzoic acid is one of the building blocks of pteroylglutamic acid, the fundamental unit of the "folic acid" family of coenzymes. These coenzymes play a key role in the biosynthesis of nucleic acids. The secret of the selective action of the sulfa drugs is, broadly speaking, that mammals cannot synthesize folic acid but must receive it in their food, as a vitamin. On the other hand, the bacteria susceptible to the action of sulfa drugs can perform this synthesis and pay for their nutritional independence by their metabolic vulnerability.

Adrien Albert, in his book on *Selective Toxicity*, rightly remarked: [1] ". . . *comparative biochemistry* is, of all branches of science, the one that holds the master key for the *logical* discovery of selectively toxic agents. It can reveal metabolic differences between the economic species which man wishes to save and the uneconomic species which he wishes to destroy. Once these metabolic peculiarities are discovered the next step is to devise selective agents which can use them to cause irreparable damage to the un-

---

[1] Reproduced by permission of John Wiley & Sons, Inc., publishers.

economic species. Unfortunately, comparative biochemistry has so far attracted few workers, although so much of selective toxicity is actually applied comparative biochemistry!"

References

Albert, A.: *Selective Toxicity,* 2d ed., Methuen & Co., Ltd., London, John Wiley & Sons., Inc., New York, 1960.

Duthie, E. S.: Molecules against Microbes, Sigma Books, London, 1946.

Fabre, R., and G. Dillemann: *Histoire de la pharmacie,* Presses Universitaires de France, Paris, 1963.

Florey, H. W., E. Chain, N. G. Heatley, M. A. Jennings, A. G. Sanders, E. P. Abraham, and M. E. Florey: *Antibiotics,* Oxford University Press, London, 1949, vol. I.

Gunther, R. T.: *The Greek Herbal of Dioscorides,* Hafner Publishing Company, Inc., New York, 1959 (republication).

Himmelweit, F. (ed.): *The Collected Papers of Paul Ehrlich,* vol. III, *Chemotherapy,* Pergamon Press, New York, 1960.

Marquardt, M.: *Paul Ehrlich,* William Heinemann, Ltd., London, 1949.

Maurois, A.: *The Life of Sir Alexander Fleming* (Translated from the French by G. Hopkins), Jonathan Cape, Ltd., London, 1959.

Papacostas, G., and J. Gaté: *Les Associations microbiennes, leurs applications thérapeutiques,* Gaston Doin et Cie, Paris, 1928.

Raper, K. B.: "A Decade of Antibiotics in America," *Mycologia,* **44:** 1–59, 1952.

Waksman, S. A.: *My Life with the Microbes,* Simon and Schuster, Inc., New York, 1954.

Waksman, S. A.: *Microbial Antagonisms and Antibiotic Substances,* The Commonwealth Fund, New York, 1947.

Waksman, S. A., and H. Lechevalier: *The Actinomycetes,* vol. III, *Antibiotics of Actinomycetes,* The Williams & Wilkins Company, Baltimore, 1962.

Genetics

12

Breeding and selection of stock desirable for one purpose or another have been carried out empirically for longer than the human race has had a written history. The phenomenon of reproduction of living organisms, however, is far from being simple, and considering the span of existence of humanity, it was only yesterday that Spallanzani was putting little pants on male frogs to make sure that they produced a seminal liquid, and Pasteur was fighting the doctrine of spontaneous generation.

It was in a little village of Moravia, then a province of Austria, that the founder of genetics, Johann Mendel, was born on July 22, 1822. The child became a man, a priest, and a genial scientist known to us as Gregor Mendel. The name under which we know him was not given at baptism but later when he became an Augustinian Father.

There were three children in the family, which was well suited to harbor the discoverer of the laws of the lottery of heredity. The older daughter looked and acted like her father (she was lean and serious); Johann was plump like his mother and was a mixture of seriousness (father) and cheerfulness (mother), whereas the younger daughter was essentially the happy image of her mother.

Johann's father was very interested in the growth of fruit trees, and under his supervision, the young boy spent many hours working in the orchard. As his love of nature burgeoned, questions must have come to his bright little mind as he implanted scions of desirable varieties in the wood of lesser trees.

When Johann reached eleven years of age, the local school teacher, who had noticed his talent, suggested that he should continue his studies at a more lavish school located in the nearest town of importance. The mother was in favor of the plan, but the father was torn between his desire to have the boy continue farming in his footsteps and the hope of seeing him escape the hard lot of the farmer. At that time, the *corvée* still existed, and Johann's father had to devote some of his time to projects which were of interest to the local lord. The desire for betterment was the strongest, and the child left for a brilliant school career.

Ready cash was not common in the pockets of the farmers of Moravia, and Johann probably did not always eat his fill during his school days. The principal of the school was an Augustinian priest, a fact which might have had some influence on the future vocation of the boy. When Johann was sixteen years old, his father suffered a serious accident while involved in a *corvée*. Three years later, as Johann started to show more interest in priest-hood than in farming, and as Mendel the elder could no longer work his farm, it was sold to the husband of the older sister. At about the same time, hardship and privations caught up with the future priest, who had to rest from his studies.

Eventually, the young man went back to complete his studies at the Olmütz Philosophical Institute. As usual he was successful, and he was accepted in 1843 as a novice at the Augustinian monastery of Saint Thomas, in Brno, a town of Czech culture. The monastery was a center of learning; almost every member of the community indulged in the arts or in sciences. Some were interested in botany and its applications. The story has reached us that one of the Augustinian botanists liked to work late at night on oenological projects. The abbot of the monastery, even though a friend of sciences and learning, was disturbed by reports permeating to his level that indicated that a certain instability might be observed in the gait of the returning botanist. One night, the abbot waited for the return of our experi-menter in order to verify the reports and to apply sanctions if needed. After a long wait, the bell was rung, and the abbot, in all his majestic austerity, opened the door. The poor botanist remained speechless, then, recovering his wits, said: "Lord, I am not fit to come into your house." So saying, he turned around to pass the rest of the night in the more understanding com-pany of the wine casks.

In 1847, Johann Mendel, now to be known as Gregor, having always lived "blamelessly, piously, and religiously," was ordained. The young priest had many good qualities but was not cut out to serve in a parish. He was too easily affected by human sufferings to console the sick and to stand at the bedside of the moribund. In addition, his knowledge of Czech was not perfect. As a result he was appointed as a substitute teacher in a high school.

Mendel had never had any trouble with his studies, but he failed an examination that would have qualified him as a regular teacher rather than as a substitute. This was not surprising since he never had had university training. Recognizing that there were deficiencies in his education, his abbot sent him to the University of Vienna. His potential value was fully recognized by this noble prelate, who stated: "I shall not grudge any expense requisite for the furtherance of this training. . . ." In 1854, Mendel was back in Brno as a substitute teacher in physics and natural history at the Modern School of Brno. He was extremely well liked by both the authorities of the school and the students. His explanations were clear, his devotion was boundless, and he was able to control the behavior of the students without recourse to any strong disciplinary measures. His kindness won him not only the affection of children and men but also that of animals. His pupils liked to visit him at the monastery to see members of his zoo. Among other creatures of God that the good father liked to have around was a hedgehog, which went to sleep one night in one of Mendel's shoes, with a distressing effect on Mendel's toes the next morning when he tried to dress. Even though a very successful teacher, Mendel never took the necessary examination that would have permitted him to become an accredited teacher, and up to the time of his election as Abbot and Prelate of his Augustinian Monastery in Brno, Mendel remained a substitute teacher. It was during his teaching years that Mendel performed the experiments on the crossing of peas and other plants that were to make him posthumously famous.

Mendel was not only interested in the breeding of plants, but also, in more general terms, he liked gardening, beekeeping, and meteorology. His interest in the latter field was in part stimulated by Pettenkofer's *Boden* theory of disease.

In 1865, Mendel presented his most important work on plant crossing before the members of the Society for the Study of Natural Science of Brno. The far-reaching value of his data was not appreciated, but they were published the following year in the transactions of the same society. Mendel tried to attract attention to his studies by sending this paper, or monograph, to some of the great scientists of his day. He especially attached much importance to the value of the judgment of Nägeli, to whom he sent not only a copy of the monograph but also lengthy letters. In his first letter to Nägeli, dated December 31, 1866, Mendel remarked that his work differed from that of others in that he had chosen to follow the fate of easily detected characteristics. He said: "Statements like . . . 'the progeny had reverted to the type of the original maternal . . . ancestor' . . . are too general, too vague to furnish a basis for sound judgment. . . . A decision can be reached only when . . . the degree of kinship between the hybrid forms and their parental species is precisely determined, rather than simply estimated from general impression. . . ."

Nägeli was not better prepared to understand the work of Mendel than the less flamboyant scientists of Brno, and the monograph fell into oblivion.

It was the custom in Austria to tax monasteries heavily at the occasion of the election of a new abbot. The frugal clergy used to select the youngest qualified man in the hope that he would have a long life, thus making the payment of the tax as seldom as possible. Mendel was young and well liked, and when the prelate of his monastery died in 1868, he was elected in his place. This gave Mendel all the space that he may have wanted in the garden for his plant crossing experiments, but it deprived him of the necessary time for experimentation. The new prelate was a kind man who enjoyed the good things in life, including jolly bowling games with friends on Sunday afternoons.

Taxation may have been one of the factors responsible for his ascension to the high post of Abbot and Prelate, but taxation also was to ruin the life of the kind priest. While he was in office, a law for the taxation of monasteries was passed. Many abbots felt that it was unconstitutional. The government, even though not willing to admit the unconstitutional aspects of the law, made it clear that it was not interested in milking the clergy. Most abbots compromised with the civil authorities with the result that the clergy kept its money and the government its prestige. Mendel, of stubborn peasant stock, was not the man to indulge in such worldly schemes, and he fought a lonely battle in defiance of the lay authorities. This miserable affair of taxation poisoned the rest of the life of the Abbot and Prelate of the Monastery of Saint Thomas, who died, embittered, in 1884. To the end, he had remained a kind friend of the poor. He now rests in Brno under a simple slab of stone that bears not only his name but also those of two other Augustinian priests.

The work of Mendel was revived in the very year that saw the birth of the twentieth century, stressing symbolically the dependence of the proud new century on the past. Three men, almost simultaneously, rediscovered the work of Mendel and found that their own fresh observations were nothing else but a duplication of what the modest Augustinian had done more than thirty years previously. In March, 1900, Hugo de Vries published two papers in which he reported that if one crossed plants having the characteristic *N* with plants having the characteristic *B*, one obtained hybrids that if permitted to autofecundate produced 25 per cent daughter plants with the characteristic *B* and 75 per cent with the characteristic *N*. He followed the fate of the progeny over several generations, observed the constant ratio of segregation, and remarked that his conclusions had been formulated long ago by Mendel.

A month later (April, 1900), Carl Correns of Tübingen published a paper that forcefully brought Mendel's work into focus by its very title: "Gregor Mendel's Rules on the Behavior of Racial Hybrids." Finally from Vienna, in June of the same year, came a paper by Erich Tschermak on the crossing of peas, further raising Mendel from oblivion.

Of the three discoverers of Mendel, Hugo de Vries is the most important figure because he introduced the notion of mutation, which was to play such an important role in the genetics of microorganisms. He was born in Haarlem in 1848, and after studying and working in Germany, he became professor at Amsterdam (1881–1918). He died at the ripe old age of eighty-seven in 1935. He was a plant physiologist, and like many men of his time, he spent much time musing over the evolutionary theory of Darwin. One fact was obvious; plant species seemed to be reasonably stable at the end of the nineteenth century with little tendency to vary. De Vries, however, thought that he had found the proper biological material to demonstrate in his days what might have taken place in the millennia of the past. He noted that the evening primrose, which was growing abundantly in the meadows surrounding Amsterdam, exhibited, apart from the typical forms, a few that were very strange. De Vries started to grow primroses and to study over numerous generations the progeny of autofecundated plants. Once in a while, from the seeds of typical plants there arose individuals that presented characteristics that were very different from the parents. These new, or odd, types had the property of reproducing their like, and De Vries called them mutants. The occurrence of mutants could indeed explain evolution. New species were formed from the mutation of old species. Mutations were more common in primeval days than in ours, speculated De Vries, and this explained how all the species known to us came into being. De Vries's theory was attacked, and doubt was cast on the value of his interpretation of the experimental work. It is now known that the odd behavior of the evening primrose can be explained in many instances on the basis of a special type of genetic recombination. However, to us, this matters very little; micro-organisms are not primroses, and they do mutate.

All sorts of mutations take place in a microbial population: a microbe sensitive to a drug may give rise to a resistant mutant; an organism able to carry out the necessary biosynthetic reactions to utilize simple chemicals as food may give birth to a more exacting strain that will grow only in presence of certain preformed metabolites that its "father" was able to make for itself. Such "biochemical mutants" were reported by G. W. Beadle and E. L. Tatum in 1941 (*Proc. Nat. Acad. Sci. U.S.*, **27**: 499–506): [1]

From the standpoint of physiological genetics the development and functioning of an organism consist essentially of an integrated system of chemical reactions controlled in some manner by genes. It is entirely tenable to suppose that these genes which are themselves a part of the system, control or regulate specific reactions in the system either by acting directly as enzymes or by determining the specificities of enzymes. Since the components of such a system are likely to be interrelated in complex ways, and since the synthesis of the parts of individual genes are presumably dependent on the functioning of other genes, it would appear that there must exist orders of directness of gene control ranging from simple one-to-one relations to relations of great complexity. In investigating the roles of genes, the physiological geneticist usually attempts to determine the physiological and biochemical bases of already known hereditary traits. This approach, as made in the study of anthocyanin pigments in plants, the fermentation of sugars by yeasts and a number of other instances, has established that many biochemical reactions are in fact controlled in specific ways by specific genes. Furthermore, investigations of this type tend to support the assumption that gene and enzyme specificities are of the same order. There are, however, a number of limitations inherent in this approach. . . . A . . . difficulty . . . is that the standard approach to the problem implies the use of characters with visible manifestations. Many such characters involve morphological variations, and these are likely to be based on systems of biochemical reactions so complex as to make analysis exceedingly difficult.

Considerations such as those just outlined have led us to investigate the general problem of the genetic control of developmental and metabolic reactions by reversing the ordinary procedure and, instead of attempting to work out the chemical bases of known genetic characters, to set out to determine if and how genes control known biochemical reactions. The ascomycete *Neurospora* offers many advantages for such an approach and is well suited to genetic studies. Accordingly, our pro-

[1] Reproduced by permission of Dr. G. Beadle and the National Academy of Sciences.

gram has been built around this organism. The procedure is based on the assumption that x-ray treatment will induce mutations in genes concerned with the control of known specific chemical reactions. If the organism must be able to carry out a certain chemical reaction to survive on a given medium, a mutant unable to do this will obviously be lethal on this medium. Such a mutant can be maintained and studied, however, if it will grow on a medium to which has been added the essential product of the genetically blocked reaction. The experimental procedure based on this reasoning can best be illustrated by considering a hypothetical example. Normal strains of *Neurospora crassa* are able to use sucrose as a carbon source, and are therefore able to carry out the specific and enzymatically controlled reaction involved in the hydrolysis of this sugar. Assuming this reaction to be genetically controlled, it should be possible to induce a gene to mutate to a condition such that the organism could no longer carry out sucrose hydrolysis. A strain carrying this mutant would then be unable to grow on a medium containing sucrose as a sole carbon source but should be able to grow on a medium containing some other normally utilizable carbon source. In other words, it should be possible to establish and maintain such a mutant strain on a medium containing glucose and detect its inability to utilize sucrose by transferring it to a sucrose medium.

Essentially similar procedures can be developed for a great many metabolic processes. For example, ability to synthesize growth factors (vitamins), amino acids and other essential substances should be lost through gene mutation if our assumptions are correct. Theoretically, any such metabolic deficiency can be "by-passed" if the substance lacking can be supplied in the medium and can pass cell walls and protoplasmic membranes.

In terms of specific experimental practice, we have devised a procedure in which x-rayed single-spore cultures are established on a so-called "complete" medium, i.e., one containing as many of the normally synthesized constituents of the organism as is practicable. Subsequently these are tested by transferring them to a "minimal" medium, i.e., one requiring the organism to carry on all the essential syntheses of which it is capable. In practice the complete medium is made up of agar, inorganic salts, malt extract, yeast extract and glucose. The minimal medium contains agar (optional), inorganic salts and biotin, and a disaccharide, fat or more complex carbon source. Biotin, the one growth factor that wild type *Neurospora* strains cannot synthesize, is supplied in the form of a commercial concentrate containing 100 micrograms of biotin per cc. Any loss of ability to synthesize an essential substance present in the complete medium and absent in the minimal medium is indicated by a strain growing on the first and failing to grow on the second medium. Such strains are then tested in a systematic manner to determine what substance or substances they are unable to synthesize. These subsequent tests include attempts to grow mutant strains on the minimal medium with (1) known vitamins added, (2) amino acids added or (3) glucose substituted for the more complex carbon source of the minimal medium. . . .

Single ascospore strains are individually derived from perithecia of *N. crassa* and *N. sitophila* x-rayed prior to meiosis. Among approximately 2000 such strains, three mutants have been found that grow essentially normally on the complete medium and scarcely at all on the minimal medium with sucrose as the carbon source. One of these strains (*N. sitophila*) proved to be unable to synthesize vitamin B_6 (pyridoxine). A second strain (*N. sitophila*) turned out to be unable to synthesize vitamin B_1 (thiamine). Additional tests show that this strain is able to synthesize the pyrimidine half of the B_1 molecule but not the thiazole half. If thiazole alone is added to the minimal medium, the strain grows essentially normally. A third strain (*N. crassa*) has been found to be unable to synthesize para-aminobenzoic acid. This mutant strain appears to be entirely normal when grown on the minimal medium to which *p*-aminobenzoic acid has been added. Only in the case of the "pyridoxinless" strain has an analysis of the inheritance of the induced metabolic defect been investigated. For this reason detailed accounts of the thiamine-deficient and *p*-aminobenzoic acid-deficient strains will be deferred.

Qualitative studies indicate clearly that the pyridoxinless mutant, grown on a medium containing one microgram or more of synthetic vitamin B_6 hydrochloride per 25 cc. of medium, closely approaches in rate and characteristics of growth normal strains grown on a similar medium with no B_6. Lower concentrations of B_6 give intermediate growth rates. . . .

In order to ascertain the inheritance of the pyridoxinless character, crosses between normal and mutant strains were made. The techniques for hybridization and ascospore isolation have been worked out and described by Dodge, and by Lindegren. The ascospores from 24 asci of the cross were isolated and their positions in the asci recorded. For some unknown reason, most of these failed to germinate. From seven asci, however, one or more spores germinated. These were grown on a medium containing glucose, malt-extract and yeast extract, and in this they all grew normally. The normal and mutant cultures were differentiated by growing them on a B_6 deficient medium. On this medium the mutant cultures grew very little, while the non-mutant ones grew normally.

The results were summarized in a table and the authors continued:

> It is clear from these rather limited data that this inability to synthesize vitamins B_6 is transmitted as it should be if it were differentiated from normal by a single gene.
>
> The preliminary results summarized above appear to us to indicate that the approach outlined may offer considerable promise as a method of learning more about how genes regulate development and function. For example, it should be possible, by finding a number of mutants unable to carry out a particular step in a given synthesis, to determine whether only one gene is ordinarily concerned with the immediate regulation of a given specific chemical reaction.
>
> It is evident, from the standpoints of biochemistry and physiology, that the method outlined is of value as a technique for discovering additional substances of physiological significance. Since the complete medium used can be made up with yeast extract or with an extract of normal *Neurospora,* it is evident that if, through mutation, there is lost the ability to synthesize an essential substance, a test strain is thereby made available for use in isolating the substance. It may, of course, be a substance not previously known to be essential for the growth of any organism. Thus we may expect to discover new vitamins, and in the same way, it should be possible to discover additional essential amino acids if such exist. We have, in fact, found a mutant strain that is able to grow on a medium containing Difco yeast extract but unable to grow on any of the synthetic media we have so far tested. Evidently some growth factor present in yeast and as yet unknown to us is essential for *Neurospora.*

These mutants played an important role in the elucidation of metabolic pathways and were most useful for studies on the biosynthesis of growth factors. Much of the earlier work was carried out on *Neurospora* mutants, but bacterial mutants were also widely used a few years later.

Mutations can be defined as "spontaneous, undirected changes that may occur in the nuclear determinants of hereditary characteristics." That changes do occur in a microbial population is not a difficult observation to make, even though, for a period of time at least, the dogmatic Koch refused to admit it. The comparatively great variation observed in microorganisms is due to their rapid rate of reproduction. Mutations are rare events usually occurring at a rate of 10^{-4} to 10^{-10} per bacterium per generation, which means that in a bacterial population, at the most, one cell in ten thousand is likely to mutate to a new type. Much more difficult than observing microbial mutants was the actual demonstration that mutations are not directed by the environment. For example, it was possible to demonstrate rather easily that a microbial population sensitive to a drug could give rise to cells resistant to the drug. What was not so simple was to demonstrate that the drug was not the cause of this change. This demonstration was first made by Luria and Delbrück in 1943, with the help of a statistical method called the "fluctuation test." Simplicity was not the main virtue of the fluctuation test. In 1949, Howard B. Newcombe published another method that disproved the directing role of agents used in the selection of bacterial mutants. Newcombe's method was simple enough to be understood without too much difficulty. His findings were published in *Nature* (**164**: 150–151, 1949) under the title of "Origin of Bacterial Variants." [1]

> Numerous bacterial variants are known which will grow in environments unfavorable to the parent strain, and to explain their occurrence two conflicting hypotheses have been advanced. The first assumes that the particular environment produces the observed change in some of the bacteria exposed to it, whereas the second assumes that the variants arise spontaneously during growth under normal conditions, the part played by the adverse environment being purely selective. These are known respectively as the "adaptation" and the "spontaneous mutation" hypotheses.
>
> In order to discriminate between them an experimental approach (the "fluctuation test") was developed by Luria and Delbrück in 1943. This test has been applied to a number of variants, some of them in widely separated strains of bacteria, and in each case the conclusion reached has been that the variant arose by spontaneous mutation. The validity of the fluctuation test has not been challenged, at least so far as I am aware; but on the other hand it has gained only limited recognition. In part this may be due to the statistical and essentially indirect nature of the argument on which it is based, and if so the more direct experimental evidence described below would seem to be of value.
>
> Bacteria of *Escherichia coli* strain B/r susceptible to phage Tl were plated on agar and incubated until a limited population increase had taken place. On alternate plates the bacteria were redistributed over the surface of the agar by spreading with 0.1 cc of sterile saline. All were then sprayed with phage T1, and counts made of the colonies of resistant survivors which developed after further incubation.
>
> On the adaptation hypothesis, the bacteria present at the end of the initial growth-period would all be phage-susceptible, and spreading

[1] Reproduced by permission of the editor of *Nature* and of Dr. H. B. Newcombe.

would serve only to redistribute the members of a homogeneous population. No striking differences in colony count between *spread* and *unspread* plates would therefore be expected.

On the alternative (spontaneous mutation) hypothesis, both susceptible and resistant cells would be present at the end of the initial growth-period wherever a sufficient end-population had been reached. Further, mutations taking place a generation or more before the cessation of growth would each be represented by a minute cluster of resistant cells all descended from the one original mutant. Where the arrangement of the bacteria is left undisturbed, a cluster would give rise to a single resistant colony after the application of phage; but where the bacteria are redistributed over the surface of the agar by spreading, a colony would develop from each resistant cell. Higher counts would thus be expected from the *spread* than from the *unspread* plates. This is, in fact, what was found, the difference being as much as fifty-fold where the end population was highest. . . .

It might be suggested that the bacteria are less likely to become "adapted" when crowded together in the developing colonies. If this were true, the proportion of bacteria becoming "adapted" should be least where microcolony size is greatest. The average number of bacteria per microcolony at the time of spraying rose with increasing period of incubation from 33 up to 54,900, but there was no corresponding decline in the proportion of resistant colonies to bacteria sprayed. Thus, crowding cannot account for the lower colony counts from unspread plates.

This experiment, therefore, confirms the conclusion drawn from the fluctuation test, namely, that phage-resistant variants arise by spontaneous change prior to contact with phage.

The passage of genetic material from one bacterial cell to another raised important questions about the actual method of this transfer. Different mechanisms were shown to be involved. One method, transformation, was discussed in 1944 (*J. Exp. Med.*, **79**: 137–157) by Oswald T. Avery, Colin M. MacLeod, and Maclyn McCarty from the Hospital of the Rockefeller Institute for Medical Research.[1]

Biologists have long attempted by chemical means to induce in higher organisms predictable and specific changes which thereafter could be transmitted in series as hereditary characters. Among microorganisms

[1] Reproduced by permission of The Rockefeller Institute Press, publishers of the *Journal of Experimental Medicine*.

the most striking example of inheritable and specific alterations in cell structure and function that can be experimentally induced and are reproducible under well defined and adequately controlled conditions is the transformation of specific types of Pneumococcus. This phenomenon was first described by Griffith, in 1928, who succeeded in transforming an attenuated and non-encapsulated (R) variant derived from one specific type into fully encapsulated and virulent (S) cells of a heterologous specific type. A typical instance will suffice to illustrate the techniques originally used and serve to indicate the wide variety of transformations that are possible within the limits of this bacterial species.

Griffith found that mice injected subcutaneously with a small amount of a living R culture derived from Pneumococcus Type II together with a large inoculum of heat-killed Type III (S) cells frequently succumbed to infection, and that the heart's blood of these animals yielded Type III pneumococci in pure culture. The fact that the R strain was avirulent and incapable by itself of causing fatal bacteremia and the additional fact that the heated suspension of Type III cells contained no viable organisms brought convincing evidence that the R forms growing under these conditions had newly acquired the capsular structure and biological specificity of Type III pneumococci.

The original observations of Griffith were later confirmed by others. [The phenomenon was also demonstrated to take place in vitro. . . .]

The present paper is concerned with a more detailed analysis of the phenomenon of transformation of specific types of Pneumococcus. The major interest has centered in attempts to isolate the active principle from crude bacterial extracts and to identify if possible its chemical nature or at least to characterize it sufficiently to place it in a general group of known chemical substances. For purposes of study, the typical example of transformation chosen as a working model was the one with which we have had most experience and which consequently seemed best suited for analysis. This particular example represents the transformation of a non-encapsulated R variant of Pneumococcus Type II to Pneumococcus Type III.

Transformation of pneumococcal types *in vitro* requires that certain cultural conditions be fulfilled before it is possible to demonstrate the reaction even in the presence of a potent extract. . . .

After having discussed in detail the conditions necessary for transformation to take place, the authors described the preparation of their active extract and showed its effectiveness. They then concluded:

The present study deals with the results of an attempt to determine the chemical nature of the substance inducing specific transformation of pneumococcal types. A desoxyribonucleic acid fraction has been isolated from Type III pneumococci which is capable of transforming unencapsulated R variants derived from Pneumococcus Type II into fully encapsulated Type III cells. Thompson and Dubos have isolated from pneumococci a nucleic acid of the ribose type. So far as the writers are aware, however, a nucleic acid of the desoxyribose type has not heretofore been recovered from pneumococci nor has specific transformation been experimentally induced *in vitro* by a chemically defined substance.

Although the observations are limited to a single example, they acquire broader significance from the work of earlier investigators who demonstrated the interconvertibility of various pneumococcal types and showed that the specificity of the changes induced is in each instance determined by the particular type of encapsulated cells used to evoke the reaction. From the point of view of the phenomenon in general, therefore, it is of special interest that in the example studied, highly purified and protein-free material consisting largely, if not exclusively, of desoxyribonucleic acid is capable of stimulating unencapsulated R variants of Pneumococcus Type II to produce a capsular polysaccharide identical in type specificity with that of the cells from which the inducing substance was isolated. Equally striking is the fact that the substance evoking the reaction and the capsular substance produced in response to it are chemically distinct, each belonging to a wholly different class of chemical compounds.

The inducing substance, on the basis of its chemical and physical properties, appears to be a highly polymerized and viscous form of sodium desoxyribonucleate. On the other hand, the Type III capsular substance, the synthesis of which is evoked by this transforming agent, consists chiefly of a non-nitrogenous polysaccharide constituted of glucose-glucuronic acid units linked in glycosidic union. The presence of the newly formed capsule containing this type-specific polysaccharide confers on the transformed cells all the distinguishing characteristics of Pneumococcus Type III. Thus, it is evident that the inducing substance and the substance produced in turn are chemically distinct and biologically specific in their action and that both are requisite in determining the type specificity of the cell of which they form a part.

The experimental data presented in this paper strongly suggest that nucleic acids, at least those of the desoxyribose type, possess different specificities as evidenced by the selective action of the transforming principle. Indeed, the possibility of the existence of specific differences in biological behavior of nucleic acids has previously been suggested but has never been experimentally demonstrated owing in part at least to the lack of suitable biological methods. The techniques used in the study of transformation appear to afford a sensitive means of testing the validity of this hypothesis, and the results thus far obtained add supporting evidence in favor of this point of view. . . .

From these limited observations it would be unwise to draw any conclusion concerning the immunological significance of the nucleic acids until further knowledge on this phase of the problem is available. Recent observations by Lackman and his collaborators (1941) have shown that nucleic acids of both the yeast and thymus type derived from hemolytic streptococci and from animal and plant sources precipitate with certain antipneumococcal sera. The reactions varied with different lots of immune serum and occurred more frequently in antipneumococcal horse serum than in corresponding sera of immune rabbits. The irregularity and broad cross reactions encountered led these investigators to express some doubt as to the immunological significance of the results. Unless special immunochemical methods can be devised similar to those so successfully used in demonstrating the serological specificity of simple non-antigenic substances, it appears that the techniques employed in the study of transformation are the only ones available at present for testing possible differences in the biological behavior of nucleic acids. . . .

In the present state of knowledge any interpretation of the mechanism involved in transformation must of necessity be purely theoretical. The biochemical events underlying the phenomenon suggest that the transforming principle interacts with the R cell giving rise to a coordinated series of enzymatic reactions that culminate in the synthesis of the Type III capsular antigen. The experimental findings have clearly demonstrated that the induced alterations are not random changes but are predictable, always corresponding in type specificity to that of the encapsulated cells from which the transforming substance was isolated. Once transformation has occurred, the newly acquired characteristics are thereafter transmitted in series through innumerable transfers in artificial media without any further addition of the transforming agent. Moreover, from the transformed cells themselves, a substance of identical activity can again be recovered in amounts far in excess of that originally added to induce the change. It is evident, therefore, that not only is the capsular material reproduced in successive generations but that the primary factor, which controls the occurrence and specificity of capsular development, is also reduplicated in the daughter cells. The induced changes are not temporary modifications but are permanent alterations which persist provided the cultural conditions are favorable for the maintenance of capsule formation. The transformed cells can be readily distinguished from the parent R forms not alone by serological reactions but by the presence of a newly formed and visible capsule which is the immunological unit of type specificity and the accessory structure essential in determining the infective capacity of the microorganism in the animal body.

It is particularly significant in the case of pneumococci that the experimentally induced alterations are definitely correlated with the development of a new morphological structure and the consequent acquisition of new antigenic and invasive properties. Equally if not more significant is the fact that these changes are predictable, type-specific, and heritable.

Various hypotheses have been advanced in explanation of the nature of the changes induced. In his original description of the phenomenon Griffith (1928) suggested that the dead bacteria in the inoculum might furnish some specific protein that serves as a "pabulum" and enables the R form to manufacture a capsular carbohydrate.

More recently the phenomenon has been interpreted from a genetic point of view. The inducing substance has been likened to a gene, and the capsular antigen which is produced in response to it has been regarded as a gene product. In discussing the phenomenon of transformation Dobzhansky (1941) has stated that "If this transformation is described as a genetic mutation—and it is difficult to avoid so describing it—we are dealing with authentic cases of induction of specific mutations by specific treatments. . . ."

Another interpretation of the phenomenon has been suggested by Stanley (1938) who has drawn the analogy between the activity of the transforming agent and that of a virus. On the other hand, Murphy (1935) has compared the causative agents of fowl tumors with the transforming principle of Pneumococcus. He has suggested that both these groups of agents be termed "transmissible mutagens" in order to differentiate them from the virus group. Whatever may prove to be the correct interpretation, these differences in viewpoint indicate the implications of the phenomenon of transformation in relation to similar problems in the fields of genetics, virology, and cancer research.

It is, of course, possible that the biological activity of the substance described is not an inherent property of the nucleic acid but is due to minute amounts of some other substance adsorbed to it or so intimately associated with it as to escape detection. If, however, the biologically active substance isolated in highly purified form as the sodium salt of desoxyribonucleic acid actually proves to be the transforming principle, as the available evidence strongly suggests, then nucleic acids of this type must be regarded not merely as structurally important but as functionally active in determining the biochemical activities and specific characteristics of pneumococcal cells. Assuming that the sodium desoxyribonucleate and the active principle are one and the same substance, then the transformation described represents a change that is chemically induced and specifically directed by a known chemical compound. If the results of the present study on the chemical nature of the transforming principle are confirmed, then nucleic acids must be regarded as possessing biological specificity the chemical basis of which is as yet undetermined.

The senior author of this most important paper, Oswald Theodore Avery, was a native of Halifax, Nova Scotia. Born in 1877, he soon left Canada to follow his father to New York. It is in this large city that Oswald Avery spent most of his long and productive life. He chose to study medicine, and in 1904 he received his medical degree from Columbia University; soon afterwards he began to specialize in bacteriology. At first he was employed in a control laboratory for dairy products. In 1906, he joined the Hoagland Laboratory of Brooklyn, which was directed by Dr. Benjamin White. Following Dr. White's confinement to the Trudeau Sanitorium for Tuberculosis a few years later, both men developed an interest in the disease. Their work culminated in the publication of three papers. One of these dealt with the extraction of a toxic substance from tubercle bacilli. From then on, the scientific interest of Dr. Avery was directed toward the relationship between the properties of microorganisms and their chemical composition.

In 1913, Dr. Avery joined the staff of the "pneumonia service" of the Rockefeller Institute. He became a leading authority on pneumococci, and while studying these organisms, he made discoveries of basic biological significance. We have already read of his fruitful collaboration with Michael Heidelberger that led to the discovery that polysaccharides could be antigens. The discovery of these nitrogen-free antigens permitted the development of quantitative immunochemistry. Avery's collaboration with René Dubos was equally productive. Dubos, a disciple of Selman A. Waksman, brought the methods of the soil microbiologist to Avery's laboratory. Dubos and Avery introduced the polysaccharide of type III pneumococci into soil. By using this soil-enrichment method, they were able in 1930 to isolate a bacterium that produced an enzyme able to hydrolyze the capsular polysaccharide. Animal experiments indicated that the enzyme might have a significant protective action in experimental infections of mice caused by type III pneumococci. The development of the sulfa drugs, followed by that of antibiotics, made unnecessary the utilization of the enzyme in medicine but did not detract from the originality of Avery and Dubos' approach to therapy.

Avery was a careful experimentor who never let fancy and speculation, of which he was fond, blur his keen critical judgment. He had but few collaborators, many of whom are now famous. Even though collaboration with him was a rewarding and pleasant experience, Avery was a proponent of individual effort in science and looked with scorn on what he called "research by squads." And indeed, the significant output of his laboratory surpassed the efforts of many squads! At the age of seventy, honored and famous, Avery left the big city and retired to Nashville, Tennessee, where he died in 1955.

We have previously noted the name of E. L. Tatum, who with Beadle had described the occurrence of biochemical mutants in *Neurospora*. That work, published in 1941, had been done in Stanford, where Beadle had organized a *Neurospora* Institute. By 1946, Tatum was at Yale University and was putting the biochemical mutants to good use. With one of his students, Joshua Lederberg, he demonstrated what seemed to be a sexual process in some strains of *Escherichia coli* (*Nature*, **158**: 558, 1946): [1]

> Analysis of mixed cultures of nutritional mutants has revealed the presence of new types which strongly suggest the occurrence of a sexual process in the bacterium, *Escherichia coli.*
>
> The mutants consist of strains which differ from their parent wild type, strain K-12, in lacking the ability to synthesize growth-factors. As a result of these deficiencies they will only grow in media supplemented with their specific nutritional requirements. In these mutants single nutritional requirements are established at single mutational steps under the influence of X-ray or ultra-violet. By successive treatments, strains with several requirements have been obtained.
>
> In the recombination studies here reported, two triple mutants have been used: Y-10, requiring threonine, leucine and thiamin, and Y-24, requiring biotin, phenylalanine and cystine. These strains were grown in mixed culture in "Bacto" yeast-beef broth. When fully grown, the cells were washed with sterile water and inoculated heavily into synthetic agar medium, to which various supplements had been added to allow the growth of colonies of various nutritional types.

[1] Reproduced by permission of the editor of *Nature* and of Dr. E. L. Tatum.

This procedure readily allows the detection of very small numbers of cell types different from the parental forms.

The only new types found in "pure" cultures of the individual mutants were occasional forms which had reverted for a single factor, giving strains which required only two of the original three substances. In mixed cultures, however, a variety of types has been found. These include wild-type strains with no growth-factor deficiencies and single mutant types requiring only thiamin or phenylalanine. In addition, double requirement types have been obtained, including strains deficient in the syntheses of biotin and leucine, biotin and threonine, and biotin and thiamin respectively. The wild-type strains have been studied most intensively, and several independent lines of evidence have indicated their stability and homogeneity.

In other experiments, using the triple mutants mentioned, except that one was resistant to the coli phage Tl (obtained by the procedure of Luria and Delbrück), nutritionally wild-type strains were found both in sensitive and in resistant categories. Similarly, recombinations between biochemical requirements and phage resistance have frequently been found.

These types can most reasonably be interpreted as instances of the assortment of genes in new combinations. In order that various genes may have the opportunity to recombine, a cell fusion would be required. The only apparent alternative to this interpretation would be the occurrence in the medium of transforming factors capable of inducing the mutation of genes, bilaterally, both to and from the wild condition. Attempts at the induction of transformations in single cultures by the use of sterile filtrates have been unsuccessful.

The fusion presumably occurs only rarely, since in the cultures investigated only one cell in a million can be classified as a recombination type. The hypothetical zygote has not been detected cytologically.

These experiments imply the occurrence of a sexual process in the bacterium *Escherichia coli;* they will be reported in more detail elsewhere.

Bacteriologists had considered for years that bacteria were asexual creatures. Lederberg and Tatum cast strong doubt on this asexuality. In so doing they had even used not an obscure bacterium but a strain of *Escherichia coli,* one of the best-known bacteria. As Lederberg and Tatum had said, the possibility was not altogether excluded that transforming factors, of the type studied by Avery and coworkers, were not involved. The final demonstration that strains of opposite "sex" had to come into intimate contact in order to obtain genetic recombination was published by Bernard

Davis in 1950 (*J. Bacteriol.*, **60**: 507–508).[1] At that time he was working at the Cornell University Medical College, in New York City:

Genetic recombination in *Escherichia coli* (Lederberg: Genetics, 32; 505, 1947; Tatum and Lederberg: J. Bact., 53; 673, 1947) appears to imply a sexual mechanism, but the evidence is incomplete since cellular fusion has not been demonstrated directly. Although the low frequency of recombination (ca. 10^{-6}) can account for this failure, a mechanism similar to the soluble pneumococcus transforming substance must also be considered. Though this hypothesis became less attractive following the demonstration of recombinant progeny involving multiple genetic factors in various combinations, one cannot rigorously exclude a complex transforming substance involving a number of genetic units, or a gamete with a full complement of genes but smaller dimensions than the bacterial cell. Since the failure of culture filtrates to produce recombination (Lederberg) could be due to unusual lability of the agent, further tests on filtrates were undertaken, deterioration being minimized by avoiding delay in transport from donor to recipient cell.

A U-tube, 25 mm in diameter, partitioned by an "ultrafine" fritted glass disk (Corning), was shown to be impermeable to the organisms under investigation. K-12 mutants Y-10 (Thr⁻, Leu⁻, Thi⁻) and 58–161 (Meth⁻), kindly furnished by J. Lederberg, were separately inoculated (1 ml of turbid culture in 10 ml of medium containing 0.5 per cent yeast extract) in the two sides of the tube. During incubation a vacuum was applied alternately to each side, approximately half the total fluid being forced through the disk in 3 cycles per hour. Accumulation of bacteria on the disk was partly prevented by shaking with glass beads. Similar cultures with a mixture of both inocula, as well as single inocula, were simultaneously incubated in test tubes.

After 4 hours all 5 cultures had reached heavy turbidity (ca. 3×10^9 cells per ml). They were washed and pour-plated in minimal medium, which would yield only recombinant prototroph colonies. The mixed tube gave 58 colonies from a 0.05-ml inoculum; the cultures from each side of the U-tube, and from separate tubes, gave none in volumes of 0.05 to 0.5 ml. Similar results were obtained in a second experiment.

Previous experience had shown that over 90 per cent of the colonies obtained under these conditions arose from recombinations occurring in the mixed tube, rather than subsequently in the plate. This experiment should therefore have detected, in the largest inoculum, U-tube recombination in $\frac{1}{500}$ the frequency of the mixed tube. Since none was

[1] Reproduced by permission of The Williams and Wilkins Company, Baltimore.

observed, recombination via a filtrable substance seems unlikely; such a mechanism could be reconciled with these observations only if the material were exceedingly unstable.

The brilliant and young Joshua Lederberg was soon demonstrating that it was not by pure chance that he had made with Tatum the previously mentioned discovery of genetic recombination in bacteria. In rapid succession he made a number of striking discoveries that soon established him as one of the leading microbiologists of our time. In 1952, in collaboration with his wife, Esther, he described a method of replicating bacterial colonies from one Petri dish to another with the help of velveteen stamps. The Lederbergs used that method to demonstrate in an elegant and convincing way what Newcombe had already done by other means, namely, that bacterial mutations were not directed by the selective agent used (*J. Bacteriol.*, **63**: 399–406, 1952).

In 1952, Lederberg himself had professorial responsibilities and was then located at the University of Wisconsin. With one of his students, Norton D. Zinder, Joshua Lederberg discovered another method of genetic exchange in bacteria. Zinder and Lederberg's paper entitled "Genetic Exchange in Salmonella" was published in the *Journal of Bacteriology* (**64**: 679–699, 1952).[1] In this the authors wrote about the transfer of genetic characteristics by small particles. They discussed their results as follows:

> Genetic exchange in *S. typhimurium* is mediated by a bacterial product which we have called FA (filtrable agent). An individual active filtrate can transfer (transduce) many hereditary traits from one strain to another. Although the total activity of this filtrate encompasses the genotype of its parental culture, each transduction transmits only a single trait per bacterium. This contrasts with genetic exchange in *E. coli*, strain K-12 where there is unrestricted recombination of the several markers that differentiate two parental lines.
>
> FA may be considered as genetic material which enters the fixed heredity of the transduced cell. We may ask whether this transfer is a simple supper-addition or a substitutive exchange and replacement of the resident genetic factors. If streptomycin resistance is a recessive mutation, as inferred from studies of heterozygous diploids in *E. coli* (Lederberg, J., J. Bact. 61; 549–550; 1951), the transduction of resistance disqualifies the simple addition mechanism.
>
> Two aspects of FA must be carefully distinguished: the biological

nature of the particles themselves and their genetic function. There is good reason to identify the particle with bacteriophage. Nevertheless, the phage particle would function as a passive carrier of the genetic material transduced from one bacterium to another. This material corresponds only to a fragment of the bacterial genotype. For example, when FA from a marked prototroph is plated with an auxotroph on minimal agar, the genotype of the presumed "donor nucleus" is not observed among the transduced prototrophs. The hypothesis of FA as a genetic complex rather than a unit might be maintained if the singular effects produced depended on a small chance of release of any particular activity from a complex particle or on some localized nonheritable happenstance in the cell that ordinarily left only one function sensitive to transduction. Still the originally singly transduced cell develops as an isolated clone. Since the clone is composed of some 10^7 bacteria, one might expect that a complex residuum of an FA particle, if viable, would transduce some one of the daughter cells for another character during the growth of the clone. However, each FA particle produces only a single transduced clone. This speaks for the simplicity of its constitution as well as of its genetic effect.

When a strain labeled LA-22 is transduced from auxotrophy (phenylalanineless and tyrosineless; tryptophanless) to prototrophy, we have an apparent dual change. If this mutant is plated on minimal agar supplemented with phenylalanine and tyrosine, it occasionally reverts to the first step auxotrophic condition. However, when LA-22 is transduced on this medium, no more first step auxotrophs are found than can be explained by spontaneous reversion. The majority of the selected colonies are prototrophs. We have not been able to affect more than one trait in any other inter- or intrastrain transductions. It seems likely that the nutrition of LA-22 was determined by two successive mutations at the same genetic site. Davis' scheme for aromatic biosynthesis corroborates this notion (J. Bact. 62; 221–230; 1951) although the mutant LA-22 can revert spontaneously to an intermediate allele, transduction brings about a substitution of the wild type gene for full synthesis.

The most plausible hypothesis for the FA granules is that they are a heterogeneous population of species each with its own competence—in other words, each carries a "single gene" or small chromosome fragment.

Regardless of the nature of the FA particles, some mechanism must be postulated for the introduction of the transduced genetic material to the fixed heredity of the recipient cell. Muller's (1947) analysis of type transformation in the pneumococcus is apropos here: ". . . there were, in effect, still viable bacterial chromosomes, or parts of chromosomes, floating free in the medium used. These might, in my opinion, have penetrated the capsuleless bacteria and in part taken root there, perhaps, after having undergone a kind of crossing-over with the chromosomes of the host."

In a preliminary report on the *Salmonella* recombination system (Lederberg *et al.*, 1951) it was suggested that FA might be related to bacterial L-forms (Klieneberger-Nobel, 1951). The occurrence of swollen "snakes," filtrable granules, and large bodies in response to certain agents is characteristic both of FA and L-forms. Except for the suggestion of viable filter passing granules we have not repeated the reported cycles. The visible agglutinable granules and the antiserum-induced swollen form are not necessary for FA activity. However, this failure to fit all of the elements to a simple scheme may be due to a system more complex than we are now aware.

The bacteriological literature has numerous reports of results which might be interpreted as transduction (see reviews by Luria, Bact. Revs., 11; 1–40; 1947 and Lederberg, J., Heredity, 2; 145–198; 1948). These experiments have been criticized or neglected because of difficulties in their reproduction and quantitization but might now be reinvestigated in light of the findings presented. A citation of some of the more pertinent ones should suffice at this time. Wollman and Wollman (1925) reported the acquisition of *Salmonella* immunological specificity by *E. coli* via filter passing material. Similar material (which can be obtained by phage lysis) has been implicated in the change of penicillin resistant staphylococci and streptococci to relative penicillin sensitivity (Voureka, 1948; George and Pandalai, 1949). *Shigella paradysenteriae* (Weil and Binder, 1947) acquired new immunological specificity when treated with extracts of heterologous types. Boivin (1947) reported a similar change in *E. coli*. Unfortunately his strains have been lost and confirmation is impossible. Bruner and Edwards (1948) in a report of variation of somatic antigens of *Salmonella* grown in the presence of specific serum commented on the possibility that bacterial products dissolved in the serum were responsible for the changes.

These systems, provocative as they are, are insufficiently documented for detailed comparison with *Salmonella* transduction. The transformations in the pneumococcus (Avery *et al.*, 1944; McCarty, 1946) and *Hemophilus influenzae* (Alexander and Leidy, 1951) have been studied more completely.

The genetic "transformation" of the capsular character of the pneumococcus depends on a specific bacterial product (pneumococcus transforming principle, PTP). Originally interpreted as a directed mutation, it is now regarded as a variety of genetic exchange (Ephrussi-Taylor,

1950). Thus far transformations have been achieved for the full capsular character (Griffith, 1928), a series of intermediate capsular characters (Ephrussi-Taylor, 1951), M protein character (Austrian and MacLeod, 1949), and penicillin resistance (Hotchkiss, 1951). As in *Salmonella* each character is transformed independently. However, there are several differences between the two systems. FA must be evoked while the PTP is extractible from healthy cells. The resistance of FA to various chemical treatments has given only negative evidence of its chemical nature. The role of desoxyribonucleic acid in the PTP was verified by its inactivation by desoxyribonucleic acidase. Retention of activity by gradocol membranes has given comparable estimates for the size (about 0.1 μ) of the FA particles affecting two different characters. On the other hand, while the particle size of the PTP has been variously estimated from an average centrifugal mass of 500,000 (Avery *et al.*, 1944) to an ionizing irradiation sensitive volume equivalent to a molecular weight of 18,000,000 with high asymmetry (Fluke *et al.*, 1951), it is considerably smaller than the FA particle. Pneumococci must be sensitized by a complex serum system for adsorption of PTP. The low but poorly determined frequency of transformations has been thought to be due to the low competence of the bacteria. In the absence of adsorption experiments a system similar to *Salmonella* has not been ruled out. Important information is still lacking in both systems and time may resolve these apparent differences.

The relationship of transduction in *Salmonella* to sexual recombination in *E. coli* is obscure. Transduction has not been found in crossable *E. coli* nor sexual recombination in *Salmonella*. These genera are extremely closely related taxonomically but seem to have entirely different modes of genetic exchange.

Sexual recombination was first demonstrated in *E. coli*, strain K-12. With the development of an efficient screening procedure, two to three per cent of *E. coli* isolates were proved to cross with strain K-12 (Lederberg, 1951). The agent of recombination in *E. coli* is almost certainly the bacterial cell. The cells apparently mate, forming zygotes from which parental and recombinant cells may emerge following meiosis, in which linkage is a prominent feature (Lederberg, 1947). The combination of genomes within a single cell has been confirmed by the exceptional occurrence of nondisjunctions which continue to segregate both haploid and diploid complements (Zelle and Lederberg, 1951). Although lysogenicity plays a critical role in transduction in *Salmonella*, all combinations of lysogenic and nonlysogenic cultures of *E. coli* cross with equal facility (Lederberg, E. M., 1951).

Owing to the lack of recombination of unselected markers, transduction is a less useful tool than sexual recombination for certain types of genetic analysis. However, as FA may correspond to extracellular genetic material, such problems as gene reproduction, metabolism, and mutation may be more accessible to attack. Sexual systems usually provide for the reassortment of genetic material and give an important source of variation for the operation of natural selection in organic evolution. Both sexual recombination and transduction, because of their low frequency, allow only limited gene interchange in bacteria. Transductive exchange is limited both in frequency and extent.

It is too early to assess the role that transduction may have played in the development of the immunologically complex *Salmonella* species. White (1926) speculated that the many serotypes evolved by loss variation from a single strain possessing all of the many possible antigens. Bruner and Edwards (1948) obtained specific examples of loss variation with contemporary species. Transduction provides a mechanism for transfer of some of the variation developed spontaneously and independently between the "descending" lines. The genus *Salmonella* includes a group of serotypes which share a receptor for *S. typhimurium* FA. Other receptor groups have yet to be sought. Within such groups it should be possible to evolve in the laboratory other new serotypes comparable to the antigenic hybrid of *S. typhi* and *S. typhimurium.*

Several different bacterial genera have been intensively studied with regard to modes of genetic exchange. Each of the several known systems differs in details that enlarge our notions of bacterial reproduction and heredity.

Joshua Lederberg's career continued to be filled with interesting investigations. He showed his versatility by not limiting himself to bacterial genetics. He contributed to the understanding of the mode of action of penicillin, and he extended his activities to space microbiology. Together with George Beadle and Edward Tatum he received the Nobel Prize for physiology and medicine in 1958.

Bacteria are able to transmit genetic elements in a number of ways. François Jacob and Elie L. Wollman, from the Pasteur Institute in Paris, tried to clarify the nomenclature of some of the genetic processes by introducing the term "episomes" for "added genetic elements" (*Compt. rend. acad. sci.,* **247**: 154–156, 1958): [1]

The determinants of classical genetics are units which are an integral part of the structure of chromosomes. Except in exceptional cases

---

[1] Translated by permission of Gauthier-Villars, publishers of the *Comptes rendus de l'académie des sciences.*

(deletions), they are always present in one or another of their allelic forms. As essential constituents of the chromosomes, they do not have, in general, an autonomous existence.

To these classical elements, one can contrast other elements which can be either present or absent, and which, in the cell, can be either independent or fixed to the genome. To these added elements, which when present can confer on the cell certain hereditary properties, we propose to give the name of *episomic elements* or *episomes*.

Bacteria, because of their haploid nature and the mechanism of their conjugation constitute a material which is particularly favorable for the study of episomic elements.

The best known case is that of temperate phages. A phagic genome introduced in a bacterium can evolve according to two distinct processes. It can multiply in an autonomous, vegetative fashion, which leads to the formation of infectious particles and to the lysis of the bacterium. It can also fix itself in the prophage state at a specific locus of the bacterial chromosome and replicate itself in harmony with the chromosome. There is every reason to think that the prophage is added to the bacterial genome and that it is neither substituted for an allelic segment nor inserted in the continuity of the chromosome. It behaves as a bacterial genetic element, but under the influence of certain agents called inductors, or under the influence of conjugation, the prophage may leave its specific locus and evolve toward the vegetative state. The phagic genome may thus exist in a bacterium in two states which are, as a matter of fact, incompatible, since the presence of a prophage inhibits the vegetative multiplication of an homologous phagic genome, it is *immunity*.

Since the genome of a temperate phage can exist in two forms, autonomous and integrated, it has all the characteristics of the elements that we propose to group under the name of episomes.

A second example is furnished by the genetic determinants for the synthesis of colicines. Even though the genetic analysis of recombinants formed during crosses does not permit their localization on the bacterial chromosome, certain methods specific for bacterial genetics permit achievement of this localization. All takes place as if during conjugation a colicinogenic determinant could leave its chromosomic locus, multiply in an autonomous fashion, then be able to integrate itself anew in the products of segregation of the zygote. Just like a prophage, a given colicinogenic determinant seems thus able to fix itself upon the chromosome of the bacterium at a specific locus or to multiply autonomously.

A third example is furnished by the sexual determinant of *Escherichia coli*. The sexual factor F is present in the donor strains, F^+ or Hfr, and

absent in the receptor strains F⁻. During conjugation, it can be transmitted very efficaciously from F⁺ bacteria to F⁻ bacteria without
apparent link with other determinants. We have observed that factor F,
no more than the colicinogenic determinant, does not segregate among
the products of a zygote isolated with a micromanipulator. One must
thus conclude that factor F is able to multiply in an autonomous
manner. In Hfr mutants from F⁺ bacteria, factor F is integrated and
can be localized on the bacterial chromosome. It is important to note
that its position on the linking group differs from one Hfr strain to
another. Let us add that the two states F⁺ and Hfr are incompatible.

Prophages, colicinogenic determinants and sexual factors seem all
three to be covered by the definition of episomic elements. . . .

The introduction of new terms, clearly defined, to refer to natural
phenomena greatly helps the development of science. However, the introduction of the word "episome" by Jacob and Wollman was not altogether
fortunate since this term had already been introduced to genetics in 1931
by D. H. Thompson. Nevertheless, this point was explained by Jacob,
Schaeffer and Wollman (*Microbial Genetics,* Tenth Symposium of the Society for General Microbiology, Cambridge University Press, 1960): [1]

In order to account for certain mutations in *Drosophila,* Thompson
advanced the hypothesis that "a gene consists of a main particle
(protosome) firmly anchored in the chromosome with varying numbers
of one or more kinds of other particles attached (episomes). Gene
mutation is due most frequently to the loss of one or more episomes
from the protosomes and less frequently to the addition of episomes.
The genetic chromosome is visualized as the conventional string of
beads, except that almost every bead bristles with side chains of
episomes." Recent analyses of the fine structure of the genetic material
make it difficult to accept the idea that a gene is composed of distinct
particles associated in a way similar to the association found between
apoenzymes and coenzymes. . . . Since there is no need for maintain-
ing the term episome in Thompson's sense, it seems that it can un-
ambiguously be applied to the kind of elements [referred to previously].

Our rapid survey of important events in microbial genetics has mainly
stressed the study of bacteria. There are important principles that can be
demonstrated with other microorganisms. The genetics of fungi, growing
under asexual conditions, already interesting in itself, started to gain im-

[1] Reproduced by permission of Cambridge University Press.

portance as *Penicillia* and *Aspergilli* were used for the production of antibiotics and other chemicals. In addition, actinomycetes, also of great industrial importance, were shown to exhibit similar mechanisms of heredity. We
shall let G. Pontecorvo, an Italian-born British geneticist from the University
of Glasgow, explain parasexual modes of transmission of genetic information
in filamentous fungi (*Caryologia,* vol. suppl., pp. 192–200, 1954): [1]

In the last ten years, processes other than standard sexual reproduction,
and yet resulting in recombination of hereditary properties, have come
to light in microorganisms of widely different groups: bacteriophages
and animal viruses, *Escherichia coli*, *Salmonella* and *Pneumococcus*.
The purpose of the present paper is to give a summary of the work
which has led, in our Laboratory, to the discovery in the filamentous
fungi of another of these mechanisms, which could be called "para-
sexual." This mechanism, based on *mitotic* segregation and recombination, occurs side by side with a standard sexual cycle in one of the
three species investigated—*Aspergillus nidulans*—and in the absence of
it in the other two—*Aspergillus niger* and *Penicillium chrysogenum*.

The implications of this "para-sexual" mechanism in the genetic
systems of filamentous fungi will be discussed. It will be clear, in
addition, that this mechanism is most relevant to two applied fields:
plant pathology and industrial fermentations. To the former, because
the knowledge that it exists permits us to look at the problems of
variation in pathogenic fungi from an additional angle. To the latter,
because it opens up the possibility of deliberate "breeding" of improved
industrial strains even in species without a sexual cycle.

The steps in the discovery of this "para-sexual" mechanism in filamentous fungi were the following. First, Roper (1952) succeeded in
synthesizing in A. *nidulans* strains carrying in their hyphae diploid
nuclei heterozygous for known markers. Second, prompted by Stern's
(1936) work on somatic crossing-over in *Drosophila*, we searched
these diploids for mitotic segregation and recombination and found it
to occur regularly and abundantly (Pontecorvo and Roper, 1952).
Third, Roper's technique was applied to a species without a normal
sexual cycle—*Aspergillus niger*—and the heterozygous diploids obtained
in this way again showed mitotic segregation and recombination (Pontecorvo, 1952). This first example of genetic analysis in a species
without a sexual cycle was extended (Pontecorvo, Roper and Forbes,
1953), and a second example with another asexual species—*Penicillium
chrysogenum*—soon followed (Pontecorvo and Sermonti, 1953). Fourth,

[1] Reproduced by permission of the editor of *Caryologia*.

investigation of the details of mitotic recombination (Pontecorvo, 1952; Pontecorvo and Roper, 1953; Pontecorvo, Tarr Gloor and Forbes, 1954) led to the conclusion that two distinct and probably independent processes operate. One is mitotic crossing-over à la Stern (1936), whereby a diploid nucleus, heterozygous at several loci, gives origin to daughter diploid nuclei, homozygous at one or more linked loci but still heterozygous at the remaining loci on the same chromosome and at all those on different chromosomes. The other is an irregular distribution of chromosomes at mitosis whereby a heterozygous diploid nucleus gives origin, in a proportion of cases, to a haploid nucleus into which each chromosome segregated as a unit, i.e., with complete linkage.

The experimental procedures have been published in full detail. It is necessary here to give only a brief summary.

Roper's technique permits the selection and recognition of strains carrying *heterozygous diploid nuclei,* which are now known to arise with exceedingly low incidence (of the order of 10^{-7}) during growth and sporulation of heterokaryons. Clearly, with an incidence as low as this, the isolation of a diploid conidium is possible only if the diploids are somehow automatically selected. The selection is based on the fact that a diploid conidium, heterozygous at two or more loci controlling growth factor requirements, will be able to multiply on a medium devoid of these growth factor(s) while haploid conidia carrying mutant alleles at any one of these loci will not. An heterokaryon is therefore synthesized between two nutritional mutant strains, differing from each other in nutritional requirements. The uninucleate conidia of the heterokaryon will have, in the overwhelming majority, the requirements of either one or the other component strain and will be unable to grow when plated, even at high densities, on agar medium devoid of the relevant growth factors. Only the rare diploid conidia among them will be able to form colonies. The use of other markers, such as colour of the conidia, gives additional clues.

By using this technique, with or without the additional help of camphor vapour treatment, a considerable number of heterozygous diploid strains have been synthesized in the species mentioned above. In every case the required diploid was obtained at the first attempt.

Colonies of heterozygous diploid strains, having the dominant phenotype in respect of all the loci at which they are heterozygous, produce, as they grow, patches of mycelium with the recessive phenotype in respect of one or more of these loci. These segregant spots can be identified by inspection for those recessive characters which are visually distinguishable (e.g., colours of the conidia) or by appropriate selective techniques for those characters which can only be distinguished by testing. Thus, for example, a colony with green conidia, doubly heterozygous for white and yellow, will produce patches of white or yellow. Isolations from each patch will lead to establish a white or yellow segregant strain.

On the other hand, a diploid colony heterozygous at a locus at which a mutant allele determines, e.g., a requirement for adenine, will not require adenine but will produce segregant patches which do require it. These patches are not visibly different from the rest of the colony. Their conidia, however, can be isolated by using, e.g., Forbes (1952) SO_2 technique, which is, in principle, the same as the penicillin technique for the selection of auxotrophic strains of bacteria.

Isolation, visual or otherwise, from segregant patches of diploids in *Aspergillus nidulans* has yielded a large number of segregant strains which could be analyzed further as to ploidy and genotype. This is, of course, the great advantage that this material has over, say, *Drosophila,* used in the pioneer work of Stern (1936) on somatic crossing-over. In *Drosophila* each spot can only be classified by phenotype; it cannot be grown into a whole fly and further analyzed.

The most complete analysis of segregant strains so far has been based on those derived from two diploids of *A. nidulans* heterozygous for the same six markers but in different coupling-repulsion arrangements (Pontecorvo, Tarr Gloor and Forbes, 1954). The six loci were distributed on three out of the four chromosomes of this species: three markers over a stretch of 25 cMo in one chromosome, two markers more than 50 cMo apart on another one, and the sixth marker on a third.

It was first found that in *A. nidulans* the diameter of the conidia is diagnostic for diploidy or haploidy. It was then found that the majority of the segregants are diploid, but about one tenth of them are haploid. In the formation of haploid segregants, recombination occurs only between non-linked genes: linked genes segregate as a unit. Finally, it was found that diploid segregants originate mainly through mitotic crossing-over with precisely the modalities inferred by Stern (1936) for somatic crossing-over in *Drosophila*.

It was also possible to estimate the frequencies with which the three events—nuclear fusion, mitotic crossing-over and haploidization —occur. A summary of the results is as follows:

1. *Nuclear fusion.* In the five cases in which it was tested, the incidence of heterozygous diploid conidia among the conidia of balanced heterokaryons was found to range from 0.25 to 3.6×10^{-7} in *A. nidulans* (Pontecorvo and Roper, 1953; Pontecorvo, Tarr Gloor and Forbes, 1954), *A. niger* (Pontecorvo, Roper and Forbes, 1953) and

Penicillium chrysogenum (Pontecorvo and Sermonti, 1954). Because of the clonal distribution of the diploid nuclei this incidence does not measure the rate at which nuclear fusion occurs. To arrive at an estimate of this, it would be necessary to evaluate the other factors leading to the equilibrium: e.g., the strength of selection for diploids during growth of a colony, the rate at which diploids break down to haploids etc. Furthermore, there is one good reason why the figures given above may be lower than those to be found in wild populations: wild balanced heterokaryons (Jinks, 1952) seem to have nuclear ratios nearer 50:50 that the artificially synthesized heterokaryons (balanced by means of induced nutritional requirements) with which we have worked.

As a provisional guess we may take it that fusion between unlike nuclei occurs at a rate of the same order as the incidence of diploid conidia, i.e., 10^{-7}.

2. *Mitotic segregation and recombination*. The incidence of conidia homozygous for one or more of the markers for which a diploid strain was heterozygous has been measured accurately in several cases in diploids of *A. nidulans* (Pontecorvo and Roper, 1953; Pontecorvo, Tarr Gloor and Forbes, 1954). Also in *A. niger* and *P. chrysogenum* the incidence of *segregant* conidia has been measured accurately in several instances (Pontecorvo, Roper and Forbes, 1953; Pontecorvo and Sermonti, 1954). Unfortunately, in these two species the analysis has not yet proceeded far enough to distinguish between segregant conidia still diploid but homozygous at one or more loci (i.e., originated through mitotic crossing-over) and haploid ones.

The data of Pontecorvo, Tarr Gloor and Forbes (1954) give an incidence of homozygosis of the order of 1:3000 for each of two distal loci examined on two chromosomes. The frequency of homozygosis resulting from mitotic crossing-over is, of course, a function of the distance from the centromere. Since only one out of two possible types of segregation (Stern, 1936) gives origin to homozygous products, and only one of the two homozygous products is detectable, we must multiply by four the above incidence, and multiply it again by four because this species has four chromosome pairs: this gives a *minimum* of about 1:200 for the incidence of mitotic chiasmata per nucleus. However, we have no clue as to how near the loci used in this work are to the distal ends of the chromosomes and selection undoubtedly acts against the segregant nuclei, because they are homozygous for recessives. All considered, the incidence of mitotic crossing-over is likely to be of the order of one "chiasma" every few ten or perhaps one hundred nuclei.

3. *Haploidisation.* Conidia with haploid nuclei carrying one or more of two distal recessive markers constitute about 15% of the segregants produced by the two diploid *A. nidulans* on which the work was carried out (Pontecorvo, Tarr Gloor and Forbes, 1954). In other words, the diploid conidia homozygous for one of these recessive markers outnumber by about 6:1 the haploid conidia carrying one of these markers. The overall incidence of haploid conidia, irrespective of whether they carry dominant or recessive alleles of either marker, can be estimated to be about 8×10^{-4}. The genotypes of the haploids are such as to leave little doubt that the process producing them has nothing to do with mitotic crossing-over: in fact, they show no recombination of linked markers, though they show random recombination between non-linked ones. Assuming that the haploids originate through some accidental failure of *regular* distribution of chromosomes at mitosis, and that the irregular distribution is random, it is possible to calculate the incidence of this failure (Pontecorvo, Tarr Gloor and Forbes, 1954). The estimates range from 1 to 6% for the proportion of nuclei in mitosis in which this failure occurs.

Mitotic crossing-over and irregular distribution of chromosomes in heterozygous diploid strains have the following consequences: mitotic crossing-over effects recombination between linked genes; irregular distribution effects recombination between non-homologous chromosomes at the same time yielding a proportion of nuclei with a balanced haploid set.

Since mitotic crossing-over, which ultimately leads to complete homozygosis, and haploidisation are enormously more frequent than fusion between unlike nuclei (say of the order of 10^{-2}, 10^{-3} and 10^{-7}, respectively), the ultimate result of growth of a heterozygous diploid strain is a variety of *haploid* strains recombining in all possible ways the genes at the heterozygous loci: thus the series of events—nuclear fusion, mitotic crossing-over, haploidisation—achieves, *without a standard sexual cycle*, the recombination of properties which were originally distributed between two or more strains.

A para-sexual cycle like the one just described, in so far as it achieves gene recombination, is *qualitatively* equivalent to a sexual cycle. Indeed it contains all the elements of a standard sexual cycle—karyogamy, recombination, haploidisation—the only difference being the absence of a precise time sequence. Crossing-over and haploidisation are not two parts of one process, following each other in that order, as they do in meiosis; they occur independently in different nuclei.

Is this para-sexual cycle an efficient substitute for the normal sexual cycle? Clearly we need much more information, especially on the genetics of wild populations of fungi, to attempt an answer. There is, of course, the easy answer, but not a serious one, that if it were not an efficient substitute, there would not be an enormous wealth of asexual filamentous fungi. In fact their existence has been for long a puzzle to the geneticist looking at the problems of evolution with neo-Darwinian field glasses. Without a diploid stage to accumulate genetic variation in the recessive condition and without recombination to offer novel genotypes to the test of selection, the neo-Darwinian scheme was left with very little. It had to fall back on a very inefficient system in which all successful genotypes arose by the chance occurrence in one clone of several mutations immediately expressed in the phenotype.

As a development of Dodge's (1942) work, it became clear about eight years ago (Pontecorvo, 1947), and it has become unquestionable recently (Jinks, 1952), that one part of this puzzle is no longer baffling. Heterokaryosis is as efficient a system as heterozygosis for storing genetic variation under the cloak of dominance. It is also a valuable mechanism for short term adaptation of a growing colony to changing conditions: the pioneer work of Hansen (1938) showed it, and the experiments of Beadle and Coonradt (1944), Pontecorvo (1947) and Jinks (1952) confirmed it. But this adaptation of a heterokaryon, based as it is on adjusting the nuclear ratios, resembles, in a sense, somatic adaptation in higher organisms, or enzyme adaptation in microorganisms: it has no *direct* effect on the production of novel genotypes, as required for variation of evolutionary significance.

It is clear now that heterokaryosis is not only a mechanism for storing genetic variability and for short term adaptation. It plays also another part of fundamental importance: it is an essential step in the para-sexual cycle described here.

As this cycle includes a diploid stage, both heterokaryosis and heterozygosis are available for storing genetic variation, and in which relative proportions, is a question to be investigated in wild populations. It seems likely that heterokaryosis will turn out to be the more important of the two in this respect, if for no other reason than isolates from nature are very often heterokaryotic (Hansen, 1938; Jinks, 1952) but diploids have not been reported so far. Perhaps they have not been recognized even if found; but the ephemeral existence of experimental diploids, demonstrated in our Laboratory, suggests that they should not be common in nature. In fact, on the basis of the figures given before haploid strains should outnumber the diploids as $10^{-3}:10^{-7}$, i.e., ten thousand haploids for one diploid, unless selection favoured the latter.

Clearly, the discovery of the mitotic recombination cycle in filamentous fungi, and of its relations to heterokaryosis, opens to investigation an enormous number of problems in the genetics of wild populations, including those of pathogenic forms. It also makes it possible to carry out purposeful breeding in asexual species of industrial importance.

The discovery that in filamentous fungi genetic recombination can occur outside, or in the absence of, the sexual cycle opens a new approach to variation in phytopathogenic fungi.

The cat-and-mouse play of host and pathogen variation, with its extraordinary specificity, is too well known to need emphasis. Part of the variation on the pathogen side has been attributed repeatedly to heterokaryosis. Now, heterokaryosis by itself can only give origin to strains with new pathogenic properties in one of two ways: (a) by dominant complementary gene action between the two or more strains from which a heterokaryon is formed; (b) by segregation of homokaryons. Heterokaryosis alone cannot give origin to new pathogenic properties by recombination. It is in this last, and most important, respect that the para-sexual cycle completes the picture.

What recombination implies can be visualized with a very crude hypothetical example. Consider two poorly pathogenic strains: one inefficient in penetrating the tissues of the host, but able to set off a serious pathological reaction inside the host; another one able to penetrate the host but inefficient in setting off the pathological reaction. With certain dominance relationships the heterokaryon will be even less pathogenic than either component. A heterozygous diploid formed by the heterokaryon will be like the heterokaryon in this respect; but some of the mitotic recombinant types, diploid or haploid, arising from the diploid may have both invasiveness and virulence.

As far back as 1947 (Pontecorvo, 1949) this kind of reasoning led to the suggestion of recombination as one of the mechanisms for the origin of "epidemic" strains of microorganisms. Clearly, the discovery of the para-sexual cycle in filamentous fungi makes now this suggestion worthy of serious experimental attack in some of the pathogenic species of this group.

Industrial fermentations based on the species of fungi in which we have already demonstrated the existence of the para-sexual cycle include those for the production of penicillin, citric acid and cortisone. No doubt others will be added to this list soon.

In all these fermentations, strain improvement has been based so far on selection, either among large collections of strains from nature, or from spontaneous or induced variation in one strain. Clearly, selection can now be applied to the variation resulting from recombination *via* the para-sexual mechanism. The use of recombination permits to introduce a certain amount of deliberate planning in the breeding programme. For instance, it may be possible to aim at recombinants associating two or more desirable properties present separately in different strains. The rational use of the para-sexual cycle may permit to introduce into the "breeding" of improved industrial strains procedures similar to those used by the plant breeder in the improvement of vegetables, crop plants and fruit trees.

Actinomycetes, which are bacteria with a fungal type of morphology, exhibit genetic phenomena that are similar to those of asexual fungi, as stated by Donald H. Braendle and Waclaw Szybalski as they sum up one of their studies (*Proc. Nat. Acad. Sci.*, **43**: 954–955, 1957): [1]

Heterokaryosis in streptomycetes denotes a co-operative venture in which genetically unlike nuclei perpetuate within a common cytoplasm as the result of hyphal anastomosis. Conidia formed by heterokaryons are of parental type only. Heterokaryosis, as defined, has been demonstrated between mutants of the same strain of *Streptomyces griseus,* *S. griseoflavus, S. fradiae, S. venezuelae, S. albus, S. sphaeroides,* and *S. coelicolor,* but not between mutants of diverse origin. Heterokaryons between amino acid-requiring mutants of the last species, and occasionally of other species were "nutritionally unbalanced," i.e., did not grow on minimal media unless at least a portion of the parental requirements was supplied.

Synkaryosis is the fusion of unlike nuclei within heterokaryons and subsequent reassortment of genetic characters to yield conidia of recombinant types. Two species, *S. coelicolor* and *S. fradiae,* have exhibited the synkaryotic type of interaction. . . .

Thus, microbial mutants, having specific properties that, as Mendel had said, " could be precisely determined," have permitted us to follow some of the mechanisms of heredity. The aim now is not only to learn about all the possible methods that permit genetic material to pass from one cell to another, but also to elucidate the very molecular mechanism that is life itself. In this endeavor, we can only be sure that asymptotically, we will come closer and closer to the Truth even though we cannot be sure whether we will ever reach it.

[1] Reproduced by permission of Dr. W. Szybalski and the *National Academy of Sciences.*

References

Adelberg, E. A.: *Papers on Bacterial Genetics,* Little, Brown and Company, Boston, 1960.

Braun, W.: *Bacterial Genetics,* W. B. Saunders Company, Philadelphia, 1953.

Iltis, H.: *Life of Mendel,* W. W. Norton & Co., Inc., New York, 1932.

Jacob, F., and E. L. Wollman: *Sexuality and the Genetics of Bacteria,* Academic Press, Inc., New York, 1961.

Lederberg, J.: *Papers in Microbial Genetics,* The University of Wisconsin Press, Madison, Wis., 1951.

Löhnis, F.: "Studies upon the Life Cycles of the Bacteria, Part I, Review of the Literature—1838–1918." *Mem. Nat. Acad. Sci.,* vol. 16, Second Memoir, Government Printing Office, Washington, D.C., 1921.

Mendel, De Vries, Correns, Tschermak (in English translation): "The Birth of Genetics," *Genetics* (suppl.), vol. 35, The Brooklyn Botanical Garden, New York, 1950.

Mendel, G.: *Experiments in Plant Hybridisation* (English translation), Harvard University Press, Cambridge, Mass., 1960.

Peters, J. A.: *Classic Papers in Genetics,* Prentice-Hall, Inc., Englewood Cliffs, N.J., 1959.

Additional References

Stent, G. S.: *Molecular Genetics; An Introductory Narrative,* W. H. Freeman, San Francisco, 1971.

Sturtevant, A. H.: *A History of Genetics,* Harper and Row, New York, 1965.

Epilogue

Man has always been subjected to the action of microorganisms. Many centuries elasped, however, before he became aware of the multitude of microbial transformations taking place under his very feet as well as within his own body. Before his eyes, animals and plants died, sometimes as a result of microbial action, and microorganisms gradually decomposed their remains, permitting the reutilization of the elements present in certain of their constituents, such as cellulose, that his own stomach was not equipped to digest. Nitrogen was fixed from the atmosphere and made biologically available. Microorganisms also played a role in the formation of peat, coal, and probably oil, all substances that man learned to use. At times, the presence of microorganisms manifested itself with fury, especially during epidemics so often associated with the miseries of war. All unknowing, man soon learned to tame some of these entities in order to protect and to feed himself; to tan leather and to brew beer, to ret flax and to leaven bread. He also learned to protect his food from microbial deterioration by smoking meat, converting milk, and salting fish. He even learned how to vaccinate.

Biologists have always been interested in probing for the basis of life. Vitalists believed that living beings were possessed of an intangible force that, although unable to explain, they were not shy to name. Under one of its most recent expressions, it is the *élan vital* of Henri Bergson. On the opposite side, the mechanists, easily inebriated by the successes of science, did not hesitate to predict that all about life would eventually be explained in measurable terms.

The investigator, however, must be guided by hypotheses without being what Pasteur would have called "a man with a system." In the study of the essence of life we find in Pasteur a model of wisdom. He had a system. His early studies on molecular dissymmetry had led him to make such pronouncements as: "The Universe is a dissymmetric complex and I am convinced that life, as it manifests itself to us, is a function of the dissymmetry of the Universe. . . ." But he did not let his system get the best of him. His efforts were guided toward the gathering of data that often had humble practical applications. In so doing, he founded microbiology and pointed to the great role of infinitesimally small beings that could cause diseases or prevent them, that could respire or ferment, that could make or break an industry.

More than a hundred years have passed since Pasteur started to investigate the secrets of the simplest forms of life, which have turned out to be so complex. Nowadays, microorganisms play an even larger role in the practical aspects of our life, and their importance as tools for basic biological and biochemical research is ever increasing. Even though we lack perspective, we can safely say that some of the most wonderful pages of the history of microbiology are now being enchased among the dry and impersonal articles found in the current issues of scientific journals.

General References

Bender, G. A., and R. A. Thom: *A History of Pharmacy in Pictures*, Parke, Davis and Co., Detroit, 1960.

Bender, G. A., and R. A. Thom: *Great Moments in Medicine*, Parke, Davis and Co., Detroit, 1961.

Brock, T. D.: *Milestones in Microbiology*, Prentice-Hall, Inc., Englewood Cliffs, N.J., 1961.

Bulloch, W.: *The History of Bacteriology*, Oxford University Press, London, 1938, reprinted 1960.

Clark, P. F.: *Pioneer Microbiologists of America*, The University of Wisconsin Press, Madison, Wis., 1961.

Clendening, L.: *Source Book of Medical History*, Dover Publications, Inc., New York, 1960 (republication).

Doetsch, R. N.: *Microbiology: Historical Contributions from 1776 to 1908*, Rutgers University Press, New Brunswick, N.J., 1960.

Gabriel, M. L., and S. Fogel: *Great Experiments in Biology*, Prentice-Hall, Inc., Englewood Cliffs, N.J., 1955.

Grainger, T. H., Jr.: *A Guide to the History of Bacteriology*, The Ronald Press Company, New York, 1958.

Leicester, H. M., and H. S. Klickstein: *A Source Book in Chemistry*, McGraw-Hill Book Company, New York, 1952.

Major, R. H.: *A History of Medicine* (2 vols.), Charles C Thomas, Publisher, Springfield, Ill., 1954.

Nordenskiöld, E.: *The History of Biology*, Tudor Publishing Company, New York, 1946 (republication).

Sarton, G.: *A Guide to the History of Science*, The Ronald Press Company, New York, 1952.

Index

A CATALOGUE OF SELECTED DOVER BOOKS
IN ALL FIELDS OF INTEREST

A CATALOGUE OF SELECTED DOVER BOOKS
IN ALL FIELDS OF INTEREST

LEATHER TOOLING AND CARVING, Chris H. Groneman. One of few books concentrating on tooling and carving, with complete instructions and grid designs for 39 projects ranging from bookmarks to bags. 148 illustrations. 111pp. 7⅞ x 10.
23061-9 Pa. $2.50

THE CODEX NUTTALL, A PICTURE MANUSCRIPT FROM ANCIENT MEXICO, as first edited by Zelia Nuttall. Only inexpensive edition, in full color, of a pre-Columbian Mexican (Mixtec) book. 88 color plates show kings, gods, heroes, temples, sacrifices. New explanatory, historical introduction by Arthur G. Miller. 96pp. 11⅜ x 8½.
23168-2 Pa. $7.50

AMERICAN PRIMITIVE PAINTING, Jean Lipman. Classic collection of an enduring American tradition. 109 plates, 8 in full color—portraits, landscapes, Biblical and historical scenes, etc., showing family groups, farm life, and so on. 80pp. of lucid text. 8⅜ x 11¼.
22815-0 Pa. $4.00

WILL BRADLEY: HIS GRAPHIC ART, edited by Clarence P. Hornung. Striking collection of work by foremost practitioner of Art Nouveau in America: posters, cover designs, sample pages, advertisements, other illustrations. 97 plates, including 8 in full color and 19 in two colors. 97pp. 9⅜ x 12¼.
20701-3 Pa. $4.00
22120-2 Clothbd. $10.00

THE UNDERGROUND SKETCHBOOK OF JAN FAUST, Jan Faust. 101 bitter, horrifying, black-humorous, penetrating sketches on sex, war, greed, various liberations, etc. Sometimes sexual, but not pornographic. Not for prudish. 101pp. 6½ x 9¼.
22740-5 Pa. $1.50

THE GIBSON GIRL AND HER AMERICA, Charles Dana Gibson. 155 finest drawings of effervescent world of 1900-1910: the Gibson Girl and her loves, amusements, adventures, Mr. Pipp, etc. Selected by E. Gillon; introduction by Henry Pitz. 144pp. 8¼ x 11⅜.
21986-0 Pa. $3.50

STAINED GLASS CRAFT, J.A.F. Divine, G. Blachford. One of the very few books that tell the beginner exactly what he needs to know: planning cuts, making shapes, avoiding design weaknesses, fitting glass, etc. 93 illustrations. 115pp.
22812-6 Pa. $1.50

CREATIVE LITHOGRAPHY AND HOW TO DO IT, Grant Arnold. Lithography as art form: working directly on stone, transfer of drawings, lithotint, mezzotint, color printing; also metal plates. Detailed, thorough. 27 illustrations. 214pp.
21208-4 Pa. $3.00

DESIGN MOTIFS OF ANCIENT MEXICO, Jorge Enciso. Vigorous, powerful ceramic stamp impressions — Maya, Aztec, Toltec, Olmec. Serpents, gods, priests, dancers, etc. 153pp. 6⅛ x 9¼.
20084-1 Pa. $2.50

AMERICAN INDIAN DESIGN AND DECORATION, Leroy Appleton. Full text, plus more than 700 precise drawings of Inca, Maya, Aztec, Pueblo, Plains, NW Coast basketry, sculpture, painting, pottery, sand paintings, metal, etc. 4 plates in color. 279pp. 8⅜ x 11¼.
22704-9 Pa. $4.50

CHINESE LATTICE DESIGNS, Daniel S. Dye. Incredibly beautiful geometric designs: circles, voluted, simple dissections, etc. Inexhaustible source of ideas, motifs. 1239 illustrations. 469pp. 6⅛ x 9¼.
23096-1 Pa. $5.00

JAPANESE DESIGN MOTIFS, Matsuya Co. Mon, or heraldic designs. Over 4000 typical, beautiful designs: birds, animals, flowers, swords, fans, geometric; all beautifully stylized. 213pp. 11⅜ x 8¼.
22874-6 Pa. $5.00

PERSPECTIVE, Jan Vredeman de Vries. 73 perspective plates from 1604 edition; buildings, townscapes, stairways, fantastic scenes. Remarkable for beauty, surrealistic atmosphere; real eye-catchers. Introduction by Adolf Placzek. 74pp. 11⅜ x 8¼.
20186-4 Pa. $2.75

EARLY AMERICAN DESIGN MOTIFS. Suzanne E. Chapman. 497 motifs, designs, from painting on wood, ceramics, appliqué, glassware, samplers, metal work, etc. Florals, landscapes, birds and animals, geometrics, letters, etc. Inexhaustible. Enlarged edition. 138pp. 8⅜ x 11¼.
22985-8 Pa. $3.50
23084-8 Clothbd. $7.95

VICTORIAN STENCILS FOR DESIGN AND DECORATION, edited by E.V. Gillon, Jr. 113 wonderful ornate Victorian pieces from German sources; florals, geometrics; borders, corner pieces; bird motifs, etc. 64pp. 9⅜ x 12¼.
21995-X Pa. $2.75

ART NOUVEAU: AN ANTHOLOGY OF DESIGN AND ILLUSTRATION FROM THE STUDIO, edited by E.V. Gillon, Jr. Graphic arts: book jackets, posters, engravings, illustrations, decorations; Crane, Beardsley, Bradley and many others. Inexhaustible. 92pp. 8⅛ x 11.
22388-4 Pa. $2.50

ORIGINAL ART DECO DESIGNS, William Rowe. First-rate, highly imaginative modern Art Deco frames, borders, compositions, alphabets, florals, insectals, Wurlitzer-types, etc. Much finest modern Art Deco. 80 plates, 8 in color. 8⅜ x 11¼.
22567-4 Pa. $3.00

HANDBOOK OF DESIGNS AND DEVICES, Clarence P. Hornung. Over 1800 basic geometric designs based on circle, triangle, square, scroll, cross, etc. Largest such collection in existence. 261pp.
20125-2 Pa. $2.50

150 MASTERPIECES OF DRAWING, edited by Anthony Toney. 150 plates, early 15th century to end of 18th century; Rembrandt, Michelangelo, Dürer, Fragonard, Watteau, Wouwerman, many others. 150pp. 8⅜ x 11¼.   21032-4 Pa. $3.50

THE GOLDEN AGE OF THE POSTER, Hayward and Blanche Cirker. 70 extraordinary posters in full colors, from Maîtres de l'Affiche, Mucha, Lautrec, Bradley, Cheret, Beardsley, many others. 9⅜ x 12¼.   22753-7 Pa. $4.95
21718-3 Clothbd. $7.95

SIMPLICISSIMUS, selection, translations and text by Stanley Appelbaum. 180 satirical drawings, 16 in full color, from the famous German weekly magazine in the years 1896 to 1926. 24 artists included: Grosz, Kley, Pascin, Kubin, Kollwitz, plus Heine, Thöny, Bruno Paul, others. 172pp. 8½ x 12¼.   23098-8 Pa. $5.00
23099-6 Clothbd. $10.00

THE EARLY WORK OF AUBREY BEARDSLEY, Aubrey Beardsley. 157 plates, 2 in color: Manon Lescaut, Madame Bovary, Morte d'Arthur, Salome, other. Introduction by H. Marillier. 175pp. 8½ x 11.   21816-3 Pa. $3.50

THE LATER WORK OF AUBREY BEARDSLEY, Aubrey Beardsley. Exotic masterpieces of full maturity: Venus and Tannhäuser, Lysistrata, Rape of the Lock, Volpone, Savoy material, etc. 174 plates, 2 in color. 176pp. 8½ x 11.   21817-1 Pa. $4.00

DRAWINGS OF WILLIAM BLAKE, William Blake. 92 plates from Book of Job, Divine Comedy, Paradise Lost, visionary heads, mythological figures, Laocoön, etc. Selection, introduction, commentary by Sir Geoffrey Keynes. 178pp. 8½ x 11.
22303-5 Pa. $3.50

LONDON: A PILGRIMAGE, Gustave Doré, Blanchard Jerrold. Squalor, riches, misery, beauty of mid-Victorian metropolis; 55 wonderful plates, 125 other illustrations, full social, cultural text by Jerrold. 191pp. of text. 8⅛ x 11.
22306-X Pa. $5.00

THE COMPLETE WOODCUTS OF ALBRECHT DÜRER, edited by Dr. W. Kurth. 346 in all: Old Testament, St. Jerome, Passion, Life of Virgin, Apocalypse, many others. Introduction by Campbell Dodgson. 285pp. 8½ x 12¼.   21097-9 Pa. $6.00

THE DISASTERS OF WAR, Francisco Goya. 83 etchings record horrors of Napoleonic wars in Spain and war in general. Reprint of 1st edition, plus 3 additional plates. Introduction by Philip Hofer. 97pp. 9⅜ x 8¼.   21872-4 Pa. $3.00

ENGRAVINGS OF HOGARTH, William Hogarth. 101 of Hogarth's greatest works: Rake's Progress, Harlot's Progress, Illustrations for Hudibras, Midnight Modern Conversation, Before and After, Beer Street and Gin Lane, many more. Full commentary. 256pp. 11 x 14.   22479-1 Pa. $7.00
23023-6 Clothbd. $13.50

PRIMITIVE ART, Franz Boas. Great anthropologist on ceramics, textiles, wood, stone, metal, etc.; patterns, technology, symbols, styles. All areas, but fullest on Northwest Coast Indians. 350 illustrations. 378pp.   20025-6 Pa. $3.50

MOTHER GOOSE'S MELODIES. Facsimile of fabulously rare Munroe and Francis "copyright 1833" Boston edition. Familiar and unusual rhymes, wonderful old woodcut illustrations. Edited by E.F. Bleiler. 128pp. 4½ x 6⅜. 22577-1 Pa. $1.00

MOTHER GOOSE IN HIEROGLYPHICS. Favorite nursery rhymes presented in rebus form for children. Fascinating 1849 edition reproduced in toto, with key. Introduction by E.F. Bleiler. About 400 woodcuts. 64pp. 6⅞ x 5¼. 20745-5 Pa. $1.00

PETER PIPER'S PRACTICAL PRINCIPLES OF PLAIN & PERFECT PRONUNCIATION. Alliterative jingles and tongue-twisters. Reproduction in full of 1830 first American edition. 25 spirited woodcuts. 32pp. 4½ x 6⅜. 22560-7 Pa. $1.00

MARMADUKE MULTIPLY'S MERRY METHOD OF MAKING MINOR MATHEMATICIANS. Fellow to Peter Piper, it teaches multiplication table by catchy rhymes and woodcuts. 1841 Munroe & Francis edition. Edited by E.F. Bleiler. 103pp. 4⅝ x 6.
22773-1 Pa. $1.25
20171-6 Clothbd. $3.00

THE NIGHT BEFORE CHRISTMAS, Clement Moore. Full text, and woodcuts from original 1848 book. Also critical, historical material. 19 illustrations. 40pp. 4⅝ x 6. 22797-9 Pa. $1.00

THE KING OF THE GOLDEN RIVER, John Ruskin. Victorian children's classic of three brothers, their attempts to reach the Golden River, what becomes of them. Facsimile of original 1889 edition. 22 illustrations. 56pp. 4⅝ x 6⅜.
20066-3 Pa. $1.25

DREAMS OF THE RAREBIT FIEND, Winsor McCay. Pioneer cartoon strip, unexcelled for beauty, imagination, in 60 full sequences. Incredible technical virtuosity, wonderful visual wit. Historical introduction. 62pp. 8⅜ x 11¼. 21347-1 Pa. $2.50

THE KATZENJAMMER KIDS, Rudolf Dirks. In full color, 14 strips from 1906-7; full of imagination, characteristic humor. Classic of great historical importance. Introduction by August Derleth. 32pp. 9¼ x 12¼. 23005-8 Pa. $2.00

LITTLE ORPHAN ANNIE AND LITTLE ORPHAN ANNIE IN COSMIC CITY, Harold Gray. Two great sequences from the early strips: our curly-haired heroine defends the Warbucks' financial empire and, then, takes on meanie Phineas P. Pinchpenny. Leapin' lizards! 178pp. 6⅛ x 8⅜. 23107-0 Pa. $2.00

WHEN A FELLER NEEDS A FRIEND, Clare Briggs. 122 cartoons by one of the greatest newspaper cartoonists of the early 20th century — about growing up, making a living, family life, daily frustrations and occasional triumphs. 121pp. 8½ x 9½.
23148-8 Pa. $2.50

THE BEST OF GLUYAS WILLIAMS. 100 drawings by one of America's finest cartoonists: The Day a Cake of Ivory Soap Sank at Proctor & Gamble's, At the Life Insurance Agents' Banquet, and many other gems from the 20's and 30's. 118pp. 8⅜ x 11¼. 22737-5 Pa. $2.50

THE BEST DR. THORNDYKE DETECTIVE STORIES, R. Austin Freeman. The Case of Oscar Brodski, The Moabite Cipher, and 5 other favorites featuring the great scientific detective, plus his long-believed-lost first adventure — 31 New Inn — reprinted here for the first time. Edited by E.F. Bleiler. USO 20388-3 Pa. $3.00

BEST "THINKING MACHINE" DETECTIVE STORIES, Jacques Futrelle. The Problem of Cell 13 and 11 other stories about Prof. Augustus S.F.X. Van Dusen, including two "lost" stories. First reprinting of several. Edited by E.F. Bleiler. 241pp.
20537-1 Pa. $3.00

UNCLE SILAS, J. Sheridan LeFanu. Victorian Gothic mystery novel, considered by many best of period, even better than Collins or Dickens. Wonderful psychological terror. Introduction by Frederick Shroyer. 436pp. 21715-9 Pa. $4.00

BEST DR. POGGIOLI DETECTIVE STORIES, T.S. Stribling. 15 best stories from EQMM and The Saint offer new adventures in Mexico, Florida, Tennessee hills as Poggioli unravels mysteries and combats Count Jalacki. 217pp. 23227-1 Pa. $3.00

EIGHT DIME NOVELS, selected with an introduction by E.F. Bleiler. Adventures of Old King Brady, Frank James, Nick Carter, Deadwood Dick, Buffalo Bill, The Steam Man, Frank Merriwell, and Horatio Alger — 1877 to 1905. Important, entertaining popular literature in facsimile reprint, with original covers. 190pp. 9 x 12. 22975-0 Pa. $3.50

ALICE'S ADVENTURES UNDER GROUND, Lewis Carroll. Facsimile of ms. Carroll gave Alice Liddell in 1864. Different in many ways from final Alice. Handlettered, illustrated by Carroll. Introduction by Martin Gardner. 128pp. 21482-6 Pa. $1.50

ALICE IN WONDERLAND COLORING BOOK, Lewis Carroll. Pictures by John Tenniel. Large-size versions of the famous illustrations of Alice, Cheshire Cat, Mad Hatter and all the others, waiting for your crayons. Abridged text. 36 illustrations. 64pp. 8¼ x 11. 22853-3 Pa. $1.50

AVENTURES D'ALICE AU PAYS DES MERVEILLES, Lewis Carroll. Bué's translation of "Alice" into French, supervised by Carroll himself. Novel way to learn language. (No English text.) 42 Tenniel illustrations. 196pp. 22836-3 Pa. $2.50

MYTHS AND FOLK TALES OF IRELAND, Jeremiah Curtin. 11 stories that are Irish versions of European fairy tales and 9 stories from the Fenian cycle — 20 tales of legend and magic that comprise an essential work in the history of folklore. 256pp. 22430-9 Pa. $3.00

EAST O' THE SUN AND WEST O' THE MOON, George W. Dasent. Only full edition of favorite, wonderful Norwegian fairytales — Why the Sea is Salt, Boots and the Troll, etc. — with 77 illustrations by Kittelsen & Werenskiöld. 418pp.
22521-6 Pa. $4.00

PERRAULT'S FAIRY TALES, Charles Perrault and Gustave Doré. Original versions of Cinderella, Sleeping Beauty, Little Red Riding Hood, etc. in best translation, with 34 wonderful illustrations by Gustave Doré. 117pp. 8⅛ x 11. 22311-6 Pa. $2.50

EARLY NEW ENGLAND GRAVESTONE RUBBINGS, Edmund V. Gillon, Jr. 43 photographs, 226 rubbings show heavily symbolic, macabre, sometimes humorous primitive American art. Up to early 19th century. 207pp. 8⅜ x 11¼.
21380-3 Pa. $4.00

L.J.M. DAGUERRE: THE HISTORY OF THE DIORAMA AND THE DAGUERREOTYPE, Helmut and Alison Gernsheim. Definitive account. Early history, life and work of Daguerre; discovery of daguerreotype process; diffusion abroad; other early photography. 124 illustrations. 226pp. 6⅙ x 9¼.
22290-X Pa. $4.00

PHOTOGRAPHY AND THE AMERICAN SCENE, Robert Taft. The basic book on American photography as art, recording form, 1839-1889. Development, influence on society, great photographers, types (portraits, war, frontier, etc.), whatever else needed. Inexhaustible. Illustrated with 322 early photos, daguerreotypes, tintypes, stereo slides, etc. 546pp. 6⅛ x 9¼.
21201-7 Pa. $5.95

PHOTOGRAPHIC SKETCHBOOK OF THE CIVIL WAR, Alexander Gardner. Reproduction of 1866 volume with 100 on-the-field photographs: Manassas, Lincoln on battlefield, slave pens, etc. Introduction by E.F. Bleiler. 224pp. 10¾ x 9.
22731-6 Pa. $5.00

THE MOVIES: A PICTURE QUIZ BOOK, Stanley Appelbaum & Hayward Cirker. Match stars with their movies, name actors and actresses, test your movie skill with 241 stills from 236 great movies, 1902-1959. Indexes of performers and films. 128pp. 8⅜ x 9¼.
20222-4 Pa. $2.50

THE TALKIES, Richard Griffith. Anthology of features, articles from Photoplay, 1928-1940, reproduced complete. Stars, famous movies, technical features, fabulous ads, etc.; Garbo, Chaplin, King Kong, Lubitsch, etc. 4 color plates, scores of illustrations. 327pp. 8⅜ x 11¼.
22762-6 Pa. $6.95

THE MOVIE MUSICAL FROM VITAPHONE TO "42ND STREET," edited by Miles Kreuger. Relive the rise of the movie musical as reported in the pages of Photoplay magazine (1926-1933): every movie review, cast list, ad, and record review; every significant feature article, production still, biography, forecast, and gossip story. Profusely illustrated. 367pp. 8⅜ x 11¼.
23154-2 Pa. $6.95

JOHANN SEBASTIAN BACH, Philipp Spitta. Great classic of biography, musical commentary, with hundreds of pieces analyzed. Also good for Bach's contemporaries. 450 musical examples. Total of 1799pp.
EUK 22278-0, 22279-9 Clothbd., Two vol. set $25.00

BEETHOVEN AND HIS NINE SYMPHONIES, Sir George Grove. Thorough history, analysis, commentary on symphonies and some related pieces. For either beginner or advanced student. 436 musical passages. 407pp.
20334-4 Pa. $4.00

MOZART AND HIS PIANO CONCERTOS, Cuthbert Girdlestone. The only full-length study. Detailed analyses of all 21 concertos, sources; 417 musical examples. 509pp.
21271-8 Pa. $4.50

THE FITZWILLIAM VIRGINAL BOOK, edited by J. Fuller Maitland, W.B. Squire. Famous early 17th century collection of keyboard music, 300 works by Morley, Byrd, Bull, Gibbons, etc. Modern notation. Total of 938pp. 8⅜ x 11.
ECE 21068-5, 21069-3 Pa., Two vol. set $14.00

COMPLETE STRING QUARTETS, Wolfgang A. Mozart. Breitkopf and Härtel edition. All 23 string quartets plus alternate slow movement to K156. Study score. 277pp. 9⅜ x 12¼. 22372-8 Pa. $6.00

COMPLETE SONG CYCLES, Franz Schubert. Complete piano, vocal music of Die Schöne Müllerin, Die Winterreise, Schwanengesang. Also Drinker English singing translations. Breitkopf and Härtel edition. 217pp. 9⅜ x 12¼.
22649-2 Pa. $4.50

THE COMPLETE PRELUDES AND ETUDES FOR PIANOFORTE SOLO, Alexander Scriabin. All the preludes and etudes including many perfectly spun miniatures. Edited by K.N. Igumnov and Y.I. Mil'shteyn. 250pp. 9 x 12. 22919-X Pa. $5.00

TRISTAN UND ISOLDE, Richard Wagner. Full orchestral score with complete instrumentation. Do not confuse with piano reduction. Commentary by Felix Mottl, great Wagnerian conductor and scholar. Study score. 655pp. 8⅛ x 11.
22915-7 Pa. $10.00

FAVORITE SONGS OF THE NINETIES, ed. Robert Fremont. Full reproduction, including covers, of 88 favorites: Ta-Ra-Ra-Boom-De-Aye, The Band Played On, Bird in a Gilded Cage, Under the Bamboo Tree, After the Ball, etc. 401pp. 9 x 12.
EBE 21536-9 Pa. $6.95

SOUSA'S GREAT MARCHES IN PIANO TRANSCRIPTION: ORIGINAL SHEET MUSIC OF 23 WORKS, John Philip Sousa. Selected by Lester S. Levy. Playing edition includes: The Stars and Stripes Forever, The Thunderer, The Gladiator, King Cotton, Washington Post, much more. 24 illustrations. 111pp. 9 x 12.
USO 23132-1 Pa. $3.50

CLASSIC PIANO RAGS, selected with an introduction by Rudi Blesh. Best ragtime music (1897-1922) by Scott Joplin, James Scott, Joseph F. Lamb, Tom Turpin, 9 others. Printed from best original sheet music, plus covers. 364pp. 9 x 12.
EBE 20469-3 Pa. $6.95

ANALYSIS OF CHINESE CHARACTERS, C.D. Wilder, J.H. Ingram. 1000 most important characters analyzed according to primitives, phonetics, historical development. Traditional method offers mnemonic aid to beginner, intermediate student of Chinese, Japanese. 365pp. 23045-7 Pa. $4.00

MODERN CHINESE: A BASIC COURSE, Faculty of Peking University. Self study, classroom course in modern Mandarin. Records contain phonetics, vocabulary, sentences, lessons. 249 page book contains all recorded text, translations, grammar, vocabulary, exercises. Best course on market. 3 12" 33⅓ monaural records, book, album. 98832-5 Set $12.50

MANUAL OF THE TREES OF NORTH AMERICA, Charles S. Sargent. The basic survey of every native tree and tree-like shrub, 717 species in all. Extremely full descriptions, information on habitat, growth, locales, economics, etc. Necessary to every serious tree lover. Over 100 finding keys. 783 illustrations. Total of 986pp.
20277-1, 20278-X Pa., Two vol. set $8.00

BIRDS OF THE NEW YORK AREA, John Bull. Indispensable guide to more than 400 species within a hundred-mile radius of Manhattan. Information on range, status, breeding, migration, distribution trends, etc. Foreword by Roger Tory Peterson. 17 drawings; maps. 540pp.
23222-0 Pa. $6.00

THE SEA-BEACH AT EBB-TIDE, Augusta Foote Arnold. Identify hundreds of marine plants and animals: algae, seaweeds, squids, crabs, corals, etc. Descriptions cover food, life cycle, size, shape, habitat. Over 600 drawings. 490pp.
21949-6 Pa. $5.00

THE MOTH BOOK, William J. Holland. Identify more than 2,000 moths of North America. General information, precise species descriptions. 623 illustrations plus 48 color plates show almost all species, full size. 1968 edition. Still the basic book. Total of 551pp. 6½ x 9¼.
21948-8 Pa. $6.00

AN INTRODUCTION TO THE REPTILES AND AMPHIBIANS OF THE UNITED STATES, Percy A. Morris. All lizards, crocodiles, turtles, snakes, toads, frogs; life history, identification, habits, suitability as pets, etc. Non-technical, but sound and broad. 130 photos. 253pp.
22982-3 Pa. $3.00

OLD NEW YORK IN EARLY PHOTOGRAPHS, edited by Mary Black. Your only chance to see New York City as it was 1853-1906, through 196 wonderful photographs from N.Y. Historical Society. Great Blizzard, Lincoln's funeral procession, great buildings. 228pp. 9 x 12.
22907-6 Pa. $6.00

THE AMERICAN REVOLUTION, A PICTURE SOURCEBOOK, John Grafton. Wonderful Bicentennial picture source, with 411 illustrations (contemporary and 19th century) showing battles, personalities, maps, events, flags, posters, soldier's life, ships, etc. all captioned and explained. A wonderful browsing book, supplement to other historical reading. 160pp. 9 x 12.
23226-3 Pa. $4.00

PERSONAL NARRATIVE OF A PILGRIMAGE TO AL-MADINAH AND MECCAH, Richard Burton. Great travel classic by remarkably colorful personality. Burton, disguised as a Moroccan, visited sacred shrines of Islam, narrowly escaping death. Wonderful observations of Islamic life, customs, personalities. 47 illustrations. Total of 959pp.
21217-3, 21218-1 Pa., Two vol. set $10.00

INCIDENTS OF TRAVEL IN CENTRAL AMERICA, CHIAPAS, AND YUCATAN, John L. Stephens. Almost single-handed discovery of Maya culture; exploration of ruined cities, monuments, temples; customs of Indians. 115 drawings. 892pp.
22404-X, 22405-8 Pa., Two vol. set $8.00

CONSTRUCTION OF AMERICAN FURNITURE TREASURES, Lester Margon. 344 detail drawings, complete text on constructing exact reproductions of 38 early American masterpieces: Hepplewhite sideboard, Duncan Phyfe drop-leaf table, mantel clock, gate-leg dining table, Pa. German cupboard, more. 38 plates. 54 photographs. 168pp. 8⅜ x 11¼.   23056-2 Pa. $4.00

JEWELRY MAKING AND DESIGN, Augustus F. Rose, Antonio Cirino. Professional secrets revealed in thorough, practical guide: tools, materials, processes; rings, brooches, chains, cast pieces, enamelling, setting stones, etc. Do not confuse with skimpy introductions: beginner can use, professional can learn from it. Over 200 illustrations. 306pp.   21750-7 Pa. $3.00

METALWORK AND ENAMELLING, Herbert Maryon. Generally coneeded best all-around book. Countless trade secrets: materials, tools, soldering, filigree, setting, inlay, niello, repoussé, casting, polishing, etc. For beginner or expert. Author was foremost British expert. 330 illustrations. 335pp.   22702-2 Pa. $3.50

WEAVING WITH FOOT-POWER LOOMS, Edward F. Worst. Setting up a loom, beginning to weave, constructing equipment, using dyes, more, plus over 285 drafts of traditional patterns including Colonial and Swedish weaves. More than 200 other figures. For beginning and advanced. 275pp. 8¾ x 6⅜ .   23064-3 Pa. $4.00

WEAVING A NAVAJO BLANKET, Gladys A. Reichard. Foremost anthropologist studied under Navajo women, reveals every step in process from wool, dyeing, spinning, setting up loom, designing, weaving. Much history, symbolism. With this book you could make one yourself. 97 illustrations. 222pp.   22992-0 Pa. $3.00

NATURAL DYES AND HOME DYEING, Rita J. Adrosko. Use natural ingredients: bark, flowers, leaves, lichens, insects etc. Over 135 specific recipes from historical sources for cotton, wool, other fabrics. Genuine premodern handicrafts. 12 illustrations. 160pp.   22688-3 Pa. $2.00

THE HAND DECORATION OF FABRICS, Francis J. Kafka. Outstanding, profusely illustrated guide to stenciling, batik, block printing, tie dyeing, freehand painting, silk screen printing, and novelty decoration. 356 illustrations. 198pp. 6 x 9.
                                                       21401-X Pa. $3.00

THOMAS NAST: CARTOONS AND ILLUSTRATIONS, with text by Thomas Nast St. Hill. Father of American political cartooning. Cartoons that destroyed Tweed Ring; inflation, free love, church and state; original Republican elephant and Democratic donkey; Santa Claus; more. 117 illustrations. 146pp. 9 x 12.
                                                       22983-1 Pa. $4.00
                                                       23067-8 Clothbd. $8.50

FREDERIC REMINGTON: 173 DRAWINGS AND ILLUSTRATIONS. Most famous of the Western artists, most responsible for our myths about the American West in its untamed days. Complete reprinting of *Drawings of Frederic Remington* (1897), plus other selections. 4 additional drawings in color on covers. 140pp. 9 x 12.
                                                       20714-5 Pa. $3.95

HOW TO SOLVE CHESS PROBLEMS, Kenneth S. Howard. Practical suggestions on problem solving for very beginners. 58 two-move problems, 46 3-movers, 8 4-movers for practice, plus hints. 171pp. 20748-X Pa. $2.00

A GUIDE TO FAIRY CHESS, Anthony Dickins. 3-D chess, 4-D chess, chess on a cylindrical board, reflecting pieces that bounce off edges, cooperative chess, retrograde chess, maximummers, much more. Most based on work of great Dawson. Full handbook, 100 problems. 66pp. 7⅞ x 10¾. 22687-5 Pa. $2.00

WIN AT BACKGAMMON, Millard Hopper. Best opening moves, running game, blocking game, back game, tables of odds, etc. Hopper makes the game clear enough for anyone to play, and win. 43 diagrams. 111pp. 22894-0 Pa. $1.50

BIDDING A BRIDGE HAND, Terence Reese. Master player "thinks out loud" the binding of 75 hands that defy point count systems. Organized by bidding problem—no-fit situations, overbidding, underbidding, cueing your defense, etc. 254pp. EBE 22830-4 Pa. $2.50

THE PRECISION BIDDING SYSTEM IN BRIDGE, C.C. Wei, edited by Alan Truscott. Inventor of precision bidding presents average hands and hands from actual play, including games from 1969 Bermuda Bowl where system emerged. 114 exercises. 116pp. 21171-1 Pa. $1.75

LEARN MAGIC, Henry Hay. 20 simple, easy-to-follow lessons on magic for the new magician: illusions, card tricks, silks, sleights of hand, coin manipulations, escapes, and more —all with a minimum amount of equipment. Final chapter explains the great stage illusions. 92 illustrations. 285pp. 21238-6 Pa. $2.95

THE NEW MAGICIAN'S MANUAL, Walter B. Gibson. Step-by-step instructions and clear illustrations guide the novice in mastering 36 tricks; much equipment supplied on 16 pages of cut-out materials. 36 additional tricks. 64 illustrations. 159pp. 6⅝ x 10. 23113-5 Pa. $3.00

PROFESSIONAL MAGIC FOR AMATEURS, Walter B. Gibson. 50 easy, effective tricks used by professionals —cards, string, tumblers, handkerchiefs, mental magic, etc. 63 illustrations. 223pp. 23012-0 Pa. $2.50

CARD MANIPULATIONS, Jean Hugard. Very rich collection of manipulations; has taught thousands of fine magicians tricks that are really workable, eye-catching. Easily followed, serious work. Over 200 illustrations. 163pp. 20539-8 Pa. $2.00

ABBOTT'S ENCYCLOPEDIA OF ROPE TRICKS FOR MAGICIANS, Stewart James. Complete reference book for amateur and professional magicians containing more than 150 tricks involving knots, penetrations, cut and restored rope, etc. 510 illustrations. Reprint of 3rd edition. 400pp. 23206-9 Pa. $3.50

THE SECRETS OF HOUDINI, J.C. Cannell. Classic study of Houdini's incredible magic, exposing closely-kept professional secrets and revealing, in general terms, the whole art of stage magic. 67 illustrations. 279pp. 22913-0 Pa. $2.50

THE MAGIC MOVING PICTURE BOOK, Bliss, Sands & Co. The pictures in this book move! Volcanoes erupt, a house burns, a serpentine dancer wiggles her way through a number. By using a specially ruled acetate screen provided, you can obtain these and 15 other startling effects. Originally "The Motograph Moving Picture Book." 32pp. 8¼ x 11.   23224-7 Pa. $1.75

STRING FIGURES AND HOW TO MAKE THEM, Caroline F. Jayne. Fullest, clearest instructions on string figures from around world: Eskimo, Navajo, Lapp, Europe, more. Cats cradle, moving spear, lightning, stars. Introduction by A.C. Haddon. 950 illustrations. 407pp.   20152-X Pa. $3.00

PAPER FOLDING FOR BEGINNERS, William D. Murray and Francis J. Rigney. Clearest book on market for making origami sail boats, roosters, frogs that move legs, cups, bonbon boxes. 40 projects. More than 275 illustrations. Photographs. 94pp.
20713-7 Pa. $1.25

INDIAN SIGN LANGUAGE, William Tomkins. Over 525 signs developed by Sioux, Blackfoot, Cheyenne, Arapahoe and other tribes. Written instructions and diagrams: how to make words, construct sentences. Also 290 pictographs of Sioux and Ojibway tribes. 111pp. 6⅛ x 9¼.   22029-X Pa. $1.50

BOOMERANGS: HOW TO MAKE AND THROW THEM, Bernard S. Mason. Easy to make and throw, dozens of designs: cross-stick, pinwheel, boomabird, tumblestick, Australian curved stick boomerang. Complete throwing instructions. All safe. 99pp.   23028-7 Pa. $1.50

25 KITES THAT FLY, Leslie Hunt. Full, easy to follow instructions for kites made from inexpensive materials. Many novelties. Reeling, raising, designing your own. 70 illustrations. 110pp.   22550-X Pa. $1.25

TRICKS AND GAMES ON THE POOL TABLE, Fred Herrmann. 79 tricks and games, some solitaires, some for 2 or more players, some competitive; mystifying shots and throws, unusual carom, tricks involving cork, coins, a hat, more. 77 figures. 95pp.   21814-7 Pa. $1.25

WOODCRAFT AND CAMPING, Bernard S. Mason. How to make a quick emergency shelter, select woods that will burn immediately, make do with limited supplies, etc. Also making many things out of wood, rawhide, bark, at camp. Formerly titled Woodcraft. 295 illustrations. 580pp.   21951-8 Pa. $4.00

AN INTRODUCTION TO CHESS MOVES AND TACTICS SIMPLY EXPLAINED, Leonard Barden. Informal intermediate introduction: reasons for moves, tactics, openings, traps, positional play, endgame. Isolates patterns. 102pp. USO 21210-6 Pa. $1.35

LASKER'S MANUAL OF CHESS, Dr. Emanuel Lasker. Great world champion offers very thorough coverage of all aspects of chess. Combinations, position play, openings, endgame, aesthetics of chess, philosophy of struggle, much more. Filled with analyzed games. 390pp.   20640-8 Pa. $3.50

SLEEPING BEAUTY, illustrated by Arthur Rackham. Perhaps the fullest, most delightful version ever, told by C.S. Evans. Rackham's best work. 49 illustrations. 110pp. 7⅞ x 10¾. 22756-1 Pa. $2.00

THE WONDERFUL WIZARD OF OZ, L. Frank Baum. Facsimile in full color of America's finest children's classic. Introduction by Martin Gardner. 143 illustrations by W.W. Denslow. 267pp. 20691-2 Pa. $2.50

GOOPS AND HOW TO BE THEM, Gelett Burgess. Classic tongue-in-cheek masquerading as etiquette book. 87 verses, 170 cartoons as Goops demonstrate virtues of table manners, neatness, courtesy, more. 88pp. 6½ x 9¼. 22233-0 Pa. $1.50

THE BROWNIES, THEIR BOOK, Palmer Cox. Small as mice, cunning as foxes, exuberant, mischievous, Brownies go to zoo, toy shop, seashore, circus, more. 24 verse adventures. 266 illustrations. 144pp. 6⅝ x 9¼. 21265-3 Pa. $1.75

BILLY WHISKERS: THE AUTOBIOGRAPHY OF A GOAT, Frances Trego Montgomery. Escapades of that rambunctious goat. Favorite from turn of the century America. 24 illustrations. 259pp. 22345-0 Pa. $2.75

THE ROCKET BOOK, Peter Newell. Fritz, janitor's kid, sets off rocket in basement of apartment house; an ingenious hole punched through every page traces course of rocket. 22 duotone drawings, verses. 48pp. 6⅞ x 8⅜. 22044-3 Pa. $1.50

PECK'S BAD BOY AND HIS PA, George W. Peck. Complete double-volume of great American childhood classic. Hennery's ingenious pranks against outraged pomposity of pa and the grocery man. 97 illustrations. Introduction by E.F. Bleiler. 347pp. 20497-9 Pa. $2.50

THE TALE OF PETER RABBIT, Beatrix Potter. The inimitable Peter's terrifying adventure in Mr. McGregor's garden, with all 27 wonderful, full-color Potter illustrations. 55pp. 4¼ x 5½. USO 22827-4 Pa. $1.00

THE TALE OF MRS. TIGGY-WINKLE, Beatrix Potter. Your child will love this story about a very special hedgehog and all 27 wonderful, full-color Potter illustrations. 57pp. 4¼ x 5½. USO 20546-0 Pa. $1.00

THE TALE OF BENJAMIN BUNNY, Beatrix Potter. Peter Rabbit's cousin coaxes him back into Mr. McGregor's garden for a whole new set of adventures. A favorite with children. All 27 full-color illustrations. 59pp. 4¼ x 5½. USO 21102-9 Pa. $1.00

THE MERRY ADVENTURES OF ROBIN HOOD, Howard Pyle. Facsimile of original (1883) edition, finest modern version of English outlaw's adventures. 23 illustrations by Pyle. 296pp. 6½ x 9¼. 22043-5 Pa. $2.75

TWO LITTLE SAVAGES, Ernest Thompson Seton. Adventures of two boys who lived as Indians; explaining Indian ways, woodlore, pioneer methods. 293 illustrations. 286pp. 20985-7 Pa. $3.00

HOUDINI ON MAGIC, Harold Houdini. Edited by Walter Gibson, Morris N. Young. How he escaped; exposés of fake spiritualists; instructions for eye-catching tricks; other fascinating material by and about greatest magician. 155 illustrations. 280pp. 20384-0 Pa. $2.50

HANDBOOK OF THE NUTRITIONAL CONTENTS OF FOOD, U.S. Dept. of Agriculture. Largest, most detailed source of food nutrition information ever prepared. Two mammoth tables: one measuring nutrients in 100 grams of edible portion; the other, in edible portion of 1 pound as purchased. Originally titled Composition of Foods. 190pp. 9 x 12. 21342-0 Pa. $4.00

COMPLETE GUIDE TO HOME CANNING, PRESERVING AND FREEZING, U.S. Dept. of Agriculture. Seven basic manuals with full instructions for jams and jellies; pickles and relishes; canning fruits, vegetables, meat; freezing anything. Really good recipes, exact instructions for optimal results. Save a fortune in food. 156 illustrations. 214pp. 6$1/8$ x 9$1/4$. 22911-4 Pa. $2.50

THE BREAD TRAY, Louis P. De Gouy. Nearly every bread the cook could buy or make: bread sticks of Italy, fruit breads of Greece, glazed rolls of Vienna, everything from corn pone to croissants. Over 500 recipes altogether. including buns, rolls, muffins, scones, and more. 463pp. 23000-7 Pa. $3.50

CREATIVE HAMBURGER COOKERY, Louis P. De Gouy. 182 unusual recipes for casseroles, meat loaves and hamburgers that turn inexpensive ground meat into memorable main dishes: Arizona chili burgers, burger tamale pie, burger stew, burger corn loaf, burger wine loaf, and more. 120pp. 23001-5 Pa. $1.75

LONG ISLAND SEAFOOD COOKBOOK, J. George Frederick and Jean Joyce. Probably the best American seafood cookbook. Hundreds of recipes. 40 gourmet sauces, 123 recipes using oysters alone! All varieties of fish and seafood amply represented. 324pp. 22677-8 Pa. $3.00

THE EPICUREAN: A COMPLETE TREATISE OF ANALYTICAL AND PRACTICAL STUDIES IN THE CULINARY ART, Charles Ranhofer. Great modern classic. 3,500 recipes from master chef of Delmonico's, turn-of-the-century America's best restaurant. Also explained, many techniques known only to professional chefs. 775 illustrations. 1183pp. 6$5/8$ x 10. 22680-8 Clothbd. $17.50

THE AMERICAN WINE COOK BOOK, Ted Hatch. Over 700 recipes: old favorites livened up with wine plus many more: Czech fish soup, quince soup, sauce Perigueux, shrimp shortcake, filets Stroganoff, cordon bleu goulash, jambonneau, wine fruit cake, more. 314pp. 22796-0 Pa. $2.50

DELICIOUS VEGETARIAN COOKING, Ivan Baker. Close to 500 delicious and varied recipes: soups, main course dishes (pea, bean, lentil, cheese, vegetable, pasta, and egg dishes), savories, stews, whole-wheat breads and cakes, more. 168pp.
USO 22834-7 Pa. $1.75

COOKIES FROM MANY LANDS, Josephine Perry. Crullers, oatmeal cookies, chaux au chocolate, English tea cakes, mandel kuchen, Sacher torte, Danish puff pastry, Swedish cookies — a mouth-watering collection of 223 recipes. 157pp.

22832-0 Pa. $2.00

ROSE RECIPES, Eleanour S. Rohde. How to make sauces, jellies, tarts, salads, pot-pourris, sweet bags, pomanders, perfumes from garden roses; all exact recipes. Century old favorites. 95pp.

22957-2 Pa. $1.25

"OSCAR" OF THE WALDORF'S COOKBOOK, Oscar Tschirky. Famous American chef reveals 3455 recipes that made Waldorf great; cream of French, German, American cooking, in all categories. Full instructions, easy home use. 1896 edition. 907pp. $6^{5}/_{8}$ x $9^{3}/_{8}$.

20790-0 Clothbd. $15.00

JAMS AND JELLIES, May Byron. Over 500 old-time recipes for delicious jams, jellies, marmalades, preserves, and many other items. Probably the largest jam and jelly book in print. Originally titled May Byron's Jam Book. 276pp.

USO 23130-5 Pa. $3.00

MUSHROOM RECIPES, André L. Simon. 110 recipes for everyday and special cooking. Champignons à la grecque, sole bonne femme, chicken liver croustades, more; 9 basic sauces, 13 ways of cooking mushrooms. 54pp.

USO 20913-X Pa. $1.25

FAVORITE SWEDISH RECIPES, edited by Sam Widenfelt. Prepared in Sweden, offers wonderful, clearly explained Swedish dishes: appetizers, meats, pastry and cookies, other categories. Suitable for American kitchen. 90 photos. 157pp.

23156-9 Pa. $2.00

THE BUCKEYE COOKBOOK, Buckeye Publishing Company. Over 1,000 easy-to-follow, traditional recipes from the American Midwest: bread (100 recipes alone), meat, game, jam, candy, cake, ice cream, and many other categories of cooking. 64 illustrations. From 1883 enlarged edition. 416pp.

23218-2 Pa. $4.00

TWENTY-TWO AUTHENTIC BANQUETS FROM INDIA, Robert H. Christie. Complete, easy-to-do recipes for almost 200 authentic Indian dishes assembled in 22 banquets. Arranged by region. Selected from Banquets of the Nations. 192pp.

23200-X Pa. $2.50

Prices subject to change without notice.
Available at your book dealer or write for free catalogue to Dept. GI, Dover Publications, Inc., 180 Varick St., N.Y., N.Y. 10014. Dover publishes more than 150 books each year on science, elementary and advanced mathematics, biology, music, art, literary history, social sciences and other areas.